ERROR-CONTROL TECHNIQUES FOR DIGITAL COMMUNICATION

ERROR-CONTROL TECHNIQUES FOR DIGITAL COMMUNICATION

ARNOLD M. MICHELSON

ALLEN H. LEVESQUE

GTE Government Systems Corporation
Needham Heights, Massachusetts

A Wiley-Interscience Publication

JOHN WILEY & SONS

New York Chichester Brisbane Toronto Singapore

Library of Congress Cataloging in Publication Data:

Michelson, Arnold M.
Error-control techniques for digital communication.
"A Wiley-Interscience publication."
Bibliography: p.
Includes index.
1. Error-correcting codes (Information theory)
2. Digital communications. I. Levesque, Allen H.
II. Title.
TK5103.7.M53 1984 621.38 84-15327
ISBN 0-471-88074-4

Printed in the United States of America

10 9 8 7 6 5 4 3 2 1

With love and affection for our families:
Karen, Joanna, and Susan Michelson
and
Barbara, Karen, Amy, and Steven Levesque

PREFACE

In the nearly four decades that have passed since the original work of Shannon, Hamming, and Golay, error-control coding has matured into an important branch of communication systems engineering. The field of coding has yielded a great many results on the mathematical structure of codes as well as a number of very efficient coding techniques. These advances, combined with the steadily decreasing costs of digital circuits, provide the communication system designer with a cost-effective means to achieve highly efficient and reliable communications. Consequently one sees today a steadily increasing use of error-control coding in a variety of digital communication systems.

However, our experience in the communication industry has shown us that many engineers called upon to lead or participate in system or product development efforts, while having a thorough understanding of the principles of digital communication, often have difficulty in grasping the concepts of error-control coding and how to apply coding in a cost-effective design. Unlike many topics in electrical and communication engineering, coding does not lend itself readily to an intuitive understanding. Furthermore, in trying to use the literature on the subject, the engineer often is dismayed by the highly axiomatic mathematical presentation, much of which is based on modern algebra. He is further perplexed by the plethora of classes and types of codes, with little or no insight provided as to which codes to use on which channels and how to select and tailor code parameters to a system at hand. The designer needs to know which coding techniques are likely to be the most useful as well as how to select a coding technique to satisfy a particular system requirement.

This book is intended to provide the communication systems engineer with a guide to the application of error-control coding. The fundamental concepts of coding theory are developed using the minimum mathematical tools possible. The book is written for the engineer who is knowledgeable in modern communications principles—digital modulation techniques, characteristics of digital communication links, and the use of computers in communications—and who wishes to be able to make intelligent choices of coding techniques for application in practical systems. The intention is to provide the background enabling the designer to address the following questions:

1. What are the important fundamentals of coding?
2. What is the role of coding in communication system design?
3. What performance gains are achievable with coding, and what is the associated complexity?
4. What codes should be used, and how are the right code parameters selected?
5. What decoding techniques should be considered, and how are they implemented?
6. How are detailed performance results obtained?

What we have attempted to do in this book is to sort out those mathematical concepts that are necessary to understand and apply the most important coding techniques, and to present them in a clear, uncluttered, and well-motivated manner. In addition to providing the reader with only the essential mathematical tools, the book develops the theory and terminology of coding so that the reader can make effective use of more rigorous textbooks and journal papers on the subject. Examples are used throughout the book to illustrate fundamental concepts and to show how particular codes, encoders, and decoders are designed. We have also chosen to emphasize performance results. A large amount of detailed performance data is included, and we show how performance results for coded systems are obtained.

Although the principal objective of this book is to describe the most useful coding techniques with a straightforward mathematical presentation, the book has a further objective—to provide the engineer with an understanding of when it makes good sense to use coding and how to select and apply the most useful coding techniques in particular types of systems. It is hoped that the book will provide the engineer with insight, not found in other books, to facilitate the proper selection and application of error-control coding techniques in communication system design. The book also provides important reference material, enabling one to quickly specify a code design. We have focused on the techniques that we have seen applied in the last 10 years. By restricting our scope to a limited set of the more important techniques, we have been able to develop certain important topics, such as alternative decoding methods, in detail.

While the book is primarily intended as a reference for the practicing engineer, it can also be used as a text or reference book in an introductory course on error-control coding in which practical applications are to be stressed. The selection of material included derives from our experience over a period of nearly twenty years in the study and design of coding systems for a wide variety of applications as well as the presentation of coding theory concepts in a number of lectures and courses prepared for engineers in industry. The contents of the book are briefly outlined below.

Chapter 1 introduces the concept of coding in the context of the overall problem of designing an efficient communication system. Key results from information theory are developed. The relationship of coding to other system functions is discussed, and the distinction is made between coding used as an applique for the correction of channel errors and coding incorporated into a combined modulation and coding design that can be optimized for maximally efficient communication performance. The theoretical and practical limits for operation of a coded communication system are identified.

Chapter 2 introduces some simple parity-check codes and develops the basic principles of parity-check redundancy as a means of detecting and correcting errors. The concepts of strictly algebraic ("hard-decision") and probabilistic ("soft-decision") decoding are introduced with some simple examples using block codes. The concept of coding gain is also introduced. This chapter is written at an elementary level, and may be omitted by the reader already familiar with the basic concepts of parity-check coding.

Chapter 3 develops the mathematical representations for binary block codes and presents some of the fundamental properties of linear codes and the important concepts of Hamming distance and the syndrome of an error pattern. The intuitively useful notion that codewords can be defined as vectors in a vector space is presented.

Chapter 4 describes the important class of binary Bose-Chaudhuri-Hocquenghem (BCH) codes, which encompasses all the important binary block codes. Essential concepts of finite fields are discussed and used to develop the theory and structure of BCH codes. The design of a BCH code for a desired amount of error correction is described in detail. Techniques for decoding binary BCH codes are given in Chapter 5, and in Chapter 6 the class of nonbinary BCH codes, which includes the Reed-Solomon (RS) codes, is described. For both the binary and nonbinary codes, numerous examples are included to illustrate particular code designs, as well as encoding and decoding algorithms.

Chapter 7 is devoted to performance analyses of block codes with bounded-distance decoding. Both bounds on code performance and exact calculations are developed for binary and nonbinary codes. Performance results for a selection of binary BCH codes and RS codes are included.

Convolutional codes are introduced in Chapter 8. Two simple yet useful decoding techniques are described, namely, feedback and threshold decoding. Viterbi (maximum likelihood) decoding of convolutional codes is treated in Chapter 9, and sequential decoding in Chapter 10. Detailed performance results for Viterbi and sequential decoding are included, and the key design issues are treated at length.

Chapter 11 deals with the important issue of selecting an error-control technique to satisfy a particular performance objective or requirement. Complex techniques that achieve highly efficient communications are presented as well as schemes that provide moderate and small improvements over uncoded operation. Finally, automatic repeat request (ARQ) and interleaving are discussed in the context of coding for compound-error channels.

ARNOLD M. MICHELSON
ALLEN H. LEVESQUE

Needham Heights, Massachusetts
January 1985

ACKNOWLEDGMENTS

It is very difficult to properly acknowledge all the individuals who have had a direct or indirect influence on the development of this book. Our colleagues and friends at GTE, other commercial organizations, and government agencies as well as the many consultants who have worked with us over the years would make a very long list, and we surely will be guilty of at least a few serious omissions. However, we wish particularly to thank Dr. B. B. Barrow, Mr. G. Blustein, Dr. E. A. Bucher, Prof. D. L. Cohn, Mr. D. R. Esten, Mr. H. M. Gibbons, Mrs. Z. McC. Huntoon, Dr. J. G. Lawton, Mr. J. H. Lindholm, Mr. H. J. Manley, Mr. J. H. Meyn, Prof. J. W. Modestino, Mr. V. Oxley, Mr. A. W. Pierce, Dr. I. Richer, Dr. S. Stein, Dr. P. J. Trafton, Mr. J. H. Wittman, and Mr. L. S. Woznak. Parts of our presentations of the basic concepts of block and convolutional codes are based on course notes prepared by Prof. J. L. Massey and Prof. J. K. Wolf. Portions of the discussion of bounds on code performance in Chapter 9 are taken from a report prepared by Prof. J. G. Proakis, who also gave us much encouragement; we owe him our sincere thanks. We would also like to thank Mr. K. G. Tong, who originally suggested that we write this book, and Prof. R. L. Pickholtz, whose interest and encouragement were instrumental in its completion.

We would like to express our special appreciation to Dr. R. J. Turyn, who read the entire manuscript very carefully and made many excellent suggestions for improvements in form and style of the presentation. A number of useful comments and suggestions for improvement were also made by Dr. D. F. Freeman and Prof. W. C. Gore.

We owe a special thanks to our secretary, Mrs. Dorothy H. Wales, for her cheerfulness, dedication, and unflagging perseverance at the typewriter and word processor throughout a seemingly endless succession of revisions of the manuscript.

We are also indebted to the management of GTE Government Systems Corporation, the Communication Systems Division, for generously providing support for the preparation of illustrations for the book. This considerable task was carried out skillfully by Mrs. Lucy O. Carroll, Technical Editor, and Mr. Frank C. Fancher and Mr. Robert W. Davis, Technical Illustrators.

Finally, we express our special appreciation to our families, for their patience, support, and love throughout this long project. They made it all worthwhile.

A.M.M.
A.H.L.

CONTENTS

CHAPTER ONE

Reliable Transmission of Digital Information

The purpose of this book is to provide the communication systems engineer with a basic understanding of error-control coding techniques and the role that coding plays in the design of efficient digital communication systems. Described in its simplest terms, error-control coding involves the addition of redundancy to transmitted data so as to provide the means for detecting and correcting errors that inevitably occur in any real communication process. Thus coding can be used to provide a desired level of accuracy in the digital data delivered to a user. There are, however, other ways to achieve accurate transmission of digital data, and this book is intended to aid the communication system designer in deciding when it makes sense to use coding and when it does not, in choosing a coding technique appropriate to the application and performance requirements at hand, and in evaluating the performance achievable with the chosen technique.

For example, in many communication systems, an alternative to the use of coding is simply to provide sufficient signal energy per unit of information to ensure that uncoded information is delivered with the required accuracy. The energy needed might be provided by setting signal power at a sufficiently high level or, if power limitations prevail, by using some form of diversity transmission and reception. However, in many cases, error-control coding can provide the required accuracy with less energy than uncoded operation and may be the economically preferred solution in spite of an increase in system complexity. Cost savings through the use of coding techniques can be dramatic when very high accuracy is needed and power is expensive. Furthermore, in some applications the savings in signal power are accompanied by important reductions in size and weight of the communication equipment.

The levels of performance that can ultimately be achieved with coded communication systems are given by the remarkable theorems of Claude Shannon, who in 1948 laid the foundation of the science of *information theory* in a famous paper entitled "A Mathematical Theory of Communication" [157]. The basic theorems of information theory not only mark the limits of efficiency in communication performance but also define the role that coding plays in achieving these limits. That is, digital codes are shown to be an efficient way of constructing the waveforms to be transmitted in order to achieve optimum communication performance for some applications.

Shannon's 1948 paper presented a statistical formulation of the communication problem, unifying earlier work by Hartley [61], Wiener [178], Rice [148], and Kotel'nikov [86]. Shannon's work sharply contradicted the long-standing intuitive but erroneous notion that noise places an inescapable limitation on the accuracy of communication. Shannon proved that the characteristics of a communication channel, namely the noise level, bandwidth, and signal power, can be used to derive a parameter C, called *channel capacity*, that gives the upper limit on the rate at which information can be transmitted through the channel and received reliably. Shannon's results showed that as long as the information transmission rate is below C, the probability of error in the information delivered can in principle be made arbitrarily low by using sufficiently long coded transmission signals. Thus noise limits the achievable information communication rate, but not the achievable accuracy. Much of the research in communication theory since the appearance of Shannon's early work has been concerned with extending and refining his basic results and with finding ways of approaching the full realization of these results in practical communication system designs. The development of error-control coding techniques has been a central element in this research.

In this book we present the most important of the error-control coding techniques that have been developed since Shannon's pioneering work. That is, we consider those techniques that have actually been used effectively in real communication systems. In this introductory chapter we begin with a description of the key elements of a modern digital communication system as well as the channel models that are used throughout the book. A heuristic discussion of information theory follows, concluding with a presentation of the key result, the channel coding theorem. It is not necessary to have a detailed understanding of information theory in order to make effective use of error-control coding techniques. However, a familiarity with the underlying principles and the meaning of the channel coding theorem is important. The fundamental limit on the efficiency of a digital communication system is given by the channel coding theorem, and this provides the gauge for measuring the overall efficiency of any given system design.

We then review the basic digital modulation and demodulation techniques. Performance curves are included that show that even the best of the practical signaling schemes fall far short of the performance limit given by the channel coding theorem. It will be seen in subsequent chapters that judicious choice of modulation and coding techniques can, in many applications, provide significant improvements in communication efficiency. The chapter concludes with a discussion of the proper way to design a digital communication system. Specifically, an analytical approach is given that permits joint optimization of the key communication functions.

1.1. THE COMMUNICATION SYSTEM DESIGN PROBLEM

We can state the task of the digital communication system designer as that of providing a cost-effective system for transmitting information from a sender to a user at the rate and level of accuracy that the user requires. The key parameters of the design are transmission bandwidth, signal power, and the complexity of the implementation chosen. The information transmission rate and accuracy of the delivered information are typically determined by the user's requirements.

The transmission bandwidth is often constrained by factors specific to the particular transmission medium used. For example, telephone circuits are separated into nominal 3-kHz bandwidth segments by longstanding engineering practice in the telephone industry. Similarly, there are standard bandwidths for individual channels on terrestrial radio circuits and satellite links due to established government regulations on spectrum utilization. In other cases, however, bandwidth constraints are not a critical issue. Examples include links to and from vehicles in deep space, where wide transmission bandwidths for a few individual links can be chosen freely without concern about possible interference with other users of the spectrum.

Finally, signal power and implementation complexity are system characteristics that are usually very much under the designer's control, and possible trade-offs between power and complexity are central issues in the design task. Both characteristics represent cost factors for the designer to consider. For example, in most systems a desired level of accuracy in the information delivered can be achieved by supplying enough power in the transmitted signal to overcome channel disturbances that produce errors. An alternative to increasing signal power is to add systematic redundancy to the transmitted information in the form of error-control coding. However, the use of coding adds complexity to the system, particularly for the implementation of the decoding operations. Since the addition of redundancy also implies the need to increase transmission bandwidth, the design trade-off must include considerations of bandwidth. In fact, in applications where bandwidth is strictly limited or very costly and when we are not permitted to lengthen message transmission time, it is difficult to use coding effectively as a means of improving information accuracy, and increasing signal power may be the only means available. These issues will be discussed in detail later in this chapter, when the fundamental results in information theory are reviewed and the implications for the design of efficient communication systems are presented.

It is clear from the outset that with any real communication system we cannot expect to receive exactly what is transmitted. At the very least, we can expect noise to be added to the transmission, causing random errors. As was stated earlier, the work of Shannon showed that channel noise limits the rate at which we can communicate reliably but not the achievable level of accuracy. The purpose of this introductory chapter is to present and illuminate this important result. It is necessary first to review the key elements of a digital communication system and the limitations imposed by the physical world. Then the role that error-control coding plays in the design of an energy-efficient communication system can be discussed in detail.

1.2. ELEMENTS OF A DIGITAL COMMUNICATION SYSTEM

The basic elements of a one-way communication system are illustrated with the general block diagram shown in Fig. 1.1. We now examine each of these elements in detail.

1.2.1. Information Source

The source of information to be transmitted may be either analog or discrete. An *analog source* produces time-continuous signals, while a *discrete source* produces sequences of discrete symbols. Examples of signals from an analog source are speech signals, radar outputs, and photographic scan data. Typical discrete sources are computer data files and messages generated at teleprinter terminals.

For analog sources, techniques are needed to efficiently represent the analog signals by sequences of digital symbols. Ordinarily this is done simply with a sampler and analog-to-digital converter. Ideally, we would like to represent the source information by the smallest number of digital samples to make the most efficient use of the digital communication system. In other words, we would like to remove as much redundancy as possible from the source information prior to transmission. Techniques for converting arbitrary information sources into digital messages efficiently are described broadly as *source coding*

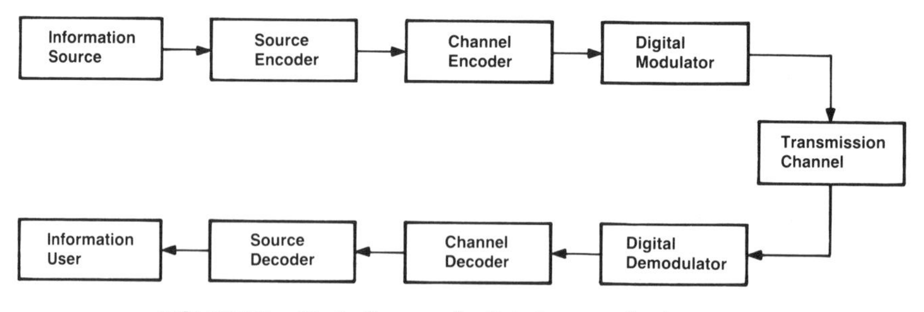

FIGURE 1.1. Block diagram of a digital communication system.

techniques. There is now a considerable body of literature on the subject of source coding, dealing with the theoretical limits on performance as well as with some very effective techniques for coding specific types of sources. To treat this subject in depth is outside the scope of this book, and we refer the interested reader to the literature on source coding [6, 175].

For our purposes in this book, we assume we are given a digital information source and our job is to provide an effective and efficient system for communicating the information it produces from one place to another. Further, we assume that the source generates a stream of statistically independent, equally likely discrete symbols. In particular, we assume that it produces binary digits, which we call *bits*, each bit occurring with equally likely binary values independent of all other bits generated. We shall say that the source produces data at a constant rate of R_s bits per second, which is defined as the *information rate of the source*. The purpose of the communication system is to deliver the source data to a user at the rate R_s and at some required level of accuracy, which is typically stated as an upper limit on acceptable *probability of bit error* or *bit-error rate* in the delivered information. It will be seen in a later discussion that the overall efficiency of the communication system is directly related to the amount of signal energy needed to deliver each information bit with the required accuracy. We denote this amount of energy as E_b and call it the required *energy per information bit*. The required signal power S is therefore given by $S = E_b R_s$.

1.2.2. Channel Encoder

The channel encoder performs all the digital operations needed to prepare the source data for modulation. We define the encoder here in a general way to encompass a variety of possible operations. In the simplest case, where no redundancy is to be added and the transmission in the physical channel is to use a binary signaling alphabet, the encoder has no function. If no redundancy is to be used but the transmission alphabet is to be nonbinary, the encoder performs the necessary binary-to-nonbinary symbol conversion. For example, if an 8-ary signaling alphabet is to be used, the encoder accepts source bits in successive blocks of three bits each and produces 8-ary symbols at a rate that is one-third of R_s.

If binary error-control coding is to be used, the encoder accepts information bits at the rate R_s and adds redundancy, producing encoded data at a higher rate R_c. For encoding with a *block code*, the encoder accepts information in successive k-bit blocks and for each k bits generates a block of n bits, where $n \geq k$. The n-bit block is called a *code block* or a *codeword*. Thus the encoder releases bits at a rate $R_c = R_s(n/k)$. We define the dimensionless ratio $R = k/n$ as the rate of the code, or simply the *code rate*. Note that we let n have values $n \geq k$ rather than simply $n > k$, to include uncoded operation as a special case. If the error-control coding is nonbinary, say M-ary where $M = 2^m$ and m is an integer greater than 1, the encoder accepts information bits in blocks of km bits each and produces encoded blocks of n M-ary symbols each. Again, the code rate is $R = k/n$.

For encoding with a *convolutional code*, the encoder accepts information bits as a continuous stream and, for a binary code, generates a continuous stream of encoded bits at a higher rate. The information stream is fed to the encoder b bits at a time, where b is typically in the range 1 to 6. The encoder operates on the current b-bit input and some number of immediately preceding b-bit inputs to produce V output bits, with $V > b$. Thus the code rate is $R = b/V$. The number of successive b-bit groups of information bits over which each encoding step operates is called the *constraint length* of the code, which we also denote by k. The encoder for a convolutional code might be thought of as a form of digital filter with memory extending $k - 1$ symbols into the past. A typical binary convolutional code will have $b = 1$, $V = 2$ or 3, and k in the range 4 to 7, although in special applications constraint lengths in the range 30 to 70 might be used.

To use a convolutional code with nonbinary transmission, each b-bit input to the encoder results in the generation of V M-ary coded symbols, where usually $M = 2^m$, and $mV > b$. A typical rate-1/2 encoder for a 16-ary ($m = 4$) transmission alphabet might have $b = 4$, $V = 2$, and $k = 2$ (four-bit symbols).

We shall not delve any further into the details of code design at this point, our immediate purpose having been served by the introduction of the concepts of code rate, block length, and constraint length. As we shall see in later discussions, these are the key parameters of a code design, since the reciprocal of the code rate gives us a measure of the required bandwidth expansion and the code block length or constraint length and rate provide a measure of the complexity of the required encoding and (more important) decoding operations. In Section 1.4 we shall see that much can be said about the communication performance achievable with well-designed codes by dealing with only these design parameters.

1.2.3. Digital Modulator

The function of the modulator is to match the encoder output to the transmission channel. The modulator accepts binary or M-ary encoded symbols and produces waveforms appropriate to the physical transmission medium, which is always analog. In many systems where coding is to be applied, the modulation and demodulation techniques and equipment are difficult or impossible to modify or replace. In other cases, the modulation technique is fixed, but changes in the method of demodulation are feasible. In yet other applications, it is possible to design the modulation and demodulation system along with the coding technique, and greatly increased latitude is provided for overall optimization of the design.

It has been conventional in much of the communication literature to define "the channel" as representing that portion of the communication system that the designer is unable or unwilling to change. In following this convention, if the modulation and demodulation equipment (usually shortened to *modem* for convenience) is not available for modification, those functions would be incorporated into the definition of the channel. However, we prefer to depart from this convention to the extent that while we shall treat the modem as part of the channel for purposes of analysis, we ask the reader to bear in mind that the design of an efficient system is best done by designing the modem functions in conjunction with the encoding and decoding functions. Section 1.6 addresses this point in more detail.

For *binary modulation*, the modulator simply converts a binary digit, 0 or 1, to a waveform, say $s_0(t)$ or $s_1(t)$, respectively, of equal duration T_s. For *M-ary modulation*, the M possible encoded symbols are converted to a corresponding set of M waveforms $s_0(t), s_1(t), \ldots, s_{M-1}(t)$. It is assumed that the reader is familiar with the common forms of digital modulation. For binary signaling, conventional modulation types include phase shift keying (PSK), differentially encoded PSK (DPSK), and frequency shift keying (FSK). Nonbinary forms of these basic modulation types are M-ary PSK (MPSK), M-ary DPSK (MDPSK), and M-ary FSK (MFSK). With the conventional forms of these modulation types, the nominal bandwidth of each waveform $s_i(t), i = 0, 1, 2, \ldots, M-1$, is approximately $1/T_s$. However, for *spread spectrum* signaling, as is implied by the name, the bandwidth of each waveform can be much wider than $1/T_s$, perhaps by as much as several orders of magnitude. For example, a spread spectrum version of binary PSK might utilize waveforms $s_0(t)$ and $s_1(t)$ in which $s_0(t)$ is a sequence of much shorter binary PSK pulses, usually called *chips*, and $s_1(t)$ is the complement of the chip sequence in $s_0(t)$. Spread spectrum signaling is used as a *multiple-access* technique and also as a means of protecting a communication system against *jamming*. For further discussion of spread spectrum systems, the reader can refer to a special *IEEE Transactions* issue [70] as well as books on the subject [31, 68].

We shall return to the modulation and demodulation functions in Section 1.5.

1.2.4. Transmission Channel

We include in the term *transmission channel* all the operations required to prepare the baseband (low-pass) modulated waveforms for transmission in the physical channel, the transmission medium itself, and the receiving operations required to bring the signals to the point just prior to demodulation. In this way, we incorporate into the transmission channel any practical limitations or impairments in the equipment. As a practical matter we are primarily concerned here with power and bandwidth limitations, which are reflected in the design of the transmitting and receiving equipment.

Transmitted signal power provides the obvious means in many systems for providing a required level of accuracy in received information. However, signal power cannot be increased arbitrarily. In telephone networks, for example, signal levels are fixed by established industry standards. In radio communication systems, there is more freedom in selecting power levels, but practical physical and economic limitations apply. An increase in power level invariably implies increases in size, weight, and cost of transmitting equipment. Even if the added cost is judged acceptable, there are some applications where mobility is very important, and strict limitations on the size and weight of the equipment must be complied with. In yet other applications, particularly in very low regions of the radio spectrum, enormous amounts of energy are required to radiate usable signals, and therefore transmitted signal power is a dominant factor in the cost of the system.

Bandwidth is the other critical design parameter governing achievable performance, since it limits the rate at which we can modulate waveforms in the channel. Restrictions on the choice of bandwidth are more or less severe depending upon the transmission medium. In telephone systems and many radio systems, where strict channelization standards are in place, bandwidth can be provided only in fixed increments, such as 3 or 6 kHz. Therefore, providing increased bandwidth in turn implies leasing additional wireline channels or acquiring added radio channel allocations and transmission equipment, the latter typically designed with standard bandwidths. In some parts of the radio spectrum, crowding is a serious problem, and it is difficult to use more bandwidth without encountering significant levels of interference with other communication signals.

Noise in received signals constitutes the most prevalent factor limiting the performance of a communication system, since noise limits the ability of the demodulator to reliably distinguish one modulated waveform from another, thereby producing errors in the demodulator output. Thermal noise is always present in electrical circuitry, for example, in the front end of the receiving equipment. However, receiver noise is not necessarily the primary concern, since in many parts of the radio spectrum other sources of noise are also significant. For example, atmospheric impulse noise due to lightning discharges can be of a sufficiently high level to be the dominant factor limiting performance. Atmospheric noise can also affect wireline communication systems, since considerable energy can couple into transmission lines during thunderstorms. Additional forms of impulsive noise affect telephone networks, arising from transients in switching equipment as well as from accidental circuit interruptions during maintenance. What distinguishes impulsive noise in its various forms from the ever-present thermal noise is a distinct difference in temporal characteristics as manifested in the ways digital errors occur. Thermal noise is broadband, is essentially steady in its power level, and has Gaussian amplitude statistics. Therefore, errors tend to occur independently from one signaling interval to the next, the rate of occurrence being derivable in a straightforward way from knowledge of the Gaussian distribution. Impulsive noise, on the other hand, is characterized by relatively long quiet intervals punctuated by short periods of intense noise. This characteristic in turn results in long error-free intervals interspersed with short *bursts* of errors.

If one takes detailed account of the different characteristics of Gaussian noise channels and burst-error channels, one is led to quite different coding techniques for each. However, what is usually done is to select the modulation and coding schemes with a view toward the limitations imposed by the Gaussian background noise and then to adapt certain features of the coding implementation to the particular characteristics of the burst-error phenomena if they are of concern for the application at hand. We shall say more on this point in Chapter 11.

Although we shall be concerned primarily with additive noise, many real communication channels exhibit other phenomena that severely limit communication performance. For example, many channels contain a sufficient amount of time dispersion that the received symbols flow into one another. This effect is commonly called *intersymbol interference*. In wireline channels, the time dispersion arises largely from significant nonlinearity of the phase characteristic within the channel bandwidth.

Intersymbol interference from any cause is a form of "self-noise" in a digital communication system, which cannot be overcome by increasing signal power. For this reason, channels affected in this way exhibit an irreducible error rate at high levels of signal-to-noise ratio (SNR), which cannot be avoided except by implementing some scheme for dealing directly with the time dispersion. Error-control coding usually does not provide an attractive approach, since intersymbol interference comes about by trying to signal as rapidly as possible within a given transmission bandwidth. It is implicit that the added bandwidth that coding would require cannot be readily provided. For this reason, intersymbol interference is treated by other techniques, for example, on telephone channels by *adaptive equalization* [98, 139].

On many radio channels, time dispersion is due directly to the *multipath* nature of signal propagation at extended distances. For example, in the high frequency (HF) band, 3 to 30 MHz, communication beyond the horizon is accomplished by refraction of signals at various layers of the ionosphere. The structure of the ionosphere causes a transmitted HF signal to arrive at a distant receiving site by a multiplicity of *propagation modes*, the modes in general having different path delays.

A direct consequence of multipath on radio channels is the phenomenon known as signal fading or simply *fading*. This comes about in ionospheric propagation due to the fact that the state of the ionosphere is dynamic. Ions within each layer are constantly in motion, and the layers also move relative to one another. As a result, the summation of several fluctuating modal components in a received multipath signal, with component signal phases randomly sliding in and out of alignment, produces the random fluctuations in received signal amplitude called fading. Associated with the amplitude variations in fading are fluctuations in the instantaneous phase of the received signal. Thus the multipath structure directly accounts for time dispersion, and the time-varying nature of the multipath accounts for fading. Radio channels that behave in this way are often called *fading multipath channels*. In addition to the HF frequency band already mentioned, examples include the VHF and UHF frequency bands (30 to 300 MHz and 300 to 3000 MHz), when used for beyond-the-horizon terrestrial communication by ionospheric and tropospheric scatter propagation, and the SHF band (3000 to 30,000 MHz), which is used for satellite communication.

There has been some success in applying adaptive equalization techniques to radio channels [3, 114]. However, the fading process greatly complicates the use of equalization, since, unlike having to adjust to a static phase characteristic, the equalizer has to accurately keep up with rapid continuous changes in signal amplitude and phase. For this reason, rather complex algorithms are used in applying these techniques to radio channels, and much of this work must still be regarded as developmental.

Where fading cannot be treated effectively by adaptive equalization, the resulting error characteristics can be dealt with by the use of efficient error-control coding techniques. On fading channels, the received errors tend to be clustered in bursts that occur in the intervals when the signal attenuation is large, that is, when the channel is going through deep fades. Since the propagation variations that produce fading are random, the durations of the error bursts and the intervening intervals of relatively error-free data are themselves random. However, the statistical parameters of the error clustering behavior can often be predicted within broad limits, given the operating frequency of a radio system and certain details of the transmission path. Some attention is given in Chapter 11 to the problem of applying error-control coding to burst-error channels.

1.2.5. Digital Demodulator

At the receiving end of the communication link, the demodulator provides the interface between the transmission channel and the functions that compute and deliver estimates of the transmitted data to the user. We include the rf receiving equipment in our definition of the transmission channel. The demodulator operates on the waveform received in each separate transmission symbol interval and produces a number or a set of numbers that represent an estimate of a transmitted binary or M-ary symbol. In some applications, the designer may choose the level of precision of this estimate.

In the simplest cases, the demodulator is designed to make a definite decision for each received symbol, that is, 0 or 1 for binary transmission or one of $0, 1, \ldots, M - 1$ for M-ary transmission. It is convenient to refer to such cases as *hard-decision demodulation*. Since the transmitted waveforms have been corrupted by the various nonidealities of the transmission channel, the symbol decisions are subject to error, and the average rate of occurrence of symbol errors, taken as a fraction of the total number of symbols received over a long period of time, is called the *symbol-error rate* or *probability of symbol error*. For binary transmission, this is the *bit-error rate* or *probability of bit error*. It is also conventional to apply the same bit-error terminology after conversion of M-ary symbol decisions to their binary representations. In a system where no error-control coding is used, these error rates are the error rates for the data delivered to the user. In a coded system the error rate at this point in the system is often called the *raw-channel error rate* or the *uncoded error rate* to make a distinction from the error statistics measured after decoding is performed.

All real transmission channels are, of course, analog channels that deliver waveforms that can in principle vary continuously over some range limited only by nonlinearities in the transmission medium and the receiving equipment. Thus the demodulator can be viewed as a form of waveform filtering followed by quantization, say to Q levels. The case of hard-decision binary demodulation thus requires quantization to $Q = 2$ levels. If the output of the binary demodulator is quantized to $Q > 2$ levels, we refer to this as *soft-decision demodulation*. In the limiting case of $Q = \infty$, the demodulator output is unquantized, corresponding to an analog matched filter output being delivered directly. For M-ary transmission, $Q > M$ constitutes soft-decision demodulation. Quantization incurs a loss of information, and thus soft-decision demodulation preserves information that can, we shall see in later discussions, be utilized profitably with an appropriate error-control decoding technique.

1.2.6. Channel Decoder

In a system using error-control coding, the decoder accomplishes the conversion of demodulator outputs into symbol decisions that reproduce, as accurately as possible, the data that was encoded by the channel encoder. For block coding, the decoder accepts consecutive blocks of n demodulator outputs and produces k decoded symbols for each block. With convolutional coding, the decoder accepts a steady stream of demodulator outputs and operates over the current received symbols and some number of previous symbols, producing b decoded outputs for each group of V received symbols. For both block and convolutional codes, the decoder attempts to make definite symbol decisions, binary or M-ary in accordance with the code design. However, the inputs need not be definite symbols.

Decoding techniques that operate on hard-decision demodulator outputs are commonly termed *hard-decision decoding* techniques. As will be seen in later chapters, they are essentially algebraic equation-solving algorithms. On the other hand, there are a number of decoding techniques, for both block and convolutional codes, that operate on soft-decision demodulator outputs and are collectively termed *soft-decision decoding* techniques. These techniques more nearly resemble signal correlation or matched-filtering operations than equation-solving routines. For many codes, practical soft-decision decoding algorithms are available that outperform the best hard-decision decoding algorithm for the same code. The range of potential performance advantage will depend to a great degree on characteristics of the transmission channel, the performance margin generally being smallest on steady-signal Gaussian noise channels.

Error probability at the output of the decoder provides an important measure of the overall performance of the communication system. In fact, for convolutional codes, performance is usually stated in terms of post-decoding bit-error rate or M-ary symbol-error rate. However, there are other measures of communication performance that are meaningful. As an example, for block-coded systems, performance can conveniently be given in terms of the probability of correctly decoded blocks. It is also possible to state the performance of block-coded systems in terms of average post-decoding bit-error rate, where the error events are averaged over all decoded k-symbol blocks. For either convolutional or block codes, other measures of performance are sometimes used as well. For example, in a system designed to transmit and receive messages of a given length, if a message decoded with one or more errors is judged to be unacceptable, the appropriate measure of performance might well be the probability of receiving an error-free message.

Thus we see that in systems designed with error-control coding there is some flexibility in the way that performance is measured. It is important to note that there is no single figure of merit that can be used to realistically compare various coded and uncoded system alternatives for an arbitrary application, since requirements vary widely. The performance afforded by many error-control coding techniques is considered in detail in Chapters 7 and 9 through 11 for several measures of communication performance.

1.2.7. Source Decoder

The final stage of processing indicated in Fig. 1.1 is source decoding. The *source decoder* accepts the sequence of symbols from the channel decoder and, in accordance with the encoding method used, attempts to reproduce the information originally generated by the analog source. Generally speaking, the output of the source decoder is an approximation to the original source output, with discrepancies due to errors in channel decoding as well as loss of detail suffered in source encoding and perhaps in decoding as well. For some analog sources, the fidelity in reproduction of the source information can be measured by a simple statistic, for example, the mean-squared error between corresponding samples of the original source output and the output of the source decoder. For some sources, however, it is very difficult to find a reasonable mathematical measure of fidelity. A good example is that of speech signals, for which statistical characterization is known to be very difficult.

We shall not pursue source coding and decoding in great detail, since we wish to concern ourselves primarily with the problem of transmitting and accurately reproducing symbols generated by a digital information source. However, we do make one final point here on the subject of source coding. In a system that uses both source coding and error-control coding, the source coding operation can be viewed as one that removes the natural, perhaps inefficient, redundancy from the source so that it can be replaced with highly structured and more efficient redundancy in the form of a well-chosen error-control code. This, in fact, is exactly what information theory requires us to do if we wish to achieve accurate transmission of the source information with the least expenditure of signal energy. These points will be made clearer in the remaining sections of this chapter.

1.3. IMPORTANT CHANNEL MODELS

A *model* of a communication system is a mathematical representation defined to realistically describe the way signals are constructed and processed and the way they are affected by the real-world communication environment. In Fig. 1.2 we show a simplified model of a digital communication system, in which the information to be transmitted is assumed to be generated by a discrete or digital information source. Using this model we next address the question of how to efficiently convey information to the user.

1.3.1. The Discrete-Time Channel

It is conventional to define a *channel model* to include the modulator, the demodulator, and all the intervening transmission equipment and media. This model, enclosed within dashed lines in Fig. 1.2, is compactly defined by the set of modulator inputs, the set of demodulator outputs, and the statistics that relate the possible outputs to each possible input. This is commonly called a *discrete-time channel model* or simply a *discrete channel*. The input-to-output statistics represent the ways in which the modulated signals are affected by amplitude and phase fluctuations, noise, interference, and equipment nonidealities and impairments. In most cases it is very difficult to define a

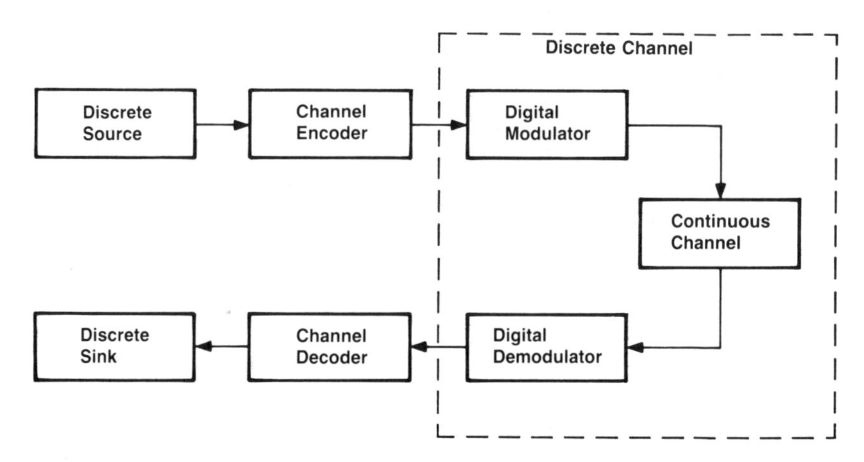

FIGURE 1.2. Model of a digital communication system.

model that thoroughly accounts for all the disturbances affecting the signals, and one must resort to reasonable approximations. However, experience has shown that even reasonably simple channel models can provide a sufficient degree of realism to enable proper design of efficient systems. Furthermore, simplified models often yield insights into underlying principles, which can be obscured by a myriad of details in more elaborate, though more accurate, models. We shall consider several channel models commonly used to analyze and design digital communication systems.

Most of the remaining discussion in this chapter will consider various forms of a channel model called the *discrete memoryless channel* (DMC), which is defined by an M-ary set of input symbols $\{x_i\}$, a Q-ary set of output symbols $\{y_j\}$, and a set of conditional probabilities, called *transition probabilities*, which we can write as

$$P(y = y_j | x = x_i) = P(y_j | x_i)$$

where $i = 0, 1, \ldots, M - 1$, and $j = 0, 1, \ldots, Q - 1$. The description of the channel as memoryless refers to the assumption that the output symbol at any instant of time depends statistically only on the input symbol at that time. The application of the DMC model to any real independent-error channel simply requires determination of the transition probabilities from definitions of the transmitted waveforms, signal power levels, and transmission channel characteristics, as well as the description of the demodulator. Examples of types of channels to which the DMC model does not apply are channels affected by atmospheric impulse noise or intersymbol interference.

1.3.2. The Binary Symmetric Channel

We now describe an important example of a DMC model. Suppose that binary modulation is used and that hard-decision demodulation is performed at the receiving end of the link. We let the modulator input x have value 0 or 1, and the demodulator output y have value 0 or 1. Let us now suppose that for either input value of x, and regardless of the transmitted or received values of any earlier bits, x is received in error with probability p or received correctly with probability $1 - p$. Using standard notation for conditional probabilities, we write this as

$$P(y = 1 | x = 0) = P(y = 0 | x = 1) = p$$

$$P(y = 0 | x = 0) = P(y = 1 | x = 1) = 1 - p$$

With these definitions we have combined the modulator, the transmission channel, and the demodulator into a compact binary-input, binary-output model depicted by the transition diagram in Fig. 1.3*a*. This simple channel model is known as the *binary symmetric channel*, usually abbreviated as BSC.

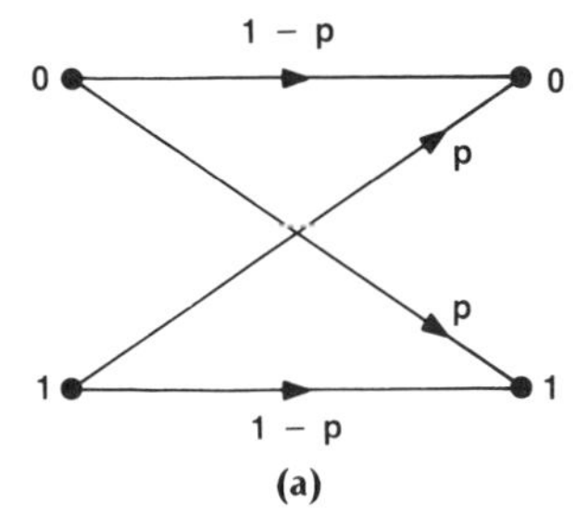

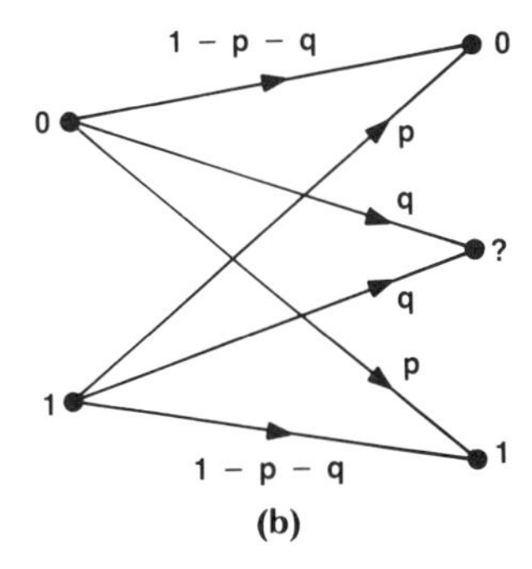

FIGURE 1.3. Discrete channel models. (*a*) Binary symmetric channel, (*b*) binary symmetric erasure channel.

1.3.3. The Binary Symmetric Erasure Channel

Another special case of the DMC is a binary-input, ternary-output channel called the *binary symmetric erasure channel* (BSEC). This channel model, depicted in Fig. 1.3*b*, includes a symmetric transition from either input symbol to an output symbol labeled ? to denote ambiguity. This model corresponds to a demodulation rule in which certain outputs are judged not to give a sufficiently reliable indication (e.g., due to a weak received signal) of which binary symbol was sent, and those outputs are erased as they leave the demodulator. A demodulation rule producing the three outputs 0, ?, and 1 is the simplest example of *soft-decision demodulation*.

1.3.4. The Additive White Gaussian Noise Channel

A channel of the DMC type having great theoretical and practical importance is one in which the output is simply the input plus broadband Gaussian noise. It is conventional to represent broadband Gaussian noise with the white Gaussian noise model. *White Gaussian noise* is defined to be a random process, each sample of which is a zero-mean Gaussian random variable and whose power spectral density is flat over the entire frequency range $-\infty \leq f \leq \infty$, with a level $N_0/2$ watts per hertz. Equivalently, the one-sided *noise spectral density* is N_0, so that, for example, a filter with a rectangular passband of width W hertz will pass N_0W watts of noise power.

The *additive white Gaussian noise* (AWGN) channel can now be described simply in terms of the input x and the output y, which are related by

$$y = x + n_G$$

where n_G is a zero-mean *Gaussian random variable* with variance σ^2 and the input x can have any one of M discrete values, where $M \geq 2$. That is, the conditional probability density function of the output y, given an input x_i, is given by

$$p(y|x = x_i) = \frac{1}{\sqrt{2\pi}\,\sigma} e^{-(y-x_i)^2/2\sigma^2}$$

The AWGN channel is an accurate model for many communication links, such as satellite and deep-space links, in which the dominant effect limiting communication performance is additive thermal or galactic noise.

1.4. INFORMATION THEORY AND CHANNEL CAPACITY

In this section we review some of the principal results of information theory and use them to provide insight into the role that coding plays, in combination with the choice of modulation technique, to achieve reliable communication with an efficient expenditure of signal energy. A heuristic explanation of Shannon's theorems on channel capacity is outlined. More complete presentations of the principles of information theory can be found in several excellent textbooks, such as Gallagher [49] or Viterbi and Omura [175].

1.4.1. Logarithmic Measures of Information

The beginnings of information theory lie in early papers by Nyquist [117] and Hartley [61], who were concerned with the achievable transmission capabilities of telegraph circuits and, more generally, with a mathematical characterization of the ultimate limitations on the amount of data that can be transmitted reliably over any given physical channel. The formulation of such problems requires an explicit mathematical measure of information. Hartley considered this question in relation to an information source producing messages, where each message is drawn with equal likelihood from a discrete set. Hartley suggested that the most natural measure of information is the logarithmic function. That is, the information content of one message drawn from a set of M equally likely messages is $\log M$, where the logarithm base is arbitrary and depends on the basic unit of information. (Note that all logarithmic measures can be related directly to one another, since $\log_a x = (\log_a b)(\log_b x)$ for any x.) Conventionally, the logarithm base is chosen to be 2, and the unit of information is called a *bit*, a contraction of "binary digit."

Note that earlier in the chapter we used the term "bit" in a more general way to mean any binary digit; both usages are common in the literature.

The logarithmic measure of information is logical and intuitively appealing, since if we equate a unit of information to a unit of storage capacity, the quantity $\log M$ corresponds exactly to the amount of storage needed to hold a representation of each of the possible messages. With binary storage, for example, we can hold a representation of one of 128 messages with seven bits, since $2^7 = 128$. Given this convention, we can define the *rate of an information source*. We assume that the source produces equally likely M-ary symbols, with the output symbols being independent from one symbol interval to the next. The rate of the source is then $\log_2 M$ bits per symbol. Throughout most of this book we assume a source that produces equally likely information bits or symbols. However, in the discussion at hand, a more general measure of information is needed for sources in which messages or symbols are not necessarily generated with equal probability.

A measure of information for general sources was provided by Shannon with the application of the concept of *entropy* to an information source. The concept of entropy has long been used in the field of physics as a measure of the randomness of a physical system whose states can be described only in statistical terms. Entropy provides an appropriate measure of the a priori uncertainty of any symbol or message to be produced by a discrete information source. Let us say that a source produces any one of M symbols, where the probabilities of occurrence are $p_1, p_2, \ldots, p_M$, with $p_1 + p_2 + \cdots + p_M = 1$. The entropy of the source is defined by

$$H = -\sum_{i=1}^{M} p_i \log p_i \tag{1.1}$$

This definition in effect averages Hartley's logarithmic information measure with respect to the set of probabilities of the individual source symbols. It is readily seen that for a set of equally likely symbols, the entropy of the source is simply $H = -\log p_i = \log M$. The entropy function H provides a measure of the average amount of information "produced" per symbol by the source. Equivalently, from the point of view of the ultimate recipient of the information, it represents the average "prior uncertainty" of the information in each symbol generated by the source. The notion of prior uncertainty will be useful when we discuss measures of information transfer through a channel. But first we want to point out certain important properties of the entropy function.

Using Eq. (1.1) we readily obtain the entropy function for a binary information source as

$$H = -p \log p - (1 - p)\log(1 - p)$$

where p is the probability of occurrence of either of the symbols. The reader should calculate and sketch a few values of H and note that the entropy is maximized when $p = 0.5$, which yields one bit of information per symbol. For an M-ary information source, it would be a simple matter to show that the entropy is maximized by having the outputs occur with equal probability, an intuitively satisfying result.

The entropy function for a discrete source generalizes readily to the case of a continuous information source, for example, a source whose output is a voltage that can have any value over a continuous range. Let us say that we have such a source and that a sample of its output has a probability density function $p(x)$. The entropy of this source is given by

$$H = -\int_{-\infty}^{\infty} p(x)\log[p(x)]\,dx$$

where the summation in Eq. (1.1) has simply been converted to an integral.

Since we want the entropy function of a continuous source to accurately describe a physical reality, such as the information content of a signal voltage, we need to place reasonable constraints on $p(x)$. It is interesting to consider several such constraints and ask what $p(x)$ maximizes the entropy. For instance, it can be shown easily that if the only constraint is to confine the output to some finite interval, the entropy is maximized by letting $p(x)$ be uniform over the given interval. This might reasonably have been guessed by generalization from the case of a discrete M-ary symbol source. Another important case is one that we consider in the following example.

EXAMPLE: A CONTINUOUS INFORMATION SOURCE WITH FIXED VARIANCE. Consider a continuous information source whose output x has the probability density function $p(x)$. We wish to find the function $p(x)$ that maximizes the entropy function given only the constraint that the variance of $p(x)$ is fixed at a given value σ^2. This is equivalent to fixing the average power of a signal voltage x. We want to maximize the integral

$$H = -\int_{-\infty}^{\infty} p(x)\log[p(x)]\,dx$$

subject to the constraints

$$\int_{-\infty}^{\infty} p(x)\,dx = 1 \tag{1.2}$$

and

$$\int_{-\infty}^{\infty} x^2 p(x)\,dx = \sigma^2 \tag{1.3}$$

This is done by straightforward application of the method of Lagrange multipliers [66]. We form the integral

$$\int_{-\infty}^{\infty} p(x)\{-\log[p(x)] + L + Mx^2\}\,dx \tag{1.4}$$

where L and M are undetermined multipliers of the integrals in Eqs. (1.2) and (1.3), which define the constraints. We want to maximize the integral in Eq. (1.4) by selecting the appropriate function $p(x)$. To do this we differentiate with respect to $p(x)$ and set the result equal to zero, producing the condition

$$-1 - \log[p(x)] + L + Mx^2 = 0$$

which in turn yields

$$p(x) = e^{L-1}e^{Mx^2} \tag{1.5}$$

The multipliers L and M are determined by substituting Eq. (1.5) into Eqs. (1.2) and (1.3), and we find

$$p(x) = \frac{1}{\sqrt{2\pi}\,\sigma} e^{-x^2/2\sigma^2}$$

which is seen to be the zero-mean Gaussian density function. Therefore, subject to a power constraint, the continuous information source having the greatest entropy is the Gaussian source. Its entropy is given by

$$\begin{aligned} H &= \log(\sqrt{2\pi e}\,\sigma) \\ &= \tfrac{1}{2}\log(2\pi e\sigma^2) \end{aligned} \tag{1.6}$$

which has units of bits per sample, and where we have used the property $\log_2 a = (\ln a)(\log_2 e)$.

We shall return to the result in Eq. (1.6) after further developing the concept of entropy as applied to a communication channel.

1.4.2. Transfer of Information Through a Channel

We have said that it is useful to view the entropy H of an information source as a measure of the prior uncertainty about the information produced by the source. Therefore, from the point of view of the intended recipient of the information, H represents the state of uncertainty before receiving information from the source. For the case of error-free discrete transmission, the user receives uncorrupted source symbols, and the uncertainty about the source information vanishes completely as symbols are received. In this ideal case, the channel transfers information from the source to the user at an average rate of H bits per symbol, exactly the amount of uncertainty prior to transmission. We can quantify this notion by observing that after receipt of each symbol, the prior probabilities of the symbols, $p_1, p_2, \ldots, p_M$, are replaced with a distribution having probability 1 for the symbol received and 0 for all other symbols. If we calculate the entropy after reception we have 0, since $\log(1) = 0$.

In all realistic cases, of course, transmission through the channel is not error-free, and thus after reception the user is left with some residual uncertainty concerning the exact identity of the transmitted information. Minimizing this residual uncertainty while making efficient use of signal energy is the essence of the communication system design problem. Given the use of the entropy function to measure a priori uncertainty of the source information, it is logical to use a corresponding function to measure the a posteriori uncertainty of the information after reception on a noisy channel. To express this more formally, we consider a discrete memoryless channel characterized by a set of inputs $x_1, x_2, \ldots, x_M$, having a priori probabilities $\{P(x_i)\}$, a set of outputs $y_1, y_2, \ldots, y_Q$, and a set of transition probabilities $\{P(y_j|x_i)\}$ specifying the probability of receiving a symbol y_j given that a symbol x_i was transmitted. This channel model is illustrated in Fig. 1.4. Using the elementary rules of probability, we can write the *a posteriori* probability of x_i given the receipt of y_j as

$$
\begin{aligned}
P(x_i|y_j) &= \frac{P(x_i, y_j)}{P(y_j)} \\
&= \frac{P(y_j|x_i)P(x_i)}{P(y_j)}
\end{aligned}
$$

Recalling that the a priori uncertainty about x_i is measured by $-\log P(x_i)$, we can define the a posteriori uncertainty about x_i given the receipt of y_j as

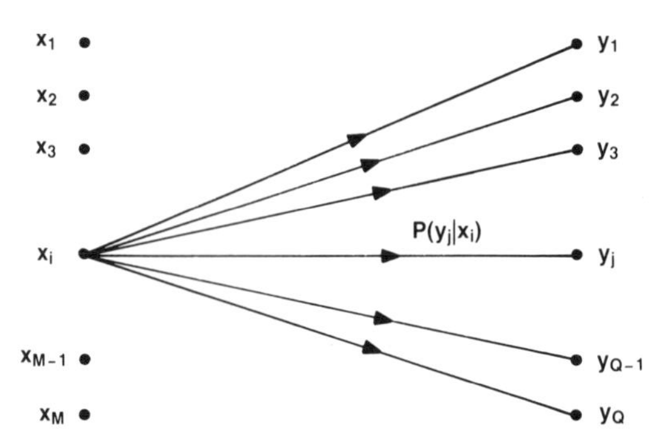

FIGURE 1.4. Model of an M-input, Q-output discrete memoryless channel.

$-\log P(x_i|y_j)$. The difference between these two values is defined as the *mutual information* associated with the transmission of x_i and reception of y_j and is given by

$$I(x_i; y_j) = \log \frac{P(y_j|x_i)}{P(y_j)}$$

We now calculate the overall rate of transfer of information through the channel by simply averaging $I(x_i; y_j)$ over all possible values of x_i and y_j, giving us the *average mutual information*, defined by

$$I(X; Y) = \sum_{i=1}^{M} \sum_{j=1}^{Q} P(x_i, y_j) \log \frac{P(y_j|x_i)}{P(y_j)} \tag{1.7}$$

which has units of bits per symbol when the logarithm base is 2. Since the output symbol probability $P(y_j)$ is given by

$$P(y_j) = \sum_{i=1}^{M} P(x_i) P(y_j|x_i)$$

it is seen that $I(X; Y)$ follows directly from specification of the probabilities of occurrence $P(x_i)$ of source symbols and the set of transition probabilities $P(y_j|x_i)$. The average mutual information is the average rate of transfer of information through the channel, given the distribution of source symbols and the channel transition probabilities. If we are given the transition probabilities defining the channel, the information transfer rate through the channel will depend upon the probabilities of occurrence of the input symbols. The *capacity* of the channel is defined as the maximum value of $I(X; Y)$ with respect to all input distributions, that is,

$$C = \max_{P(x_i)} I(X; Y) \tag{1.8}$$

given in bits per transmitted channel symbol. (Channel capacity can be expressed with various units of measurement, and the separate forms are useful in different contexts. We shall be careful to specify the different forms in our discussions.)

Thus we see that given any channel that is defined by the input-to-output transition probabilities, one can determine the maximum achievable information transfer rate through the channel by performing the indicated maximization over all possible input distributions. Let us consider a simple example.

EXAMPLE: CAPACITY OF THE BINARY SYMMETRIC CHANNEL. The binary symmetric channel was discussed in Section 1.3 and is described by the transition diagram in Fig. 1.3*a*. It is readily seen that the channel is completely characterized by the crossover probability p. If we denote the input symbols as $x = 0, 1$ and the corresponding output symbols as $y = 0, 1$, the transition probabilities $P(y_j|x_i)$ are simply

$$P(0|0) = P(1|1) = 1 - p$$

$$P(1|0) = P(0|1) = p$$

The information transfer rate of this simple channel is maximized when the input symbols are equally likely, that is $P(0) = P(1) = 0.5$, which results in the following expression for capacity:

$$C = 1 + p \log p + (1 - p)\log(1 - p)$$

A plot of this expression is shown in Fig. 1.5. Note that for $p = 0$, corresponding to error-free transmission, C equals one bit per transmitted symbol, which is the entropy of the source with $P(0) = P(1) = 0.5$. For $p = 0.5$, either value of y is received with the same probability regardless of which value of x is transmitted. Thus $C = 0$, and the channel fails to transmit any information about the source symbols to the user. Note also that the capacity curve for the BSC is symmetric about $p = 0.5$. The portion of the curve for $p > 0.5$ has the same meaning as the region $p < 0.5$ if we interchange the assignments of the values 0 and 1 for the output symbols. Therefore, in using the binary symmetric channel model we need only consider values of p in the range $0 \leq p \leq 0.5$.

It is useful to write the expression for average information transfer given in Eq. (1.7) in either of the forms

$$I(X; Y) = H(X) - H(X|Y)$$

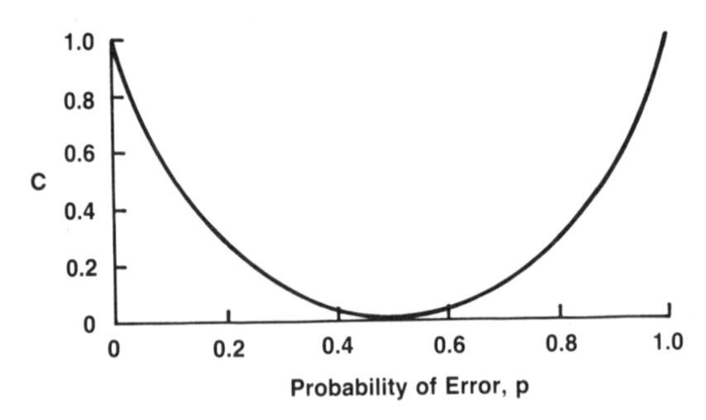

FIGURE 1.5. Capacity of the binary symmetric channel. The units of C are bits per transmitted symbol.

or

$$I(X;Y) = H(Y) - H(Y|X) \tag{1.9}$$

That is, $I(X;Y)$ can be expressed in terms of entropy functions calculated with unconditional or conditional probability distributions of input and output symbols. The second form, Eq. (1.9), is particularly instructive, with

$$H(Y) = -\sum_{j=1}^{Q} P(y_j)\log P(y_j) \tag{1.10}$$

and

$$H(Y|X) = -\sum_{i=1}^{M}\sum_{j=1}^{Q} P(x_i, y_j)\log P(y_j|x_i) \tag{1.11}$$

What is important about the use of Eqs. (1.9) to (1.11) for expressing average information transfer is that for certain types of channels the determination of channel capacity is relatively straightforward. In particular, there are channels described as *uniform from the input*, defined by the property that the sets of transition probabilities from the various channel inputs are permutations of the same set of numbers $p_1, p_2, \ldots, p_Q$. The importance of this property is that for any channel of this type the conditional entropy $H(Y|x_i)$ is given by

$$H(Y|x_i) = -\sum_{j=1}^{Q} P(y_j|x_i)\log P(y_j|x_i)$$

$$= -\sum_{j=1}^{Q} p_j \log p_j$$

which has the same value for any channel input x_i. Therefore, the average conditional entropy given by Eq. (1.11) is independent of the distribution of input symbols and may be written as

$$H(Y|X) = -\sum_{j=1}^{Q} p_j \log p_j$$

The practical interpretation of this model is that when a channel is uniform from the input, the transmission of any of the input symbols is disturbed to the same extent by channel noise. Thus, if Eq. (1.9) is used to derive the capacity, we see that only the first term, $H(Y)$, varies with the input symbol distribution, and we may write the definition of capacity using Eqs. (1.8) and (1.9) as

$$C = \max_{P(x_i)} [H(Y)] - H(Y|X) \tag{1.12}$$

We can now see that for any channel of this type, the information transfer rate is maximized by choosing the input symbol distribution to maximize the entropy of the set of output symbols. From Fig. 1.3 it is readily seen that the BSC and BSEC models are uniform from the input. The capacity of the BSC has already been given. For the BSEC, the reader should verify that $H(Y)$ is maximized when the two input symbols are equally likely, and show that the capacity is

$$C = (1 - q)[1 - \log(1 - q)] + (1 - p - q)\log(1 - p - q) + p\log p$$

For discrete channels with larger numbers of input and output symbols, the uniform distribution of input symbols does not necessarily achieve capacity, even in cases where there is uniformity from the input. However, for a type of channel called a *doubly uniform channel*, capacity is always achieved with a uniform input distribution. A doubly uniform channel is not only uniform from the input but also *uniform from the output*. By the latter property we mean that for all the Q output symbols the sets of transition probabilities from the M inputs are permutations of the same set of M numbers. For any channel of this type it can be shown that capacity is given by

$$C = \log Q + \sum_{j=1}^{Q} p_j \log p_j$$

where the $\{p_j\}$ are the transition probabilities conditioned on any one of the M input symbols. It should be noted that the BSC is a doubly uniform channel, but the BSEC is uniform only from the input.

For channels with continuous rather than discrete input or output signals, the summations in Eqs. (1.10) and (1.11) are converted to appropriate integrals involving probability density functions, and equivalent arguments apply.

1.4.3. Capacity of the Continuous AWGN Channel

We now discuss a channel in which communication is accomplished by transmission of continuous waveforms rather than discrete symbols. In particular, consider a band-limited AWGN channel, and let the bandwidth be denoted by W, given in hertz (Hz). We would like to know the maximum information transfer rate for the channel, that is, the channel capacity, and also what transmission waveforms should be used to achieve capacity. The only constraint we place on the allowable signals is that they have some finite average power. We shall provide a brief outline of the derivation of capacity for this channel, following a similar presentation given by Woodward [183]. More complete and rigorous derivations may be found in a number of texts on information theory, including references already cited in this chapter.

By well-known principles of sampling theory, the signal and noise waveforms received in the band-limited channel can be represented by independent samples taken at the *Nyquist rate* $2W$, where W is the channel bandwidth. Let us denote the transmitted waveform as $x(t)$, the additive white Gaussian noise as $n(t)$, and the noisy received waveform as $y(t) = x(t) + n(t)$. Also let us denote the average power of a received signal or noise sample as, respectively,

$$\overline{x}^2 = S \qquad \text{and} \qquad \overline{n}^2 = N$$

and let us suppose that these average power levels are fixed. Since we are considering continuous rather than discrete signals, we can modify the expressions developed in Section 1.4.2 by replacing the input and output symbol distributions $\{P(x_i)\}$ and $\{P(y_j)\}$ with probability density functions $p(x)$ and $p(y)$, respectively. We also replace the transition probability $P(y_j|x_i)$ with the conditional probability density function $p(y|x)$. Now, for an amplitude-continuous additive noise channel, it is clear that $p(y|x)$ depends on y and x only through their difference $y - x$, and for the case of Gaussian noise we have

$$p(y|x) = \frac{1}{\sqrt{2\pi N}} e^{-(y-x)^2/2N}$$

where N is the average power or variance of a noise sample. Noise additivity is a property of continuous channels that corresponds directly to the property of uniformity from the input discussed in the previous section for discrete channels. Therefore, we can derive channel capacity using Eq. (1.12), where the output entropy $H(Y)$ is calculated for sequences of samples of the received noisy signal $y(t)$, and the conditional entropy $H(Y|X)$ depends only on the statistical distribution of the noise samples, that is, sequences of samples of $n(t)$. We rewrite Eq. (1.12) as

$$C = \max_{p(x)} [H(\mathbf{y})] - H(\mathbf{n})$$

where $\mathbf{y}$ denotes a sequence or vector of received noisy signal samples and $\mathbf{n}$ denotes a sequence or vector of noise samples. Now we can compactly define the capacity of the continuous-waveform AWGN channel as the maximum entropy of the total received signal minus the entropy of the noise. The maximization is to be done by appropriate selection of the statistics of the transmitted waveforms $x(t)$.

First we consider $H(\mathbf{n})$. Recall from the discussion leading to Eq. (1.6) that the entropy of a sample drawn from a Gaussian distribution is $\frac{1}{2}\log(2\pi e\sigma^2)$, where σ^2 is the variance of the distribution. Thus, here the entropy of a sample of the Gaussian noise is $\frac{1}{2}\log(2\pi eN)$, where N is the variance, or average power, of a noise sample. The entropy of a sequence of $2WT$ noise samples taken in an interval of time equal to T seconds is then

$$H(n) = WT \log(2\pi eN)$$

We now consider $H(y)$ and recall from the same discussion leading to Eq. (1.6) that if we constrain a continuous random variable to have a given fixed variance, the entropy of the variable is maximized by letting its probability density function be Gaussian. Therefore, since we have constrained the signal plus noise $y(t)$ to have average power $S + N$, it follows that the received signal entropy $H(y)$ will be maximized by letting $y(t)$ have the statistics of Gaussian noise, each sample having variance $S + N$. We can make this happen by transmitting signals that are Gaussian noise waveforms, and by reapplication of Eq. (1.6) we have for a sequence of $2WT$ samples,

$$\max_{p(x)} H(y) = WT \log[2\pi e(S + N)]$$

Therefore the maximum information transfer rate is

$$I_{\max} = WT \log\left(\frac{S+N}{N}\right)$$

which is measured in bits transmitted per T seconds. Equivalently, capacity per unit time is given by

$$C = W \log\left(\frac{S+N}{N}\right) = W \log\left(1 + \frac{S}{N}\right) \text{ bits/s} \tag{1.13}$$

This is Shannon's capacity formula for the band-limited continuous AWGN channel. We shall see shortly that the practical implication of this and another closely related result is that the most efficient use of the channel is achieved by setting up a suitable correspondence between long sequences of source information digits and long noiselike signaling waveforms $x(t)$ for transmission on the channel. The most remarkable of Shannon's results, which we shall not derive here, is the following:

If we take increasingly long sequences of source digits and map them into correspondingly long transmission waveforms, the error rate in the data delivered can be brought arbitrarily close to zero, as long as we do not attempt to transmit data at a rate higher than C. Therefore, at any nonzero level of channel signal-to-noise ratio S/N, there is some nonzero information transfer rate below which arbitrarily accurate communication can in principle be achieved.

The essence of Shannon's result, which is called the *channel coding theorem*, is that noise in the channel does not inherently limit the accuracy with which communication can be achieved, but only the rate at which information can be reliably transmitted [158,159]. We discuss this result in more detail in the following section.

1.4.4. Channel Coding Theorem

The channel coding theorem states that every channel has a channel capacity C, and that for any information transfer rate $R < C$ there exist codes of block length n and rate R having probability of incorrect decoding $P(E)$ bounded by

$$P(E) \leq 2^{-nE_b(R)} \tag{1.14}$$

where the exponent $E_b(R)$ is a positive function of R for $R < C$ and is determined solely by the characteristics of the channel. The implication of the bound in Eq. (1.14) is that for any information rate less than C, the error probability can be made arbitrarily small by increasing the code block length n while holding the code rate constant. A similar bound can be written for convolutional codes, where n is replaced by k, the code constraint length.

Shannon's derivation of the exponential error bound formula was based upon an analysis called the *random coding argument*. The bound is obtained by averaging the error probability over an ensemble of randomly selected codes. Since some codes in the ensemble must perform better than the average, the coding theorem guarantees the existence of codes capable of achieving the bound in Eq. (1.14).

Therefore, Shannon's work shows that it is not really necessary to transmit Gaussian noise waveforms in order to achieve capacity; rather, well-chosen codes can be used to produce the same result. For a coded communication system, sequences of information bits are mapped into long codewords by the error-control encoder and then into long digital waveforms by the modulator. The demodulator and decoder then utilize all the received signal energy during the transmission of a codeword in the decision-making process. Several practical issues need to be dealt with, however. First, the coding theorem provides no means for constructing effective codes. Second, requirements for very low error probabilities will compel the use of very long codes, and this in turn will lead to very complex decoding operations.

Because of the issues just outlined, much of the research in the coding field over the last three decades has dealt with two key problems: finding classes of codes that yield good performance over wide ranges of lengths, and designing decoding algorithms that realize the intrinsic code performance without prohibitive complexity. Chapters 2 through 11 will present some of the more important and useful results that have been obtained in addressing these questions.

However, no practical means of achieving channel capacity has yet been found. In Section 1.6 we discuss other implications of the channel capacity formula and outline a valuable concept closely related to channel capacity that provides a calculable measure of modulation and coding performance achievable with practical implementations.

1.5. MODULATION PERFORMANCE ON THE AWGN CHANNEL

In this section we summarize formulas giving the performance that can be achieved in additive white Gaussian noise with several of the commonly used forms of digital modulation. Detailed derivations of these formulas can be found in a number of references, including Arthurs and Dym [4], Viterbi [172], and Proakis [139].

1.5.1. Phase-Shift Keying

Let us first consider binary signaling in the AWGN channel, so that we write the received signal as

$$r(t) = s_i(t) + n(t), \qquad 0 \le t \le T$$

where $s_i(t)$, with $i = 0$ or 1, is the transmitted waveform and $n(t)$ represents the white Gaussian noise waveform in the signaling interval T. For *phase-shift keying* (PSK), the two possible waveforms are chosen to be *antipodal*, so that

$$s_0(t) = -s_1(t), \qquad 0 \le t \le T$$

where $s_0(t)$ and $s_1(t)$ are sinusoids of the same frequency with fixed phases 180° apart. Binary PSK is sometimes called *biphase* modulation.

The transmission channel is assumed to have finite bandwidth, which places a practical upper limit on the pulse transmission rate. Let us say that PSK pulses are transmitted at the rate B pulses per second, which is equal to or less than the upper limit. If the average transmitted signal power is S, then the energy per PSK pulse is $E_b = S/B = ST$. Note that E_b was defined earlier to be the average energy transmitted per bit of source information. For uncoded binary transmission the signal energy per pulse equals the energy per source bit, and so we shall use the symbol E_b in the error-probability formulas that follow.

Optimum detection of binary PSK signals is done with a *matched filter* followed by a sampler, the filter being matched to either $s_0(t)$ or $s_1(t)$. Assuming perfect phase coherence, analysis of matched filter detection for PSK shows the probability of error p to be

$$p = \frac{1}{\sqrt{2\pi}} \int_{\sqrt{2E_b/N_0}}^{\infty} e^{-x^2/2}\, dx$$

which can be written more compactly as

$$p = \frac{1}{2}\operatorname{erfc}\left(\sqrt{A_b}\right) \tag{1.15}$$

where $\operatorname{erfc}(x)$ is the complementary error function, defined by

$$\operatorname{erfc}(x) = \frac{1}{\sqrt{\pi}}\int_x^\infty e^{-t^2}\,dt$$

and $A_b = E_b/N_0$ is the *signal-to-noise ratio* (SNR) per bit.

1.5.2. Differential PSK

Coherent PSK demodulation requires the use of a phase reference. This is usually done by extracting the carrier phase from the received signal with a *phase-locked loop*. In some systems, however, it is difficult to obtain a reliable phase reference due to characteristics of the transmission medium or the demodulation equipment itself. In such cases, PSK signaling can still be used if the signals to be transmitted are mapped into successive phase differences. The resulting modulated waveform is known as *differentially encoded PSK* or simply *differential PSK*, abbreviated as DPSK. In binary DPSK a 1 is transmitted by sending the pulse waveform which is 180° out of phase with the pulse sent in the previous interval, while a 0 is transmitted by sending a pulse in phase with the previous pulse. This is called *differential encoding* or *differential precoding*. *Differentially coherent demodulation* of DPSK signals is implemented by detection of the phase difference between received pulses in successive signaling intervals. For steady-signal reception of binary DPSK signals on an AWGN channel, the average probability of error is given by

$$p = \tfrac{1}{2}e^{-A_b} \tag{1.16}$$

where $A_b = E_b/N_0$ is the SNR per bit. The bit-error probability for DPSK was derived originally by Lawton [87].

Because of the partial overlap of signal and noise components involved in successive binary decisions, there is a tendency for errors in differentially coherent demodulation of DPSK to occur in clusters of two. This occurs because, especially under high-SNR, low-error-rate operation, an error is likely to be caused by a momentarily high noise level associated with a single pulse completely distorting its phase. Since any such single pulse is involved in two successive binary decisions, both may have a high likelihood of error. Such a clustering of errors may be of major significance when error-control coding is used, since the coding must then be designed to cope with clusters of errors rather than with statistically independent errors. Numerical results on error clustering behavior with DPSK are given in a paper by Salz and Saltzberg [152].

There is another form of DPSK that is sometimes used in systems where coherent demodulation can be done but where there is an unresolvable 180° ambiguity in the carrier phase. Here a DPSK signal is transmitted, and at the receiving end coherent PSK demodulation is performed on individual pulses, followed by *differential decoding* of the stream of demodulated bits. This scheme does not perform as well as strict-sense coherent PSK signaling and reception, since a single error leaving the PSK demodulator produces two errors after differential decoding. The probability of bit error for differentially encoded coherent PSK lies between that of strict-sense coherent PSK and that of DPSK; at high SNR, the probability of bit error is approximately twice that of coherent PSK. A detailed analysis of the performance of coherent demodulation of DPSK, including a table of computed bit-error probabilities, can be found in Lindsey and Simon [97].

1.5.3. Coherent Frequency Shift Keying

Frequency shift keying (FSK) is a special case of *orthogonal signaling*. A set of signals is said to be orthogonal over an interval T if any pair of different signals, say $s_i(t)$ and $s_j(t)$, have zero crosstalk, that is, if

$$\int_0^T s_i(t)s_j(t)\,dt = 0, \qquad i \neq j$$

With FSK a symbol is transmitted by sending one of a set of tones, where the tone frequencies are chosen so that any pair of tones at different frequencies are orthogonal over the signaling interval T. In binary FSK, two tones are used, and if the tones are both generated and demodulated with known phases it can be shown that orthogonality is obtained with any tone spacing equal to an integer multiple of $1/2T$. In practice, the tone spacing is usually chosen as $1/2T$ in the interest of minimizing bandwidth. Optimum demodulation is done with a pair of coherent matched filters. The bit-error probability for coherent FSK detection in AWGN is

$$p = \tfrac{1}{2}\operatorname{erfc}\left(\sqrt{\frac{A_b}{2}}\right) \tag{1.17}$$

where $A_b = E_b/N_0$ is the SNR per bit.

If we compare this expression with Eq. (1.15) for coherent PSK, we see that equal bit-error probabilities are obtained with 3 dB greater SNR in the case of coherent FSK. In other words, binary coherent FSK has 3 dB poorer performance than binary coherent PSK.

If, in generating the coherent FSK signal, phase continuity is maintained from one pulse to the next, the resulting modulation is called *continuous-phase FSK* (CPFSK). In many applications CPFSK is an attractive form of modulation, because the phase continuity results in a signal spectrum that rolls off more rapidly than spectra for other forms of modulation. There is an important special case of CPFSK modulation and demodulation that achieves performance identical to coherent PSK. The scheme is called *minimum-shift keying* (MSK). In MSK, the two tones have a frequency separation equal to $1/2T$, hence the term "minimum shift." Optimum performance for MSK is achieved by exploiting a special property of the MSK waveform, namely that it can be constructed as two binary phase-modulated pulse streams or "subchannels" in phase quadrature, each subchannel carrying sinusoidally shaped pulses of duration $2T$, with the two subchannels offset by T seconds. That is, MSK can be viewed as a particular way to implement antipodal signaling on two orthogonal subchannels. The optimum demodulator performs coherent matched-filter detection in each of the quadrature subchannels, producing bit decisions at times $0, 2T, 4T, \ldots$ in one subchannel and at times $T, 3T, 5T, \ldots$ in the other subchannel. It can be shown that the signal energy usable in each subchannel demodulation is E_b, so that the average probability of bit error is the same as for coherent PSK and thus is given by Eq. (1.15).

As with PSK, the data bits may be differentially precoded prior to quadrature-channel MSK modulation, with the objective of again providing resistance to channel phase inversions. This produces CPFSK, the frequencies of the transmitted pulses having a one-to-one correspondence to the source data bits. (This is not the case if differential precoding is omitted.) This form of MSK modulation has also been called *fast frequency-shift keying* (FFSK). If this signal is demodulated with a coherent matched filter correlating over one T-second pulse at a time, the probability of bit error is that already given for coherent FSK in Eq. (1.17). However, the signal can be optimally demodulated using the coherent quadrature channel method outlined earlier, but followed now by differential decoding. With this modulation and demodulation scheme, the bit-error probability is the same as that observed with differentially encoded, coherently detected PSK, discussed at the end of Section 1.5.2.

MSK has found application in a number of systems for which efficient spectral utilization and low crosstalk are important requirements. The interested reader may refer to a tutorial paper on MSK by Pasupathy and other references cited there [126].

1.5.4. Noncoherent Binary FSK

In most applications, binary FSK signals are demodulated noncoherently, that is, with a detector that operates without knowledge of the received carrier phase. Then it is necessary that the tones be spaced by an integer multiple of $1/T$ Hz, which ensures orthogonality even if the phases are arbitrary. Demodulation is usually done with two pairs of quadrature filters, one pair matched to each frequency, followed by a comparison of envelopes or squared envelopes of the outputs of the two filter pairs. For steady-signal reception in AWGN, the bit-error probability is

$$p = \tfrac{1}{2} e^{-A_b/2}$$

where $A_b = E_b/N_0$ is the SNR per bit.

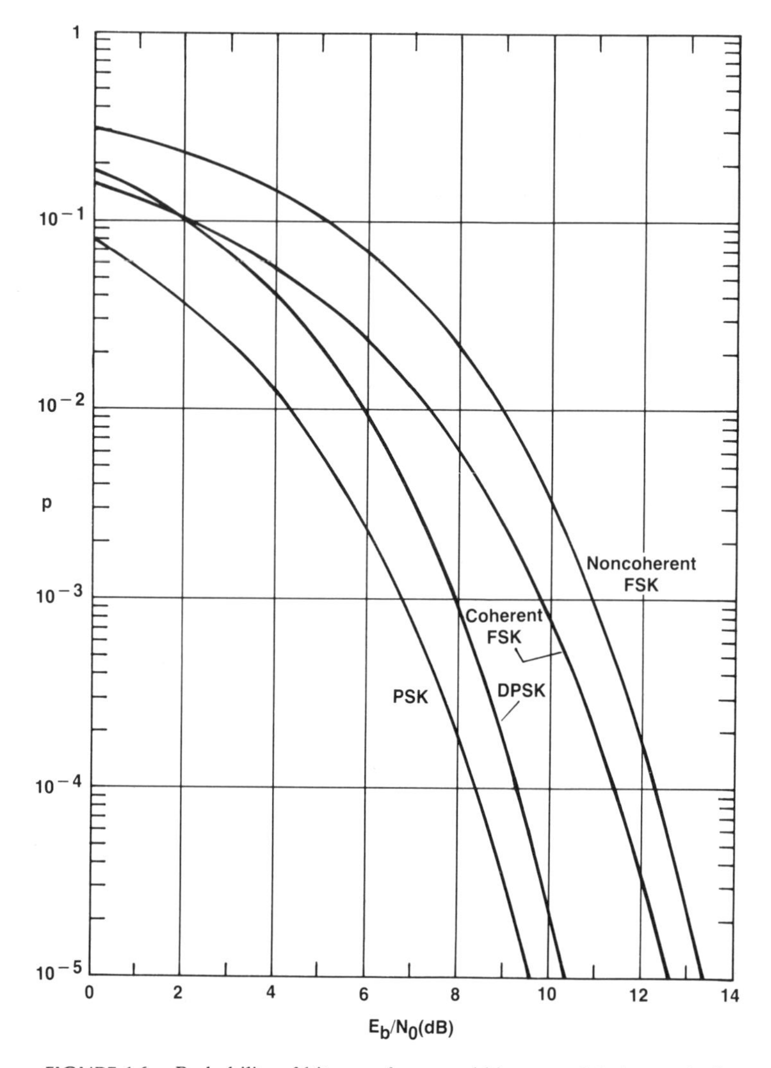

FIGURE 1.6. Probability of bit error for several binary modulation methods.

Comparing the above expression with Eq. (1.16), one sees that the performance of noncoherent FSK is 3 dB poorer than that of DPSK.

Figure 1.6 shows the probability of bit error versus SNR per bit for coherent PSK, DPSK, coherent FSK, and noncoherent FSK.

1.5.5. *M*-ary Signaling on the Gaussian Noise Channel

The basic types of signaling waveforms outlined above, PSK and FSK, are readily generalized from binary to M-ary forms by using M signal phases or frequencies, respectively. For M-ary PSK transmission, the set of transmitted waveforms consists of sinusoidal carriers with relative phases $0°, 360°/M, 2 \cdot 360°/M, \ldots, (M-1) \cdot 360°/M$. In describing the performance of M-ary modulation techniques, we need to distinguish between the *signal energy per channel symbol*, which we denote by E_s, and the energy per bit E_b, already defined. In most applications, M is a power of 2, say $M = 2^m$, so that each M-ary symbol represents m channel bits and therefore $E_s = mE_b$. Performance is most readily derived in terms of the M-ary *symbol-error probability*. For four-phase PSK with coherent detection and AWGN, the 4-ary symbol error probability can be shown to be

$$P_4 = \operatorname{erfc}\left(\sqrt{A_b}\right)\left[1 - \tfrac{1}{4}\operatorname{erfc}\left(\sqrt{A_b}\right)\right] \tag{1.18}$$

where $A_b = E_b/N_0$ is the SNR per bit. Note that since $M = 4$, $A_b = 0.5A_s$, where $A_s = E_s/N_0$. For values of M other than 2 or 4, the error probability formulas cannot be presented in such compact forms as Eqs. (1.15) and (1.18), but rather the symbol-error probability in each case must be obtained by numerical integration. However, it can be shown that for $A_b \gg 1$ the M-ary symbol-error probability is given approximately by

$$P_M \approx \operatorname{erfc}\left(\sqrt{mA_b}\sin\frac{\pi}{M}\right) \tag{1.19}$$

where $m = \log_2 M$.

The relationship between M-ary symbol-error probability and bit-error probability in the corresponding m-bit groups depends upon the assignment of bit groups to symbols, which is up to the designer. The preferred mapping in most cases is *Gray coding*, in which the bit groups assigned to adjacent phases differ by only one bit. For example, a Gray code for 4-ary PSK would assign the bit groups 00, 01, 11, and 10 to successive phases. With the use of a Gray code, and given a practical level of SNR, symbol errors will be predominantly transitions to adjacent phases, and therefore the bit-error probability corresponding to a given level of symbol-error probability is reasonably well approximated by

$$p \approx \frac{1}{m}P_M$$

Curves of P_M versus SNR per bit are shown in Fig. 1.7 for several values of M.

For M-ary DPSK, the error-probability formulas are rather tedious to derive, and even for the relatively simple case of $M = 4$, the error probability must be expressed in terms of higher transcendental functions. We do not wish to take the space to present these formulas here, but instead refer the interested reader to reference [139] or [4].

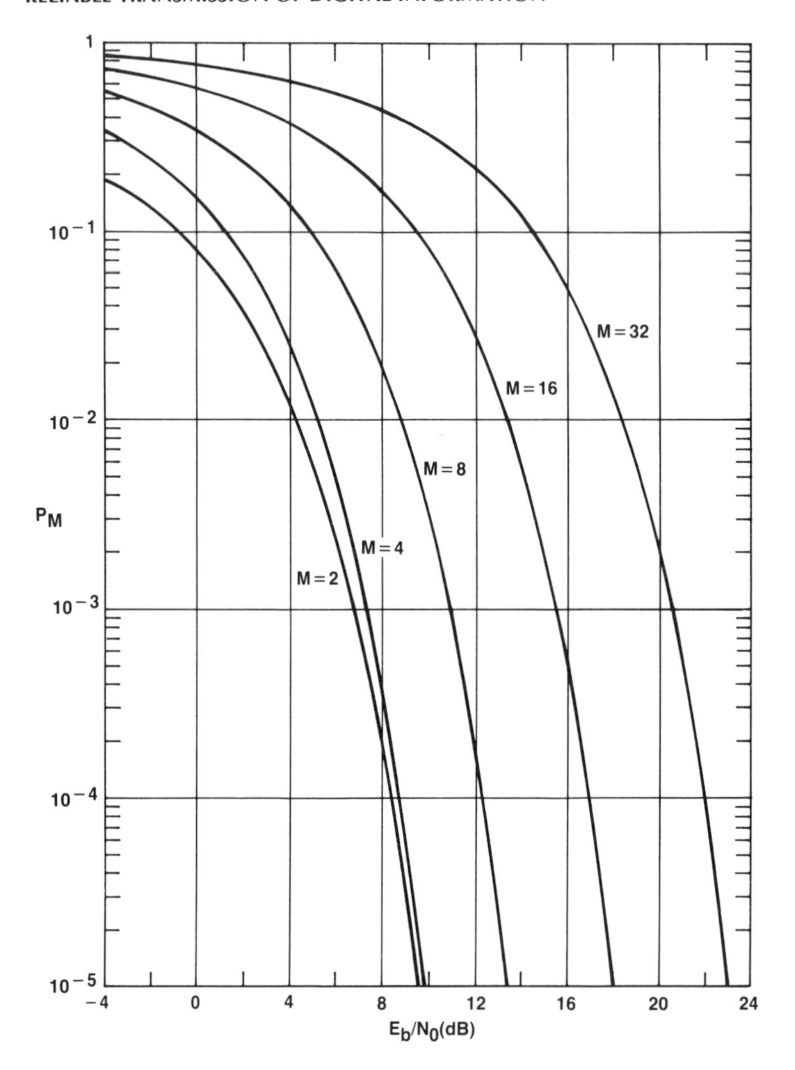

FIGURE 1.7. Probability of symbol error for M-ary PSK modulation.

For coherent detection of M-ary FSK in AWGN, the probability of M-ary symbol error is given by

$$P_M = \frac{1}{\sqrt{2\pi}} \int_{-\infty}^{\infty} \left\{1 - \left[1 - \frac{1}{2}\operatorname{erfc}\left(\frac{y}{\sqrt{2}}\right)\right]^{M-1}\right\} e^{-(y-\sqrt{2mA_b})^2/2}\, dy$$

where $A_b = E_b/N_0$ is the SNR per bit, $M = 2^m$, and y is the variable of integration. Curves of P_M versus E_b/N_0 are shown in Fig. 1.8 for several values of M.

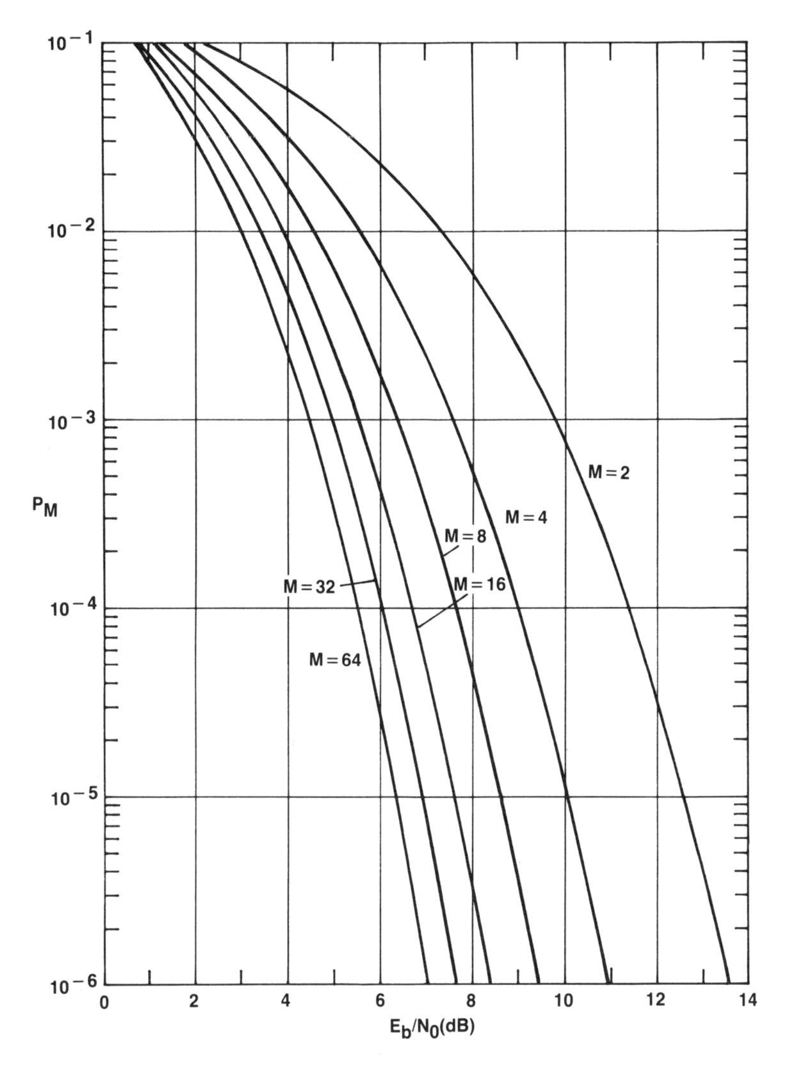

FIGURE 1.8. Probability of symbol error for coherent detection of M-ary FSK signals.

As in the case of binary FSK, M-ary FSK signals may be generated with phase continuity from pulse to pulse. The resulting modulation is called *M-ary CPFSK*. As a consequence of the phase continuity being maintained, the M-ary CPFSK signal has memory that in general can extend over a number of pulses. While it is possible to ignore this inherent memory in the detection process, a demodulator that makes use of this memory performs better than one that does not. See, for example, Schonhoff [156].

The error probability P_M for M-ary symbols can be converted to a bit-error probability p for the equivalent m-bit groups by assuming that when an M-ary symbol is in error, each of the $2^m - 1$ incorrect m-bit patterns is equally likely.

It can be shown easily that this leads to the relationship

$$p = \frac{2^{m-1}}{2^m - 1} P_M$$

For noncoherent detection of M-ary FSK in AWGN, the probability of symbol error is given by

$$P_M = \sum_{i=1}^{M-1} (-1)^{i+1} \binom{M-1}{i} \frac{1}{i+1} e^{-imA_b/(i+1)}$$

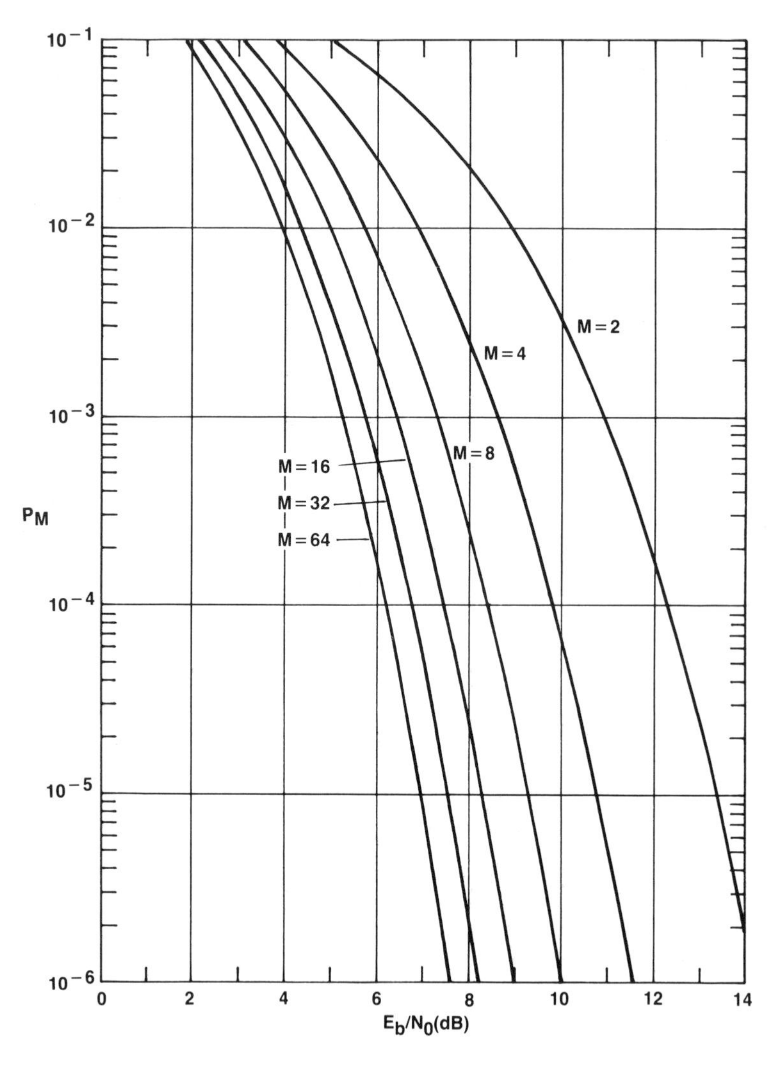

FIGURE 1.9. Probability of symbol error for noncoherent detection of M-ary FSK signals.

Symbol-error probability P_M is plotted in Fig. 1.9 as a function of SNR per bit for several values of M.

Tables of computed error probabilities for all the forms of binary and M-ary modulation described here can be found in Lindsey and Simon [97]. Tables of error probabilities for coherent demodulation of M-ary orthogonal signals can also be found in Viterbi [171].

1.5.6. Comparison of Binary Modulation Techniques

In the chapters that follow, the performance of many errror-control techniques will be presented. Particularly for the block-coded systems, it will be convenient to give results as a function of p, the bit-error rate in the channel, or P_{CE}, the symbol-error rate. These results can be related to channel SNR by inspecting the applicable modulation performance curve. A word of caution is in order, though, since the error-control results assume an independent-error channel. This assumption is valid for all the channel models considered here except binary and nonbinary DPSK and CPFSK. Some degradation can be expected due to the correlation described previously, but this degradation can be eliminated with interleaving (see Chapter 11).

There is an important observation to be made in comparing Figs. 1.7 and 1.8, which give error probabilities for M-ary PSK and M-ary coherent FSK, respectively. Note that for M-ary PSK, the error-rate curves shift to the right (increasing SNR per bit) with increasing M, while for M-ary coherent FSK the curves shift to the left (decreasing SNR per bit) with increasing M. Therefore we see that by providing an expansion of bandwidth, as we must do to increase the library of FSK tones, we can achieve improvements in communication efficiency as measured by the SNR per bit, E_b/N_0. That is, for a desired quality of service, measured here as the probability of a symbol error, the E_b/N_0 required is reduced. In fact, it can be shown that for coherent M-ary FSK signaling, as $M \to \infty$, arbitrarily small error rates can be provided for any $E_b/N_0 \geq \ln 2$, which is equal to -1.6 dB.

Unfortunately, the use of extremely large FSK tone libraries cannot be considered a practical design approach. To see this we consider a parameter called the *bandwidth expansion factor*, B_e, which is defined for any digital modulation scheme as

$$B_e = \frac{W}{R}$$

where W is the overall bandwidth of the set of modulation waveforms measured in hertz and R is the information rate of the modulation in bits per second. For M-ary FSK with tones spaced at $1/2T$ Hz, the overall required bandwidth is approximately $M/2T$, where T is the FSK pulse duration, and R is m/T, where $M = 2^m$. Therefore we see that for M-ary FSK signaling, the bandwidth expansion factor is

$$B_e = \frac{W}{R} = \frac{M}{2\log_2 M}$$

Thus as $M \to \infty$, the bandwidth expansion required also goes to infinity. In addition, coherent reception of a very large number of orthogonal signals would be required, leading to a very complex design. Nonetheless, the asymptotic MFSK result demonstrates that utilizing additional bandwidth can provide highly efficient and reliable communication. However, by the use of simpler modulation schemes and error-control coding, one can construct waveforms that perform as well as orthogonal waveforms while requiring smaller bandwidth expansions. In other words, coded waveforms can be designed that are more efficient than orthogonal waveforms in their use of bandwidth. The coded waveforms are nonorthogonal, in general.

1.6. COMBINED MODULATION AND CODING FOR EFFICIENT SIGNAL DESIGN

It is becoming common in the field of digital communication to refer to the set of modulation, coding, and decoding techniques used in a system design by the overall name *signal design*. The use of this terminology reflects a growing awareness that the design of an efficient digital communication system is best approached by the selection of modulation and coding techniques jointly as part of an integral design. In this section we present this viewpoint by discussing certain implications of Shannon's capacity formula as well as a closely related capacity-like performance measure called R_0("R zero").

1.6.1. Implications of the Capacity Formula

The underlying principles of efficient signal design can best be seen by examining the capacity formula for the AWGN channel, which we rewrite from Eq. (1.13) in the form

$$\frac{C}{W} = \log\left(1 + \frac{S}{N_0 C}\frac{C}{W}\right) \text{bits/Hz}$$

where N_0 is the one-sided power spectral density of the noise.

This form describes channel capacity in terms of two convenient normalized parameters, S/N_0C and C/W. For transmission at capacity, the first parameter becomes

$$\frac{S}{N_0 C} = \frac{(E_b)_{\min}}{N_0}$$

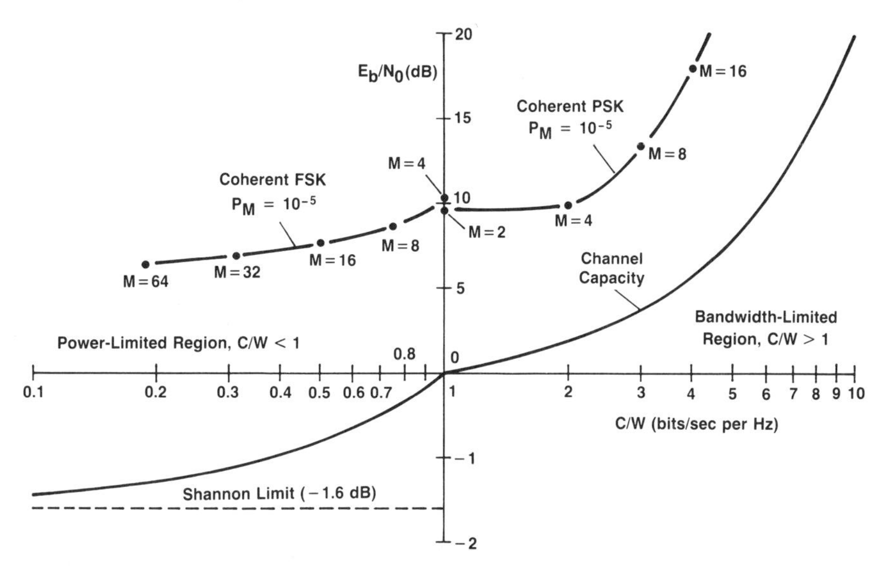

FIGURE 1.10. Channel capacity and a comparison of several modulation methods at symbol-error probability equal to 10^{-5}. Note that the vertical scale is expanded below the origin.

where $(E_b)_{\min}$ is the minimum energy per transmitted source information bit required for reliable communication. The second parameter C/W simply normalizes channel capacity with respect to an arbitrary bandwidth and is seen to be the reciprocal of the bandwidth expansion for operation at capacity.

Figure 1.10 shows the relationship between E_b/N_0 and C/W as given by the capacity formula. Note that the lower portion of the vertical scale is expanded for convenience in drawing the figure. It is conventional to call the region of $C/W < 1$ the *power-limited region* of operation and the region of $C/W > 1$ the *bandwidth-limited region*. As can be seen from the figure, the greatest energy efficiency is achieved when the bandwidth can be made large by comparison with the information rate. In the limiting case, very large bandwidths, C/W approaching zero, E_b/N_0 approaches ln 2 or -1.6 dB, which is called the *Shannon limit*. Thus, by invoking the coding theorem, we can say that for any SNR per information bit E_b/N_0 equal to or greater than -1.6 dB, the probability of error in delivered information can theoretically be made arbitrarily close to zero by use of a suitably chosen error-control code.

It is seen that the Shannon limit, -1.6 dB, is identical to the minimum level of E_b/N_0 achievable with coherent *M-ary FSK* as $M \to \infty$ (see Section 1.5.6). The use of error-control coding, however, has promise of less complex transmission and reception equipment, since long codewords are easier to generate and decode than long orthogonal M-ary waveforms. With M-ary FSK, energy efficiency is realized entirely with the modem and the complexity of demodulation grows very rapidly with M. Furthermore, M-ary FSK waveforms are relatively extravagent in their use of bandwidth when compared with well-chosen codes as we noted in Section 1.5.6. A key point here is that the bandwidth consumed by MFSK waveforms provides orthogonality for every pair of waveforms in the set, but the achievement of energy-efficient communication does not require orthogonality. Design points for several MFSK modulations at symbol error probability 10^{-5} are shown in Fig. 1.10. Note that for practical values of M, $M \leq 32$, significant gains in communication efficiency can be obtained by increasing M.

The assumptions of severely limited power and practically unlimited bandwidth are valid in some applications, for example in the design of data links for space probes and some satellite systems. In these cases the relationship shown in Fig. 1.10 suggests the use of highly redundant coding in the power-limited region, where E_b/N_0 is to be minimized. For operation in this region, highly efficient communication systems have been designed using binary PSK modulation and low-rate block and convolutional codes decoded by powerful decoding algorithms, which will be described in later chapters. While the Shannon limit cannot be achieved in practice, there are practical schemes that can approach a computation-limited information transfer rate, termed R_0, which is exactly 3 dB above the Shannon limit for the AWGN channel. This point is addressed further in Section 1.6.2, and the coding schemes that we refer to are described in Chapters 10 and 11.

Many important communication channels can be characterized as bandwidth-limited — for example, wireline telephone circuits and radio channels in crowded regions of the radio spectrum. It is instructive, therefore, to examine the capacity formula in the bandwidth-limited region, which in turn implies high SNR if capacity is to be achieved.

It will be useful here to use another normalized form of the capacity formula:

$$\frac{C}{2W} = \frac{1}{2}\log\left(1 + \frac{S}{N_0W}\right)\text{bits/symbol}$$

which expresses capacity in bits per transmission symbol, assuming signaling at the Nyquist rate of $2W$ symbols per second.

At high levels of SNR we see that this formula gives

$$C \cong \frac{1}{2}\log\left(\frac{S}{N_0W}\right)\ \text{bits/symbol}$$

The need for operation at high SNR in limited bandwidth suggests the use of nonbinary modulation, and so we consider the asymptotic performance of M-ary coherent PSK modulation in Gaussian noise as an example. An approximation to the probability of symbol error, P_M, obtained from Eq.

(1.19), is

$$P_M \cong \operatorname{erfc}\left(\sqrt{\frac{E_s}{N_0}}\sin\frac{\pi}{M}\right)$$

where E_s is the energy per symbol, that is, per M-ary PSK pulse, and erfc(x) is the complementary error function. If $E_s/N_0 \gg 1$, this expression gives a very accurate approximation to P_M for arbitrary M. For large M and E_s/N_0, we have, using the small-angle formula,

$$P_M \cong \operatorname{erfc}\left(\sqrt{\frac{E_s}{N_0}}\frac{\pi}{M}\right)$$

Therefore, as M becomes large, the SNR per pulse required to maintain a given probability of symbol error must increase as the square of the number of modulation states; that is, we have, for large E_s/N_0 and large M,

$$\frac{E_s}{N_0} \propto M^2$$

Since the information conveyed per transmission symbol is $\log_2 M$ bits, the information rate needed to maintain constant P_M is

$$R = \log_2 M \propto \log\left(\frac{E_s}{N_0}\right) \text{bits/symbol}$$

which in turn implies

$$R \propto \log\left(\frac{S}{N_0 W}\right) \text{bits/symbol}$$

Therefore the information rate that can be achieved for high-order PSK exhibits asymptotically the same function of SNR as channel capacity.

Design points for several uncoded modulation methods operating at symbol error probability 10^{-5} are shown in Fig. 1.10. Similar asymptotic behavior can also be shown for amplitude-modulated (AM) signals and combined AM/PSK signals.

It is seen, therefore, that for operation in the high-SNR, bandwidth-limited region, channel capacity can be approached by simply using large multiphase or multiamplitude alphabets without a need for coding. This underlines the point that as greater levels of signal power can be provided, system operating points move from the strictly power-limited region, where coding and bandwidth expansion must be employed to achieve reliable transmission at the highest possible rate with the least signal energy, toward the bandwidth-limited region, where coding offers little or no performance gain with respect to uncoded high-order signaling alphabets. For operating points in the latter region, emphasis is placed on the design of bandwidth-efficient signals and the effects of bandwidth limitations on system performance.

The above discussion serves to outline some of the fundamental principles of modulation and coding design for digital communication channels. In the following section we describe a useful analytical methodology for arriving at specific modulation and coding designs that optimize the efficiency of a digital communication system.

1.6.2. The R_0 Criterion for Modulation and Coding Design

In section 1.6.1 we examined the capacity formula for the AWGN channel with a view toward understanding the conditions under which coding promises to be useful in the design of an energy-efficient communication system. However, we have already pointed out that there are practical limitations in trying to achieve capacity in an actual design and that in fact it has not been done. This would seem to leave us at an impasse, with informative theoretical bounds on information rate and error probability but no guidance for the means to approach the theoretical limits to some reasonable degree and with practical designs. As it happens, there is a line of analysis that provides exactly the design guidance that we would like to have. This analysis revolves around a parameter called the *cutoff rate* of the channel, denoted as R_0 ("R zero"). The cutoff rate R_0 is a capacity-like quantity defined for any discrete memoryless channel, whose value is always less than the channel capacity C. R_0 gives the practical limit on the rate at which information can be reliably transmitted through the channel.

It has been shown [173] that for a system using a convolutional code of constraint length k, the post-decoding error probability is bounded as

$$P_e < C_R 2^{-kR_0}, \qquad R \leq R_0 \tag{1.20}$$

where R is the code rate and C_R is a small constant usually determined experimentally. The corresponding bound for block coding is

$$P_e < C_R 2^{-nR_0}, \qquad R \leq R_0 \tag{1.21}$$

where n is the code block length. Together with these results it has been shown that for rates $R < R_0$ codes with long constraint length k or block length n can be decoded without suffering an unbounded growth in the number of decoding computations. Thus R_0 at once provides both an error-bound exponent and a practical limit on information transfer rate. Comparison of Eqs. (1.20) and (1.21) with Eq. (1.14) shows that R_0 is an exponential-bound parameter of nearly the form of $E_b(R)$. In fact, R_0 was first derived as a lower bound on $E_b(R)$ (see Gallager [49] and Wozencraft and Kennedy [186]). In the development and analysis of powerful decoding algorithms for long-constraint-length convolutional codes, R_0 has been given a more descriptive name, the *computational cutoff rate*, denoted by R_{COMP} (see Chapters 10 and 11).

As we said earlier, R_0 is a quantity similar to capacity that can be evaluated for any given channel. Mathematically, the cutoff rate for an M-input, Q-output discrete memoryless channel is given by

$$R_0 = -\log_2\left\{ \min_{P(x_i)} \sum_{j=0}^{Q-1} \left[\sum_{i=0}^{M-1} \sqrt{P(y_j|x_i)}\, P(x_i) \right]^2 \right\} \tag{1.22}$$

where the $\{P(y_j|x_i)\}$ denote the channel transition probabilities, and the minimization is taken over all possible probability distributions $P(x_i)$ on the channel input symbols. The cutoff rate has the same units as capacity — that is, in the form given above, source information bits transferred per symbol—and it is obtained by a process of optimization as in the derivation of capacity. However, unlike capacity, R_0 is optimized not only with respect to the distribution of symbols entering the channel, but also with respect to the boundaries between decision regions at the output of the demodulator. That is, for a given number Q of demodulator outputs (quantization intervals), the decision boundaries are chosen so as to maximize the value of R_0.

Thus we see from Eqs. (1.20) and (1.21) that maximizing R_0 serves two purposes. The practical limit on the rate of information transfer is maximized, and the bound on the probability of error is minimized. The complexity of the modulation and demodulation strategy is established as part of the optimizations of input symbol set and output decision regions.

Massey and other researchers have proposed a unified theory of modulation and coding design based upon R_0 as the fundamental channel parameter and have defined an approach to the design of optimum modulation systems to be used in conjunction with efficient decoders [106]. This approach has come to be known as designing according to the "R_0 criterion." The point of this approach is that a modulation system should be designed to achieve the highest possible value of R_0 rather than the lowest value of post-demodulation error probability.

It is instructive to calculate R_0 for a relatively simple case that is of considerable practical importance, that of binary antipodal signaling on the AWGN channel. Let us say that the two channel inputs are $x_0 = -a$ and $x_1 = a$ and that they occur with equal probability $P(x_0) = P(x_1) = \frac{1}{2}$. If we assume that the channel adds a sample of Gaussian noise with variance σ^2 to each transmitted digit and that the channel outputs are unquantized, then Eq. (1.22) is rewritten as

$$R_0 = -\log \int_{-\infty}^{\infty} \left(\frac{1}{2}\sqrt{p(y|x_0)} + \frac{1}{2}\sqrt{p(y|x_1)} \right)^2 dy \tag{1.23}$$

where the transition probabilities are given by the Gaussian conditional probability density functions

$$p(y|x_0) = \frac{1}{\sqrt{2\pi}\,\sigma} e^{-(y+a)^2/2\sigma^2} \tag{1.24}$$

and

$$p(y|x_1) = \frac{1}{\sqrt{2\pi}\,\sigma} e^{-(y-a)^2/2\sigma^2} \tag{1.25}$$

Because of the symmetry in this simple binary case, the minimization over $P(x)$ is eliminated. Also, the demodulation decision boundary to maximize R_0 is obvious: $y = 0$. The integration in Eq. (1.23) is straightforward, yielding

$$R_0 = -\log\frac{1}{2}\left(1 + e^{-a^2/2\sigma^2}\right)$$

$$= 1 - \log\left(1 + e^{-a^2/2\sigma^2}\right)$$

If we now relate this channel model to the case of binary PSK signaling in AWGN, with unquantized coherent matched-filter detection, the quantity $a^2/2\sigma^2$ corresponds to E_s/N_0, where E_s is the signal energy per PSK pulse and N_0 is the one-sided noise spectral density. We can then write the cutoff rate for this channel as

$$R_0 = 1 - \log_2(1 + e^{-E_s/N_0}) \tag{1.26}$$

which has units of bits per channel symbol. Note that code rate is measured in the same units.

We would now like to determine the form of R_0 in the power-limited region (Fig. 1.10), where, with unlimited bandwidth, coding can be used to reduce the required signal energy to the lowest possible level. To do this we assume the use of a code with rate equal to R_0 and rewrite E_s/N_0 as

$$\frac{E_s}{N_0} = R_0\frac{E_b}{N_0}$$

where E_b is the signal energy per source information bit. Now, letting the rate R_0 become very small, we rewrite Eq. (1.26) as

$$R_0 = 1 - \log_2(1 + e^{-R_0E_b/N_0}),$$

$$\cong 1 - \log_2(2 - R_0E_b/N_0)$$

$$\cong \frac{R_0}{2\ln 2}\left(\frac{E_b}{N_0}\right) \tag{1.27}$$

Solving Eq. (1.27) for E_b/N_0, we find that the minimum value of E_b/N_0 to achieve the cutoff rate R_0 is

$$\frac{E_b}{N_0} = 2\ln 2 \qquad \text{or} \qquad \frac{E_b}{N_0} = 1.4 \text{ dB}$$

Referring to the discussion in Section 1.6.1, we see that this is exactly 3 dB above the Shannon limit (ln 2 or -1.6 dB), the minimum value of E_b/N_0 needed to achieve channel capacity on the AWGN channel.

The reader may question this comparison, since R_0 was derived here for a binary-input AWGN channel, while capacity C (Eq. 1.13) was derived for a channel whose inputs were unconstrained except for a limit on the average signal power. If we were to compute C for the binary-input AWGN channel, we would convert Eq. (1.7) to an integral form and then use Eq. (1.8) to give

$$C = \frac{1}{2}\sum_{i=0}^{1}\int_{-\infty}^{\infty} p(y|x_i)\log\frac{p(y|x_i)}{p(y)}\,dy \tag{1.28}$$

with $p(y|x_0)$ and $p(y|x_1)$ given by Eqs. (1.24) and (1.25) and

$$p(y) = \frac{1}{2}p(y|x_0) + \frac{1}{2}p(y|x_1)$$

In this calculation, C has units of bits per channel symbol. For this case it is not possible to carry out the integral in Eq. (1.28) in closed form. However, it can be shown that for code rates R approaching zero, channel capacity can be accurately approximated by

$$C \cong \frac{R}{\ln 2}\left(\frac{E_b}{N_0}\right)$$

Again we set $C = R$ and find that the minimum SNR per information bit needed to achieve capacity is

$$\frac{E_b}{N_0} = \ln 2 \qquad \text{or} \qquad \frac{E_b}{N_0} = -1.6 \text{ dB}$$

This is the Shannon limit, the minimum value of E_b/N_0 found for the continuous AWGN channel. Thus in the power-limited region of operation (arbitrarily large bandwidths and code rates approaching zero), capacity is in principle achievable with binary modulation and coding.

To develop fully the concept of the R_0 criterion and discuss the extensive implications is beyond the scope of this book. However, it has served our purpose to state that by determining R_0 for any given channel, we establish an upper bound on the information transfer rate that can be achieved with practical coding implementations and also obtain an exponential bound on post-decoding error probability. In Chapters 10 and 11 we shall see examples of error-control coding schemes that operate reliably at information transfer rates remarkably close to R_0. For schemes that operate at rates above R_0 but less than capacity, the computational complexity increases drastically. We shall also see that the particular figure of communication system performance used is not important; that is, low post-decoding error rates or high probability of correct decoding can both be provided for rates $R \leq R_0$.

1.7. SUMMARY AND CONCLUSIONS

In this introductory chapter we have outlined the problem of designing an efficient digital communication system and have reviewed the principal results of information theory in order to provide a framework for the remainder of the book. At this point it is useful to summarize the major points that have been made and to place them in their proper perspective for the sequel.

We can describe the communication system design problem succinctly as follows. We first want to remove all redundancy from the source information, so that the amount of data to be transmitted is minimized, and we also want to communicate this information reliably with the smallest possible expenditure of signal energy. The two key parameters here are the information rate R_s of the source, which is the minimum number of bits per second needed to represent the output of the source, and the channel capacity C, the maximum rate at which information can be transmitted through the channel and received reliably. The channel coding theorem provides us with the important result that if the source rate R_s does not exceed the channel capacity C, it is possible to deliver the source information with arbitrarily low probability of error. We do not concern ourselves with details of the source-coding function, that is, the reduction of a source output to a stream of bits occurring at the rate R_s. This is a large subject unto itself and is treated extensively by other authors. Rather, we assume that the information source produces a sequence of information bits completely free of redundancy and concentrate on the problem of communicating this information as reliably and efficiently as possible.

The central idea of efficient channel coding is to transform long sequences of source data bits into even longer coded channel sequences or signaling waveforms. That is, we must put well-structured redundancy back into the source data for transmission. The amount of redundancy required depends on the quality of the channel (the channel SNR or BER) and the desired level of reliability in the delivered information. At the receive side we use the known structure of the possible transmitted signals to detect and decode the output of the channel and to deliver a representation of the source data.

Information theory provides the fundamental limits on the reliability and efficiency of a digital communication system. For the important case of the AWGN channel, communication efficiency is measured by the SNR per source bit, E_b/N_0. For operation in the power-limited region, where we assume that large bandwidth expansion can be used, the Shannon limit tells us that for all $E_b/N_0 > -1.6$ dB, arbitrarily reliable communication can theoretically be provided. Furthermore, very reliable communication at low SNRs can best be achieved by using an crror-control code to construct the long channel sequences prior to modulation. On the other hand, for operation in the bandwidth-limited region, highly efficient operation can be obtained without resorting to coding, that is, by the use of complex modems that implement high-order modulation alphabets.

Thus we shall be concerned primarily with operation of a digital communication system in the power-limited region, and much of this book is devoted to the solution of two basic problems. The first is finding the best ways to transform the source information into the redundant channel sequences (the code design problem) and the second is finding ways to invert that transformation that are not unduly complex (the decoding problem). Thus, we shall be concerned primarily with finding good code designs and efficient decoding algorithms.

While channel capacity establishes the theoretical limit on the performance that can be achieved with a digital communication system, under conditions of severe signal power limitations operation at or near channel capacity may require an unacceptably complex system design. Channel capacity provides little in the way of practical design guidance. We have asserted that the SNR corresponding to the computational cutoff rate of a channel, R_0, gives us a limit that can be approached with a system design that is not unduly complex. For the case of antipodal signaling with coherent reception in AWGN, this corresponds to an SNR that is 3 dB above the Shannon limit, namely $E_b/N_0 = 1.4$ dB. In fact, in Chapters 10 and 11 we shall see examples of powerful code designs that come close to achieving this limit.

Although we have stressed the view that an improvement in communication efficiency is directly related to a reduction in the required level of signal power, there are other important practical interpretations. We have assumed that the time required for transmission of a message is held fixed, and consequently for coded operation, bandwidth expansion is needed. However, in many applications, both the available signal power and bandwidth are limited. In such situations, if highly reliable and efficient operation is to be provided, a lengthening of message transmission time is inescapable for both coded and uncoded systems. When transmission time is an adjustable parameter, comparison of alternative designs on the basis of communication efficiency is equally valid. Improvements in communication efficiency can in this case be related directly to a reduction in the required level of signal power, a reduction in transmission time, or a simultaneous reduction in both. In addition, for some applications, a reduction in required power levels to achieve the desired performance can be used to increase the effective range of the communication system.

For most applications encountered, operation near channel capacity or computational cutoff is not needed, and considerably less severe requirements apply. We shall see in this book that error-control coding can also be used effectively at higher SNRs. There are, in fact, a number of coding techniques that provide sufficient improvements in communication efficiency to justify the added implementation complexity associated with coding. To apply error-control coding in a cost-effective design for any application, it is clearly necessary to determine both the complexity of the coding technique considered and the performance improvements that accrue. Therefore, the remainder of this book is concerned with the details of code design, encoding and decoding techniques, and the evaluation of performance of error-control-coded systems.

1.8. NOTES

The mathematical formulation of error-correcting codes was founded by R. W. Hamming, whose early work was cited in Shannon's 1948 paper [157] on the mathematical theory of communication. Apparently delayed because of patent considerations at Bell Telephone Laboratories [10], Hamming's own paper appeared in 1950 [60]. The papers of Hamming and Shannon represent, respectively, an essentially combinatorial discipline termed coding theory and an essentially statistical discipline known as information theory or *Shannon theory*.

An important area of information theory, and in particular an important aspect of the source coding problem, is that of *rate-distortion theory*. This theory addresses the question of what accuracy must be sacrificed in the delivery of information from a source to a user when the capacity of the intervening communication channel is less than the minimum information rate needed to completely reconstruct the output of the source. Much research has been devoted to this interesting subject, but a thorough discussion is beyond the scope of this book. The interested reader will find detailed treatments of the subject in books by Berger [6], Gallager [49], and Viterbi and Omura [175].

Many of the key papers in the development of information theory have been collected into an IEEE Reprint Series volume edited by Slepian [163]. Additional bibliographies and surveys of published research are also cited there. An excellent survey of the development of coding theory is provided in a companion volume edited by Berlekamp [10]. Other surveys of the field are also cited there. A book by MacWilliams and Sloane [109] provides an almost exhaustive treatment of the theory of block codes with many references to the literature up to 1981. A brief introduction to the theory of error-correcting block codes is the subject of a recent book by Pless [134]. Convolutional codes and sequential decoding are treated in detail in a text on information theory and coding by Gallager [49] as well as an earlier text by Wozencraft and Jacobs [185].

■ ■ CHAPTER TWO

Some Fundamentals and Simple Block Codes

Many of the fundamentals of error-control coding can be illustrated with elementary examples and a minimum of mathematics. In this chapter we present some basic concepts and definitions relating to parity-check codes, describe modulo-2 arithmetic, and discuss the structure and properties of some simple block codes. Both hard- and soft-decision decoding techniques are considered. The parity-check matrix and the syndrome are introduced. The important issue of the overall efficiency of a coded communication system is discussed, and coding gain is defined. The chapter concludes with a description of the binary Hamming codes.

2.1. PARITY-CHECK CODES

The fundamental concepts of parity-check block codes can be described succinctly as follows: The *encoder* accepts k *information digits* from the information source and appends a set of r *parity-check digits*, which are derived from the information digits in accordance with a prescribed encoding rule. The encoding rule determines the mathematical structure of the code. The information and parity digits are transmitted as a block of $n = k + r$ digits on the communication channel. It is customary to call the code an (n,k) *block code*. The n-bit block is called the *code block* or *codeword*, and n is called the *block length* of the code. The *code rate* is defined as the ratio $R = k/n$.

The received version of the codeword is called simply the *received word* to allow for the possibility that it may not be the intended codeword or even a valid codeword if errors occurred during transmission. The *decoder* operates on the received word to determine whether the information and check digits satisfy the encoding rules, and uses any observed discrepancy to detect and possibly correct errors that have occurred in transmission. The decoder performs either the first or all three of the following functions:

1. Reapply the encoding rules to the received word to determine whether the parity-check relationships are satisfied. Discrepancy in the parity-check relationships indicates the presence of one or more errors in the word. If only *error detection* is to be performed, the decoding function is completed with an announcement either that the received word is a codeword or that errors have been detected. If *error correction* is to be performed, the next two steps are required.
2. The parity-check discrepancy is used to derive an estimate of the *error pattern* contained in the received word. For any given code, there are often several alternative algorithms that may be used to accomplish this step of decoding. This is usually the most complicated of the decoder's functions.
3. When an estimate of the error pattern has been determined, the errors are corrected, the check digits removed from the codeword, and the decoded information digits delivered to the information user.

A central problem in coding theory is to find code designs for which the r check digits can be used as effectively as possible in the detection and correction of errors, and much of the coding literature has been concerned with research on this problem. In addition, considerable attention has been given to the development of computationally efficient algorithms for decoding block codes.

Several points need to be made in reference to the general problem of designing effective error-control codes. First, a given code may be utilized for error detection only or for error detection and correction, commonly designated by the acronym *EDAC*. This was mentioned in the outline of the decoding functions. It will be seen in later discussions that the principle governing the design of a code for error detection is essentially the same as that for effective error correction. Second, no code can be designed that will successfully correct all error patterns to which the transmitted codeword can be subjected. For example, an error pattern that changes one codeword into a different codeword is always undetectable. Rather, codes are designed to correct *the most likely error patterns*. The reader may quite correctly infer that the effectiveness of a code on a given channel will depend in part upon the characteristics of the error patterns produced by the channel. We shall see in a later chapter that given a code design and a decoding algorithm, it is possible to specify the error patterns that will be correctable, those that will be detectable but not correctable, and those that will escape successful detection and correction. It will also be seen that, given an accurate model for channel-error behavior, it is possible to make precise calculations of the performance achievable with a selected code.

2.2. MODULO-2 ARITHMETIC

The codes most widely used in practical systems are binary codes, that is, codes defined on the two-letter alphabet commonly denoted as 0 and 1. Binary codes are structured and implemented according to the algebraic rules of *modulo-2 arithmetic*, which ensures that all mathematical operations performed on code digits, including the generation of redundant check digits from information digits, always result in digits that remain within the binary alphabet.

The rules for modulo-2 arithmetic are completely described by just two operations, addition and multiplication, which follow rules very similar to those of ordinary arithmetic. The modulo-2 addition and multiplication rules are defined as follows:

Modulo-2 Addition			Modulo-2 Multiplication		
+	0	1	•	0	1
0	0	1	0	0	0
1	1	0	1	0	1

Note that the modulo-2 arithmetic rules are the same as those of ordinary arithmetic except that $1 + 1 = 0$. That is, for addition, summations are reduced modulo 2 (often abbreviated *mod 2*), which makes all odd sums equal to 1 and all even sums equal to 0. Thus when the modulo-2 addition rule is applied to find the sum of a number of binary digits, the result is 0 or 1—0 when the number of ones among the digits is even and 1 when odd. This is called a *parity-check sum* and is a common operation in the manipulation of binary codes.

It should be noted that the addition table also gives the rules for subtraction, since when we write $z = x - y$, we are asking: "What z must be added to y in order that the sum be equal to x?" If we use this interpretation to construct a modulo-2 subtraction table, we find that the subtraction table is identical to the addition table, thus showing that addition and subtraction are identical operations. Division can also be performed in the binary alphabet, with division by a symbol being allowed only if the symbol is nonzero, as in ordinary arithmetic. Since the binary alphabet contains only one nonzero symbol, the division operation is trivial ($1 \div 1 = 1$).

The binary alphabet, with its accompanying rules for modulo-2 addition, subtraction, multiplication, and division, is the simplest form of an algebraic system called a *finite field*. Much of the work in coding theory rests upon the mathematical foundation of finite fields, of which more will be said in Chapters 3 and 4. In this chapter, however, we confine our attention to simple concepts of binary codes, which can be discussed in terms of modulo-2 arithmetic. Careful examination of the properties of even simple codes provides a great deal of insight into the properties of block codes in general.

2.3. SINGLE-PARITY-CHECK CODES

A very simple parity-check code is one that uses a single *parity-check bit*, or simply *parity bit* or *check bit*, appended to a block of k *information bits*, the check bit being chosen to satisfy an overall parity rule for the codeword. Note that here we use the term "bit" to designate a binary digit equal to 0 or 1, as opposed to the fundamental theoretical measure of information. Encoding of the *single-parity-check code* is described by the equation

$$i_1 + i_2 + \cdots + i_k + p_1 = 0 \qquad \text{mod } 2$$

where $i_1, i_2, \ldots, i_k$ are arbitrary information bits, and p_1 is the parity bit. The equation specifies that the parity bit be chosen so that the codeword has *even parity*, that is, an even number of ones. We might just as easily have specified an *odd-parity* rule for setting the check bit, and the properties of the code would have been exactly the same. For consistency, we use an even-parity rule throughout the book. Most, though not all, coded systems in operation today use the even-parity rule.

The structure of the single-parity-check code can be illustrated by use of an example. We consider the (4,3) binary code whose codewords are listed as follows:

The (4,3) Binary Single-Parity-Check Code

Information Sequence	Codeword
0 0 0	0 0 0 0
0 0 1	0 0 1 1
0 1 0	0 1 0 1
0 1 1	0 1 1 0
1 0 0	1 0 0 1
1 0 1	1 0 1 0
1 1 0	1 1 0 0
1 1 1	1 1 1 1

Note that the fourth bit in each codeword is the parity-check bit.

2.3.1. Error-Detection Decoding

At the decoder, the received word, possibly containing transmission error is checked to determine whether or not the parity relationship is satisfied. This is done by calculating from the received word, which we represent as $r_1, r_2, \ldots, r_n$, the sum

$$S = r_1 + r_2 + \cdots + r_n \qquad \text{mod } 2 \tag{2.1}$$

and observing whether S is 0 or 1. If $S = 0$, the decoder assumes that the codeword was received correctly, removes the parity-check bit, and releases the $n - 1$ information bits. If $S = 1$, the decoder declares the received word to be in error.

This simple example can be used to make several points that are pertinent to all block codes. First, the single parity check cannot detect all possible error patterns, although it can detect any error pattern containing an odd number of errors. This is easily verified by noting that since every valid codeword contains an even number of ones, the occurrence of any odd number of errors will cause the word to be received with an odd number of ones, resulting in a calculated parity-check sum $S = 1$. We term this decoding outcome *error detection* or *decoding failure*. By the same token, the decoder will not detect any error pattern containing an even number of errors, but for each such error pattern will calculate $S = 0$ and decide, incorrectly, that the word was received without error. We denote this type of decoding outcome as an *incorrect decoding*. The point to be emphasized here is that a single binary parity check can provide only one binary-valued indicator of the condition of the received word, namely that the prescribed parity relationship among a set of bits is or is not satisfied. We summarize this situation for the single parity check code as follows:

Decoding Outcomes for the $(n, n-1)$ Single-Parity-Check Code

Actual Number of Errors	Value of S(mod 2)	Decoding Decision
$0, 2, 4, 6, \ldots$	0	"No error"
$1, 3, 5, 7, \ldots$	1	"Errors detected"

It is instructive to examine the performance of the decoding rule for the $(n, n-1)$ code on the binary symmetric channel. In particular, consider the error patterns that preserve even parity ($S = 0$) in the received word, patterns having an even number of errors. On the binary symmetric channel, the probability of occurrence of any particular pattern of i errors is simply

$$\text{Prob}(i\text{-error pattern}) = p^i(1-p)^{n-i}$$

where p is the bit-error probability in the channel ($p < \frac{1}{2}$). If we compare the probabilities of zero errors and a two-error pattern, we see that for $p < \frac{1}{2}$,

$$(1 - p)^n > p^2(1 - p)^{n-2}$$

since

$$(1 - p)^2 > p^2$$

Thus, given $S = 0$, the probability of no errors in the received word is greater than the probability of any particular two-error pattern. In addition, the reader can easily verify that given $S = 0$, the probability of zero errors is greater than that of any error pattern containing an even number of errors. We can now state that the interpretation of $S = 0$ as the occurrence of "no error" and $S = 1$ as errors detected is in fact a *maximum likelihood decoding rule* for the binary symmetric channel. A decoder that selects the word of largest probability conditioned on transmission of that word is called a *maximum likelihood decoder*.

Knowing the decoding outcomes associated with the various possible error patterns, it is a simple matter to calculate the probabilities of all the decoding outcomes for the single-parity-check code. We have

$$\text{Prob(correct decoding)} = (1 - p)^n$$

$$\text{Prob(error detection)} = \sum_{\substack{i=1 \\ i \text{ odd}}}^{n} \binom{n}{i} p^i (1 - p)^{n-i}$$

$$\text{Prob(incorrect decoding)} = \sum_{\substack{i=2 \\ i \text{ even}}}^{n} \binom{n}{i} p^i (1 - p)^{n-i}$$

In these expressions $\binom{n}{i}$ is the binomial coefficient, the number of combinations of n objects taken i at a time, which is calculated as

$$\binom{n}{i} = \frac{n!}{(n - i)!\,i!}$$

The discussion of the single-parity-check code has shown that the single check bit permits a decoder merely to classify received words into two categories, "no error" and "errors detected." Later in this chapter we shall examine some codes that utilize multiple parity checks to accomplish more than simple error detection. Before doing this, however, we describe one way in which a single parity check can be used with information derived in the receiver or demodulator to do more than simple error detection.

2.3.2. Erasure Filling

We pointed out in Chapter 1 that coding systems can be implemented to perform either *hard-decision* or *soft-decision* decoding, depending on the amount of information conveyed to the decoder with each demodulated symbol. One very simple scheme for using soft-decision symbol information in conjunction with a single-parity-check code has been applied with success in several existing communication systems. The scheme, sometimes called *Wagner coding* [161], operates as follows. The demodulator makes a hard binary decision and at the same time a *bit-quality* measurement on each received bit. The bit-quality estimate might be determined, for example, by measuring the matched-filter output for the signal arriving in the bit interval. The bit-quality information is conveyed to the decoder along with each corresponding bit decision. When all the bits in the received word have been entered, the decoder computes the overall parity-check sum given by Eq. (2.1). If the parity check is satisfied, the codeword is accepted. If the parity check fails, the received bit having the poorest quality measure is inverted, forcing the parity check to be satisfied. The information bits are then released by the decoder. This soft-decision decoding procedure is clearly more powerful than hard-decision error-detection decoding, since many erroneously received words can be decoded correctly.

As discussed in Chapter 1, the process of flagging a received symbol as unreliable is known as symbol *erasure*. The decoding procedure of adjusting erased symbols so as to satisfy one or more parity checks in a codeword is called *erasure filling*, and the Wagner scheme is the simplest example of this procedure. We shall have more to say later about erasure filling as a means of enhancing the error-correction power of a code. For the present, however, we simply state that erasure filling is a very simple form of *probabilistic decoding*, a designation covering a variety of techniques in which the decoder utilizes some form of channel statistics to aid in the decoding process. Probabilistic decoding techniques are widely used with convolutional codes, which are treated in detail in Chapters 8, 9, and 10.

2.4. PRODUCT CODES

While the single-parity-check code is limited to error detection or erasure filling, there is a way of combining two single-parity-check codes to permit error correction with relatively simply decoding procedures. This scheme uses a structure called a *product code*, which is shown in Fig. 2.1. A product code applies two block codes, (n_1,k_1) and (n_2,k_2), to the rows and columns, respectively, of a two-dimensional array of information digits, as shown in Fig. 2.1*a*. In general, the two block codes may be the same or different. It can be shown that for any given array of information digits the check digits in the lower right corner are the same whether they are constructed using the row code or the column code.

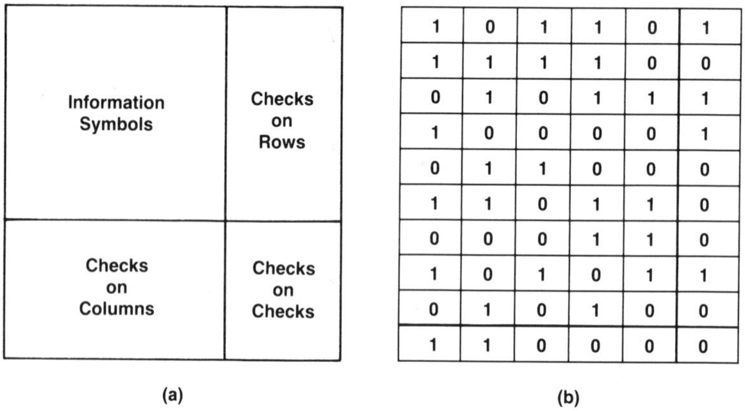

Information Symbols	Checks on Rows
Checks on Columns	Checks on Checks

(a)

1	0	1	1	0	1
1	1	1	1	0	0
0	1	0	1	1	1
1	0	0	0	0	1
0	1	1	0	0	0
1	1	0	1	1	0
0	0	0	1	1	0
1	0	1	0	1	1
0	1	0	1	0	0
1	1	0	0	0	0

(b)

FIGURE 2.1. Two-dimensional product codes. (*a*) General code structure, (*b*) a code designed with single parity checks.

A product code may be generalized to higher dimensions as well. In three dimensions the code block can be visualized as a box with dimensions $n_1 \times n_2 \times n_3$, while in four dimensions the code block is n_4 such three-dimensional boxes. This generalization suggests the name *iterated code*, which in fact was coined for this coding scheme by its inventor, Elias [32].

Our interest here is confined to two-dimensional codes, in particular simple codes that use a single binary parity check in each dimension. An example is shown in Fig. 2.1*b*. It is left as an exercise for the reader to prove that the parity bit in the lower right corner must be the same whether it is generated as a row or a column check.

Since a product code is a block code, it can be referred to compactly as an (n,k) code where $n = n_1 n_2$ and $k = k_1 k_2$.

2.4.1. Single-Error Correction

In Fig. 2.2*a* we show the (9,4) product code, which is constructed using the (3,2) single-parity-check code in each dimension. Clearly the code is capable of correcting any single error occurring in the code block, since an error causes the parity checks to fail in the row and column containing the error. This is shown in Fig. 2.2*b*, where the marginal ones indicate the parity checks that fail ($S = 1$). It is seen that the row and column positions of the parity checks that fail serve as coordinates of the single error. This suggests a simple decoding procedure in which the nine parity-check failure patterns corresponding to single-error events are stored for comparison against the pattern found for a received block.

Some useful insights can be gained by further examination of this simple example. Consider the fact that the five parity checks can produce 32 different binary combinations, of which one (all $S = 0$) indicates an error-free code block and nine indicate the single-error patterns. This leaves 22 parity-check combinations unaccounted for, which raises a question. Can the extra combinations be used to correctly identify the error patterns containing more than one error?

This can be answered immediately with simple arithmetic. Suppose, for example, that we wanted to correct all the two-error patterns. We have $\binom{9}{2} = 36$ two-error patterns, and there are clearly not enough available parity-check outcomes to uniquely identify all these patterns. For example, Fig. 2.2*c* shows a two-error pattern that is detectable but not correctable. One readily sees that the pattern is not correctable, since its row location is invisible, all three row checks being satisfied. It is also readily seen that all two-error patterns are detectable, since to escape detection a pattern must have an even number of errors in each row and column occupied. A product code of this type can, in fact, detect all the three-error patterns as well. However, if single-error correction is being done, some three-error patterns will cause incorrect decodings. One such pattern is shown in Fig. 2.2*d*, where the parity-check outcomes are the same as for the single-error pattern in Fig. 2.2*b*. In this case, the decoder will invert bit i_1, converting the three-error pattern to

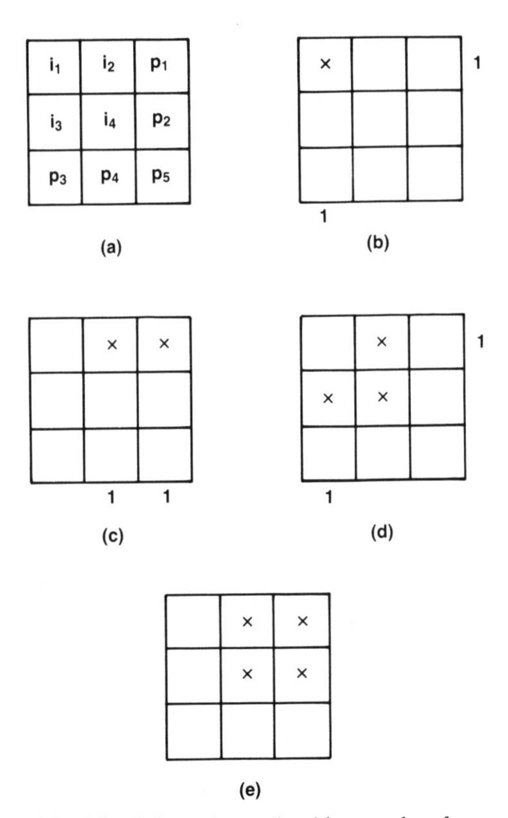

FIGURE 2.2. The (9,4) product code with examples of error patterns.

a four-error pattern, and an incorrect decoding will occur. The four-error pattern in Fig. 2.2*e* is undetectable, since all parity checks are satisfied. There are, of course, many four-error patterns that are detectable though not correctable. We could continue with further examples to show that the (9,4) product code is capable of assured detection of all five-, seven-, and nine-error patterns, while the sets of six- and eight-error patterns will each include some that are detectable, some that are undetectable, and some that result in incorrect decoding after false (but unavoidable) interpretation of the parity check outcomes as indicators of single-error events.

In summary, we have seen that a product code constructed with two single-parity-check codes is able to correct any single error in a received code block and has sufficient parity-check redundancy to provide detection, without correction, of a great many error patterns containing larger numbers of errors. It is conventional to describe the capability of a block code in terms of the maximum number of errors *guaranteed correctable* in a code block. Thus we can refer to the $(n_1, n_1 - 1) \times (n_2, n_2 - 1)$ product code as a *single-error-correcting code* of block length $n_1 n_2$.

2.4.2. Soft-Decision Decoding

It was shown earlier in this chapter how statistical measurements made on received bits can be used to enhance the power of a single-parity-check code by using the erasure-filling scheme termed Wagner coding. The same concept can be applied to the decoding of a product code if a measure of reliability is provided for each received bit as part of the demodulation function. A particularly effective algorithm for decoding a product code with the use of reliability information has been given by Chase [23]. It is convenient to describe this decoding algorithm, called *rank decoding*, with the example of the (9,4) product code.

We begin by assigning symbols to the information and parity bits as in Fig. 2.2*a*. The parity bits are formed by taking modulo-2 sums as follows: $p_1 = i_1 + i_2$, $p_2 = i_3 + i_4$, $p_3 = i_1 + i_3$, $p_4 = i_2 + i_4$, and $p_5 = p_1 + p_2 = p_3 + p_4 = i_1 + i_2 + i_3 + i_4$.

The nine received bits are individually detected and also ranked according to their relative reliability. The bits are labeled $1, 2, \ldots, 9$, where 9 indicates the most reliable bit and 1 the least reliable. Consider the example shown in Fig. 2.3*a*, where the reliability ranking numbers are entered into their respective bit locations. The **x**'s indicate the locations of three errors (the most that the decoder can find and correct in this case) and the f's indicate flags for the three least reliable bits.

The decoder begins by checking the parity of each row and column of the received word in Fig. 2.3*a*. A check (✓) indicates that the parity check is satisfied, and a crossed check (✓̸)indicates a failure. Note that a parity equation will check if there are no errors or two errors in a row or column. The

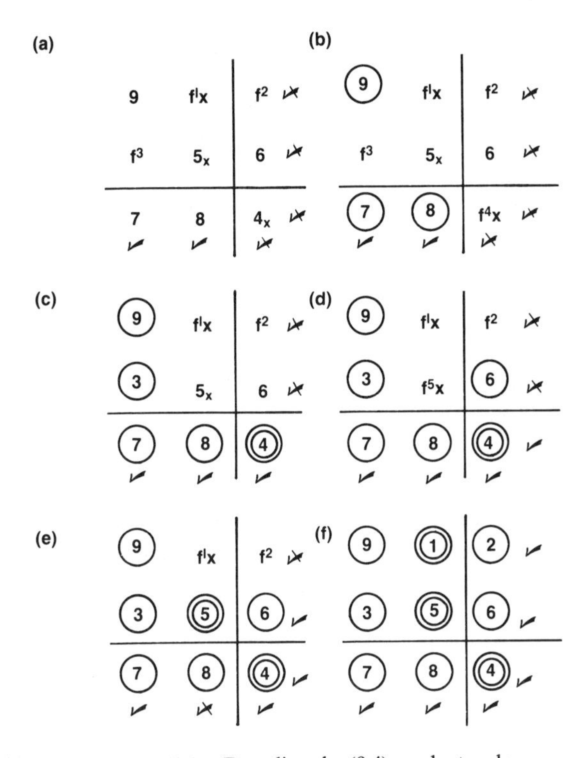

FIGURE 2.3. Decoding the (9,4) product code.

remainder of the algorithm can be stated in three main parts:

1. Examine the highest ranked unflagged undecoded bit.
   a. If both parity equations check, decode the bit as received.
   b. If both equations fail to check, flag the bit.
   c. If only one equation checks, decode the bit as received, provided that its rank is larger than the lowest ranked undecoded bit in each crossed-checked equation. Flag the lowest ranked undecoded bit in each crossed-checked equation, if it is not already flagged.
2. After decoding a bit, if there remains only one undecoded bit in any equation, decode the remaining bit by forcing parity to check.
3. When all remaining undecoded bits are flagged, decode the highest ranked flagged bit and go to step 2.

Figures 2.3*b* to *f* trace the decoding steps for the example given. We begin with bit 9 and notice that the vertical parity check is satisfied but the horizontal check fails. Invoking step 1c, we decode bit 9 and circle it to indicate this. Bits 7 and 8 are similarly decoded by step 1c (Fig. 2.3*b*). As soon as bit 7 is decoded, though, the first column and third row have only one remaining undecoded bit. Step 2 requires that we now decode bits 3 and 4 by forcing the parity in the first column and third row to check. Thus, in Fig. 2.3*c*, bits 3 and 4 are decoded and circled. Bit 4 is circled twice to indicate that it had to be complemented for the third row parity to check. At this point we have corrected our first error (Fig. 2.3*c*). The highest-ranked undecoded bit (bit 6) is next decoded by step 1*c*, as shown in Fig. 2.3*d*, leaving bit 5 the only undecoded bit in the second row. Step 2 again requires that bit 5 be complemented and decoded to assure correct parity on the second row. We have just found and corrected the second error (Fig. 2.3*e*). Finally bit 2 is decoded by step 2, enabling bit 1 also to be decoded properly by step 2, correcting the third error and completing the decoding process (Fig. 2.3*f*).

The rank decoding technique has been implemented with the (25,16) product code in an experimental HF modem called CODEM I. This modem, designed primarily to provide reliable data communication in signal fading, has exhibited as much as one to two orders of magnitude improvement in bit-error probability relative to uncoded transmission on HF radio circuits [23].

2.5. BINARY REPETITION CODES

The simplest type of block code allowing a variable amount of redundancy is the repetition code. With this code a single information bit is encoded into a block of n identical bits, producing an $(n,1)$ code. The code contains only two codewords, the all-zeros word and the all-ones word.

The repetition code can be discussed without reference to error-control coding concepts by observing that the repeated transmission of the information symbol is a form of *time-diversity signaling*. The received n-bit word, detected one bit at a time, may be operated on by the *majority-voting rule*. (We assume n is odd.) That this is a *maximum likelihood* rule can be simply shown with the following example.

Consider the transmission of a codeword from the (5,1) code. Suppose that the received word is 01001. Obviously there are errors in the word, since only 00000 or 11111 could have been sent. Two interpretations are possible: either five zeros were transmitted and two errors occurred, or five ones were transmitted and three errors occurred. If we compare the conditional probabilities of these two error events for the binary symmetric channel, we find that

$$p^2(1-p)^3 > p^3(1-p)^2$$

for $p < \frac{1}{2}$. Thus the two-error pattern is the event of greater likelihood. The corresponding analysis can be made of each possible received word to show that the majority-voting rule is a maximum likelihood decision rule for the binary symmetric channel. It can be seen that all single- and double-error patterns are correctable with the majority-voting rule.

The generalization of these results to a repetition code of arbitrary length should be clear to the reader. That is, the majority-voting rule decides between two explanations for the received word by favoring the one that assumes the smaller number of channel errors. It is easy to see that this is a maximum likelihood decoding rule by comparing the likelihoods of an i-error and a j-error event, where $i < j$. For $p < \frac{1}{2}$, the likelihoods obey the inequality

$$p^i(1-p)^{n-i} > p^j(1-p)^{n-j}, \qquad i < j$$

since

$$(1-p)^{j-i} > p^{j-i}, \qquad i < j$$

For a repetition code with even block length, a received word containing an equal number of ones and zeros does not permit a maximum likelihood decision. In this case the receiver can do no better than announce that the n-bit word was received in error.

2.5.1. The Repetition Code as a Parity-Check Code

We next discuss the repetition code as a parity-check code. The discussion will serve to introduce certain principles of parity-check coding that underlie all commonly used block coding techniques.

The $(n,1)$ repetition code can be described as a parity-check code in which $n-1$ parity bits are generated from one information bit by the following set of equations:

$$\begin{array}{llllll} i_1 & +p_1 & & & & = 0 \\ i_1 & & +p_2 & & & = 0 \\ i_1 & & & +p_3 & & = 0 \\ \vdots & & & & & \vdots \\ i_1 & & & & +p_{n-1} & = 0 \end{array} \tag{2.2}$$

These simultaneous equations are called *parity-check equations* and can be generalized to describe encoding rules for any parity-check block code. It is readily seen that this set of equations produces a codeword that is a block of n replicas of the information bit.

Many of the commonly used algorithms for decoding parity-check codes begin by testing the parity-check relationships in the received word $r_1, r_2, \ldots, r_n$ as the first step in decoding. For the $(n,1)$ repetition code, this step can be written as follows:

$$
\begin{aligned}
r_1 + r_2 \quad\quad\quad\quad &= S_1 \\
r_1 \quad + r_3 \quad\quad &= S_2 \\
r_1 \quad\quad + r_4 \quad &= S_3 \\
\vdots \quad\quad\quad\quad\quad &\quad \vdots \\
r_1 \quad\quad\quad + r_n &= S_{n-1}
\end{aligned}
\tag{2.3}
$$

It is obvious from examination of the parity-check equations, Eq. (2.2), that if the received word $r_1, r_2, \ldots, r_n$ is error-free, Eq. (2.3) yields $n - 1$ parity check sums $S_1, S_2, \ldots, S_{n-1}$ all equal to zero. Let us now examine the set of check sums, that is, the S-vector, resulting from various received error patterns for the example of the (5,1) code. We recall from previous discussion that the (5,1) code is capable of correcting any pattern of up to two errors in a received word. Therefore, in Table 2.1 we list all such correctable error patterns (including the zero-error pattern) together with the parity-check result for each pattern. The parity-check results may be verified from Eq. (2.3) by noting that a single error in one of the two bits checked in any parity-check equation will produce odd parity, while errors in both bits will yield even parity.

The key point to be observed in Table 2.1 is that each of the correctable error patterns results in a unique S-vector in the parity-check test. Thus the S-vector, which is called the *syndrome vector* or simply the *syndrome*, provides

TABLE 2.1. Parity-Check Tests for the (5,1) Repetition Code

Received Error Pattern					Parity Test Results			
1	2	3	4	5	S_1	S_2	S_3	S_4
0	0	0	0	0	0	0	0	0
X	0	0	0	0	1	1	1	1
0	X	0	0	0	1	0	0	0
0	0	X	0	0	0	1	0	0
0	0	0	X	0	0	0	1	0
0	0	0	0	X	0	0	0	1
X	X	0	0	0	0	1	1	1
X	0	X	0	0	1	0	1	1
X	0	0	X	0	1	1	0	1
X	0	0	0	X	1	1	1	0
0	X	X	0	0	1	1	0	0
0	X	0	X	0	1	0	1	0
0	X	0	0	X	1	0	0	1
0	0	X	X	0	0	1	1	0
0	0	X	0	X	0	1	0	1
0	0	0	X	X	0	0	1	1

a unique indication for each error pattern that can be corrected by the code. We see, therefore, that as an alternative to a majority-voting rule on the received bits the syndrome may be used in a decoding procedure to determine which of the set of correctable error patterns has actually occurred.

In order to discuss parity checks and syndromes more conveniently, we rewrite the equations in matrix notation, as is done with simultaneous equations in ordinary arithmetic. (Appendix A provides a brief review of standard matrix notation and terminology.)

To rewrite the parity-check equations given in Eq. (2.2), we represent the codeword as a row vector $\mathbf{c} = i_1, p_1, p_2, \ldots, p_{n-1}$ and the matrix of coefficients of the parity-check equations as $\mathbf{H}$, a matrix having $n - 1$ rows and n columns. We call $\mathbf{H}$ written as

$$\mathbf{H} = \begin{bmatrix} 1 & 1 & 0 & 0 & 0 \\ 1 & 0 & 1 & 0 & 0 \\ 1 & 0 & 0 & 1 & 0 \\ 1 & 0 & 0 & 0 & 1 \end{bmatrix} \tag{2.4}$$

the *parity-check matrix* for the (5,1) repetition code. We can now write the parity-check equations in the form of a single compact matrix equation,

$$\mathbf{cH}^T = \mathbf{0}$$

where $\mathbf{H}^T$ denotes the transpose of $\mathbf{H}$ and $\mathbf{0}$ is the row vector of $n - 1$ zeros. Similarly, we can rewrite the set of syndrome equations, Eq. (2.3), as the single matrix equation

$$\mathbf{S} = \mathbf{rH}^T$$

where $\mathbf{r}$ is the row vector representing the received word $r_1, r_2, \ldots, r_n$, and $\mathbf{S}$ is the resulting syndrome row vector. (Note: We use these matrix forms of the parity-check and syndrome equations for typographical convenience, so that we always write codewords, received words, and syndromes as row vectors. Had we chosen to use column vectors instead, the matrix equations would be written $\mathbf{Hc} = \mathbf{0}$ and $\mathbf{S} = \mathbf{Hr}$.)

The key point to be observed in examining the matrix form of the syndrome equations is the following. The syndrome $\mathbf{S}$ resulting from multiplication of the received vector $\mathbf{r}$ by $\mathbf{H}^T$ is the sum of the columns of $\mathbf{H}$ corresponding to the locations of the errors in $\mathbf{r}$, transposed into a row vector. The reader should verify, for example, that regardless of which codeword is transmitted, 00000 or 11111, the error pattern 0000X will produce $\mathbf{S} = 0001$, the fifth column of $\mathbf{H}$ transposed, and the error pattern 0XX00 will produce $\mathbf{S} = 1100$, which corresponds to the sum of the second and third columns of $\mathbf{H}$ transposed. Therefore, the columns of the parity-check matrix $\mathbf{H}$ serve as *location vectors* for the bit positions where errors occurred in a received word, and the syndrome vector $\mathbf{S}$ is simply the sum of the location vectors transposed. The task of the decoder then is to determine which column of **H** or which sum of columns of **H** produced the observed syndrome. This states succinctly the task that must be performed by the decoder for any parity-check block code. We shall have more to say on this point in Section 2.6.

A further point to be made with regard to Table 2.1 is that the number of possible four-bit syndrome vectors, $2^4 = 16$, is exactly equal to the number of error patterns that can be corrected with the (5,1) repetition code:

$$\binom{5}{0} + \binom{5}{1} + \binom{5}{2} = 1 + 5 + 10 = 16$$

Therefore, the set of possible syndrome vectors are fully utilized as indicators of the correctable error patterns. It is easy to show that for any binary repetition code of odd block length n, the number of parity checks, $n - 1$, is the minimum number required to enumerate all the correctable error patterns, that is, all patterns having from 0 to $(n - 1)/2$ errors. This exact sufficiency of parity checks for the required syndromes occurs only in special cases, and the codes for which this relationship holds are referred to as *perfect codes*. (See Section 3.8.) As we shall see shortly, though, a perfect code is not necessarily a "good" code in the sense of communication efficiency.

2.5.2. Performance of Binary Repetition Codes

Knowing that the $(n,1)$ repetition code with n odd can be used to correct up to $(n - 1)/2$ errors, we can calculate the probability of correctly decoding a received word on the binary symmetric channel as

$$\text{Prob(correct decoding)} = \sum_{i=0}^{(n-1)/2} \binom{n}{i} p^i (1 - p)^{n-i}$$

For n odd, all patterns of $(n + 1)/2$ or more errors will be decoded incorrectly, so we can also write

$$\text{Prob (incorrect decoding)} = 1 - \text{prob(correct decoding)}$$

Table 2.2 provides comparisons of performance for uncoded binary transmission and for transmission with three different repetition codes. Performance parameters are shown in the table for a delivered (post-decoding) bit-error rate (BER) equal to 10^{-5}. The second column gives the channel BER for which each code yields prob(incorrect decoding) $= 10^{-5}$, which is equal to the delivered BER, since there is one information bit per codeword. It is seen that the use of increasing amounts of repetition-code redundancy permits steadily higher levels of channel BER while maintaining a desired level of delivered BER. The third column makes the same set of comparisons in terms of channel SNR (E_s/N_0) for binary PSK transmission on the AWGN channel, as given by Eq. (1.1).

TABLE 2.2. Use of Uncoded and Repetition-Coded Binary PSK Signaling at Delivered BER = 10^{-5}

Code (n, k)	Channel BER	E_s/N_0 (dB)	E_b/N_0 (dB)
(1,1)	1.0×10^{-5}	9.6	9.6
(3,1)	1.8×10^{-3}	6.2	11.0
(5,1)	1.0×10^{-2}	4.3	11.3
(7,1)	2.3×10^{-2}	2.9	11.4

It is seen that for comparisons made on the basis of channel BER or E_s/N_0, repetition coding achieves consistently better performance than uncoded transmission, the improvement increasing with n. This type of comparison does not, however, tell a complete story with respect to the efficiency of repetition coding as a communication technique. The comparison is useful in the design of a system where the modulation scheme (the modem) and the channel signaling rate are fixed and the code redundancy may be varied at will to accommodate various levels of channel SNR. However, if the channel signaling rate is fixed, the use of an $(n,1)$ repetition code causes a reduction of the overall information transmission rate for the system by the ratio $1/n$. If channel SNR is also fixed, this reduction in information rate is accompanied by an n-fold increase in E_b, the signal energy transmitted per source information bit, which we first discussed in Chapter 1.

Therefore it is important to consider the overall communication efficiency of the repetition codes as measured by the energy per source bit required for various levels of performance. This is done by showing values of E_b/N_0 in the fourth column of Table 2.2. It is seen in the table that when compared on the basis of E_b/N_0 the repetition codes are less efficient than uncoded transmission, the loss in efficiency increasing with block length n. Thus repetition coding may serve well as a simple applique for improving error-rate performance in a system where the modulation and demodulation parameters must remain fixed and a reduction in the system information rate can be tolerated. However, the technique is not satisfactory if the objective is to increase overall communication efficiency.

This simple set of calculations serves to make an important point: the use of coding is not guaranteed to improve overall communication efficiency. The central idea here is that simple repetition coding does not represent an efficient way to use redundancy. The reader will recall that in Chapter 1, where Shannon's capacity formula and coding theorem were discussed, it was pointed out that the key element in providing efficient reliable communication is the transformation of long strings of source bits into long coded waveforms. Simple repetition coding contradicts this prescription by inserting only a single information bit into each codeword. In fact, efficiency is lost because the binary demodulation decision is converted into n separate binary decisions, each having to operate with available signal energy reduced by $1/n$.

The efficiency lost with use of the repetition code is in fact salvageable if the demodulation function can be modified appropriately. Specifically, if the n-bit block is coherently detected as a single waveform, there is no loss in communication efficiency relative to uncoded transmission. There is no gain either. Later in the book, we shall present many efficient code designs that reduce output bit-error probabilities and simultaneously achieve improvements in communication efficiency as measured by a reduction in the required E_b/N_0 relative to uncoded transmission.

2.6. PROPERTIES OF THE SYNDROME

Certain key properties of parity-check block codes, seen thus far only in terms of simple examples, will now be described with somewhat more mathematical detail. For compactness of presentation, we shall use vector and matrix notation.

We consider an arbitrary binary (n, k) parity-check block code. Since the k information-bit positions may be filled with any combination of ones and zeros, the code will contain a total of 2^k codewords. We can denote the set of codewords as $\{\mathbf{c}_i\}$, where $i = 0, 1, 2, \ldots, 2^k - 1$. However, for the present discussion, we denote an arbitrary codeword simply as $\mathbf{c}$.

When a particular codeword $\mathbf{c}$ is transmitted, we receive a word $\mathbf{r}$ which may differ from $\mathbf{c}$ in one or more bit positions due to channel-induced errors. We can write this quite generally as

$$\mathbf{r} = \mathbf{c} + \mathbf{e}$$

where $\mathbf{e}$, which is called the *error vector* or *error pattern*, has zeros in those positions where $\mathbf{r}$ and $\mathbf{c}$ agree and ones in those positions where $\mathbf{r}$ and $\mathbf{c}$ disagree. In other words, the ones in $\mathbf{e}$ mark the positions where errors have occurred. The decoding problem, then, is to attempt to determine $\mathbf{e}$ so that the intended codeword may be recovered by forming $\hat{\mathbf{c}} = \mathbf{r} - \hat{\mathbf{e}}$ where $\hat{\mathbf{e}}$ is an estimate of the error pattern and $\hat{\mathbf{c}}$ is the decoded word. Note that when $\hat{\mathbf{e}} = \mathbf{e}$, we have $\hat{\mathbf{c}} = \mathbf{c}$, which means that decoding successfully reproduces the transmitted codeword.

The algorithms commonly used for decoding parity-check block codes start with the calculation of an $(n - k)$-element vector called the *syndrome* (a name invented by Hagelbarger [59]), introduced earlier in discussion of the repetition code. The syndrome derives its name from the fact that it is symptomatic of the error vector that actually occurred. We use the term "symptomatic" quite deliberately, since the syndrome does not uniquely specify the true error vector. What is important, though, is that the syndrome can be used to identify the complete set of error vectors that could have occurred. The decoder must then make the best selection from that set.

Consider the transmission of a codeword $\mathbf{c}$ satisfying the matrix equation $\mathbf{cH}^T = \mathbf{0}$ and the reception of the vector $\mathbf{r}$. The syndrome corresponding to $\mathbf{r}$ has been defined as $\mathbf{S} = \mathbf{rH}^T$. Important general properties of the syndrome for parity-check codes will now be given.

1. The syndrome $\mathbf{S}$ is independent of the transmitted codeword $\mathbf{c}$ and is a function only of the error vector $\mathbf{e}$, specifically, $\mathbf{S} = \mathbf{eH}^T$.

PROOF.

$$\mathbf{S} = \mathbf{rH}^T = (\mathbf{c} + \mathbf{e})\mathbf{H}^T = \mathbf{cH}^T + \mathbf{eH}^T = \mathbf{0} + \mathbf{eH}^T = \mathbf{eH}^T$$

Note that when no errors have occurred, the syndrome is all zeros. The property $\mathbf{S} = \mathbf{eH}^T$ is of central importance in the decoding of parity-check codes. It is seen, in the proof just given, to come directly from the relationship $\mathbf{cH}^T = \mathbf{0}$, which will be treated in greater detail in Chapter 3.

2. For any error vector $\mathbf{e}$, the 2^k distinct vectors $\mathbf{e}_i$ defined as

$$\mathbf{e}_i = \mathbf{e} + \mathbf{c}_i, \qquad i = 0, 1, 2, \ldots, 2^k - 1$$

all have the same syndrome.

PROOF.

$$\mathbf{e}_i\mathbf{H}^T = (\mathbf{e} + \mathbf{c}_i)\mathbf{H}^T = \mathbf{eH}^T$$

$$\mathbf{e}_j\mathbf{H}^T = (\mathbf{e} + \mathbf{c}_j)\mathbf{H}^T = \mathbf{eH}^T$$

therefore, $\mathbf{e}_i\mathbf{H}^T = \mathbf{e}_j\mathbf{H}^T$

Property 2 substantiates our earlier statement that the syndrome does not uniquely specify the actual error pattern $\mathbf{e}$. In fact, for a specific received vector $\mathbf{r}$, there are 2^k error patterns that could have produced the syndrome $\mathbf{rH}^T$. Therefore, the decoding problem is to devise algorithms to find the one of the 2^k candidate error patterns that is most likely to have occurred.

3. The syndrome $\mathbf{S}$ is the sum of the rows of $\mathbf{H}^T$ corresponding to the bit positions in which the errors have occurred.

PROOF. First we write the parity-check matrix in the form

$$\mathbf{H}^T = \begin{bmatrix} \mathbf{h}_1 \\ \mathbf{h}_2 \\ \mathbf{h}_3 \\ \vdots \\ \mathbf{h}_n \end{bmatrix}$$

where $\mathbf{h}_j$ is the jth row of $\mathbf{H}^T$. The syndrome can then be written as

$$\mathbf{S} = \mathbf{e}\mathbf{H}^T = \sum_{j=1}^{n} e_j \mathbf{h}_j$$

where e_j is the jth element of the error vector $\mathbf{e}$. However, $e_j = 1$ if an error has occurred in the jth bit position, and $e_j = 0$ otherwise. Therefore, $\mathbf{S}$ is a summation of those rows of $\mathbf{H}^T$ whose row positions are error locations in $\mathbf{e}$.

The properties of the syndrome as well as the decoding problem that has been alluded to can be illustrated by reference to the simple codes we have discussed thus far in this chapter. In the case of the single-parity-check code, the parity-check-sum outcome $\mathbf{S} = 0$ indicates not only the reception of an error-free codeword, but also the reception of a codeword plus an error pattern equal to any codeword, which in turn means any error pattern with an even number of ones. By the same token, the check-sum outcome $\mathbf{S} = 1$ indicates the presence of one error or any other odd number of errors in the received word. For the decoding rule that accepts words received with $\mathbf{S} = 0$, the assumption is made that the most likely situation is that the word was received without error. This is seen to conform to the principle of maximum likelihood decoding as discussed in Section 2.5.

In the case of the product code constructed with two single-parity-check codes, the syndrome is simply a vector of $k_1 + k_2 + 1$ bits giving the outcomes of the corresponding parity-check sums calculated for a received block. The all-zeros syndrome indicates either an error-free block or reception with any error pattern identical to a codeword such as the square pattern of four errors shown in Fig. 2.2e. The syndrome that indicates the single error shown in Fig. 2.2b also indicates any of a number of other patterns, including the three-error pattern shown in Fig. 2.2d. (Note that the three-error pattern can be interpreted as a single error superimposed on the codeword with four ones at the upper left corner of the code array.) The single-error-correction decoding rule, which inverts the bit at the intersection of two parity checks that fail, is a sensible decoding rule, since it selects the error pattern that contains the smallest number of errors. It is, in fact, a maximum likelihood decoding rule for the independent-error channel.

For the $(n, 1)$ repetition codes, the situation can also be described simply. The all-zeros syndrome indicates no error or n errors, the former being more likely. Each nonzero syndrome with t ones indicates t errors or $n - t$ errors, and the maximum likelihood decoding rule is to decide in favor of the smaller number of errors, which must be $(n - 1)/2$ or less.

In general, hard-decision decoding algorithms for block codes are designed to interpret the syndrome as arising from the error pattern having the smallest number of errors, given the set of 2^k error patterns that could have produced the syndrome. However, for an arbitrary parity-check code, it is not a simple matter to determine which combination of rows of $\mathbf{H}^T$ corresponds to that error pattern. Algorithms for decoding important binary block codes are treated in detail in Chapter 5.

2.7. BINARY HAMMING CODES

Thus far we have discussed several techniques for using single-parity-check codes and repetition codes for error detection or correction. For most applications, however, these simple forms of coding lack sufficient power to provide good overall communication efficiency. However, more complex code designs do provide the desired efficiency. Some of the concepts introduced in the earlier discussions will provide a basis for understanding the structure of more powerful block codes.

Key concepts introduced thus far include the description of a parity-check block code in terms of its parity-check matrix $\mathbf{H}$ and the interpretation of the syndrome as the sum of those columns of the parity-check matrix (or rows of $\mathbf{H}^T$) corresponding to the locations of errors in the received word. Parity-check matrices are shown below for two simple codes, a single-parity-check code with $n = 7$ and a repetition code with $n = 3$.

$$[1 \quad 1 \quad 1 \quad 1 \quad 1 \quad 1 \quad 1] \qquad \begin{bmatrix} 1 & 1 & 0 \\ 1 & 0 & 1 \end{bmatrix}$$

With the single-parity-check code, any odd number of errors produces a syndrome equal to 1, and since all columns of $\mathbf{H}$ for the code are equal to 1, the nonzero syndrome can yield no information on locations of the errors. However, the parity-check matrix for the repetition code has three distinct column vectors, and any single error will produce a syndrome that uniquely identifies one of the columns of $\mathbf{H}$ and in turn the position of the single error in the received word.

Thus we begin to see how the error-detection and error-correction properties of a code are related to the structure of its parity-check matrix $\mathbf{H}$. Every single-error pattern in a received word will be detectable as long as no column of $\mathbf{H}$ is all zeros, and every single-error pattern will be correctable if no two columns of $\mathbf{H}$ are the same. In other words, a single-error-correcting code of block length n is one whose parity-check matrix has n unique nonzero columns. Therefore, we can construct a single-error-correcting code by specifying an $(n - k) \times n$ parity-check matrix with unique nonzero columns. This can always be done as long as $n \leq 2^{n-k} - 1$. The limitation reflects the fact that there can be no more than $2^{n-k} - 1$ unique nonzero binary vectors of length $n - k$. It can be seen, therefore, that the longest single-error-correcting binary block code that can be constructed with r parity checks has block length $n = 2^r - 1$. A code with these parameters is called a *binary Hamming code* [60].

Below we show two parity-check matrices for the (7, 4) binary Hamming code.

$$\begin{bmatrix} 0 & 0 & 0 & 1 & 1 & 1 & 1 \\ 0 & 1 & 1 & 0 & 0 & 1 & 1 \\ 1 & 0 & 1 & 0 & 1 & 0 & 1 \end{bmatrix} \qquad \begin{bmatrix} 1 & 1 & 1 & 0 & 1 & 0 & 0 \\ 0 & 1 & 1 & 1 & 0 & 1 & 0 \\ 1 & 1 & 0 & 1 & 0 & 0 & 1 \end{bmatrix}$$

On the left the columns of **H** are arranged as the binary-coded numbers 1 to 7. Any rearrangement of the columns of **H** will preserve the single-error-correction capability (uniqueness of columns), and thus the ordering of the columns may be chosen in accordance with convenience of implementation for encoding or decoding. In the matrix on the right, if the parity-check positions are identified with the three rightmost columns, the parity bits are represented individually as functions of the four information bits. This form of a parity-check matrix is called the *reduced echelon form* or the *systematic form* [131]. It is characterized by the partitioned structure $\mathbf{H} = \mathbf{P}^T; \mathbf{I}_r$, where $\mathbf{I}_r$ is the $r \times r$ identity matrix, and $\mathbf{P}^T$ is an $r \times k$ submatrix that defines the essential structure of the code. The submatrix **P**, the transpose of $\mathbf{P}^T$, is sometimes called the *parity matrix*. A block code constructed with a parity-check matrix of this form is called a *systematic* code. This point is treated in greater detail in Section 3.4.

Two codes are said to be *equivalent* if they differ only in the ordering of their symbols. Thus the codes formed by the two parity-check matrices considered above are equivalent Hamming codes. We can generalize by saying that all binary Hamming codes of a given length are equivalent. As an exercise, the reader should construct a parity-check matrix for the (15, 11) Hamming code and arrange the columns into the systematic form.

It is instructive to analyze the performance of a Hamming code for a simple case. This is done in the following example.

Example. Determine the channel BER and SNRs E_s/N_0 and E_b/N_0 at which the (7, 4) binary Hamming code will provide a delivered (post-decoding) BER equal to 10^{-5} on an AWGN channel with coherent PSK signaling. Assume hard-decision single-error-correction decoding.

We begin by noting that any syndrome computed from a received word will be decoded, since it will either be all zeros or equal one of the seven columns of **H**. This in turn means that all received words will be decoded by the single-error-correction procedure outlined earlier. All words received with no errors or one error will be decoded correctly, and therefore all other received words will be decoded incorrectly. Thus, on a channel where errors are statistically independent and occur with probability p, we can calculate the probability of incorrect decoding as simply $1 - [(1 - p)^7 + 7p(1 - p)^6]$. We next note that for reasonably small values of p, the probability of incorrect decoding will be dominated by the occurrence of error patterns containing two errors out of seven received bits. In the syndrome calculation, each such two-error pattern will cause two different columns of **H** to be added together to produce a third column of **H**. The resulting syndrome will indicate a fictitious error in the location of that third column, and that bit will be inverted by the decoder. The result, therefore, is that a two-error pattern in the received word is changed into a three-error pattern by the decoder. Again we see that incorrect decoding can introduce additional errors into an error pattern that exceeds the correction capability of the code.

Therefore, in this example we can estimate the post-decoding BER as 3/7 times the probability of incorrect decoding. For a delivered bit error rate of 10^{-5}, the probability of incorrect decoding must be about 2.33×10^{-5}. Using the expression written earlier for the probability of incorrect decoding, one finds that the required value of channel BER is 1.06×10^{-3}. The per-pulse SNR in the PSK channel is found from Eq. (1.1) to be $E_s/N_0 = 6.74$ dB. Comparison with Table 2.2 shows a saving of about 2.9 dB in channel SNR relative to uncoded PSK transmission. However, in order to assess the overall efficiency of communication with the Hamming code we must examine E_b/N_0, which is greater than E_s/N_0 by $10 \log_{10}(7/4)$ due to the transmission of check bits. The result is $E_b/N_0 = 9.2$ dB, which provides a gain of 0.4 dB relative to uncoded transmission.

The saving in energy per source bit of information for a coded system relative to an uncoded system, both operating at the same delivered BER, is called *coding gain* [73]. Coding gain is the parameter commonly used for evaluating the effectiveness of a code and for comparing the communication efficiencies of different code designs. The analysis of the performance of some complex block codes is treated in detail in Chapter 7.

In this chapter we have considered some very simple codes in order to introduce fundamental concepts and the terminology of error-control coding with a minimum of mathematical details. We have also examined the performance achievable with two types of codes in order to illustrate the important concept of overall communication efficiency, as measured by E_b/N_0, and how it relates to the design of an error-control code.

We saw an example of a type of code, the repetition code, that results in a loss of communication efficiency and another, the Hamming code, that yields a gain, although a modest one. Much of the remainder of this book is devoted to error-control techniques that provide substantial coding gains. To provide large coding gains, more complex codes are needed that have more mathematical structure than the codes discussed thus far. The presentation of these more complex techniques requires a more general mathematical formulation, which is the central topic of the next chapter.

2.8. NOTES

Single-parity-check coding is incorporated into a number of binary transmission alphabets widely used in the data communication industry. The

American Standard Code for Information Interchange, commonly referred to as the *ASCII code*, uses eight-bit characters, with seven bits (also called "levels") used to convey information and an eighth bit available as an overall odd or even parity check over the character [47]. The ASCII code is the most extensively used transmission code in the United States and is gaining wide acceptance elsewhere in the world. The CCITT No. 5 alphabet is proposed in CCITT Recommendations V.3 and V.4 as a new international standard for information interchange. It is also a seven-bit information code with an eighth bit added for parity. CCITT recommends the odd-parity convention under some circumstances and even parity for others [35].

The Extended Binary Coded Decimal Interchange Code (EBCDIC), another widely used code, evolved over a period of years from the original binary-coded decimal codes designed for strictly numerical data. EBCDIC is a true eight-bit code capable of accommodating 256 binary combinations, leaving no bits for parity checking [10]. As a result, many users have developed variations of EBCDIC that use bits for odd or even parity. These variations are not necessarily compatible with one another. EBCDIC is widely used in systems manufactured by IBM Corporation.

Other simple block codes have been used extensively in data transmission systems. In some systems, single parity checks across characters are used in conjunction with a parity-check character constructed by forming a single parity check on each bit position in a block of data or message characters. This is often called *longitudinal redundancy checking*, and the resulting parity character is called a *block-check character* [10]. In typical systems using such a scheme, failure of any of the character checks or the block check causes the entire block to be deleted and retransmitted.

A number of commercial and military communication systems have used various forms of repetition coding as an expedient way of overlaying an error-control capability onto an existing communication system. Where the extension of transmission time has not been of critical concern, this has been an attractive way of providing a modest error-control capability with a simple and inexpensive implementation.

Single-parity-check codes have been used for a number of years to protect data recorded on magnetic tapes. Typically, tapes have nine recording tracks, and characters are written across the tape, with eight bits in each character devoted to data and one to a parity check (often referred to as *vertical redundancy check*). More complex codes are usually used to provide longitudinal parity checking over the length of a data file recorded on the tape [11, 25, 94]. Hamming codes and other, more complex codes are used for error control in computer memories and other mass storage systems. In the last several years, semiconductor manufacturers have been providing an increasing number of LSI memory chips having built-in error detection and, in some cases, correction. Several of these devices use the (15, 11) Hamming code and perform automatic single-error correction.

■ ■ CHAPTER THREE

Algebra of Linear Block Codes

In this chapter we describe a large and important class of error-control codes, called linear block codes. We shall be concerned principally with binary codes, but nonbinary codes will be considered as well. Linear codes are parity-check codes, and almost all the block codes used in error-control systems belong to this class. It is in this chapter that we begin to develop the important mathematical background necessary to describe encoding and decoding of powerful block codes.

The chapter begins with an introduction to three algebraic systems that are fundamental to the theory of error-detecting and -correcting codes: groups, fields, and vector spaces. We show that a linear block code can be defined as a set of vectors in an n-dimensional vector space over a finite field. This description is related to the generator matrix and parity-check matrix for a code. The important concepts of Hamming distance, minimum distance, and codeword weight distribution are presented. The minimum distance of a code is then discussed as a measure of error-correction or error-detection capability. The chapter concludes with a discussion of code design and decoding techniques in terms of a geometric representation of linear codes.

3.1. GROUPS

The most basic algebraic system used in specifying the properties of error-control codes is a *group*. A group is a system of elements upon which one mathematical operation and its inverse are defined, such as addition and subtraction or multiplication and division. If the defining operation is addition, the system is called an *additive group*, and if multiplication, a *multiplicative group*. In this discussion we concentrate on additive groups, and we define a group more formally as follows.

Definition. A group G is a set of elements and an operation, say $+$, with the following properties:

1. G is *closed* under the defining operation, that is, if a and b are in G, $a + b$ is also in G.
2. The *associative law* of ordinary arithmetic holds, that is, for any a, b, and c in G, $(a + b) + c = a + (b + c)$.
3. G has an *identity element* 0 such that for any a in G, $a + 0 = a$.
4. Every element a in G has an *inverse element*, written as $-a$, such that $a + (-a) = 0$.

That the identity element 0 is unique can be proved as follows.

PROOF. Let there be two additive identity elements, 0 and Z. Then we have $0 + Z = 0$ and $Z + 0 = Z$, so that $0 = Z$.

Using property 2, it is a simple matter to show that the inverse of each element in G is unique.

PROOF. Let the element a have two inverses, $-a$ and d. Then we have

$$-a = -a + 0 = -a + (a + d) = (-a + a) + d = 0 + d = d$$

which says that the two assumed inverses are in fact the same element.

If $a + b = b + a$ for any two elements a and b, the group is called a *commutative* or *Abelian* group.

The number of elements in a group is called the *order* of the group, and a group may be of finite or infinite order. Examples of infinite groups include the set of all real numbers under ordinary addition and the set of all positive and negative integers, and 0, under addition. Note that the set of all nonnegative integers does not constitute an additive group, since the set does not include the inverses. The set of all real numbers excluding 0 is an infinite group under ordinary multiplication; the multiplicative identity element is 1. However, the set of all integers excluding 0 does not constitute a group under multiplication, since the fractional numbers needed for multiplicative inverses are missing.

The concept of a group is quite general and is not restricted to systems of numbers. For example, a discrete set of equally spaced phase positions in the xy plane is a finite group under the operation of rotation. The identity element is identified by arbitrarily assigning one of the phase positions a null value, say zero degrees. Closely related in principle to this example is a type of finite group that will be of interest to us, the set of integers $0, 1, 2, \ldots, M - 1$ under *modulo-M addition*. Modulo-M addition means that the ordinary sum of a set of integers is divided by M and the remainder is saved as the result. A group can be formed in this way with any positive integer M, and it is conventional to refer to the elements of the group as simply the integers modulo M. The simplest example of this type of group is the integers modulo 2, containing only 0 and 1, which was introduced in Chapter 2. Another simple example is the set of integers modulo 3, which gives the elements $0, 1$, and 2. Note here that 1 and 2 are inverses of each other, since $1 + 2 = 2 + 1 = 3 = 0$ modulo 3. As an exercise the reader should construct the group of integers under addition modulo 7 and determine the additive inverse for each nonzero element. In the same sense that the integers modulo 2 represent symbols in a binary code, the integers modulo M can (for certain values of M) be used to represent symbols in codes constructed on M-ary alphabets.

Many groups are contained in larger groups. For example, one can easily visualize a set of four uniformly spaced phase rotations as a finite group contained within larger groups consisting of $8, 16, 32, \ldots$ rotations. Also, the infinite group of even integers is contained within the group of all integers. These examples suggest the concept of a subgroup. A subset S of the elements in a group G is called a *subgroup* of G if S itself is a group with respect to the operation defined on G.

We are principally interested here in finite groups and subgroups. An important theorem (due to Lagrange) states that the number of elements in a finite group, its order, is a multiple of the order of each of its subgroups. We illustrate this with a simple example.

EXAMPLE. Let us consider a group G consisting of the four bit pairs $00, 01, 10$, and 11, where the defining operation is modulo-2 addition, applied bit by bit. That is, $10 + 11 = 01$, and so forth. The reader can easily verify that the four elements constitute a group under the stated operation. Since the order of G is 4, Lagrange's theorem tells us that if there are any nontrivial subgroups of G they must have order 2. (We ignore the subgroup of order 1, containing only 00.) There are, in fact, three subgroups of order 2, $(00, 01)$, $(00, 10)$, and $(00, 11)$, each constituting a group under the rule of bit-by-bit modulo-2 addition.

The type of group that is of central interest to us is an additive group composed of elements that are codewords. Before developing this idea, however, we must introduce two more mathematical structures, fields and vector spaces.

3.2. FIELDS

A *field* is described simply as a set of elements with two operations defined, addition $(+)$ and multiplication $(\cdot)$. Two further operations, subtraction and division, are implied by the existence of inverse elements under each of the defining operations. Stated more completely, the elements in a field F, taken together with the operations $+$ and $\cdot$, must satisfy the following conditions:

1. F is closed under the two operations, that is, the sum or product of any two elements in F is also in F.
2. For each operation, the associative and commutative laws of ordinary arithmetic hold, so that for any elements u, v, and w in F,

$$(u+v)+w=u+(v+w)$$

$$u+v=v+u$$

$$(u\cdot v)\cdot w=u\cdot(v\cdot w)$$

$$u\cdot v=v\cdot u$$

3. Connecting the two operations, the distributive law of ordinary arithmetic holds, so that

$$u\cdot(v+w)=u\cdot v+u\cdot w$$

for any u, v, and w in F.

4. F contains a unique additive identity element 0 and a unique multiplicative identity, different from 0 and written as 1, such that

$$u+0=u$$

$$u\cdot 1=u$$

for any element u in F. The two identity elements are the minimum elements that any field must contain. We call these two elements simply zero and unity.

5. Each element u in the field has a unique additive inverse, denoted by $-u$, such that

$$u+(-u)=0$$

and, for $u\neq 0$, a unique multiplicative inverse, denoted by u^{-1}, such that

$$u\cdot u^{-1}=1$$

From the above, the inverse operations subtraction $(-)$ and division $(\div)$ are defined by

$$u-v=u+(-v),\qquad \text{any } u, v \text{ in } F$$

$$u\div v=u\cdot(v^{-1}),\qquad v\neq 0$$

where $-v$ and v^{-1} are the additive and multiplicative inverses, respectively, of v.

Thus we see that a field provides the four elementary operations and all the familiar rules of ordinary arithmetic. Common examples of fields are the set of all real numbers and the set of all rational numbers under ordinary addition and multiplication. The set of all real numbers equal to or greater than zero does not constitute a field under the rules of ordinary arithmetic, since the set does not include additive inverses for nonzero numbers. Similarly, the set of all integers under ordinary arithmetic is not a field, since integers other than 1 do not have multiplicative inverses in the set.

The number of elements in a field, called the *order* of the field, may be finite or infinite, but for our purposes we are interested only in fields having finite numbers of elements. A field having a finite number of elements is called a *finite field* and is denoted by $GF(q)$, where q is the number of elements in the field. The notation is related to the designation *Galois field*, which is used interchangeably with "finite field" in the literature.* A finite field $GF(p^m)$ exists for any p^m, where p is a prime and m is an integer. The simplest example of a finite field is a *prime field*, $GF(p)$, consisting of the set of all integers modulo p, where p is any prime number greater than 1 and the addition and multiplication operations are addition and multiplication modulo p. The simplest prime field is $GF(2)$, which contains only the zero and unity elements 0 and 1; it was introduced and discussed in Chapter 2. As another example, the addition and multiplication tables for $GF(5)$ are shown below.

$GF(5)$ Addition

+	0	1	2	3	4
0	0	1	2	3	4
1	1	2	3	4	0
2	2	3	4	0	1
3	3	4	0	1	2
4	4	0	1	2	3

$GF(5)$ Multiplication

•	0	1	2	3	4
0	0	0	0	0	0
1	0	1	2	3	4
2	0	2	4	1	3
3	0	3	1	4	2
4	0	4	3	2	1

The reader should verify the uniqueness of the identity elements 0 and 1 and the fact that each element in $GF(5)$ has a unique additive inverse and (except for 0) a unique multiplicative inverse.

The reader can also verify that adding the multiplicative identity element for $GF(5)$, namely 1, to itself 5 times (5 addends) produces the additive identity

*Galois fields are named for the French mathematical prodigy Évariste Galois (1811–1832), who by the age of 17 had established the branch of mathematics called group theory. Galois died following a duel with a political opponent.

element 0. That is, $1 + 1 + 1 + 1 + 1 = 0$ modulo 5. The smallest number of times for which the multiplicative identity can be summed to produce the additive identity is called the *characteristic* of the field. Clearly, the characteristic of $GF(5)$ is 5, and the characteristic of $GF(2)$ is 2. We shall subsequently see that the characteristic of $GF(p^m)$ is p.

Just as groups can be contained in larger groups, fields can be contained in larger fields. In fact, the notation $GF(p^m)$ for a general finite field alludes to fields that are constructed as *extensions* of prime fields. We do not need to delve into this concept at this point, but we shall return to it in Chapter 4.

3.3. VECTOR SPACES

One of the most important algebraic concepts used in the mathematical description of codes is that of a *vector space*. The reader is assumed to be familiar with the interpretation of a vector as a directed line and the writing of a vector $\mathbf{v}$ as an enumeration of its coordinates $v_1, v_2, \ldots, v_m$. In two- or three-dimensional Euclidean space, the coordinates are simply the projections of $\mathbf{v}$ onto coordinate axes. We also assume familiarity with the rule for addition of two vectors, that is, addition of corresponding coordinates of the two vectors, and the multiplication of a vector by a real number, which is done simply by multiplying each coordinate by the number. These concepts are readily generalized to n dimensions, although for $n > 3$ it is difficult to visualize the vectors geometrically. However, the familiar properties of geometric vectors in ordinary coordinate systems provide an intuitive foundation for the discussion that follows.

The concept of a *vector space V over a field F* brings together a set of vectors and a set of field elements, called *scalars*, under the operations of addition and multiplication in a mathematical system very much like a system of geometric vectors, real numbers, and ordinary algebra. Vectors can be added (*vector addition*), and a scalar can multiply a vector (*scalar multiplication*). The addition of any two vectors $\mathbf{u}$ and $\mathbf{v}$ in V, written as $\mathbf{u} + \mathbf{v}$, produces a vector that is also in V. The multiplication of a scalar a in F by a vector $\mathbf{v}$ in V, written as $a\mathbf{v}$, also yields a vector in V. (For typographical convenience we have omitted the use of a multiplication symbol.) One should note the similarity of the properties stated above to the closure properties of groups and fields. This is made concrete by the following formal definition.

Definition. A vector space V over a field F is a set of elements, called *vectors*, forming a commutative additive group satisfying the following additional conditions for all vectors $\mathbf{u}$ and $\mathbf{v}$ in V and all scalars a and b in F:

1. The distributive laws apply, so that $a(\mathbf{u} + \mathbf{v}) = a\mathbf{u} + a\mathbf{v}$, and $(a + b)\mathbf{u} = a\mathbf{u} + b\mathbf{u}$.
2. The associative law applies, so that $(ab)\mathbf{u} = a(b\mathbf{u})$.
3. For the multiplicative identity element 1 in F, $1\mathbf{u} = \mathbf{u}$.

Note that there are two additive identity elements to deal with here. Since the vectors in V form a commutative group under addition, there is an identity element, called the *zero vector* and written as $\mathbf{0}$, such that $\mathbf{v} + \mathbf{0} = \mathbf{0} + \mathbf{v} = \mathbf{v}$ for all vectors $\mathbf{v}$ in V. This is not to be confused with 0, the zero scalar element in the field F. The two identity elements are closely connected, as will now be shown. From the two distributive laws we have, for all a in F and all $\mathbf{v}$ in V,

$$a\mathbf{v} + 0\mathbf{v} = (a + 0)\mathbf{v} = a\mathbf{v} = a\mathbf{v} + \mathbf{0}$$

$$a\mathbf{v} + a\mathbf{0} = a(\mathbf{v} + \mathbf{0}) = a\mathbf{v} = a\mathbf{v} + \mathbf{0}$$

and canceling $a\mathbf{v}$ from both sides (using the cancellation law of addition), we have

$$0\mathbf{v} = \mathbf{0} \qquad \text{for all } \mathbf{v} \text{ in } V$$

$$a\mathbf{0} = \mathbf{0} \qquad \text{for all } a \text{ in } F$$

Finally, the additive inverse of the unity scalar element in F, written as -1, serves to provide the additive inverse of any vector $\mathbf{v}$ in V, since

$$\mathbf{v} + (-1)\mathbf{v} = [1 + (-1)]\mathbf{v} = 0\mathbf{v} = \mathbf{0}$$

and therefore the additive inverse of any vector $\mathbf{v}$ in V is $(-1)\mathbf{v}$, written simply as $-\mathbf{v}$. It should be evident to the reader that the set of all vectors in Euclidean space is a simple example of a vector space.

We now describe the type of vector space of special interest to us in defining error-control codes. Consider an ordered set of n elements $u_1, u_2, \ldots, u_n$, where each element u_i belongs to the field F. This is called an *n-tuple over F*. Next consider the totality of all n-tuples over F, and let us define the addition of any two n-tuples as the element-by-element addition

$$(u_1, u_2, \ldots, u_n) + (v_1, v_2, \ldots, v_n) = (u_1 + v_1, u_2 + v_2, \ldots, u_n + v_n)$$

where each addition $u_i + v_i$ is performed in F. Due to the closure property of addition in F, we immediately see that the addition of two n-tuples over F produces another n-tuple over F. Now define the multiplication of an element from F by an n-tuple over F as the element-by-element multiplication

$$a(u_1, u_2, \ldots, u_n) = (au_1, au_2, \ldots, au_n)$$

where each multiplication au_i is done in F. We see that the result is again an n-tuple over F, and if $a = 1$ the result of the multiplication is simply the original n-tuple itself. It would be straightforward to show that the addition and multiplication rules just given satisfy the distributive and associative laws, and that in fact the set of all n-tuples over F, taken together with the operations of addition and multiplication in F, constitutes a vector space over F, with the all-zeros n-tuple being the identity element, or zero vector, in the additive group of the vector space. A vector space can be constructed in this manner with any field F, but our primary interest is in vector spaces over finite fields, in particular where the scalars represent code symbols and the vectors (n-tuples) represent codewords.

3.3.1. Linear Operations in a Vector Space over a Field

We now outline three key concepts involving vector spaces: linear combinations of vectors, basis vectors, and the dimensionality of a vector space. The reader familiar with the theory of simultaneous linear equations in ordinary algebra [66] will find that theory to be closely paralleled by the concepts presented here, the distinction being the use of algebraic operations in a finite field.

If $\mathbf{v}_1, \mathbf{v}_2, \ldots, \mathbf{v}_k$ are vectors in a vector space V over a field F, then a *linear combination* of $\mathbf{v}_1, \mathbf{v}_2, \ldots, \mathbf{v}_k$ is any sum of the form

$$a_1\mathbf{v}_1 + a_2\mathbf{v}_2 + \cdots + a_k\mathbf{v}_k \tag{3.1}$$

where each a_i is in the field F. The given set of vectors $\{\mathbf{v}_i\}$ is said to *span* the vector space V if any vector in V can be generated by a linear combination of the form of Eq. (3.1).

A set of k vectors $\mathbf{v}_1, \mathbf{v}_2, \ldots, \mathbf{v}_k$ is said to be *linearly independent* if no set of scalars $a_1, a_2, \ldots, a_k$ (except all $a_i = 0$) exists such that

$$a_1\mathbf{v}_1 + a_2\mathbf{v}_2 + \cdots + a_k\mathbf{v}_k = \mathbf{0} \tag{3.2}$$

where $\mathbf{0}$ denotes the zero vector. If Eq. (3.2) can be satisfied for at least one set of scalars not all equal to zero, the vectors $\mathbf{v}_1, \mathbf{v}_2, \ldots, \mathbf{v}_k$ are said to be *linearly dependent*. For example, the vectors $(1, 0, 0), (0, 1, 0)$, and $(0, 0, 1)$ are linearly independent over any field. However, the vectors $(1, 1, 0)$, $(0, 1, 1)$, and $(1, 0, 1)$ are linearly dependent over $GF(2)$ since they sum to the zero vector.

A subset S containing at least one vector in a vector space V is called a *subspace* of V if S itself has all the properties of a vector space with respect to the operations of addition and scalar multiplication in V. For example, it is simple to show that if we take a subset of vectors $\mathbf{v}_1, \mathbf{v}_2, \ldots, \mathbf{v}_r$ from V over F, the set S of all vectors formed by linear combinations of $\mathbf{v}_1, \mathbf{v}_2, \ldots, \mathbf{v}_r$ over F constitutes a vector space.

PROOF.

1. Closure is satisfied, that is, for any vectors $\mathbf{v}_a$ and $\mathbf{v}_b$ in S, $\mathbf{v}_a + \mathbf{v}_b$ is in S. We let $\mathbf{v}_a = a_1\mathbf{v}_1 + a_2\mathbf{v}_2 + \cdots + a_r\mathbf{v}_r$ and $\mathbf{v}_b = b_1\mathbf{v}_1 + b_2\mathbf{v}_2 + \cdots + b_r\mathbf{v}_r$ so that $\mathbf{v}_a + \mathbf{v}_b = (a_1 + b_1)\mathbf{v}_1 + (a_2 + b_2)\mathbf{v}_2 + \cdots + (a_r + b_r)\mathbf{v}_r = c_1\mathbf{v}_1 + c_2\mathbf{v}_2 + \cdots + c_r\mathbf{v}_r$. Because of the closure property of F under addition, the scalars $c_1, c_2, \ldots, c_r$ are seen to be in F, and therefore $\mathbf{v}_a + \mathbf{v}_b$ is in S.
2. We also require that for any b in F and $\mathbf{v}$ in S, $b\mathbf{v}$ be in S. If we let $\mathbf{v} = a_1\mathbf{v}_1 + a_2\mathbf{v}_2 + \cdots + a_r\mathbf{v}_r$, then, using distributive and associative laws, we have $b\mathbf{v} = ba_1\mathbf{v}_1 + ba_2\mathbf{v}_2 + \cdots + ba_r\mathbf{v}_r = c_1\mathbf{v}_1 + c_2\mathbf{v}_2 + \cdots + c_r\mathbf{v}_r$. Because of

the closure property of F under multiplication, $c_1, c_2, \ldots, c_r$ are in F and therefore $b\mathbf{v}$ is in S.

From part 2 of the proof we see that S must contain the zero vector, since $0\mathbf{v}_a = \mathbf{0}$, where $\mathbf{v}_a$ is any vector in S and 0 is the zero scalar in F. The vector $\mathbf{v}_a$ might be the zero vector itself. If so, S contains only the zero vector. We can also see that each vector $\mathbf{v} = a_1\mathbf{v}_1 + a_2\mathbf{v}_2 + \cdots + a_r\mathbf{v}_r$ in S has an additive inverse given by $-\mathbf{v} = (-a_1)\mathbf{v}_1 + (-a_2)\mathbf{v}_2 + \cdots + (-a_r)\mathbf{v}_r$, where $-a_i$ is the additive inverse of a_i in F. The commutative, associative, and distributive laws carry over from V into S, since the vectors in S are contained in V.

We said earlier that a set of vectors that generates a vector space V by linear combinations is said to span V. The same terminology applies to any subspace S of V. For example, the vectors $\mathbf{v}_1, \mathbf{v}_2, \ldots, \mathbf{v}_r$ just discussed are said to span the subspace S.

EXAMPLE. Consider the three binary vectors $(1,0,0)$, $(0,1,0)$ and $(0,0,1)$. These vectors are said to span the vector space, call it V_3, composed of all eight binary vectors $(0,0,0,)$, $(0,0,1), \ldots,$ $(1,1,1)$ formed by linear combinations of the three spanning vectors over $GF(2)$. If we use the two vectors $(1,0,0)$ and $(0,1,0)$ to form linear combinations over $GF(2)$, the resulting vectors $(0,0,0)$, $(1,0,0)$, $(0,1,0)$, and $(1,1,0)$ constitute a vector space, say V_2, which is a subspace of V_3. The vectors $(1,0,0)$ and $(0,1,0)$ are said to span V_2.

It is possible to show that in any vector space there is at least one set of linearly independent vectors that spans the space. Any such set is said to be a *basis* of the vector space. In the example just given, the spanning vectors $(1,0,0)$, $(0,1,0)$, and $(0,0,1)$ are linearly independent and therefore constitute a basis of V_3. The reader will recognize the direct parallel to the familiar x, y, and z axes of 3-dimensional coordinate systems in Euclidean space. The basis vectors discussed above for V_3 represent a very natural choice for a basis, being a set of vectors each having 1 in one position and 0 in all other positions, no two vectors having 1 in the same position. Such vectors are called *unit vectors*, and a basis consisting of unit vectors is called a *normal orthogonal* or *orthonormal* basis. However, the unit vectors are not the only basis vectors for V_3. For example, $(1,0,0)$, $(0,1,0)$, and $(0,1,1)$ also form a basis of V_3. It is left as an exercise for the reader to show that these three vectors are linearly independent and that they generate all the vectors in V_3.

Relative to the previous example, the vectors $(1,0,0)$ and $(1,1,0)$ form an alternative basis of the subspace V_2. As a further example, the vector $(1,1,1)$ is a basis for another subspace, call it V_1, containing only the vectors $(0,0,0)$ and $(1,1,1)$. It is possible to show that all bases of a given vector space contain the same number of vectors; the number is called the *dimension* of the vector space. Therefore the dimension of a vector space is the number of vectors in any linearly independent set of vectors that can be used to generate the space by forming linear combinations. In the examples given above, V_3, V_2, and V_1 have dimension 3, 2, and 1, respectively.

3.3.2. Matrix Representation of a Vector Space

We saw in the preceding discussion that a vector space composed of n-tuples need not have dimension n but may instead have dimension less than n. Consider, for example, the vector space generated by linear combinations of the three 7-tuples (0, 0, 0, 1, 1, 1, 1), (0, 1, 1, 0, 0, 1, 1), and (1, 0, 1, 0, 1, 0, 1) over $GF(2)$. The linear independence of the vectors is easily verified by noting that elements in the first, second, and fourth positions of any linear combination cannot all be zero unless all three vectors are given the scalar multiplier zero. Therefore, the three vectors generate a vector space of dimension 3, and this is a subspace of the vector space of all 7-tuples over $GF(2)$.

It is convenient to write the linear combination of basis vectors in a shorthand matrix notation. Operations with matrices follow rules much like those of ordinary algebra. Matrices can be added and multiplied, given certain conditions regarding their dimensions. These operations obey the distributive and associative laws of ordinary arithmetic; however, matrix multiplication is not commutative in general. Vectors can be defined as matrices, and this provides for multiplication of matrices by vectors. Matrices can also be multiplied by scalars.

A linear combination $\mathbf{c}$ of the three 7-tuples is written as

$$\mathbf{c} = a_1(0, 0, 0, 1, 1, 1, 1) + a_2(0, 1, 1, 0, 0, 1, 1) + a_3(1, 0, 1, 0, 1, 0, 1)$$

$$= (a_1 a_2 a_3)\begin{bmatrix} 0 & 0 & 0 & 1 & 1 & 1 & 1 \\ 0 & 1 & 1 & 0 & 0 & 1 & 1 \\ 1 & 0 & 1 & 0 & 1 & 0 & 1 \end{bmatrix}$$

We write this more compactly as

$$\mathbf{c} = \mathbf{aT}$$

where the matrix multiplication of the 1×3 row vector $\mathbf{a}$ times the 3×7 matrix $\mathbf{T}$ yields a 1×7 row vector $\mathbf{c}$. This matrix multiplication constitutes a *linear transformation* of a 3-tuple $\mathbf{a}$ into a 7-tuple $\mathbf{c}$, where the nature of the transformation is determined by the elements of $\mathbf{T}$. Stated a little differently, the matrix $\mathbf{T}$ maps 3-tuples into 7-tuples. Since the rows of $\mathbf{T}$ are basis vectors in a space of dimension 3, if the 3-tuples $\mathbf{a}$ are allowed to take on all eight binary combinations, the transformation generates the entire vector space spanned by the rows of $\mathbf{T}$. To generalize this, a $k \times n$ matrix $\mathbf{T}$ having k linearly independent rows generates a k-dimensional vector space C_k, where the vectors in C_k are a subspace of the vector space of all n-tuples over the scalar field.

3.4. BINARY LINEAR BLOCK CODES

We return now to the central purpose of the chapter, which is to provide a general mathematical framework for the description of binary block codes. We begin by defining a linear block code in terms of vector spaces. We then show how this leads to a simple rule for encoding k bits of source information into an n-bit redundant codeword where $k < n$. We do this by using the close relationship between vector spaces and linear block codes. For simplicity, we confine our attention to binary codes, whose symbols are drawn from the finite field $GF(2)$.

Definition. An (n,k) *linear block code* over a field F is a k-dimensional vector subspace of the space of all n-tuples over F, where $k < n$. The term linear refers to the formation of a vector space by linear combinations of basis vectors. Thus we can state that the set of all n-bit vectors formed by linear combinations over $GF(2)$ of k linearly independent basis vectors $\mathbf{g}_1, \mathbf{g}_2, \ldots, \mathbf{g}_k$ is a binary (n,k) linear block code C. That is, a vector $\mathbf{c}$ is an (n,k) codeword in C if and only if it lies in the k-dimensional vector space spanned by the $\{\mathbf{g}_i\}$. Equivalently, if the $\{\mathbf{g}_i\}$ are arranged as rows of a $k \times n$ matrix $\mathbf{G}$, a codeword $\mathbf{c}$ can be expressed as

$$\mathbf{c} = (i_1, i_2, \ldots, i_k)\begin{bmatrix} \mathbf{g}_1 \\ \mathbf{g}_2 \\ \vdots \\ \mathbf{g}_k \end{bmatrix} = \mathbf{iG} \tag{3.3}$$

where $\mathbf{i}$ is a k-bit information vector and $\mathbf{G}$ is called the *generator matrix* of the code. The full set of codewords, referred to simply as the code, is generated by letting $\mathbf{i}$ range through the set of all 2^k binary k-tuples.

Since the code is a vector space, by definition the 2^k codewords constitute a group under vector addition. In fact, it is possible to define a binary (n,k) linear code as a subgroup of order 2^k taken from the group of all binary n-tuples, and therefore binary linear block codes are sometimes called *group codes*. Certain obvious properties of linear block codes follow from the group property: the sum of two codewords is a codeword, and every linear code contains a zero vector, the all-zeros n-tuple. These properties are also readily seen from Eq. (3.3), since $\mathbf{0G} = \mathbf{0}$ and, for any two codewords $\mathbf{c}_j$ and $\mathbf{c}_k$ in C,

$$\mathbf{c}_j + \mathbf{c}_k = \mathbf{i}_j\mathbf{G} + \mathbf{i}_k\mathbf{G} = (\mathbf{i}_j + \mathbf{i}_k)\mathbf{G} = \mathbf{i}_l\mathbf{G}$$

The vector space consisting of all vectors formed by $\mathbf{iG}$ is called the *row space* of the matrix $\mathbf{G}$. Therefore, the code generated by $\mathbf{G}$ can also be described as the *row space of* $\mathbf{G}$.

The simple codes discussed in Chapter 2 are linear codes and are therefore vector spaces over $GF(2)$. For example, consider the triple-repetition code, for which Eq. (3.3) gives

$$\mathbf{c} = (i_1, i_1, i_1) = i_1 \mathbf{G}$$

where the generator matrix is

$$\mathbf{G} = (1 \quad 1 \quad 1)$$

For the single-parity-check code we have

$$\mathbf{c} = [i_1, i_2, i_3, \ldots, i_k, (i_1 + i_2 + \cdots + i_k)]$$

which we write in matrix form as

$$\mathbf{c} = (i_1, i_2, \ldots, i_k) \begin{bmatrix} 1 & 0 & 0 & \cdots & 0 & 1 \\ 0 & 1 & 0 & \cdots & 0 & 1 \\ 0 & 0 & 1 & \cdots & 0 & 1 \\ \vdots & & & & & \vdots \\ 0 & 0 & 0 & \cdots & 1 & 1 \end{bmatrix} = \mathbf{iG}$$

Finally, for the (9,4) product code (see Fig. 2.2), we have

$$\mathbf{c} = (i_1, i_2, i_3, i_4) \begin{bmatrix} 1 & 0 & 0 & 0 & 1 & 0 & 1 & 0 & 1 \\ 0 & 1 & 0 & 0 & 1 & 0 & 0 & 1 & 1 \\ 0 & 0 & 1 & 0 & 0 & 1 & 1 & 0 & 1 \\ 0 & 0 & 0 & 1 & 0 & 1 & 0 & 1 & 1 \end{bmatrix} = \mathbf{iG} \qquad (3.4)$$

Note that each of these codes is structured so that k of the bits of each codeword vector are the source information bits unaltered. That is, the generator matrix $\mathbf{G}$ is of the form

$$\mathbf{G} = [\mathbf{I}_k; \mathbf{P}]$$

where $\mathbf{I}_k$ is the $k \times k$ identity matrix and $\mathbf{P}$ is the $k \times (n - k)$ parity matrix whose composition determines how the parity-check bits are related to the information bits. Codes constructed with distinct information bits and check bits in each codeword are called *systematic codes*. (The systematic form for block codes was introduced in Chapter 2, where code structure was discussed in terms of parity-check matrices.) The use of *nonsystematic* block codes is generally avoided, since encoding and decoding involve the steps of converting the information vectors to and from the nonsystematic code vectors. Furthermore, any nonsystematic linear block code is equivalent to a systematic code, which can be shown as follows.

Given a nonsystematic generator matrix $\mathbf{G}''$ for a linear block code, it is possible to convert $\mathbf{G}''$ to a systematic generator matrix $\mathbf{G}$ by a straightforward procedure involving column interchanges and *elementary row operations* on $\mathbf{G}''$. For this purpose consider three types of elementary row operations for matrices:

1. Interchanging any two rows.
2. Multiplying a row by a nonzero scalar.
3. Adding a scalar multiple of one row to another row.

The procedure for converting $\mathbf{G}''$ to $\mathbf{G}$ is the same as procedures for diagonalizing matrices in ordinary linear algebra. It should be clear that the code generated by the systematic generator matrix $\mathbf{G}$ must contain exactly the same set of codewords as does $\mathbf{G}''$, except for a possible rearrangement of bit positions in codewords. This can be seen as follows. Let the conversion from $\mathbf{G}''$ to $\mathbf{G}$ be done in two steps, $\mathbf{G}''$ to $\mathbf{G}'$ and $\mathbf{G}'$ to $\mathbf{G}$, where the first step consists only of elementary row operations. Obviously, the row operations of types 1 and 2 do not change the vector space generated by $\mathbf{G}''$. For type 3 operations, a row constructed in $\mathbf{G}'$ is a linear combination of rows in $\mathbf{G}''$ and thus is contained in the vector space generated by $\mathbf{G}''$. However, any type 3 operation used in going from $\mathbf{G}''$ to $\mathbf{G}'$ can be undone by applying a type 3 operation to $\mathbf{G}'$, and so the vector space generated by $\mathbf{G}'$ is contained in the vector space generated by $\mathbf{G}''$, proving that $\mathbf{G}''$ and $\mathbf{G}'$ generate the same vector space. The interchange of columns in going from $\mathbf{G}'$ to $\mathbf{G}$ can, of course, change the vectors in the vector space. However, this is simply a reordering of bits in the codewords, which does not change the error-correction power of the code. This procedure is described in detail using an example in the book by Pless [134].

3.5. THE PARITY-CHECK MATRIX REVISITED

Consider the generator matrix $\mathbf{G}$

$$\mathbf{G} = \begin{bmatrix} \mathbf{g}_1 \\ \mathbf{g}_2 \\ \mathbf{g}_3 \\ \mathbf{g}_4 \\ \vdots \\ \mathbf{g}_k \end{bmatrix} = \begin{bmatrix} 1 & 0 & 0 & 0 & \cdots & 0, & \mathbf{p}_1 \\ 0 & 1 & 0 & 0 & \cdots & 0, & \mathbf{p}_2 \\ 0 & 0 & 1 & 0 & \cdots & 0, & \mathbf{p}_3 \\ 0 & 0 & 0 & 1 & \cdots & 0, & \mathbf{p}_4 \\ \vdots & & & & & & \vdots \\ 0 & 0 & 0 & 0 & \cdots & 1, & \mathbf{p}_k \end{bmatrix} = [\mathbf{I}_k;\mathbf{P}]$$

shown here for an arbitrary (n,k) systematic linear block code, where $\mathbf{p}_j$ denotes the jth row in the $k \times (n-k)$ matrix $\mathbf{P}$. Since every row in $\mathbf{G}$ is in the vector space that defines the code, every row is itself a codeword. It is seen then that the jth row of $\mathbf{G}$ is the codeword that corresponds to an information vector having a one in the jth position and zeros elsewhere. It is also clear from the form of the encoding transformation $\mathbf{c} = \mathbf{iG}$ that a codeword with an arbitrary k-bit information vector is obtained by forming a linear combination of the vectors $\{\mathbf{g}_i\}$. Specifically, the codeword associated with an information vector $\mathbf{i}$ is the sum of the rows of $\mathbf{G}$ whose row numbers correspond to positions of ones in $\mathbf{i}$.

As we saw in Chapter 2, the relationship between the information bits and parity bits can be written in another useful way. For example, let $r = n - k$ and consider the $r \times n$ matrix $\mathbf{H}$ given by

$$\mathbf{H} = \left[\mathbf{P}^T;\mathbf{I}_r\right]$$

where $\mathbf{P}^T$ is the transpose of the parity matrix $\mathbf{P}$ and $\mathbf{I}_r$ is the $r \times r$ identity matrix. By multiplication of partitioned matrices we then have

$$\begin{aligned} \mathbf{GH}^T &= [\mathbf{I}_k;\mathbf{P}]\begin{bmatrix}\mathbf{P}\\ \mathbf{I}_r\end{bmatrix} \\ &= \mathbf{P} + \mathbf{P} = \mathbf{0} \end{aligned}$$

where $\mathbf{0}$ represents the $k \times r$ all-zeros matrix. Therefore we have

$$\mathbf{cH}^T = \mathbf{iGH}^T = \mathbf{0} \tag{3.5}$$

and the code vectors $\mathbf{c}$ are said to be in the *null space* of $\mathbf{H}$. The matrix $\mathbf{H}$ is the parity-check matrix of the code, introduced in Chapter 2, and the equations specified by Eq. (3.5) are the parity-check equations. Recalling our earlier description of the code generated by $\mathbf{G}$ as the row space of $\mathbf{G}$, we can now say that the row space of $\mathbf{G}$ is in the null space of $\mathbf{H}$.

For the (9,4) product code (see Fig. 2.2), the reader should verify that the parity-check matrix is

$$\mathbf{H} = \begin{bmatrix} 1 & 1 & 0 & 0 & 1 & 0 & 0 & 0 & 0 \\ 0 & 0 & 1 & 1 & 0 & 1 & 0 & 0 & 0 \\ 1 & 0 & 1 & 0 & 0 & 0 & 1 & 0 & 0 \\ 0 & 1 & 0 & 1 & 0 & 0 & 0 & 1 & 0 \\ 1 & 1 & 1 & 1 & 0 & 0 & 0 & 0 & 1 \end{bmatrix}$$

and use the generator matrix given in Eq. (3.4) to show that $\mathbf{GH}^T = \mathbf{0}$.

In Section 2.7 the parity-check matrix for the systematic (7,4) binary Hamming code was given. Interpreting $\mathbf{H}$ as $(\mathbf{P}^T;\mathbf{I}_3)$, we can write the corresponding generator matrix $\mathbf{G} = (\mathbf{I}_4;\mathbf{P})$ as

$$\mathbf{G} = \begin{bmatrix} 1 & 0 & 0 & 0 & 1 & 0 & 1 \\ 0 & 1 & 0 & 0 & 1 & 1 & 1 \\ 0 & 0 & 1 & 0 & 1 & 1 & 0 \\ 0 & 0 & 0 & 1 & 0 & 1 & 1 \end{bmatrix}$$

The reader should verify again that $\mathbf{GH}^T = \mathbf{0}$.

Thus we see the close connection between the generator matrix and the parity-check matrix of a linear block code. The relationship $\mathbf{GH}^T = \mathbf{0}$ can be interpreted as the requirement that each row of $\mathbf{G}$, being a codeword, must satisfy all the parity-check equations. This in turn means that every codeword, a sum of rows in $\mathbf{G}$, must likewise satisfy all the parity-check equations. We see, therefore, that codewords may be generated by either of two operations. Information vectors may be transformed into codewords by the operation $\mathbf{c} = \mathbf{iG}$, which forms sums of rows of the generator matrix. Alternatively, parity bits may be generated by applying the parity-check equations $\mathbf{cH}^T = \mathbf{0}$, setting parity bits for even parity. As a practical matter, techniques equivalent to the use of $\mathbf{G}$ are usually employed for encoding, while the parity-check relationships defined by $\mathbf{H}$ are typically used in decoding. This is discussed in detail in Chapters 4 and 5.

3.6. DUAL CODES

Another important relationship between the generator matrix $\mathbf{G}$ and the parity-check matrix $\mathbf{H}$ can be illustrated by transposing the two sides of the matrix equation $\mathbf{GH}^T = \mathbf{0}$, which yields

$$\mathbf{HG}^T = \mathbf{0}^T$$

We have transposed the zero matrix simply to emphasize the fact that since $\mathbf{0}$ has k rows and $n - k$ columns, $\mathbf{0}^T$ has $n - k$ rows and k columns.

Now, in the spirit of the discussions in Sections 3.4 and 3.5, we can let $\mathbf{i}$ be an $(n - k)$-element binary vector and observe that the product $\mathbf{iH}$ represents linear combinations of rows in $\mathbf{H}$. By letting $\mathbf{i}$ range through all 2^{n-k} binary combinations, we generate all 2^{n-k} linear combinations of rows of $\mathbf{H}$. Then by writing $\mathbf{iHG}^T = \mathbf{0}$ we see that all the linear combinations of rows in $\mathbf{H}$ are in the null space of the matrix $\mathbf{G}$. This suggests that here $\mathbf{H}$ is serving as a generator matrix and $\mathbf{G}$ as a parity-check matrix for a code with 2^{n-k} codewords. This in fact is true, and to prove it we would first have to show that the null space of $\mathbf{G}$ constitutes a vector space. This is easily shown by demonstrating closure. That is, let $\mathbf{i}_j\mathbf{H} = \mathbf{v}_j$ and $\mathbf{i}_k\mathbf{H} = \mathbf{v}_k$, which are both in the null space of $\mathbf{G}$, since $\mathbf{v}_j\mathbf{G}^T = \mathbf{0}$ and $\mathbf{v}_k\mathbf{G}^T = \mathbf{0}$. We then use the distributive law and find $(\mathbf{v}_j + \mathbf{v}_k)\mathbf{G}^T = \mathbf{0}$, which says that $(\mathbf{v}_j + \mathbf{v}_k)$ is also in the null space of $\mathbf{G}$, proving that the vectors generated by $\mathbf{iH}$ constitute a vector space. It remains to be shown that the dimensionality of the vector space is $n - k$. We do not prove this but rather invoke a theorem from linear algebra. The theorem states that if the dimension of a subspace of n-tuples is m, the dimension of the null space is $n - m$. This means that if the rows of a $k \times n$ matrix $\mathbf{G}$ generate a vector space of dimension k, the null space of $\mathbf{G}$ is a vector space of dimension $n - k$. Therefore, in our case the set of all n-tuples $\mathbf{v}$ satisfying $\mathbf{vG}^T = \mathbf{0}$ is a vector space of dimension $n - k$, or in other words a linear code with 2^{n-k} codewords. Alternatively, we could have invoked the theorem to show that **H**, which has a null space of dimension k, must generate a vector subspace of dimension $n - k$. This subspace is therefore a code generated by the matrix **H**, called the *dual* of the code generated by **G**. Therefore, every (n,k) linear binary block code with generator matrix **G** and parity-check matrix **H** has a dual, with parameters $(n,n - k)$, generator matrix **H**, and parity-check matrix **G**.

For example, the dual of the (5,4) single-parity-check code is the (5,1) repetition code, and the (7,4) Hamming code has a dual that is a (7,3) code.

3.7. HAMMING DISTANCE AND THE WEIGHT DISTRIBUTION

Before considering the error-correction power of linear block codes, it is first necessary to define some key concepts.

The *Hamming distance* between two vectors (or n-tuples) having the same number of elements is defined as the number of positions in which the elements differ. For example, the Hamming distance between the vectors (1, 0, 1) and (0, 1, 1) is 2.

The *Hamming weight* of a vector is the number of nonzero elements in the vector. Stated differently, the Hamming weight of a vector is equal to the Hamming distance between that vector and the all-zeros vector. Thus the Hamming distance between two vectors $\mathbf{v}_i$ and $\mathbf{v}_j$ is the Hamming weight of $\mathbf{v}_i - \mathbf{v}_j$.

The *minimum distance d* of a linear block code C is the smallest of the Hamming distances between pairs of different codewords in the code. It follows that the minimum distance is the smallest Hamming weight of $\mathbf{c}_i - \mathbf{c}_j$, where $\mathbf{c}_i$ and $\mathbf{c}_j$ are any two different codewords in C. Since the difference (or sum for a binary code) of a pair of codewords is another codeword, the minimum distance of a linear code is the minimum Hamming weight of the nonzero codewords.

We next establish a key relationship between the minimum distance of a linear block code C and the structure of its parity-check matrix **H**. We begin by writing **H** in the form

$$\mathbf{H} = [\mathbf{h}_1, \mathbf{h}_2, \ldots, \mathbf{h}_n]$$

where $\mathbf{h}_i$ is the ith column of **H**. We know we can define C as the set of all n-tuples or vectors for which $\mathbf{c}\mathbf{H}^T = \mathbf{0}$. However, we can restate this by saying that a codeword in C is a vector having ones in positions such that the corresponding columns of **H** sum to the zero vector **0**. If the codeword has w ones it is said to be a *codeword of weight w*. Therefore, the minimum weight of any codeword in C is the minimum number of distinct columns of **H** that sum to **0**. Since the minimum codeword weight is equal to the minimum Hamming distance between any pair of different codewords, the minimum number of columns of **H** summing to **0** also defines the minimum distance of the code.

We consider a few examples. The single-parity-check code introduced in Chapter 2 has a parity-check matrix that is simply a row vector of n ones. The minimum number of columns that sum to zero is two, and thus $d = 2$. For the $(n,1)$ repetition code, $\mathbf{H}$ has a single column of all ones, and the remaining columns form an $(n - 1) \times (n - 1)$ identity matrix. Therefore, all n columns must be added to produce the zero vector, and so the minimum distance is $d = n$. The parity-check matrix for the (7,4) Hamming code has columns that are the seven nonzero binary 3-tuples. Since no column is repeated, no two columns can sum to $\mathbf{0}$, and thus d must be at least 3. However, any two columns must sum to a third column in $\mathbf{H}$, and thus by adding that third column into the sum we produce $\mathbf{0}$. Therefore, we conclude that the minimum distance of the (7,4) Hamming code (or, by exactly the same argument, any Hamming code) is $d = 3$.

By combinatorial analysis one can enumerate the combinations of w columns of $\mathbf{H}$ that sum to $\mathbf{0}$, thereby determining the number of codewords of weight w in C. The full enumeration of the number of codewords of every possible Hamming weight is called the *weight distribution* or *weight spectrum* of the code, denoted by $\{A_i\}$, $i = 0, 1, 2, \ldots, n$. A_0 must always be 1, since the all-zeros codeword is contained in every linear code. For the single-parity-check code, the weight distribution is simply $A_i = \binom{n}{i}$ for i even and $A_i = 0$ for i odd. For the $(n,1)$ repetition code, $A_0 = A_n = 1$. For any (n,k) binary code, of course, we must have

$$\sum_{i=0}^{n} A_i = 2^k$$

For the (7,4) Hamming code, the weight distribution is easily found. We start with $A_0 = 1$. Since $d = 3$, we know that $A_1 = A_2 = 0$. We next calculate A_3 by noting that the selection of any two columns of $\mathbf{H}$ will uniquely determine the third column to be summed to produce $\mathbf{0}$. The total number of such selections is $\binom{7}{2}$, or 21. However, this set of selections comprises three permutations of each unique three-column combination $[(\mathbf{a} + \mathbf{b}) + \mathbf{c}, (\mathbf{b} + \mathbf{c}) + \mathbf{a}, (\mathbf{c} + \mathbf{a}) + \mathbf{b}]$, so that $A_3 = 21/3 = 7$. [For an arbitrary binary Hamming code, this generalizes to $A_d = n(n - 1)/6$.] Next note from $\mathbf{H}$ that the information bits 1111 produce parity bits 111 so that $A_7 = 1$. Finally, we note that the addition of each of the weight-3 codewords to the all-ones codeword produces a weight-4 codeword, and thus $A_4 = 7$. Since $1 + 7 + 7 + 1 = 16 = 2^4$, the calculation is complete, that is, all other A_i's are zero.

The weight distribution of a code is important in the calculation of its performance, as will be seen in Chapter 7. For codes more complex than those considered above, calculation of weight distributions is considerably more laborious than was indicated by the simple examples. In many cases weight distributions have been generated by means of computer searches aided by ancillary combinatorial analysis. A general formula for the weight distribution of a binary Hamming code of any length n is known, however. The number of codewords of weight i is found as the coefficient of z^i in the polynomial

$$A(z) = \frac{1}{n+1}\left[(1+z)^n + n(1+z)^{(n-1)/2}(1-z)^{(n+1)/2}\right]$$

Polynomials giving weight distribution values are usually called *weight enumerators*, but sometimes they are referred to by the more general name *generating function* [109, 131].

A widely used modification of a binary Hamming code, usually called an *extended Hamming code*, is a distance-4 code constructed by adding an overall parity check to each codeword of a Hamming code. The result is an (n,k) code with block length $n = 2^m, m$ any integer ≥ 2, and with $k = n - m - 1$ information bits per codeword. The weight enumerator for an extended Hamming code of block length n is

$$A(z) = \frac{1}{2n}\left[(1+z)^n + (1-z)^n + 2(n-1)(1-z^2)^{n/2}\right]$$

Partial and complete weight distributions for many other codes can be found in the literature [7, 130].

3.8. CODE GEOMETRY AND ERROR-CORRECTION CAPABILITY

The minimum distance d of a linear block code is the key property of a code that determines its capability to detect and correct errors. Specifically, a code is capable of correcting up to t errors in any received word if the minimum distance d between codewords is at least $2t + 1$. This is best understood by thinking of transmitted codewords and received words as points in geometric space. Then visualize each codeword of the code at the center of a *sphere of radius t*, where the sphere is composed of the set of all received words at Hamming distance t or less from the codeword. We now say that we would like the code to have a guaranteed capability for *t-error correction*, and we can relate this to our geometric model as follows. We say that if a codeword $\mathbf{c}_i$ is transmitted and is received with no more than t errors, the received word can be unambiguously identified as "belonging to" codeword $\mathbf{c}_i$ as long as the word lies in only one radius-t sphere surrounding $\mathbf{c}_i$. Clearly, no ambiguity can arise as long as the radius-t spheres centered on all the codewords in the code are *disjoint*, that is, no two spheres intersect. Since the smallest of the distances between centers of pairs of spheres is d, the minimum distance of the code, we see that received words with up to t errors can be correctly decoded as long as $d \geq 2t + 1$.

These ideas are illustrated in Fig. 3.1, where we depict a pair of codewords from each of four simple codes discussed in Chapters 2 and 3. In each diagram

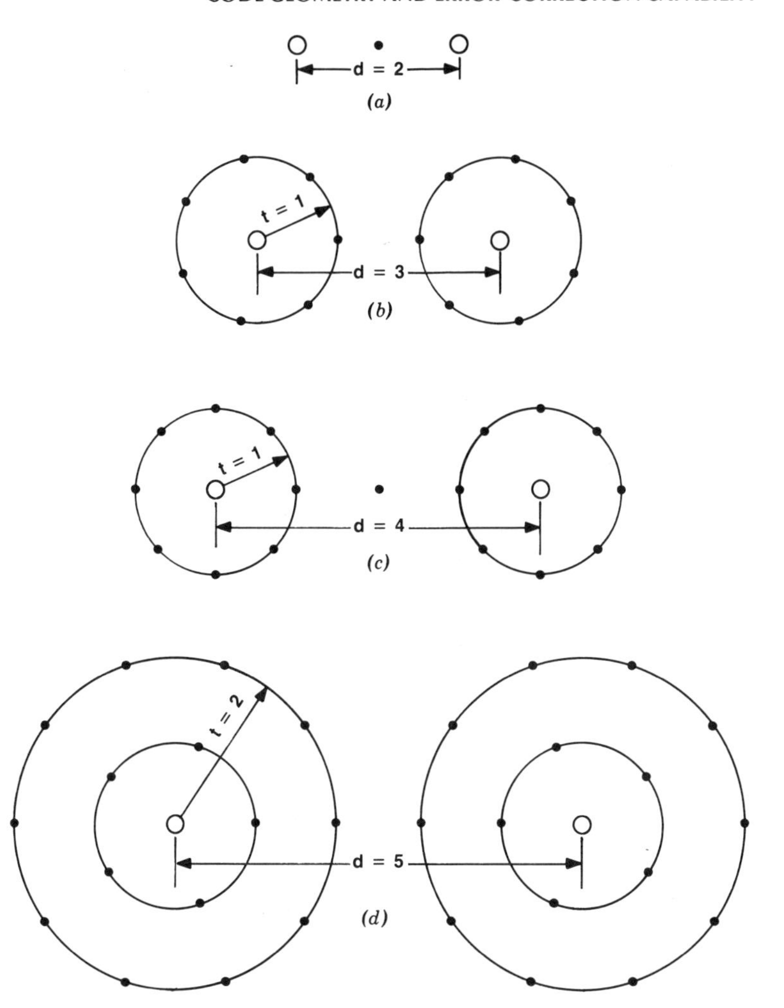

FIGURE 3.1. Representations of several binary codes.

we use small circles to represent codewords and solid dots to represent words received with one or more errors. In Fig. 3.1*a* we indicate a single-parity-check code, which has only even-weight codewords and minimum distance 2. The solid dot in the center of the diagram represents a word received with a single error and therefore at distance 1 from each of two valid codewords. The single error can be detected but not corrected. In the remaining three diagrams we use circles to suggest spheres of radius t around codewords.

Figure 3.1*b* represents a (7,4) binary Hamming code, which has minimum distance 3. On each circle we show seven solid dots representing words received each with a single error, which can be correctly decoded to the intended codeword at the center of the circle. Figure 3.1*c* represents the (8,4) extended Hamming code, which has minimum distance 4. Each circle in this diagram contains eight received words separated from a codeword by a single error. We also see in this diagram that the spheres cannot be enlarged to have radius 2, since the received word at the center of the diagram would then lie simultaneously in two spheres. However, the error pattern contained in that word is detectable, since it is not a codeword. Furthermore, a decoder will know that it contains more than one error, since even if a bit is changed by the decoder in an attempt at single-error correction, the result will not be a codeword, because the received word is at distance 2 from the nearest codewords.

Finally, in Fig. 3.1*d* we indicate the (5,1) repetition code, which has minimum distance 5 and is therefore capable of correcting one or two errors in every received word. Here, two "orbits" encircle each codeword, the inner orbit accounting for the five correctable single-error patterns in received words and the outer orbit for the 10 correctable double-error patterns. As an exercise the reader should sketch a diagram for the (6,1) repetition code, similarly accounting for all the correctable error patterns.

These examples serve to illustrate the type of decoding rule that has been assumed in all our discussions; the rule is defined as follows:

Definition. A decoding procedure that corrects all error patterns of weight l or less and no others, where $l \le t$ and t is the largest integer equal to or less than $(d - 1)/2$, is called *bounded-distance decoding*.

The parameter t is called the *error-correction limit* of the code. It is conventional to refer to a code with odd minimum distance d and $t = (d - 1)/2$ as a *t-error-correcting code*. If the code's minimum distance is even, so that $t = (d - 2)/2$, the code is referred to as a *t-error-correcting*, $(t + 1)$-*error-detecting code*. For example, the (7,4) Hamming code is a single-error-correcting code, while the (8,4) extended Hamming code is a single-error-correcting, double-error-detecting code.

All practical bounded-distance decoders use some form of syndrome decoding, in which the syndrome computed from the received word is interpreted as an indicator of either some specific error pattern of weight l or less or the occurrence of some unidentifiable error pattern that has caused the received word to lie outside any sphere of radius l centered on a codeword. Thus we might describe a t-error-correcting code as a code that provides enough syndromes (or, equivalently, enough parity-check combinations) to uniquely identify every possible error pattern of weight t or less. It can also be seen that since no error pattern of weight less than d can produce the all-zeros syndrome, a distance-d code is guaranteed capable of detecting any number of errors up to and including $d - 1$. This is done with a bounded-distance decoder implemented with $l = 0$.

3.8.1. Complete and Incomplete Decoding

The foregoing discussion leads us to a useful and intuitively appealing way to visualize a code design, which is to think of the set of received n-tuples as 2^n points in an n-dimensional geometric space where $GF(2)$ arithmetic prevails. The 2^k codewords in a code are then a subspace of all the points in the space. For example, in Fig. 3.2 we attempt a somewhat distorted two-dimensional projection of an n-dimensional space. The diagram represents an (n,k) code with minimum distance $d = 2t + 1$. Each sphere is centered on a codeword and contains

$$\binom{n}{0} + \binom{n}{1} + \cdots + \binom{n}{t}$$

points or possible received n-tuples. The bounded-distance decoding rule requires decoding any n-tuple that lies within a radius-t sphere of a codeword to the word at the center of that sphere. No attempt is made to correct more than t errors. In the case suggested by the figure, bounded-distance decoding is an *incomplete* decoding procedure, in that there are received n-tuples that cannot be decoded since they are not contained within any radius-t sphere surrounding a codeword, that is, they are not within distance t of any codeword. When such an n-tuple is received, decoding fails and error detection is announced. Thus, as we have observed previously, three kinds of decoding outcomes are possible: (1) correct decoding, (2) incorrect decoding, and (3)

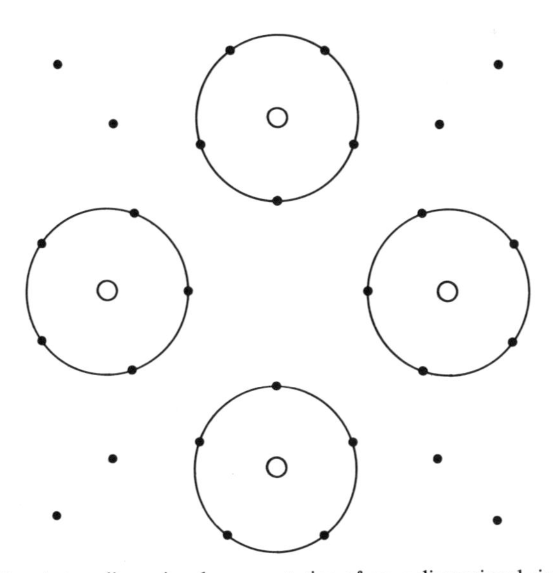

FIGURE 3.2. A two-dimensional representation of an n-dimensional signal space.

error detection without correction. The last event is often called *decoding failure*.

Figure 3.2 has been drawn in a way that provides an exact accounting of all the 5-tuples transmitted and received for a (5,2) single-error-correcting code. It is left as an exercise for the reader to construct a parity-check matrix for this code (several are possible) and to make the appropriate interpretation of the 32 points shown in the figure.

For many of the block codes used in error-control systems, the 2^k radius-t spheres around codewords do not account for all received words, and therefore bounded-distance decoding is very often an incomplete decoding algorithm. A *complete* decoding algorithm is one in which every received word is decoded into one of the possible transmitted codewords. A reasonable rule for doing this would be to decode each received word into the codeword closest to it, as measured by Hamming distance. This is called the *nearest-neighbor decoding rule*. It is also the *maximum likelihood* decoding rule for the binary symmetric channel, since it assumes that the most likely transmitted word is that obtained from the received word by correction of the smallest number of errors. In nearest-neighbor decoding, there are only two kinds of decoding events, correct and incorrect decoding. For many codes of interest, nearest-neighbor decoding is not a practical strategy, since it requires comparing received words with a great many of the set of possible codewords, a procedure usually involving a great amount of computation. This is why bounded-distance decoding is used in most applications of block codes.

There are important cases, however, where bounded-distance decoding is a complete decoding algorithm. We have already encountered such cases in the discussions in Chapter 2 of decoding the repetition codes and the Hamming codes. We shall have more to say about this.

3.8.2. Code Design and Sphere Packing

At this point it should be apparent to the reader that the error-correction capability of an (n,k) linear block code is directly related to the defining property of the code as a k-dimensional subspace of the space of all n-tuples over a field. The way the subspace is constructed determines the minimum distance between pairs of codewords, and this in turn determines how many errors can be corrected. In using a code, the "leftover" n-tuples, the n-tuples that are not in the code subspace, are identified (by the syndrome calculation) as codewords corrupted by errors. In bounded-distance decoding, the problem is determining the radius-t sphere to which the received n-tuple belongs. The details of how the k-dimensional code subspaces are constructed to facilitate decoding are presented in Chapter 4. However, it is instructive to do some simple arithmetic that illustrates the central characteristic of a well-designed code.

Let us say we wish to have an (n,k) code capable of correcting up to t errors in each code block by the use of bounded-distance decoding. Assume that we have chosen k and t and we want the block length n to be as small as possible in the interest of overall communication efficiency. Since a received word is decoded only if it lies within a radius-t sphere centered on a codeword, we would ideally like every received word to fall within one of the spheres. That is, by minimizing the number of n-tuples outside the spheres we are minimizing n. This means that if we compute the volume of each sphere and multiply by the number of codewords, we would like the result to exactly equal the total number of n-tuples, as given by the following equation for a binary code.

$$\left[1 + \binom{n}{1} + \cdots + \binom{n}{t}\right]2^k = 2^n \tag{3.6}$$

We have already noted that the spheres cannot overlap, for there would then be an ambiguity as to the decoding of a received word lying simultaneously in more than one sphere.

Equation (3.6) is the generalization for an arbitrary t-error-correcting binary code of a calculation done in Chapter 2 as part of the discussion of the (5,1) repetition code. A binary code for which Eq. (3.6) is satisfied is called a *perfect code*. All the $(n,1)$ repetition codes of odd block length are perfect codes (the proof is very simple and is left to the reader). The binary Hamming codes, introduced in Chapter 2, are also perfect codes, which can be shown as follows. The binary Hamming codes are defined to have block length

$$n = 2^{n-k} - 1$$

so that all words within spheres of radius $t = 1$ are accounted for by

$$(1 + n)2^k = 2^n$$

which satisfies Eq. (3.6).

Other than these two classes of codes, there exists only one other perfect binary code, the three-error-correcting (23,12) Golay code (see Section 4.3). For this code we have

$$\left[1 + \binom{23}{1} + \binom{23}{2} + \binom{23}{3}\right]2^{11} = 2^{23}$$

which again satisfies Eq. (3.6). It has been proved that no other perfect binary codes exist [168]. For binary codes in general, the $\leq$ sign must be substituted for the $=$ sign in Eq. (3.6), thus defining a bound relating n, k, and t that is satisfied with equality only for the perfect codes. The bound can be generalized from the binary case to an arbitrary (n,k) t-error-correcting code over a q-ary alphabet, resulting in the *sphere-packing bound*, given by

$$\left[1 + (q-1)\binom{n}{1} + \cdots + (q-1)^t\binom{n}{t}\right]q^k \leq q^n$$

The form of the bound is easily discerned by noting that there is only one way of receiving a codeword correctly, $\binom{n}{1}$ ways of having one wrong symbol in the block with $(q - 1)$ possibilities for the wrong symbol value, and so forth. The total volume of all the spheres cannot be greater than the number of received words, q^n. This bound is also called the *Hamming bound*. A nonbinary code that satisfies the bound with equality is a perfect code.

In the realm of nonbinary codes, the simple repetition codes are again perfect codes, as are the infinite class of single-error-correcting nonbinary Hamming codes. Beyond these, there is only one other perfect code, again due to Golay, the two-error-correcting (11,6) code on a ternary alphabet. The sphere-packing bound is satisfied by this code as follows:

$$\left[1 + 2\binom{11}{1} + 4\binom{11}{2}\right]3^6 = 3^{11}$$

In many applications, one desires codes that come as close as possible to achieving the bound. For example, given k and t, block length n should be as small as possible in order to minimize the required redundancy. Or, given n and k, we want the error-correction limit t, and hence the minimum distance d, to be as large as possible. The desired relationship among n, k, and t is equally appropriate whether the code is to be used for error correction or only for error detection. In either case the power of a code is directly related to minimum distance, which should be made as large as possible. It is customary to refer to codes as *tightly packed* or *loosely packed* in accordance with how well or how poorly they approach the sphere-packing bound.

Finally, we point out that for a perfect code with minimum distance $d = 2t + 1$, a bounded-distance decoding algorithm implemented to correct up to $l = t$ errors is a complete decoding procedure.

3.9. NOTES

Complete and rigorous treatments of groups, fields, and vector spaces can be found in a number of texts on modern algebra. Birkhoff and MacLane [12] is an excellent comprehensive reference. A short monograph by Artin [5] is also recommended.

While linear, that is, parity-check, codes are preferred in most practical applications due to their relative convenience of implementation, there are some *nonlinear* codes having advantages of their own. For example, if one wants a code of a particular block length n and error-correction limit t, one can sometimes find a nonlinear code with more codewords than the largest linear code. Nonlinear codes cannot be constructed with linear parity-check equations.

An important class of codes, some of which are nonlinear, is the *Hadamard codes*. A Hadamard code is constructed by selecting as codewords rows of a

Hadamard matrix. A Hadamard matrix of order n is an $n \times n$ matrix of $+1$s and -1s, where all pairs of rows are orthogonal. Hadamard matrices exist only for orders 1, 2, and multiples of 4. That they exist for all multiples of 4 has been conjectured but not yet proved. If the matrix elements are replaced with ones and zeros, respectively, a Hadamard matrix has one row of all zeros, and the rest have an equal number of ones and zeros. Linear Hadamard codes can be constructed for block lengths $n = 2^m$, where m is any integer, all nonzero codewords having weight $n/2$, and all pairs of codewords being separated by Hamming distance $n/2$. Hadamard codes of other lengths are possible, but they are nonlinear.

A code derived from the 32×32 Hadamard matrix was used on a telemetry link from the *Mariner '69* spacecraft to transmit high quality photographs of the planet Mars. The code was decoded with a fast Fourier transform decoder that was named the Green Machine [136]. In another development, Hadamard codes were found to be advantageous for implementation with binary on-off signaling where it was desired to transmit equal-energy codewords [132]. The book by MacWilliams and Sloane includes a chapter on nonlinear codes, including a detailed discussion of Hadamard codes [109].

A very simple form of nonlinear coding has been used for years in data transmission alphabets. The technique consists simply in transmitting coded characters with a fixed number of ones and zeros ("marks" and "spaces") per character. Teleprinter systems operating on HF radio circuits have since the late 1930s used automatic-repeat-request (ARQ) systems, which operate with a full duplex, synchronous protocol. Characters are formed by treating the START and STOP bits in each seven-element code group (CCITT Alphabet No. 2) as character code bits and encoding the seven bits with a 3-out-of-7 *constant-ratio code* [115]. This provides 35 valid bit combinations out of 128 possible received combinations. When a character is received with other than three ones, an error signal is sent to the other end of the link, where the transmission is stopped and repeated, starting with the erroneous character. It is easy to see that this is an error-detection code, but the modulo-2 bit-by-bit addition of two 3:7 characters cannot yield another 3:7 character, and thus the code is nonlinear.

Other constant-ratio character transmission codes have also been used. A common example is a 4-out-of-8 error-detection code widely used for data transmission over telephone circuits. This code, with 70 allowable bit combinations, has typically been used to provide error protection for data already coded with a six-bit alphabet (64 allowed combinations). With a scheme of this type, code conversions are required at both ends of a link. The six extra characters provided by the 4-out-of-8 code are used as transmission control characters. Descriptions of other error-control schemes used in the telecommunication industry can be found in Martin [101] and Freeman [46].

CHAPTER FOUR

Binary Cyclic Codes and BCH Codes

Most of the block codes that have proved to be useful in practical applications belong to a class of codes called cyclic codes, first studied by Prange [137]. Cyclic codes are easy to encode and have well-defined algebraic structure, which has led to the development of very efficient decoding schemes. Cyclic codes include a broad and important subclass called Bose-Chaudhuri-Hocquenghem codes, or simply BCH codes. The BCH codes are efficient multiple-error-correcting codes that include codes defined on both binary and nonbinary symbol alphabets. The Hamming single-error-correcting codes and all the known perfect codes can be described as BCH codes.

In order to discuss the structure of cyclic codes it is necessary to further develop the algebraic concepts introduced in Chapter 3. We begin with a discussion of more general finite fields and the representation of field elements as polynomials. Straightforward ways to implement finite field arithmetic will follow. We then define cyclic codes and give their key properties. Finally, we consider the binary BCH codes in some detail. Decoding binary BCH codes is treated in Chapter 5 and nonbinary codes in Chapter 6. Many of the mathematical tools developed here will carry over to the case of nonbinary codes.

4.1. REPRESENTATIONS OF FINITE FIELDS

In Chapter 3 we discussed the ways in which linear block codes can be described with simple matrix equations. We saw that codewords can be defined by $\mathbf{c} = \mathbf{iG}$ or $\mathbf{cH}^T = \mathbf{0}$, and we discussed an important property of the syndrome, namely $\mathbf{S} = \mathbf{eH}^T$. In order that we may more fully describe the properties of cyclic codes and the procedures for determining error patterns from syndromes, we need to have more powerful mathematical tools than have been described thus far. In particular, we need a convenient algebraic system for manipulating vectors of scalars, where the vectors will be columns in a parity-check matrix or syndromes that are equivalent to sums of columns in a parity-check matrix. The individual scalars are elements in the field from which code symbols are drawn. In order to describe the way cyclic codes are constructed and decoded, it will be necessary to do addition, subtraction, multiplication, and division of vectors of finite field elements. We are thus concerned with obtaining consistent sets of rules for manipulating individual field elements and sets of field elements treated as vectors. We begin by describing a representation for field elements that is useful for doing multiplication.

4.1.1. The Primitive Element of a Finite Field

An important property of finite fields is that every finite field $GF(q)$ contains at least one *primitive element*, called α, which has the property that the $q - 1$ powers of α are the $q - 1$ nonzero elements of the field. This means that the nonzero field elements can be represented as $\alpha, \alpha^2, \ldots, \alpha^{q-1}$. Because of this property, α is sometimes called the *generator* of the multiplicative group of $GF(q)$.

If we take an arbitrary nonzero element β in the field and raise it to successive powers, we eventually arrive at some exponent e such that $\beta^e = 1$. For an arbitrary β in the field, the smallest positive integer e such that $\beta^e = 1$ is called the *order of the element*. (This is not to be confused with the *order of the field*, defined as the number of elements in the field, which is equal to q in the present discussion.) In the generation of the nonzero elements of $GF(q)$ as powers of a primitive element α, we always find that $\alpha^{q-1} = \alpha^0 = 1$, but no smaller power of α equals 1, so that the order of a primitive element is $q - 1$. In general, the various elements of the field can have different orders, but there is a theorem (the same theorem, due to Lagrange, that we used in Section 3.1) stating that the order of an arbitrary element must be either $q - 1$ or a divisor of $q - 1$.

For example, consider the prime field $GF(5)$, which consists of the integers modulo 5. Since $q - 1 = 4$, we anticipate that the orders of various elements can be 1, 2, or 4. The element 1 has order 1. Taking successive powers of 2, we find $2^1 = 2$, $2^2 = 4$, $2^3 = 8 = 3 \bmod 5$, $2^4 = 6 = 1 \bmod 5$, and thus 2 has order 4. One also finds that 3 has order 4, and 4 has order 2. Therefore 2 and 3 are primitive elements of $GF(5)$, while 1 and 4 are *nonprimitive* elements. By the same token, we know that nonzero elements in $GF(7)$ must be of order 1, 2, 3, or 6, elements in $GF(11)$ of order 1, 2, 5, or 10, and so forth. Since it is known that $GF(q)$ must have at least one element of order $q - 1$, a primitive element in $GF(q)$ can also be defined as an element of maximum order. Since the order of any element β divides $q - 1$, it follows that $\beta^{q-1} = 1 = \beta^0$.

It is simple to show that the first $q-1$ powers of a primitive element, $\alpha, \alpha^2, \ldots, \alpha^{q-1}$, are all distinct and nonzero, as has been asserted. To see this, let $\alpha^i = \alpha^j$ for $0 < j < i \leq q-1$. For the equality to hold, we must have $\alpha^{i-j} = 1$. However, for i and j unequal and both greater than zero and smaller than q, the exponent $i-j$ must be $q-2$ or smaller, which contradicts the definition of α as an element of order $q-1$. Thus α^i and α^j are distinct. Further, α^i cannot equal the zero element for $0 < i \leq q-1$, since if it did, $\alpha^{i+1}, \alpha^{i+2}, \ldots, \alpha^{q-1}$ would all equal zero and again α would not have order $q-1$. Therefore, the first $q-1$ powers of the primitive element are all of the nonzero field elements.

From the property that $\alpha, \alpha^2, \ldots, \alpha^{q-1}$ are all distinct, it follows that the $q-1$ products $\alpha^i\alpha^j$, for i fixed and $0 < j \leq q-1$, are also distinct. This is readily shown by factoring $\alpha^{q-1} = \alpha^0 = 1$ out of each product in the sequence, which then leaves the same set of exponents of α, but with a shift in their numerical order.

It is also easy to show that each nonzero element in a finite field has a unique multiplicative inverse. For example, suppose the element β has two inverses, γ and γ'. Then

$$\beta\gamma = \beta\gamma' = 1$$

and consequently

$$\gamma(\beta\gamma') = \gamma$$

$$(\gamma\beta)\gamma' = \gamma$$

so

$$\gamma' = \gamma$$

The uniqueness of the inverse of a nonzero field element is also assured by the property that the products $\alpha^i\alpha^j$ are unique and nonzero for i fixed and $0 < j \leq q-1$.

Next, having seen that all the nonzero elements in a field can be expressed as the first $q-1$ powers of a primitive element α, we note that we can represent the field elements in terms of their exponents. The exponents are in effect *logarithms to the base* α. As in ordinary arithmetic, the logarithm of zero is undefined, though for convenience the notation $0 = \alpha^{-\infty}$ is often used. Below we show the logarithm tables for $GF(5)$ formed with $\alpha = 2$ and $\alpha = 3$.

β	$\log_2 \beta$	$\log_3 \beta$
0	$-\infty$	$-\infty$
1	0	0
2	1	3
3	3	1
4	2	2

As an exercise the reader should find the primitive elements of $GF(7)$ (there are two) and construct the respective logarithm tables.

Just as in ordinary arithmetic, multiplication of field elements can be done by adding logarithms. For example, in $GF(5)$, using $\alpha = 2$ we can multiply 2 times 4 by adding the logarithms $(1 + 2 = 3)$ and looking up the resulting element (3 mod 5) in an antilogarithm table. In coding implementations, finite field multiplications are often done with logarithm and antilogarithm tables. More will be said about this later.

4.1.2. Vectors of Field Elements

We next return to the matter of providing an algebraic system for doing calculations with vectors, or m-tuples, of finite field elements and a representation for field elements that is convenient for implementation in a digital machine. First we note that we can enumerate all the m-tuples of elements in a field $GF(q)$, q^m in number, and we recall from Section 3.3 that they in fact constitute an m-dimensional vector space over $GF(q)$. Thus we can add and subtract vectors, using vector (element-by-element) addition and subtraction in $GF(q)$, and the result in every case is another vector in the vector space. However, we shall also want to do multiplication and division of vectors. To accomplish this we associate each vector with a polynomial having coefficients corresponding to the elements in the vector. For example, the set of four 2-tuples on $GF(2)$ can be represented by 0, 1, x, and $x + 1$, corresponding to 00, 01, 10, and 11, respectively. Clearly we can do term-by-term addition of the polynomials just as we would add the vectors. All we have done is replace the set of all 2-tuples defined on $GF(2)$ with the set of all degree-1 polynomials defined on $GF(2)$.

However, just as we have closure with addition of vectors, we must also have closure under multiplication. In fact, if we can find a way to multiply the polynomials that conforms to all the properties of multiplication in a finite field, we will have constructed a finite field with q^m elements. First we want the product of any two polynomials in the set to be another polynomial in the set (closure). This is no problem if the product is a polynomial of degree $m - 1$ or less. But what do we do with a polynomial product of degree m or greater? Clearly we can reduce the product by taking its remainder with respect to a fixed polynomial of degree m. The remainder will always be of degree $m - 1$ or less, and closure is achieved. However, we need to know what sort of polynomial to use in this reduction so that the other properties of a finite field are assured.

We can gain some insight into this question by observing that the product of any two nonzero field elements must be nonzero. For example, let two nonzero elements α^i and α^j be represented by $a(x)$ and $b(x)$, respectively, each of degree $m - 1$ or less. Then, assuming a *reduction polynomial* $p(x)$ of degree m, we can write the product $\alpha^i\alpha^j$ as

$$\alpha^i\alpha^j = a(x)b(x) \bmod p(x)$$

Now, let us set this product equal to zero and see what type of reduction polynomial would allow this to happen. That is, we write

$$a(x)b(x) \bmod p(x) = 0$$

or equivalently

$$a(x)b(x) = c(x)p(x) \tag{4.1}$$

which says that the left-hand side of Eq. (4.1) must be evenly divisible by $p(x)$. Now, if $p(x)$ is *factorable*, that is, expressible as the product of two or more polynomials of degree $m - 1$ or less, there may well be polynomials $a(x)b(x)$ that are evenly divisible by $p(x)$. However, if $p(x)$ is chosen to be a degree-m *irreducible polynomial* (a polynomial that cannot be factored), then $p(x)$ must be a factor of either $a(x)$ or $b(x)$. We can readily see that neither factoring is possible, since the polynomials $a(x)$ and $b(x)$ are each of degree $m - 1$ or less and $p(x)$ is of degree m. We therefore conclude that if $p(x)$ is chosen to be an irreducible polynomial of degree m, the equality in Eq. (4.1) cannot be satisfied unless $a(x)$ or $b(x)$ equals zero, in which case $c(x) = 0$. By similar arguments we could show that the requirement for uniqueness of the products $a(x)b(x)$, and hence the uniqueness of the inverse for each polynomial, again results in choosing the reduction polynomial $p(x)$ to be an irreducible (nonfactorable) degree-m polynomial in $GF(q)$.

While the discussion just presented does not constitute a rigorous proof, it does serve to illustrate a key relationship between polynomials and fields: associated with every degree-m irreducible polynomial with coefficients in a finite field $GF(q)$ there is at least one finite field containing q^m elements, denoted by $GF(q^m)$. More will be said about this in later sections.

Returning now to the simple example of the four 2-tuples defined on $GF(2)$, we can use $p(x) = x^2 + x + 1$, since $x^2 + x + 1$ cannot be factored into any lower-degree polynomials on $GF(2)$. (The only candidates for factors are x and $x + 1$, and the reader can easily verify that none of the products of these two polynomials equals $x^2 + x + 1$.) With $p(x)$ chosen, we can now write the multiplication table for the degree-1 binary polynomials as follows:

$\cdot$	0	1	x	$x+1$
0	0	0	0	0
1	0	1	x	$x+1$
x	0	x	$x+1$	1
$x+1$	0	$x+1$	1	x

We see from the multiplication table that each nonzero polynomial has a unique multiplicative inverse, x being the inverse of $x + 1$ and vice versa, while 1 is its own inverse, as always. Thus we see that by using a degree-2 irreducible polynomial on $GF(2)$ as a reduction polynomial, we have constructed the multiplicative group with four elements, and taking this together with the term-by-term addition rule for polynomials, we have defined a finite field with four elements, which we denote by $GF(4)$.

Finally, we complete this example by describing $GF(4)$ in terms of a primitive element. We can test for a primitive element simply by taking a nonzero element other than 1 (which we know has order $e = 1$) and raising it to successive powers until we find its order. For example, testing x we have $x^1 = x$, $x^2 = x + 1$, $x^3 = x^2 + x = (x + 1) + x = 1$, where calculation of x^2 and x^3 required reduction modulo $x^2 + x + 1$. We therefore see that x has order $e = 3$, and since $q - 1 = 3$, x is a primitive element. It is left for the reader to test $x + 1$ and show that it is primitive also. Having found two primitive elements in $GF(4)$, we can now use either one to generate a list of the nonzero field elements as powers of α. This is shown here with a table of field elements for each primitive element. The table also shows a representation of the four elements in $GF(4)$ that is convenient for implementation in a digital machine. With each polynomial we associate a binary 2-tuple, for example, $0 = 00$ and $1 + x = 11$. Addition of the digital representations of field elements is then conveniently implemented with the exclusive-OR operation.

Representations for Field Elements

$\alpha = x$			$\alpha = x + 1$		
$\alpha^{-\infty}$	$= 0$	$= 00$	$\alpha^{-\infty}$	$= 0$	$= 00$
α^0	$= 1$	$= 01$	α^0	$= 1$	$= 01$
α^1	$= x$	$= 10$	α^1	$= x + 1$	$= 11$
α^2	$= x + 1$	$= 11$	α^2	$= x$	$= 10$

Thus we have a complete representation for elements in $GF(4)$ and a consistent set of operations for addition and multiplication of elements. For multiplication, the appropriate logarithm table can be used. In Subsection 4.1.3, we generalize these results and define somewhat more formally the properties of fields constructed with m-tuples of field elements.

4.1.3. Extension Fields and Primitive Polynomials

In general, a finite field $GF(p^m)$ exists for any number p^m, where p is a prime and m is a positive integer. For $m = 1$ we have the prime fields $GF(p)$ already considered. The fields $GF(p^m)$ for $m > 1$ are commonly called *prime-power fields*, where p is the *characteristic* of the field. That is, p is the smallest integer such that

$$\sum_{i=1}^{p} \alpha^0 = 0$$

where α^0 is the multiplicative identity element. For fields of characteristic 2, each element is its own additive inverse and a minus sign is unnecessary; that is, addition and subtraction are the same operations.

The relationship between $GF(p)$ and $GF(p^m)$ is such that $GF(p)$ is a *subfield* of $GF(p^m)$; that is, the elements of $GF(p)$ are a subset of the elements in $GF(p^m)$, the subset itself having all the properties of a finite field. Equivalently, $GF(p^m)$ is called an *extension field*, or simply an *extension*, of $GF(p)$.

A familiar example of an extension of an infinite field is the set of all complex numbers, which contains the real numbers as a subfield. The concept of complex numbers came about after years of struggling with the question of how to characterize the roots of the equation $x^2 = -1, x^2$ never being negative. The hurdle was overcome by inventing an "imaginary" number j, satisfying $j^2 = -1$ and otherwise conforming to the laws of ordinary algebra. This device (*deus ex machina*!) neatly skirted the question of exactly what the roots of the equation were, but the implication $j = \sqrt{-1}$ was held in great suspicion by many mathematicians for a long time. The notation $x + yj$, x and y real, evolved as a means of providing for addition and multiplication according to the laws of ordinary algebra. In the modern notation and terminology of fields, we would simply describe the field of complex numbers as an extension of the field of real numbers formed by taking the real numbers modulo $x^2 + 1$, and would denote a complex number by the *couple* x, y.

The procedure followed in Section 4.1.2 for constructing $GF(4)$ from $GF(2)$ serves as an example of how one constructs an extension field from a subfield. The procedure generalizes in a straightforward way to any extension field $GF(p^m)$. That is, we represent elements in $GF(p^m)$ as the p^m polynomials of degree $m - 1$ or lower with coefficients in $GF(p)$. Polynomials are added by adding coefficients of corresponding powers of x, addition being done in $GF(p)$. To define multiplication, a degree-m irreducible polynomial over $GF(p)$ is selected and a primitive element α is found. Then the polynomials corresponding to the $p^m - 1$ distinct powers of α are constructed. We see that the irreducible polynomial $p(x)$ provides the key link between the addition and multiplication tables and thus fixes the structure that allows us to define the two arithmetic operations and their inverses in a consistent way. Thus we can say that the set of all polynomials in $GF(p)$ reduced with respect to a degree-m irreducible polynomial over $GF(p)$ forms the field $GF(p^m)$. The role of the irreducible polynomial is seen to be directly analogous to the use of a prime number p to define the finite field $GF(p)$.

We now examine irreducible polynomials a little more closely with another example. A polynomial $p(x)$ of degree m with coefficients in $GF(p)$ is said to be irreducible if it is not divisible by any polynomial with coefficients in $GF(p)$ of degree less than m and greater than zero. For example, consider the polynomial $p(x) = x^3 + x + 1$ having degree 3 and coefficients in $GF(2)$. We can quickly convince ourselves that $x^3 + x + 1$ is not factorable in $GF(2)$, as follows. If it is factorable, it must have at least one factor of degree 1. Of course x is not a factor of $p(x)$, since the lowest order term in $p(x)$ is $x^0 = 1$. Thus, the only candidate is $x + 1$, but if this were a factor, then $x = 1$ would be a root of $p(x)$. It is easily verified that this is not the case, since $p(x)$ has an odd number of terms and therefore $p(x)$ evaluated at $x = 1$ sums to $1 \bmod 2$. Therefore, $p(x) = x^3 + x + 1$ is irreducible in $GF(2)$. [We point out in passing that an irreducible polynomial on $GF(2)$ must have an odd number of terms and lowest order term equal to 1. These are necessary but not sufficient conditions for irreducibility of binary polynomials.]

Now we are able to generate the 3-tuples representing elements of $GF(2^3)$ simply by listing all 2^3 polynomials of the form $a(x) = a_2x^2 + a_1x + a_0$ and taking each 3-tuple as the vector of coefficients a_2, a_1, a_0. It is convenient to list the polynomials $a(x)$ in a sequence that automatically provides a consecutive ordering by logarithms. This can be done here by using the polynomial $a(x) = x$ as the primitive element, multiplying repeatedly by x, and reducing the result modulo $p(x)$. This is shown in Table 4.1. We see from the table that by forming successive powers of x, reduced modulo $x^3 + x + 1$, we obtain all the polynomials defining the nonzero elements of $GF(2^3)$. In order for the procedure to generate the full list of 3-tuples, it is necessary that x be a primitive element, which is clearly the case in this example.

Although $p(x) = x^3 + x + 1$ is an irreducible binary polynomial and consequently has no roots in $GF(2)$, it does have roots defined in an extension field. In fact, it is a simple matter to find one of its roots, since from Table 4.1 we see that we could as easily have generated the table using powers of α, letting $\alpha = x$ and $\alpha^3 = \alpha + 1$, and therefore α is a root of $p(x)$. An irreducible polynomial having a primitive element as a root is called a *primitive irreducible polynomial* or simply a *primitive polynomial.* While an irreducible polynomial with coefficients in $GF(p)$ has no roots in $GF(p)$, it has roots in the extension field $GF(p^m)$. In fact, the degree-m polynomial $p(x)$ must have exactly m roots in the extension field $GF(p^m)$.

The idea that an irreducible polynomial has roots in an extension field has a direct counterpart in ordinary complex arithmetic. For example, the polynomial

TABLE 4.1. A Representation of $GF(2^3)$ Generated from $x^3 + x + 1$

Zero and Powers of x		Polynomials over $GF(2)$		Vectors over $GF(2)$
0	=	0	=	000
x^0	=	1	=	001
x^1	=	x	=	010
x^2	=	x^2	=	100
x^3	=	$x + 1$	=	011
x^4	=	$x^2 + x$	=	110
x^5	=	$x^2 + x + 1$	=	111
x^6	=	$x^2 \quad + 1$	=	101

$x^2 - 2x + 2$ has no real-valued roots. Its roots, which we call complex, are $1 \pm j$, where j is the complex operator.

It is important to note that not all irreducible polynomials are primitive. In the example given above, with $p = 2$ and $m = 3$, we did not have to concern ourselves with this question because of a well-known property of irreducible polynomials, that for $2^m - 1$ prime, every degree-m polynomial irreducible in $GF(2)$ is primitive. Since $2^3 - 1 = 7$ is prime, we conclude that $x^3 + x + 1$ is primitive. This property is a special case of a more general property of any degree-m irreducible polynomial on $GF(2)$, that its roots are all of the same order and the order must divide $2^m - 1$. When $2^m - 1$ is prime, it has no divisors other than itself and unity, and thus all the roots must be primitive. However, if $2^m - 1$ is factorable, the order of the roots can equal $2^m - 1$ or some factor of $2^m - 1$.

For example, it can be shown that the polynomial $x^4 + x^3 + x^2 + x + 1$ is irreducible and nonprimitive and has roots of order 5, since, for example from the identity

$$x^5 + 1 = (x + 1)(x^4 + x^3 + x^2 + x + 1)$$

we know that all roots of $x^5 + 1$ satisfy $x^5 = 1$ and therefore must be of order 5. As an exercise, the reader should attempt to generate the elements of $GF(2^4)$ from this polynomial, using successive powers of x as we did in constructing Table 4.1. It will be found that after five lines the list of polynomials repeats itself, which means that x has order 5. However, we know that the field must contain at least one primitive element, and thus polynomials other than x must be tried. Raising $x + 1$ to successive powers is found to generate all 15 nonzero polynomials of degree ≤ 3, and thus $\alpha = x + 1$ is a primitive element of $GF(2^4)$, when generated by $x^4 + x^3 + x^2 + x + 1$. We see, therefore, that an irreducible polynomial need not be primitive to serve as a generator of an extension field. However, it is convenient to use a primitive polynomial, since the field elements can be generated with powers of x, and it is not necessary to search for a primitive element. As a practical matter, tables of irreducible polynomials with primitive polynomials identified are available in the literature. Such a table of polynomials on $GF(2)$ is provided in Appendix B.

In summary, to construct a representation of $GF(p^m)$, we go to a table of irreducible polynomials on $GF(p)$ and find a polynomial $p(x)$ (preferably primitive) of degree m. We then generate the list of p^m polynomials modulo $p(x)$ and take the vectors of polynomial coefficients as m-tuples representing the elements of $GF(p^m)$. Consistent addition and multiplication tables can then be constructed for $GF(p^m)$. The addition table is formed by adding corresponding elements in m-tuples, modulo p. The multiplication table can be formed by addition of exponents of α. The addition and multiplication tables for $GF(2^3)$, formed with the use of Table 4.1, are shown in Table 4.2. Note that we constructed the addition and multiplication tables using powers of α although we expressed field elements using polynomials in x in Table 4.1. However, since $x^3 + x + 1$ is a primitive polynomial, it has α as a root, so that

TABLE 4.2. Addition and Multiplication Tables for $GF(2^3)$

$+$	0	1	α	α^2	α^3	α^4	α^5	α^6	$\cdot$	0	1	α	α^2	α^3	α^4	α^5	α^6
0	0	1	α	α^2	α^3	α^4	α^5	α^6	0	0	0	0	0	0	0	0	0
1	1	0	α^3	α^6	α	α^5	α^4	α^2	1	0	1	α	α^2	α^3	α^4	α^5	α^6
α	α	α^3	0	α^4	1	α^2	α^6	α^5	α	0	α	α^2	α^3	α^4	α^5	α^6	1
α^2	α^2	α^6	α^4	0	α^5	α	α^3	1	α^2	0	α^2	α^3	α^4	α^5	α^6	1	α
α^3	α^3	α	1	α^5	0	α^6	α^2	α^4	α^3	0	α^3	α^4	α^5	α^6	1	α	α^2
α^4	α^4	α^5	α^2	α	α^6	0	1	α^3	α^4	0	α^4	α^5	α^6	1	α	α^2	α^3
α^5	α^5	α^4	α^6	α^3	α^2	1	0	α	α^5	0	α^5	α^6	1	α	α^2	α^3	α^4
α^6	α^6	α^2	α^5	1	α^4	α^3	α	0	α^6	0	α^6	1	α	α^2	α^3	α^4	α^5

$\alpha^3 + \alpha + 1 = 0$. Thus Table 4.1 might as easily have been written with α replacing x, as we observed earler.

It should be noted that while the multiplication table is most easily constructed by addition of exponents of α, it can also be constructed by multiplying the polynomial representations of two elements and reducing the result modulo $p(x)$. For example, we can use Table 4.1 to calculate

$$\begin{aligned} \alpha^2\alpha^4 &= x^2(x^2 + x) \bmod x^3 + x + 1 \\ &= x^4 + x^3 \bmod x^3 + x + 1 \\ &= x^2 + x + x + 1 \\ &= x^2 + 1 \\ &= \alpha^6 \end{aligned}$$

TABLE 4.3. A Representation of $GF(2^4)$ Generated from $\alpha^4 + \alpha + 1$

Zero and Powers of α		Polynomials over $GF(2)$		Vectors over $GF(2)$
0	$=$	0	$=$	0000
α^0	$=$	1	$=$	0001
α^1	$=$	α	$=$	0010
α^2	$=$	α^2	$=$	0100
α^3	$=$	α^3	$=$	1000
α^4	$=$	$\alpha + 1$	$=$	0011
α^5	$=$	$\alpha^2 + \alpha$	$=$	0110
α^6	$=$	$\alpha^3 + \alpha^2$	$=$	1100
α^7	$=$	$\alpha^3 + \alpha + 1$	$=$	1011
α^8	$=$	$\alpha^2 + 1$	$=$	0101
α^9	$=$	$\alpha^3 + \alpha$	$=$	1010
α^{10}	$=$	$\alpha^2 + \alpha + 1$	$=$	0111
α^{11}	$=$	$\alpha^3 + \alpha^2 + \alpha$	$=$	1110
α^{12}	$=$	$\alpha^3 + \alpha^2 + \alpha + 1$	$=$	1111
α^{13}	$=$	$\alpha^3 + \alpha^2 + 1$	$=$	1101
α^{14}	$=$	$\alpha^3 + 1$	$=$	1001

TABLE 4.4. A Representation of $GF(2^5)$ Generated from $\alpha^5 + \alpha^2 + 1$

Zero and Powers of α		Polynomials over $GF(2)$		Vectors over $GF(2)$
0	=	0	=	00000
α^0	=	1	=	00001
α^1	=	α	=	00010
α^2	=	α^2	=	00100
α^3	=	α^3	=	01000
α^4	=	α^4	=	10000
α^5	=	$\alpha^2 + 1$	=	00101
α^6	=	$\alpha^3 + \alpha$	=	01010
α^7	=	$\alpha^4 + \alpha^2$	=	10100
α^8	=	$\alpha^3 + \alpha^2 + 1$	=	01101
α^9	=	$\alpha^4 + \alpha^3 + \alpha$	=	11010
α^{10}	=	$\alpha^4 + 1$	=	10001
α^{11}	=	$\alpha^2 + \alpha + 1$	=	00111
α^{12}	=	$\alpha^3 + \alpha^2 + \alpha$	=	01110
α^{13}	=	$\alpha^4 + \alpha^3 + \alpha^2$	=	11100
α^{14}	=	$\alpha^4 + \alpha^3 + \alpha^2 + 1$	=	11101
α^{15}	=	$\alpha^4 + \alpha^3 + \alpha^2 + \alpha + 1$	=	11111
α^{16}	=	$\alpha^4 + \alpha^3 + \alpha + 1$	=	11011
α^{17}	=	$\alpha^4 + \alpha + 1$	=	10011
α^{18}	=	$\alpha + 1$	=	00011
α^{19}	=	$\alpha^2 + \alpha$	=	00110
α^{20}	=	$\alpha^3 + \alpha^2$	=	01100
α^{21}	=	$\alpha^4 + \alpha^3$	=	11000
α^{22}	=	$\alpha^4 + \alpha^2 + 1$	=	10101
α^{23}	=	$\alpha^3 + \alpha^2 + \alpha + 1$	=	01111
α^{24}	=	$\alpha^4 + \alpha^3 + \alpha^2 + \alpha$	=	11110
α^{25}	=	$\alpha^4 + \alpha^3 + 1$	=	11001
α^{26}	=	$\alpha^4 + \alpha^2 + \alpha + 1$	=	10111
α^{27}	=	$\alpha^3 + \alpha + 1$	=	01011
α^{28}	=	$\alpha^4 + \alpha^2 + \alpha$	=	10110
α^{29}	=	$\alpha^3 + 1$	=	01001
α^{30}	=	$\alpha^4 + \alpha$	=	10010

This is analogous to generating the multiplication table for a prime field $GF(p)$ by multiplying integers and reducing the product modulo p.

Representations for $GF(16)$ and $GF(32)$ are shown in Tables 4.3 and 4.4, respectively. Note that these two tables are written in terms of powers of α.

4.1.4. Relationship to Maximum-Length Sequences

It is useful at this point to show the relationship between the field elements generated by an irreducible polynomial and the sequences generated by a

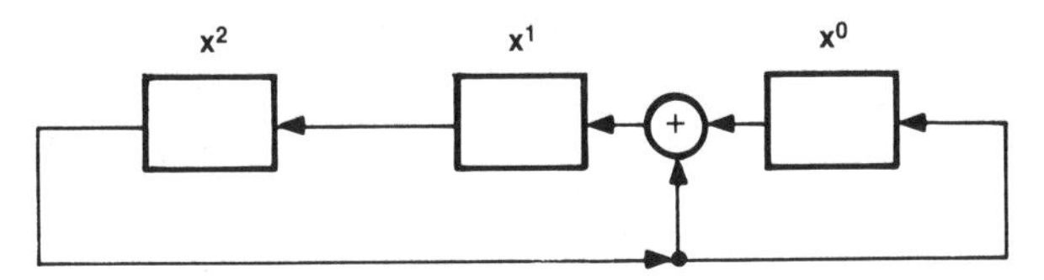

FIGURE 4.1. Sequence generator for the polynomial $x^3 + x + 1$.

feedback shift register corresponding to the same polynomial. Figure 4.1 shows such a shift register configured for the polynomial $x^3 + x + 1$, where the feedback tap connections form the relationship $x^3 = x + 1$ and the additions are done modulo 2. Let the initial state of the shift register be a one in the rightmost stage of the register, with zero in each of the other two stages, that is, 001 reading left to right. As the shift register is clocked to the left, the successive shift register states are seen to correspond exactly to the representations of field elements in $GF(2^3)$ as shown in Table 4.1. That is, each shift to the left corresponds to multiplication of the current state by x, and the feedback connections provide a reduction of the new state modulo $x^3 + x + 1$ by setting $x^3 + x + 1 = 0$ or $x^3 = x + 1$. [Again, in $GF(2)$, addition and subtraction are equivalent.] Thus the circuit generates the successive powers of x in the field $GF(2^3)$. Furthermore, it can be shown that the sequence of bits in any stage of the shift register is in fact the length-$(2^3 - 1)$ *maximum-length shift-register sequence* [54], or *m-sequence*, generated by the primitive irreducible polynomial $x^3 + x + 1$ on $GF(2)$, cyclically repeated. This generalizes in a straightforward way, and we can summarize by stating that the generation of field elements as consecutive powers of α in $GF(p^m)$ is equivalent to the generation of the m-sequence of length $p^m - 1$ from a degree-m primitive irreducible polynomial with coefficients in $GF(p)$. Therefore we see that an m-sequence generator is a circuit for recursively multiplying by α, a primitive element in $GF(p^m)$.

4.2. THE STRUCTURE OF BINARY CYCLIC CODES

We next turn our attention to the structure of binary cyclic codes and a large and important subclass, binary BCH codes. Much of the early literature on coding typically discussed the theory of code structure as well as encoding and decoding techniques in terms of matrix operations on code vectors. However, as the subject has matured, more effective ways of characterizing cyclic codes have been developed, using a polynomial representation for codewords and manipulations of polynomials for encoding and decoding operations. The polynomial representation makes it convenient to draw upon the rich theory of finite fields and their relationship to irreducible polynomials. It also provides for realization of coding operations with shift register circuits, which can be used to implement polynomial multiplication and division.

We begin by summarizing certain key properties of irreducible polynomials and their roots in extension fields. We then present and discuss the central idea in the construction of a cyclic code, namely that a cyclic code is designed to have prescribed roots in an extension of the field of code symbols. These concepts will then be utilized to define binary BCH codes.

4.2.1. Key Properties of Irreducible Polynomials

In Section 4.1 we utilized certain properties of irreducible polynomials to provide a consistent set of rules for performing addition and multiplication in a finite field $GF(p^m)$. It will now be necessary to present further details on the properties of these polynomials, which form the basis for describing the structure of cyclic codes. At this point, for simplicity of presentation, we confine our attention to fields of characteristic 2, $GF(2^m)$. The generalizations to $GF(p^m)$ are discussed in Chapter 6, where nonbinary codes are treated.

Our discussion here concentrates on polynomials that are irreducible in $GF(2)$, that is, degree-m binary polynomials that have no factors of degree less than m and greater than 0. It has already been stated that every degree-m polynomial $f(x)$ on $GF(2)$ has m roots (as in ordinary arithmetic), and if $f(x)$ is irreducible all m roots are in the extension field $GF(2^m)$. The properties of these roots in extension fields are of central importance in the theory of cyclic codes, and thus we summarize, without proof, key points required in subsequent discussion. For convenience of presentation, certain points made earlier in Chapter 4 are repeated in this summary:

1. Given a polynomial $f(x)$ with coefficients in $GF(2)$, we say that β is a root of $f(x)$ if and only if $f(\beta) = 0$, where β is an element of $GF(2)$ or some extension $GF(2^m)$. The multiplications and additions required for the evaluation of the polynomial can be performed in the consistent arithmetic system $GF(2^m)$, since $GF(2)$ is contained in any of its extensions.
2. Every polynomial of degree m has exactly m roots, some of which may be repeated.
3. For any m, there is at least one degree-m polynomial on $GF(2)$ that is irreducible.
4. If $f(x)$ is a degree-m irreducible polynomial ($m \geq 2$) on $GF(2)$, it has no roots in $GF(2)$, but all its roots lie in some extension of $GF(2)$. If $f(x)$ has a root that is a primitive element of $GF(2^m)$, $f(x)$ is called a *primitive irreducible polynomial*, or simply a *primitive polynomial.* Since it can be shown that all the roots of an irreducible polynomial are of the same order, all the roots of a primitive polynomial are primitive. For any m, there is at least one irreducible polynomial on $GF(2)$ that is primitive.
5. For every element β in an extension field $GF(2^m)$, there is a polynomial on $GF(2)$, called the *minimal polynomial* of β, that is the lowest degree

monic (highest order term has coefficient 1) polynomial having β as a root. Of course, all polynomials defined on $GF(2)$ are monic. Minimal polynomials, sometimes called *minimum functions*, have an important place in the design of cyclic codes, and we shall have more to say about them.

6. If $f(x)$ is an irreducible degree-m polynomial on $GF(2)$ and has a root β, then $\beta, \beta^2, \beta^4, \beta^8, \ldots, \beta^{2^{m-1}}$ are all the roots of $f(x)$. This is an important property relating to the structure of cyclic codes.

4.2.2. Minimal Polynomials

Associated with every element β in an extension field $GF(2^m)$ is its minimal polynomial $m_\beta(x)$ with coefficients in $GF(2)$. The minimal polynomial has the property that it is the monic polynomial of lowest degree having β as a root. There is a minimal polynomial for every element in the field, even if the element lies in $GF(2)$ itself. The important properties of minimal polynomials are summarized as follows:

1. The minimal polynomial $m_\beta(x)$ of any field element β must be irreducible. If this were not the case, one of the factors of $m_\beta(x)$ would have β as a root and would be of lower degree than $m_\beta(x)$ and contradict the definition.
2. The minimal polynomial of β is unique, that is, for every β there is one and only one minimal polynomial of β. However, different elements of $GF(2^m)$ can have the same minimal polynomial. (See Section 4.2.1, property 6.)
3. For every element in $GF(2^m)$, the degree of the minimal polynomial over $GF(2)$ is at most m.
4. The minimal polynomial of a primitive element of $GF(2^m)$ has degree m and is a primitive polynomial.

A very simple example of a minimal polynomial in the more familiar context of the infinite field of complex numbers is the polynomial $x^2 + 1$ with coefficients in the field of real numbers. This is the minimal polynomial of the numbers $+j$ and $-j$, the square roots of -1.

Consider again the case of the extension field $GF(2^2)$, which we represent as the polynomials of degree 1 or less, modulo the irreducible polynomial $y^2 + y + 1$. We have

$$\beta_0 = 0$$

$$\beta_1 = 1$$

$$\beta_2 = y$$

$$\beta_3 = y + 1$$

The minimal polynomials of β_0 and β_1 are simply

$$m_{\beta_0}(x) = x \qquad \text{and} \qquad m_{\beta_1} = x + 1$$

To find the minimal polynomial of $\beta_2 = y$, we use property 6 in Section 4.2.1, which tells us that the irreducible polynomial having β_2 as a root has β_2 and β_2^2 as roots, and no others. Therefore we can write

$$\begin{aligned} m_{\beta_2}(x) &= (x - y)(x - y^2) \\ &= (x + y)(x + y + 1) \\ &= x^2 + xy + x + yx + y^2 + y \\ &= x^2 + x + 1 \end{aligned}$$

where we use $y^2 = y + 1$ to reduce powers of y greater than unity. Similarly,

$$\begin{aligned} m_{\beta_3}(x) &= (x + y + 1)(x + y^2 + 1) \\ &= x^2 + xy^2 + x + yx + y^3 + y + x + y^2 + 1 \\ &= x^2 + x + 1 \end{aligned}$$

Thus we see that β_2 and β_3 have the same minimal polynomial. (We could have shown this directly by noting that $\beta_2^2 = \beta_3$.) Sets of elements having this property are called *conjugates*. In the field of complex numbers, $+j$ and $-j$ are conjugates having the same minimal polynomial $x^2 + 1$.

These examples are given only to provide a clearer explanation of the concept of a minimal polynomial. We shall see that, fortunately, it is not necessary to derive minimal polynomials in most cases of binary code design, since they are available in published lists.

We finish our description of minimal polynomials with a property of polynomials that have minimal polynomials as factors.

Let $\beta_1, \beta_2, \dots, \beta_L$ be elements in some extension field of $GF(2)$, and let the minimal polynomials of these elements be $m_{\beta_1}(x), m_{\beta_2}(x), \dots, m_{\beta_L}(x)$. Then the smallest degree monic polynomial with coefficients from $GF(2)$ having $\beta_1, \beta_2, \dots, \beta_L$ as roots, say $g(x)$, is given by

$$g(x) = \text{LCM}\left[m_{\beta_1}(x), m_{\beta_2}(x), \dots, m_{\beta_L}(x)\right]$$

where LCM denotes the *least common multiple*.

We might well refer to $g(x)$ as the minimal polynomial of the set of elements $\beta_1, \beta_2, \dots, \beta_L$. If the minimal polynomials of these elements are distinct (recall that different field elements can have the same minimal polynomial), then $g(x)$ is simply

$$g(x) = \prod_{i=1}^{L} m_{\beta_i}(x)$$

The proof of these relationships is straightforward. Since $g(x)$ is to have β_i as a root, $m_{\beta_i}(x)$ must divide $g(x)$ for $i = 1, 2, \ldots, L$. The smallest degree monic polynomial having $\beta_1, \beta_2, \ldots, \beta_L$ as roots is then the smallest degree polynomial having $m_{\beta_1}(x), m_{\beta_2}(x), \ldots, m_{\beta_L}(x)$ as factors, that is, the least common multiple of these minimal polynomials. If these minimal polynomials are distinct, the least common multiple is simply their product.

4.2.3. A Heuristic Description of Binary Cyclic Codes

We next describe the important class of binary cyclic block codes. A binary code is said to be cyclic if the following two properties hold:

1. The code is linear, that is, the bit-by-bit addition of two codewords in $GF(2)$ is again a codeword.
2. Any cyclic ("end-around") shift of a codeword is also a codeword.

The first property means that cyclic codes can be described as parity-check codes, as were all the simple codes discussed so far. The second property simply means that if $\mathbf{c} = (c_0, c_1, c_2, \ldots, c_{n-1})$ is a codeword, then so are all cyclic shifts, that is, $(c_{n-1}, c_0, c_1, c_2, \ldots, c_{n-2})$, $(c_{n-2}, c_{n-1}, c_0, c_1, \ldots, c_{n-3})$, and so forth, are all codewords.

Before proceeding with a detailed mathematical description of cyclic codes, it will be instructive to consider an example. Since a cyclic code (call it C) is a linear code, it can be described as the set of all n-tuples generated by the matrix equation $\mathbf{c} = \mathbf{iG}$, where $\mathbf{i}$ is any length-k binary vector and $\mathbf{G}$ is a $k \times n$ binary matrix called the generator matrix, as discussed in Section 3.4. Consider the case of the (7,4) binary Hamming code, for which the generator matrix (see Section 3.5) can be written as

$$\mathbf{G} = \begin{bmatrix} \mathbf{g}_1 \\ \mathbf{g}_2 \\ \mathbf{g}_3 \\ \mathbf{g}_4 \end{bmatrix} = \begin{bmatrix} 1 & 0 & 0 & 0 & 1 & 0 & 1 \\ 0 & 1 & 0 & 0 & 1 & 1 & 1 \\ 0 & 0 & 1 & 0 & 1 & 1 & 0 \\ 0 & 0 & 0 & 1 & 0 & 1 & 1 \end{bmatrix}$$

Note that $\mathbf{G}$ is shown in a systematic form, with the four leftmost positions being information bit positions. The fourth row of the matrix, $\mathbf{g}_4$, is seen to be the codeword generated from the information vector 0001.

We shall now show that $\mathbf{g}_4$ specifies $\mathbf{G}$ entirely. First note that $\mathbf{g}_3$ is simply $\mathbf{g}_4$ shifted once to the left. Next see that $\mathbf{g}_2$ is obtained by shifting $\mathbf{g}_3$ once to the left and adding $\mathbf{g}_4$ to it. In a similar manner, $\mathbf{g}_1$ is obtained by shifting $\mathbf{g}_2$ once to the left and adding $\mathbf{g}_4$. Therefore, specifying the single row $\mathbf{g}_4$ provides sufficient information to construct the entire generator matrix $\mathbf{G}$. The three cyclic shifts of $\mathbf{g}_4$ simply serve to align $\mathbf{g}_4$ with each of the first three information bit positions. The row additions are needed to create the 4×4 identity matrix in the four leftmost columns of $\mathbf{G}$. This is a simple application of elementary row operations, discussed in Section 3.4. Therefore, since the code C consists of all 16 possible linear combinations of the rows of $\mathbf{G}$, we can say that the row $\mathbf{g}_4$ completely specifies the code. We can also say that all codewords are sums of cyclic shifts of $\mathbf{g}_4$.

It is convenient to discuss these ideas in terms of polynomials rather than vectors. For example, $\mathbf{g}_4$ is associated with the degree-3 polynomial

$$g(x) = x^3 + x + 1$$

Since $g(x)$ can be used to form all the codewords in the code, it is called the *generator polynomial* of the code. The cyclic shift operation can be represented mathematically as multiplication by x modulo $x^7 - 1$. For example,

$$x(x^6 + x^5 + x^3 + 1) \bmod x^7 - 1 = x^7 + x^6 + x^4 + x \bmod x^7 - 1$$

$$= x^6 + x^4 + x + 1$$

For an arbitrary (n,k) cyclic code, we have the following properties:

1. $g(x)$ has degree $n - k = r$.
2. $g(x)$ must be of the form

$$g(x) = x^r + \sum_{i=1}^{r-1} g_i x^i + 1, \qquad g_i = 0 \text{ or } 1$$

PROOF. Otherwise a single cyclic shift to the right would result in all zero information bits and nonzero parity-check bits, whereas the relationship $\mathbf{c} = \mathbf{iG}$ requires the codeword to be all zeros.

3. Every codeword is a multiple of the generator polynomial.

PROOF. We have seen that a code polynomial $c(x)$ can be represented as sums of cyclic shifts of $g(x)$, and since $g(x)$ divides each term in the sum evenly it must divide into the sum evenly.

As a consequence of property 3, encoding can be implemented as polynomial multiplication of a k-bit [degree-$(k-1)$] information polynomial by the generator polynomial. In practice this leads to a nonsystematic code and is avoided. However, encoding can also be thought of as a division operation. For example, suppose we shift the degree-$(k-1)$ polynomial $i(x)$ left r places and divide by $g(x)$. The remainder of that division when added to $x^r i(x)$ is certainly divisible by $g(x)$ and is therefore a codeword. Specifically,

$$c(x) = x^r i(x) + [x^r i(x) \bmod g(x)]$$

This produces a systematic code structure. It is convenient nonetheless to visualize the code polynomials as the product of information polynomials with the generator polynomial when describing the structure and properties of cyclic codes.

4.2.4. A Polynomial Description of Cyclic Codes

We now present a description of cyclic block codes using the algebra of polynomials. For simplicity of presentation we confine our treatment to binary codes, nonbinary block codes being treated separately in Chapter 6. There are two principal motivations for the emphasis placed on the polynomial characterization of cyclic codes. The first is the insight gained into the structure of these codes by using the theory of polynomial algebra, and the second is the relative advantage of the polynomial characterization in devising efficient encoding and decoding algorithms. We begin by presenting the basic definitions of cyclic codes and discussing some of their properties.

In describing cyclic block codes, we represent a codeword of length n as a polynomial of degree $n-1$ with coefficients that are symbols from the code alphabet. Thus we may represent the codeword as a *codeword polynomial*,

$$c(x) = c_0 + c_1 x + c_2 x^2 + \cdots + c_{n-1} x^{n-1}$$

where in this discussion the components of the codeword, $c_0, c_1, c_2, \ldots, c_{n-1}$ are symbols in $GF(2)$. In order to identify information bits and parity bits in a codeword, we let k be the number of information bits and $r = n - k$ the number of parity bits, and we can write, for a systematic code,

$$c(x) = i_1 + i_2 x + \cdots + i_k x^{k-1} + p_1 x^k + \cdots + p_r x^{n-1}$$

or

$$c(x) = p_1 + p_2 x + \cdots + p_r x^{r-1} + i_1 x^r + \cdots + i_k x^{n-1}$$

where the sets of i's are information bits and the p's are parity bits. The two representations are, of course, equivalent, but the second, with parity bits in the lowest-order positions in $c(x)$, generally proves to be more convenient in the manipulation of polynomials, and it is assumed in the following discussions.

Turning again to the two defining properties of cyclic codes (linearity and the cyclic-shift property), we see first that two codeword polynomials are added by simply adding, in $GF(2)$, coefficients of corresponding terms of each power of x. We can describe the second property with polynomial notation by stating that if $c(x)$ is a codeword polynomial, then

$$x^j c(x) \bmod x^n - 1$$

is also a codeword polynomial for any cyclic shift j. This is true since multiplication of the codeword polynomial by x^j, setting $x^n = 1$, is equivalent to a cyclic shift of the codeword. [Note that we write $x^n - 1$ instead of $x^n + 1$, although we stated earlier that we would use only + signs in polynomials over $GF(2)$, since addition and subtraction are the same. It is conventional to use $x^n - 1$, as it is a general form applicable with polynomials over any finite field.]

These definitions provide a compact description of a cyclic code but do not make clear how such codes are constructed. As we have observed, all the codeword polynomials of a cyclic code must be multiples of a generator polynomial $g(x)$, where $g(x)$ is the codeword polynomial of lowest degree. That is, a cyclic code with block length n can be represented as all the polynomials of the form

$$c(x) = a(x)g(x) \bmod x^n - 1$$

where $g(x)$ is the generator polynomial of the code. We next show that a cyclic code of block length n is formed from any polynomial $g(x)$ that divides $x^n - 1$. That is, the generator polynomial must be such that

$$x^n - 1 = g(x)h(x)$$

We can verify that a code generated in this manner is cyclic, as follows. We wish to prove that a cyclic shift of a codeword,

$$x^j c(x) \bmod x^n - 1 = x^j a(x)g(x) \bmod x^n - 1$$

is also a codeword. If it is, the polynomial $x^j c(x) \bmod x^n - 1$ must be divisible by $g(x)$, that is, we must have

$$\left[x^j a(x)g(x) \bmod x^n - 1\right] \bmod g(x) = 0$$

Now, if $g(x)$ is a factor of $x^n - 1$, then $[b(x) \bmod x^n - 1] \bmod g(x)$ is simply $b(x) \bmod g(x)$, so that we can write

$$\left[x^j a(x)g(x) \bmod x^n - 1\right] \bmod g(x) = x^j a(x)g(x) \bmod g(x) = 0$$

showing that the cyclically shifted codeword is divisible by $g(x)$ and is therefore itself a codeword.

Let r be the degree of some generator polynomial that divides $x^n - 1$. It is now a simple matter to show that the resulting cyclic code has 2^k codewords, where $k = n - r$. This follows from the fact that all polynomials $a(x)$ of degree less than k produce distinct codewords, since the products $g(x)a(x)$ must have degree less than n, which in turn means that each product modulo $x^n - 1$ will simply be the polynomial $a(x)g(x)$ itself. Since there are 2^k distinct polynomials of degree $k - 1$ or less, there must be 2^k distinct codewords in the code. Therefore, using the notation adopted in earlier chapters, we say that the vectors of coefficients of the codeword polynomials generated by $g(x)$, where $g(x)$ divides $x^n - 1$, form an (n,k) cyclic block code. Henceforth, for convenience, we shall use the term codeword interchangeably with codeword polynomial.

An important property of a code generated with a polynomial that divides $x^n - 1$ is that for any codeword $c(x)$ in the code,

$$c(x)h(x) = 0 \bmod x^n - 1$$

where $g(x)h(x) = x^n - 1$. This can be shown simply as follows. Since every codeword is of the form $c(x) = a(x)g(x) \bmod x^n - 1$, let us multiply a codeword through by $h(x)$, so that we have

$$\begin{aligned} c(x)h(x) &= a(x)g(x)h(x) \bmod x^n - 1 \\ &= 0 \bmod x^n - 1 \end{aligned}$$

since $a(x)g(x)h(x)$ is a multiple of $x^n - 1$.

We now see that there are two relationships for codewords, $c(x) = a(x)g(x) \bmod x^n - 1$ and $h(x)c(x) = 0 \bmod x^n - 1$, which exactly parallel the code matrix equations $\mathbf{c} = \mathbf{iG}$ and $\mathbf{cH}^T = \mathbf{0}$ discussed in Chapters 2 and 3. In fact, the generator polynomial $g(x)$ is equivalent to the generator matrix $\mathbf{G}$ as a representation of the code, and the polynomial $h(x)$, which we call the *parity-check polynomial*, is equivalent to the parity-check matrix $\mathbf{H}$. To complete this equivalence, we note that the relationship $g(x)h(x) = x^n - 1$ corresponds to $\mathbf{GH}^T = \mathbf{0}$ in the matrix representation of linear parity-check codes.

An important distinction between these two fomulations does exist, however, in that the matrix representation is applicable to any linear parity-check code, whether cyclic or noncyclic, while the polynomial representation applies only to cyclic codes. That is, for any degree-r polynomial $g(x)$ that divides $x^n - 1$, there is an equivalent $k \times n$ generator matrix $\mathbf{G}$ where $k = n - r$, but the reverse is true only if $\mathbf{G}$ represents a cyclic code. We now establish this relationship in more detail with the aid of an example.

The polynomial $x^7 - 1$ has the following factorization into irreducible polynomials:

$$x^7 - 1 = (x + 1)(x^3 + x + 1)(x^3 + x^2 + 1)$$

Let us now take $g(x) = x^3 + x + 1$ as the generator polynomial for a code of block length 7, which means that the parity-check polynomial is

$$h(x) = (x + 1)(x^3 + x^2 + 1)$$

$$= x^4 + x^2 + x + 1$$

Therefore, $n = 7$, $r = 3$, $k = 4$, and the code consists of the $2^4 = 16$ polynomials formed by $a(x)(x^3 + x + 1)$, where we let $a(x)$ be in turn all of the 16 binary polynomials of degree less than 4. The generation of the list of codewords is shown in Table 4.5. The reader should note that the codewords in the table are not generated in a systematic form, that is, with information and check bits in segregated positions. Encoding of this code in a systematic form will be discussed in Section 4.4.

At this point it should be evident to the reader that the identification of $g(x)$ and $h(x)$ in a factorization of $x^n - 1$ is arbitrary. That is, given a factorization into two polynomials $p_1(x)$ and $p_2(x)$, we might choose either to be a generator polynomial $g(x)$. If we choose $p_1(x)$, then $p_2(x)$ becomes the corresponding parity-check polynomial $h(x)$. But if we choose $p_2(x)$ as $g(x)$, $p_1(x)$ becomes $h(x)$. The codes corresponding to these two sets of assignments are *duals* of each other. This directly parallels the description of dual codes presented in Section 3.6, where we showed that we could transpose $\mathbf{GH}^T = \mathbf{0}$ into $\mathbf{HG}^T = \mathbf{0}$, thereby interchanging the roles of the generator and parity-check matrices. In the example given above, we might as easily have

TABLE 4.5. Generation of a (7,4) Binary Cyclic Code from $g(x) = x^3 + x + 1$

$a(x)$	$a(x)g(x)$	*Code Vector*
0	0	0000000
1	$x^3 + x + 1$	0001011
x	$x^4 + x^2 + x$	0010110
$x + 1$	$x^4 + x^3 + x^2 + 1$	0011101
x^2	$x^5 + x^3 + x^2$	0101100
$x^2 + 1$	$x^5 + x^2 + x + 1$	0100111
$x^2 + x$	$x^5 + x^4 + x^3 + x$	0111010
$x^2 + x + 1$	$x^5 + x^4 + 1$	0110001
x^3	$x^6 + x^4 + x^3$	1011000
$x^3 + 1$	$x^6 + x^4 + x + 1$	1010011
$x^3 + x$	$x^6 + x^3 + x^2 + x$	1001110
$x^3 + x + 1$	$x^6 + x^2 + 1$	1000101
$x^3 + x^2$	$x^6 + x^5 + x^4 + x^2$	1110100
$x^3 + x^2 + 1$	$x^6 + x^5 + x^4 + x^3 + x^2 + x + 1$	1111111
$x^3 + x^2 + x$	$x^6 + x^5 + x$	1100010
$x^3 + x^2 + x + 1$	$x^6 + x^5 + x^3 + 1$	1101001

taken $g(x) = x^4 + x^2 + x + 1$ as the generator polynomial, which would have resulted in a (7,3) code with eight codewords.

Since each codeword in a cyclic code contains $g(x)$ as a factor, each code polynomial will have roots [from solution of $c(x) = 0$] that must include the roots of $g(x)$. It then follows that since the cyclic code is completely described by $g(x)$, we may define the code by specifying the roots of $g(x)$. For this example, recall from our earlier discussion of irreducible polynomials that the $g(x)$ we have chosen, $x^3 + x + 1$, is irreducible in $GF(2)$ and therefore has all its roots in an extension field of $GF(2)$. Furthermore, as we pointed out earlier, $x^3 + x + 1$ is primitive, so that its roots are all primitive elements of the extension field $GF(2^3)$. Therefore, having chosen $g(x)$ to be irreducible in $GF(2)$, we can describe the code succinctly by specifying the roots of $g(x)$ in $GF(2^3)$. Since we know from property 6 of Section 4.2.1 how the three roots of $x^3 + x + 1$ are related to one another, it is in fact necessary only to specify one of its roots in $GF(2^3)$ in order to describe the code completely. Let us simply say, then, that this code is a (7,4) binary cyclic code in which every codeword must have as a root α a primitive element in $GF(2^3)$. That is, for every word $c(x)$ in the code,

$$c(x) = 0 \qquad \text{for } x = \alpha$$

or simply

$$c(\alpha) = 0$$

It is important to note that in this example the polynomial $x^3 + x + 1$ plays two roles. First, since it divides $x^7 - 1$, it is used as a generator polynomial, and second, since it is primitive, it is used to provide a representation for $GF(2^3)$.

Let us now see how we would carry out the arithmetic needed to determine whether a given polynomial $c(x)$ satisfies the above equation, $c(\alpha) = 0$. First we write the expression out as

$$c_0\alpha^0 + c_1\alpha^1 + c_2\alpha^2 + \cdots + c_{n-1}\alpha^{n-1} = 0$$

Clearly we need to make use of multiplications and additions in $GF(2^3)$, which includes $GF(2)$, the field of the coefficients $\{c_i\}$, as a subfield. We can utilize the polynomial representation of $GF(2^3)$ generated in Table 4.1, where each power of α is represented by a binary polynomial of degree less than 3. The binary coefficients $\{c_i\}$ are simply 0 or $\alpha^0 = 1$, and addition is done using term-by-term modulo-2 addition of the polynomials (or the equivalent length-3 vectors) corresponding to the field elements. We can verify, for example, that the code polynomial $c(x) = 1 + x^2 + x^6$ satisfies $c(\alpha) = 0$ by using the table to write

$$\begin{aligned} \alpha^0 + \alpha^2 + \alpha^6 &= 001 + 100 + 101 \\ &= 000 \end{aligned}$$

Since we are using a set of binary vectors to represent the powers of α, it is convenient to write $c(\alpha) = 0$ in an equivalent matrix form as

$$(c_0 \quad c_1 \quad c_2 \quad \ldots \quad c_{n-1})[\alpha^0 \quad \alpha^1 \quad \alpha^2 \quad \ldots \quad \alpha^{n-1}]^T = \mathbf{0}$$

or

$$\mathbf{c}\begin{bmatrix} 0 & 0 & 1 & 0 & 1 & 1 & 1 \\ 0 & 1 & 0 & 1 & 1 & 1 & 0 \\ 1 & 0 & 0 & 1 & 0 & 1 & 1 \end{bmatrix}^T = \mathbf{0}$$

where the representations of powers of α are written as column vectors, the polynomial $c(x)$ is represented by a row vector, and $\mathbf{0}$ is a 3×1 column vector of all zeros. We now observe that the last expression is of the form $\mathbf{cH}^T = \mathbf{0}$, which is the parity-check matrix equation for a linear parity-check code. Furthermore, we see by referring to the discussion in Section 2.6 that the parity-check matrix constructed in this example has as its columns all $2^3 - 1 = 7$ nonzero binary vectors of length 3 and therefore represents a binary Hamming code of block length 7. Since we constructed this code using the cyclic formulation, it is in fact an example of a binary *cyclic Hamming code*.

From the discussion in Section 2.6, we know that rearrangement of the columns of a parity-check matrix does not change the essential properties of the code, the minimum distance and the number of codewords of a given weight. Two codes that differ only in the order of their symbols are said to be *equivalent* codes. Therefore the Hamming code constructed in this example is equivalent to any other (7,4) binary Hamming code. However, constructing the code with the polynomial formulation has produced a cyclic version of the code, that is, a particular ordering of the columns of the parity-check matrix that causes any cyclic shift of a codeword to also be a codeword. As we shall see later, a cyclic version of a linear block code has advantages for encoding and decoding.

In the example given above, we could have taken the other degree-3 irreducible factor of $x^7 - 1$, namely, $x^3 + x^2 + 1$, as a generator polynomial. This polynomial generates another (7,4) cyclic block code. Since $x^3 + x^2 + 1$ is also primitive, another $\mathbf{H}$ matrix with seven distinct nonzero columns results; in other words, another binary cyclic Hamming code is formed. It is equivalent to the code formed with $g(x) = x^3 + x + 1$.

In general, since it is known that for any integer m there is at least one primitive irreducible polynomial of degree m, we know that we can always construct at least one binary cyclic Hamming code of block length $n = 2^m - 1$. Therefore, we can say that any primitive irreducible polynomial of degree m generates a binary cyclic Hamming code of block length $n = 2^m - 1$. For a given value of m, there may be more than one such code, but if so, they are all equivalent. They all have distance $d = 3$, and they may differ only in a reordering of the columns of the parity-check matrix.

The foregoing discussion of cyclic codes constructed from generator polynomials and their relationship to the more general parity-check codes illustrates the usefulness of polynomial algebra in representing block codes. The factorizations of $x^n - 1$ provide a number of generator polynomials for cyclic codes with block length n, the degree r of the generator determining the number of parity check bits, and $k = n - r$ the number of information bits in the code. The description of codewords as multiples of the generator polynomial provides a characterization of the codewords as the set of polynomials whose roots are the roots of the generator polynomial. Finally, this characterization leads directly to the construction of the parity-check matrix, which determines the weight distribution of the code. The great value of this approach in describing cyclic codes has been to enable coding theorists to draw upon the extensive body of mathematical theory on the algebra of polynomials and their roots in finite fields. It is not our purpose here to describe the theory of cyclic codes in detail. Rather, we have simply stated key properties of cyclic codes and their generator polynomials. In the next section we turn our attention to the most important class of multiple-error-correcting cyclic codes, the BCH codes.

4.3. BINARY BCH CODES

The *Bose-Chaudhuri-Hocquenghem codes*, usually referred to as *BCH codes*, are an infinite class of cyclic block codes that have capabilities for *multiple-error detection and correction* [16,67]. In this chapter we restrict our discussion to binary BCH codes. Nonbinary BCH codes are discussed in Chapter 6.

4.3.1. Primitive BCH Codes

For any positive integers m and $t < n/2$, there exists a binary BCH code with block length $n = 2^m - 1$ and minimum distance $d \geq 2t + 1$ having no more than mt parity check bits. Each such code can correct up to t random errors per codeword and thus is a *t-error-correcting code*.

A BCH code is a cyclic code and thus can be defined in terms of its generator polynomial $g(x)$. Let α be a primitive element of the extension field $GF(2^m)$. The generator polynomial for a t-error-correcting BCH code is chosen so that $2t$ consecutive powers of α, such as $\alpha, \alpha^2, \alpha^3, \ldots, \alpha^{2t}$, are roots of the generator polynomial and consequently are also roots of each codeword.

This defines a somewhat restricted subclass of binary BCH codes, called *primitive BCH codes* because the roots are specified to be consecutive powers of a primitive element of $GF(2^m)$. The block length of a BCH code is the order of the element used in defining the consecutive roots. Since α is a primitive element in $GF(2^m)$, the block length of a primitive BCH code must be $2^m - 1$. To generalize the definition, if $2^m - 1$ is factorable, $2t$ consecutive powers of some nonprimitive element β of $GF(2^m)$ may instead be specified as roots of

TABLE 4.6. Binary BCH Codes Generated by Primitive Elements of Order Less Than 2^{10}

n	k	t	n	k	t	n	k	t
7	4	1	255	239	2	511	421	10
				231	3		412	11
15	11	1		223	4		403	12
	7	2		215	5		394	13
	5	3		207	6		385	14
				199	7		376	15
31	26	1		191	8		367	16
	21	2		187	9		358	18
	16	3		179	10		349	19
	11	5		171	11		340	20
	6	7		163	12		331	21
				155	13		322	22
63	57	1		147	14		313	23
	51	2		139	15		304	25
	45	3		131	18		295	26
	39	4		123	19		286	27
	36	5		115	21		277	28
	30	6		107	22		268	29
	24	7		99	23		259	30
	18	10		91	25		250	31
	16	11		87	26		241	36
	10	13		79	27		238	37
	7	15		71	29		229	38
				63	30		220	39
127	120	1		55	31		211	41
	113	2		47	42		202	42
	106	3		45	43		193	43
	99	4		37	45		184	45
	92	5		29	47		175	46
	85	6		21	55		166	47
	78	7		13	59		157	51
	71	9		9	63		148	53
	64	10					139	54
	57	11					130	55
	50	13	511	502	1		121	58
	43	14		493	2		112	59
	36	15		484	3		103	61
	29	21		475	4		94	62
	22	23		466	5		85	63
	15	27		457	6		76	85
	8	31		448	7		67	87
				439	8		58	91
255	247	1		430	9		49	93
511	40	95	1023	698	35	1023	338	89

TABLE 4.6. ***(Continued)***

n	k	t	n	k	t	n	k	t
	31	109		688	36		328	90
	28	111		678	37		318	91
	19	119		668	38		308	93
	10	127		658	39		298	94
				648	41		288	95
1023	1013	1		638	42		278	102
	1003	2		628	43		268	103
	993	3		618	44		258	106
	983	4		608	45		248	107
	973	5		598	46		238	109
	963	6		588	47		228	110
	953	7		578	49		218	111
	943	8		573	50		208	115
	933	9		563	51		203	117
	923	10		553	52		193	118
	913	11		543	53		183	119
	903	12		533	54		173	122
	893	13		523	55		163	123
	883	14		513	57		153	125
	873	15		503	58		143	126
	863	16		493	59		133	127
	858	17		483	60		123	170
	848	18		473	61		121	171
	838	19		463	62		111	173
	828	20		453	63		101	175
	818	21		443	73		91	181
	808	22		433	74		86	183
	798	23		423	75		76	187
	788	24		413	77		66	189
	778	25		403	78		56	191
	768	26		393	79		46	219
	758	27		383	82		36	223
	748	28		378	83		26	239
	738	29		368	85		16	247
	728	30		358	86		11	255
	718	31		348	87			
	708	34						

Source: Peterson, W. W., and E. J. Weldon, Jr., *Error-Correcting Codes*, 2nd ed., MIT Press, Cambridge, MA, 1972. © 1972 by the Massachusetts Institute of Technology. Reprinted by permission.

the codewords. The resulting code will be a *nonprimitive BCH code* and will have a block length that divides $2^m - 1$. Here we confine most of the discussion to primitive codes, but a brief discussion of nonprimitive codes is given in Section 4.3.3.

In general the definition of a BCH code allows the powers of the roots to range over any interval of consecutive values, say $m_0, m_0 + 1, \ldots, m_0 + 2t - 1$. The parameter m_0 is usually chosen to be zero or one. For the present discussion we let $m_0 = 1$, but we shall return to this point in Section 4.3.2.

Since the BCH codes are cyclic, codewords are assured of having the desired set of roots by choosing the generator polynomial so that it has $\alpha, \alpha^2, \ldots, \alpha^{2t}$ as roots. This is done by choosing $g(x)$ as the least common multiple of the minimal polynomials of $\alpha, \alpha^2, \ldots, \alpha^{2t}$, that is, we write

$$g(x) = LCM\left[m_{\alpha^1}(x), m_{\alpha^2}(x), \ldots, m_{\alpha^{2t}}(x)\right] \tag{4.2}$$

Now, we know from property 6 in the earlier discussion of irreducible polynomials (Section 4.2.1) that if a binary irreducible polynomial has β as a root, where β is an element of an extension field of $GF(2)$, then it also has β^2 as a root. Therefore, in Eq. (4.2) each even-power element α^{2i} and the corresponding element α^i are roots of the same minimal polynomial, $m_{\alpha^i}(x)$, and we can condense the sequence of minimal polynomials and write instead

$$g(x) = LCM\left[m_{\alpha}(x), m_{\alpha^3}(x), \ldots, m_{\alpha^{2t-1}}(x)\right] \tag{4.3}$$

Since we know from property 3 of Section 4.2.2 that the minimal polynomial of an element in $GF(2^m)$ will have degree no greater than m, we know that $g(x)$ will have degree no greater than mt, and thus the number of parity bits $r = n - k$ will be $\leq mt$, as stated at the beginning of this section. Table 4.6 gives the parameters of all primitive BCH codes of length $n = 2^m - 1$ up to $n = 1023$. It can be seen from the table that for all 1-error- and 2-error-correcting codes, $n - k$ is exactly equal to mt, and as t is increased $n - k$ can be smaller than mt. The single-error-correcting primitive BCH codes, $n = 2^m - 1$, $n - k = m$, are the cyclic Hamming codes.

The quantity $2t + 1$ used in specifying the generator polynomial of a BCH code is called the *design distance* of the code, but the true minimum distance will in some cases be greater, that is, $d \geq 2t + 1$. The true minimum distance for an arbitrary BCH code cannot be readily given, as this general problem is as yet unsolved. However, there are many results in the literature giving the true minimum distance for various specific cases. For example, see Chapter 9 of reference 109. For a great many cases of practical interest, the true minimum distance is equal to the design distance.

After selecting a BCH code with parameters that are useful for our purposes, we need to determine the generator polynomial for the code. Tables of generator polynomials for many codes are available in the literature, but a simple example will be included here to show how a generator is obtained.

Let us suppose that we require a three-error-correcting BCH code of block length $n = 31$. The generator polynomial may then, for example, have α, α^3, and α^5 as roots, where α is a primitive element of $GF(32)$. Therefore, we form $g(x)$ by multiplying together the minimal polynomials of α, α^3, and α^5 in $GF(32)$. A table of polynomials set up for our purpose is given in Appendix B. (This table was originally published by Peterson [129] in the first book devoted to the subject of error-correcting codes. It also appears in Peterson and Weldon [131], which is an expanded and updated edition of the original text. For convenience we refer to the table, reproduced here with permission, as the *Peterson table*). In the fourth line of the table we find:

DEGREE 5   1 45E   3 75G   5 67H

The three entries on this line give, respectively, the minimal polynomials of α, α^3, and α^5, with α being a primitive element of $GF(32)$. The polynomials are given in octal notation, right-justified with low-order bits to the right, so that we may write

$$m_\alpha(x) = 45_8 = x^5 + x^2 + 1$$

$$m_{\alpha^3}(x) = 75_8 = x^5 + x^4 + x^3 + x^2 + 1$$

$$m_{\alpha^5}(x) = 67_8 = x^5 + x^4 + x^2 + x + 1$$

The notations E, G and H in the table entry refer in part to the identification of these polynomials as being primitive; the reader will find a more complete explanation of the table in Appendix B.

We now enumerate all the roots of each of these minimal polynomials to determine whether or not any one polynomial has more than one of the required three roots. This is done with the aid of property 6 in Section 4.2.1. The roots of the three minimal polynomials are as follows:

Roots of $m_\alpha(x)$: $\alpha, \alpha^2, \alpha^4, \alpha^8, \alpha^{16}$
Roots of $m_{\alpha^3}(x)$: $\alpha^3, \alpha^6, \alpha^{12}, \alpha^{24}, \alpha^{48} = \alpha^{17}$
Roots of $m_{\alpha^5}(x)$: $\alpha^5, \alpha^{10}, \alpha^{20}, \alpha^{40} = \alpha^9, \alpha^{18}$

From this enumeration, we see that α, α^3, and α^5 are roots of three distinct polynomials, and therefore the required generator polynomial is simply the product of the three minimal polynomials just found, that is,

$$g(x) = (x^5 + x^2 + 1)(x^5 + x^4 + x^3 + x^2 + 1)(x^5 + x^4 + x^2 + x + 1)$$

$$= x^{15} + x^{11} + x^{10} + x^9 + x^8 + x^7 + x^5 + x^3 + x^2 + x + 1$$

From the enumeration of all the roots of $m_\alpha(x)$, $m_{\alpha^3}(x)$, and $m_{\alpha^5}(x)$, the reader can also verify that each of the three minimal polynomials must be of degree 5, as we found directly by use of the table of polynomials. This distance-7 code therefore has 15 parity bits and 16 information bits in each codeword.

Note from Table 4.6 that if we next design a length-31 code with five more parity bits, that is, a degree-20 generator polynomial, the resulting (31,11) code can correct five errors rather than only four. This is readily seen by first enumerating the roots of the minimal polynomial of α^7, namely

$$\text{Roots of } m_{\alpha^7}(x): \qquad \alpha^7, \alpha^{14}, \alpha^{28}, \alpha^{56} = \alpha^{25}, \alpha^{50} = \alpha^{19}$$

and noting that the collective set of roots of the minimal polynomials of α, α^3, α^5, and α^7 includes 10 consecutive powers of α, that is, α through α^{10}. Therefore, the design distance of the (31,11) code is 11.

We also point out a simple example from Table 4.6 in which the number of parity bits is less than mt, namely the (15,5) distance-7 code. The roots of the minimal polynomials of α, α^3, and α^5, where α is a primitive element of $GF(16)$, are as follows:

$$\text{Roots of } m_{\alpha}(x): \qquad \alpha, \alpha^2, \alpha^4, \alpha^8$$
$$\text{Roots of } m_{\alpha^3}(x): \qquad \alpha^3, \alpha^6, \alpha^{12}, \alpha^{24} = \alpha^9$$
$$\text{Roots of } m_{\alpha^5}(x): \qquad \alpha^5, \alpha^{10}$$

It is seen that the LCM of this set of minimal polynomials is of degree 10 and has roots containing six consecutive powers of α, so that a (15,5) code with design distance 7 is produced.

A table of generator polynomials for primitive BCH codes of block length up to 255 has been published by Stenbit [166] and is reproduced here as Table 4.7. (A table of generator polynomials for BCH codes with lengths up to 1023 is given in a text by Lin and Costello [94].) It can be verified that the above generator polynomial is included in Table 4.7, identified as $(107657)_8$.

Note that the Stenbit table (Table 4.7) gives only one code generator for $n = 7$, namely, $g(x) = (13)_8 = x^3 + x + 1$, although $x^7 - 1$ has two factors of degree 3, as we pointed out in Section 4.2.4. The other degree-3 factor, $x^3 + x^2 + 1$, is the *reciprocal polynomial* of $x^3 + x + 1$, where the reciprocal polynomial of a polynomial $f(x)$ of degree m is defined as $x^m f(1/x)$. It is easy to see from the definition that the roots of the reciprocal polynomial are the reciprocals of the roots of the original polynomial. Also, the reciprocal polynomial of an irreducible polynomial is irreducible, and the reciprocal polynomial of a primitive polynomial is primitive. The Stenbit table is based on the irreducible polynomials given in the Peterson table (Appendix B), and as the introduction to Appendix B explains, of any pair consisting of a polynomial and its reciprocal polynomial, only one is listed in the table. (Some irreducible polynomials are identical to their reciprocal polynomials, for example, $x^2 + x + 1$ and $x^4 + x^3 + x^2 + x + 1$.)

TABLE 4.7. Generator Polynomials for Primitive BCH Codes

n	k	t	$g(x)$
7	4	1	13
15	11	1	23
	7	2	721
	5	3	2467
31	26	1	45
	21	2	3551
	16	3	107657
	11	5	5423325
	6	7	313365047
63	57	1	103
	51	2	12471
	45	3	1701317
	39	4	166623567
	36	5	1033500423
	30	6	157464165547
	24	7	17323260404441
	18	10	1363026512351725
	16	11	6331141367235453
	10	13	472622305527250155
	7	15	5231045543503271737
127	120	1	211
	113	2	41567
	106	3	11554743
	99	4	3447023271
	92	5	624730022327
	85	6	130704476322273
	78	7	26230002166130115
	71	9	6255010713253127753
	64	10	1206534025570773100045
	57	11	335265252505705053517721
	50	13	54446512523314012421501421
	43	14	17721772213651227521220574343
	36	15	3146074666522075044764574721735
	29	21	403114461367670603667530141176155
	22	23	123376070404722522435445626637647043
	15	27	22057042445604554770523013762217604353
	8	31	7047264052751030651476224271567733130217
255	247	1	435
	239	2	267543
	231	3	156720665
	223	4	75626641375
	215	5	23157564726421
	207	6	16176560567636227
	199	7	7633031270420722341
	191	8	2663470176115333714567
	187	9	52755313540001322236351
	179	10	22624710717340432416300455
255	171	11	15416214212342356077061630637
	163	12	7500415510075602551574724514601
	155	13	3757513005407665015722506464677633
	147	14	1642130173537165525304165305441011711
	139	15	461401732060175561570722730247453567445
	131	18	215713331471510151261250277442142024165471
	123	19	120614052242066003717210326516141226272506267
	115	21	60526665572100247263636404600276352556313472737
	107	22	22205772322066256312417300235347420176574750154441
	99	23	10656667253473174222741416201574332252411076432303431
	91	25	6750265030327444172723631724732511075550762720724344561
	87	26	110136763414743236435231634307172046206722545273311721317
	79	27	66700035637657500020270344207366174621015326711766541342355
	71	29	24024710520644321515554172112331163205444250362557643221706035
	63	30	10754475055163544325315217357707003666111726455267613656702543301
	55	31	7315425203501100133015275306032054325414326755010557044426035473617
	47	42	2533542017062646563033041377406233175123334145446045005066024552543173
	45	43	15202056055234161131101346376423701563670024470762373033202157025051541
	37	45	5136330255067007414177447245437530420735706174323432347644354737403044003
	29	47	3025715536673071465527064012361377115342242324201174114060254757410403565037
	21	55	1256215257060332656001773153607612103227341405653074542521153121614466513473725
	13	59	464173200505256454442657371425006600433067744547656140317467721357026134460500547
	9	63	15726025217472463201031043255355134614162367212044074545112766115547705561677516057

Source: Stenbit, J. P., *IEEE Trans. Inf. Theory*, IT-10, 390–391(1964). © 1964 IEEE. Reprinted by permission.

4.3.2. BCH Codes with $m_0 = 0$

At the beginning of the discussion of binary BCH codes, we stated that the most general definition permits defining a code as having an arbitrary consecutive sequence of roots, which we now write as $\alpha^{m_0}, \alpha^{m_0+1}, \alpha^{m_0+2}, \ldots, \alpha^{m_0+d-2}$, where d is the design distance of the code. The true minimum distance can be equal to or greater than d. We first considered the case in which the parameter m_0 is chosen to be 1, so that the sequence of roots is $\alpha, \alpha^2, \ldots, \alpha^{2t}$, where $2t = d - 1$. Therefore t is the maximum error-correction capability of the code if d is the true minimum distance. We shall now see what happens if instead we let $m_0 = 0$.

With $m_0 = 0$, a BCH code of design distance d consists of words having as roots the elements $\alpha^0, \alpha^1, \alpha^2, \ldots, \alpha^{d-2}$. That is, the consecutive sequence of roots now starts with $\alpha^0 = 1$ instead of α, and the design distance of the code is one greater than the number of roots in the sequence, as was the case with $m_0 = 1$. It is now a simple matter to show that with $m_0 = 0$ the design distance d must be even. We begin by assuming that we have chosen d to be odd. If d is odd, the highest exponent in the sequence of roots, say L, must also be odd. Then, if L is odd, there is a root of power $j = (L + 1)/2$ at a lower position in the sequence, for $d \geq 1$. However, we know that for codewords on $GF(2)$, if α^j is a root, α^{2j} is also a root, where here $2j = L + 1$, an even number. We conclude, therefore, that if the consecutive sequence of roots extends to the root α^L, with L odd, the element α^{L+1} is also a root, and the highest exponent in the sequence is actually even rather than odd, which is to say that $d - 2$ is even, and thus d is even. Therefore, we conclude that the highest exponent in the sequence of roots is always even.

This can be made clear by considering an example. Let us first specify the roots of a primitive two-error-correcting code, using $m_0 = 1$. The roots are $\alpha, \alpha^2, \alpha^3$, and α^4. The design distance of this code is $d = 5$, that is, one greater than the number of roots in the sequence. If we now modify this code by appending α^0 to the sequence of roots, the sequence becomes α^0, α^1, α^2, α^3, and α^4, yielding a design distance of 6 instead of 5. Note that we have added one, and only one, root to the sequence, since $(\alpha^0)^2 = \alpha^0$. Therefore, using $m_0 = 0$ produces codes with even design distance, and appending α^0 to a code having a sequence of roots already starting with α simply increases the design distance by 1. This, of course, does not increase the error-correction power of the original code; only the residual error detection power is increased. For example, with the original distance-5 code, certain three-error patterns can lead to incorrect decodings, but all three-error patterns are detectable though not correctable with the distance-6 code.

Construction of the generator polynomial $g(x)$ for a BCH code designed with $m_0 = 0$ proceeds in the same manner as with $m_0 = 1$, except that the defining equation, Eq. (4.2), is rewritten as

$$g(x) = LCM\left[m_{\alpha^0}(x), m_{\alpha^1}(x), m_{\alpha^2}(x), \ldots, m_{\alpha^{2t}}(x)\right]$$

where $m_{\alpha^0}(x)$ is the minimal polynomial of $\alpha^0 = 1$. Given that we are concentrating on binary codes, we can condense the sequence of minimal polynomials, as we did earlier in forming Eq. (4.3), and define $g(x)$ as

$$g(x) = LCM\left[m_{\alpha^0}(x), m_{\alpha^1}(x), m_{\alpha^3}(x), \ldots, m_{\alpha^{2t-1}}(x)\right]$$

The simplest generator polynomial for a binary BCH code is that for a distance-2 code with $g(x) = m_{\alpha^0}(x)$. Since the minimal polynomial of α^0 in any finite field is simply $x + 1$, we have for the distance-2 code $g(x) = x + 1$. Since the degree of $g(x)$ is 1, the code has $r = 1$ check bit and $k = n - r = n - 1$ information bits. The code is, in fact, simply the set of all n-bit vectors with an even number of ones, since it is required only that $c(\alpha^0) = 0$ for each codeword. We can easily satisfy ourselves that the code is cyclic in that $x + 1$ is a factor of $x^n - 1$, that is, $x = 1$ is a root of $x^n - 1$ for any block length n. The parity-check matrix **H** for the distance-2 code is the row vector having every column equal to $\alpha^0 = 1$, which is to say that **H** is simply the $1 \times n$ matrix

$$H = (1 \quad 1 \quad 1 \quad \cdots \quad 1)$$

It should now be clear by generalization from this simple case that choosing $m_0 = 0$ in constructing a binary BCH code results in the inclusion of $x + 1$ as a factor in $g(x)$, and this in turn means that the all-ones row vector is one of the rows of the parity-check matrix. Stated a little differently, if we append α^0 to the sequence of roots already defined for a t-error-correcting binary BCH code, the set of information bits is reduced by one and the set of parity bits is increased by one, the added parity bit serving to remove from the original code those words (exactly half) that do not have overall even parity.

A set of codes widely used for error detection with long data packets and files in communication network and computer applications is called the *cyclic redundancy check* (CRC) codes [110]. A CRC code is a distance-4 binary BCH code whose generator polynomial is formed by multiplying the generator polynomial of a Hamming code by $x + 1$. In some applications CRC codes are modified so that the all-zeros information set is not associated with the all-zeros check set in order to detect certain types of hardware failures.

4.3.3. Nonprimitive BCH Codes

Another generalization of the BCH codes mentioned at the beginning of the discussion permits specifying a consecutive sequence of roots that can be powers of any element of $GF(2^m)$. That is, with $m_0 = 1$, the sequence of roots can be selected as

$$\beta^1, \beta^2, \beta^3, \ldots, \beta^{2t}$$

where β need not be a primitive element. If $2^m - 1$ is not prime, some of the elements of $GF(2^m)$ will be nonprimitive, and the order of each such element divides $2^m - 1$. For example, $GF(2^4)$ contains elements of order 3, 5, and 15; $GF(2^6)$ contains elements of order 3, 7, 9, 21, and 63; and so on. BCH codes generated from nonprimitive field elements are called *nonprimitive BCH codes*. Each such code, defined as having roots that are $d - 1$ consecutive powers of an element β, will have design distance d and block length equal to the order of β. We illustrate with an example given by Peterson and Weldon [131].

Let β equal α^3, where α is a primitive element of $GF(2^6)$. We see that β has order 21 as follows:

$$\beta^0 = \alpha^0 = 1, \qquad \beta^1 = \alpha^3, \qquad \beta^2 = \alpha^6, \qquad \ldots, \qquad \beta^{21} = \alpha^{63} = \alpha^0 = 1$$

Let us now design a distance-5 two-error-correcting BCH code having β, β^2, β^3, and β^4 as roots. The generator polynomial will be the least common multiple of $m_\beta(x)$ and $m_{\beta^3}(x)$. From property 6 of Section 4.2.1, we can write all the roots of each minimal polynomial:

Roots of $m_\beta(x)$: $\quad \beta, \beta^2, \beta^4, \beta^8, \beta^{16}, \beta^{32} = \beta^{11}$

Roots of $m_{\beta^3}(x)$: $\quad \beta^3, \beta^6, \beta^{12}$

We therefore see that $m_\beta(x)$ must have degree 6 and $m_{\beta^3}(x)$ must have degree 3, so that the code will have nine parity checks. The block length is 21, the order of β, and thus there are 12 information bits in the code.

The generator polynomial for this code can be constructed with the use of the table of irreducible polynomials in Appendix B. The appropriate line in the table shows

DEGREE 6 1 103F 3 127B 5 147H 7 111A 9 015

The required polynomials are 127 and 015, the minimal polynomials of $\beta = \alpha^3$ and $\beta = \alpha^9$, respectively, which yield

$$g(x) = (x^6 + x^4 + x^2 + x + 1)(x^3 + x^2 + 1)$$

$$= x^9 + x^8 + x^7 + x^5 + x^4 + x + 1$$

This construction procedure does not tell us the actual minimum distance of the code, which in fact is equal to the design distance, 5. In general, a nonprimitive BCH code can have an actual minimum distance greater than the design distance. An important example is the three-error-correcting (23,12) binary Golay code [52], which may be constructed as a nonprimitive BCH code with roots in $GF(2^{11})$. Since $2^{11} - 1 = 23 \times 89$, $GF(2^{11})$ has elements of order 23 and 89 as well as 2047. By selecting $\beta = \alpha^{89}$, we can construct a code of length 23. Consider the design of a single-error-correcting code, which we specify as having roots β and β^2. Using property 6 of Section 4.2.1 once again, we enumerate the roots of the minimal polynomial of β as

$$\beta, \beta^2, \beta^4, \beta^8, \beta^{16}, \beta^9, \beta^{18}, \beta^{13}, \beta^3, \beta^6, \beta^{12}$$

where $\beta^{23} = \alpha^{2047} = 1$ is used to reduce powers of β greater than 22. Notice that the sequence is found to include four consecutive powers of β, namely β, β^2, β^3, and β^4. Therefore, we see that by using the minimal polynomial $m_\beta(x)$ as a BCH code generator polynomial, with the intention of designing a distance-3 code, we have "discovered" two additional consecutive roots and thus have actually constructed a code with design distance 5. The code has 11 check bits, and its generator polynomial can be found in Appendix B, listed as 89 5343B, which yields

$$g(x) = x^{11} + x^9 + x^7 + x^6 + x^5 + x + 1$$

[The reciprocal of this polynomial, $x^{11} + x^{10} + x^6 + x^5 + x^4 + x^2 + 1$, may also be used as a generator for the code, as it also has four consecutive roots in $GF(2^{11})$.] What is very important about this code, which we do not prove here, is that the true minimum distance is 7 rather than 5, and the code is in fact the three-error-correcting (23,12) binary Golay code.

TABLE 4.8. Some Binary BCH Codes Generated by Nonprimitive Elements

n	k	t	n	k	t	n	k	t
21	12	2	45	11	4	65	41	2
	6	3		7	7		29	3
	4	4		5	10		17	5
23	12	2	47	24	2		5	6
25	5	2	49	7	3	69	36	2
27	3	4		4	10		14	7
33	13	2	51	35	2		3	11
35	11	2		27	4	71	36	3
	8	3		19	5	73	55	2
	4	7		11	8		46	4
39	15	3		9	9		37	5
	3	6	55	15	2		28	6
43	15	3		5	5		19	8
45	29	2	57	21	2		10	12
	23	3		3	9			

Source: Peterson, W. W., *Error-Correcting Codes*, MIT Press, Cambridge, MA, 1961. © 1961 by The Massachusetts Institute of Technology. Reprinted by permission.

The generator matrix for the (23,12) Golay code can be constructed directly from $g(x)$ by the method outlined in Section 4.2.3. That is, we can use the vector equivalent of $g(x)$ as the bottom row $\mathbf{g}_1$ of the generator matrix $\mathbf{G}$, and then with 11 successive shifts and vector additions of $\mathbf{g}_1$, produce a systematic generator matrix in which the 12 leftmost positions are information bit positions and the 11 rightmost positions are parity-check positions. The parity-check matrix $\mathbf{H}$ can in turn be constructed as $\mathbf{H} = [\mathbf{P}^T; \mathbf{I}_{11}]$, where $\mathbf{G} = [\mathbf{I}_{12}; \mathbf{P}]$. Alternatively, $\mathbf{H}$ can be constructed first, by writing its columns as $\mathbf{H} = (\beta^0; \beta; \beta^2; \ldots; \beta^{22})$, where β is a root of $g(x) = x^{11} + x^9 + x^7 + x^6 + x^5 + x + 1$. In other words, we would construct a list of 11-tuples using $g(x)$ as a reducing polynomial, just as Tables 4.1, 4.3, and 4.4 were constructed. The list repeats after 23 lines, and the 23 distinct 11-tuples are used as the columns of $\mathbf{H}$. Then $\mathbf{G}$ can be obtained from $\mathbf{H}$.

A short table of the parameters n, k, and t for some nonprimitive binary BCH codes is given in Table 4.8.

4.3.4. Shortening and Extending BCH Codes

In applications of binary block codes, it is often necessary to provide a code with a block length that does not correspond exactly to one of the strict-sense BCH codes. This can usually be accomplished by choosing a BCH code with block length greater than the required length and *shortening* the code by an appropriate amount. The shortening is most readily done by setting a number of the information bits equal to zero in the encoding operation. The number of codewords that can be generated is reduced accordingly, and since the reduced set of codewords is a subset of the codewords in the unshortened code, the minimum distance of the shortened code must be at least as great as that of the unshortened code. Depending upon the amount of the shortening and which particular bits are omitted, the minimum distance may be unchanged or it may increase. For relatively short codes, the minimum distance of the shortened code can be determined by direct examination of the list of resulting codewords.

In general, a shortened BCH code may or may not be cyclic, depending upon which particular information bits are omitted. There is no general theory available to give guidance about which bits are best omitted for a required amount of shortening. Typically, the shortening is done in the most convenient manner, which is to set a string of consecutive information bits equal to zero.

Another commonly used code modification is the *extension* of a code of odd minimum distance by addition of a single overall parity check. Since this modification causes the weight of any odd-weight codeword in the original code to be increased by 1, the minimum distance of the original code is also increased by 1. It should be noted that this modification is not the same as inserting the factor $x - 1$ into the generator polynomial as was discussed in Section 4.3.2, although minimum distance $2t + 2$ results in both cases. In the earlier case, while the parity set was increased by one bit, the information set was simultaneously decreased by one bit, so that the block length was unchanged. The resulting code was also cyclic. With the extension being described here, the information set remains unchanged and the block length is increased by one bit. Furthermore, the code obtained by this one-bit extension is not a cyclic code.

A frequently used extended code is the (24,12) distance-8 code, often called the *extended Golay code*, which is obtained by appending an overall parity check bit to the (23,12) distance-7 Golay code. The extended code is attractive in many applications because the rate k/n is exactly equal to 0.5. The codeword weight distributions for the (23,12) Golay code and the (24,12) extended Golay code are as follows:

Hamming Weight	0	7	8	11	12	15	16	23	24
Words in (23,12) code	1	253	506	1288	1288	506	253	1	—
Words in (24,12) code	1	0	759	0	2576	0	759	0	1

The relationship between the two weight distributions is readily explained by noting that the added parity-check bit increases the weight of any odd-weight codeword by 1 but leaves the weight of any even-weight codeword unchanged.

Given the $r \times n$ parity-check matrix for a distance-$(2t + 1)$ BCH code, the check matrix for the code as extended by one check bit is found as follows:

First append a column of r zeros to the original check matrix, and then append a row of $n + 1$ ones to produce the new $(r + 1) \times (n + 1)$ parity-check matrix.

4.4. ENCODING BINARY BCH CODES

As we have pointed out, a BCH code or any other cyclic code can be encoded by using the generator polynomial $g(x)$ in the manner indicated by the basic definition of a cyclic code, namely,

$$c(x) = i(x)g(x)$$

That is, we associate a polynomial $i(x)$ of degree $k - 1$ with the set of k information bits to be transmitted and multiply by the degree-r polynomial $g(x)$ forming the degree-$(n - 1)$ code polynomial $c(x)$. However, this results in a nonsystematic code structure, and therefore it is preferred instead to form codewords using

$$c(x) = [x^r i(x) \bmod g(x)] + x^r i(x)$$

It is seen that this encoding operation places the k information bits in the k highest order terms of the code polynomial, while the parity-check bits, represented by $x^r i(x) \bmod g(x)$, are confined to the r lowest order terms. That is, this operation generates codewords of the form

$$c(x) = p_1 x^0 + p_2 x + \cdots + p_r x^{r-1} + i_1 x^r + i_2 x^{r+1} + \cdots + i_k x^{n-1}$$

The encoding of any binary cyclic code can be done in a straightforward manner using a linear feedback shift register. An encoding circuit using a shift register with $r = n - k$ stages is shown in Fig. 4.2. Each box in the circuit is a binary storage device. The additions indicated are done modulo 2, and the tap connections are specified by the coefficients of the generator polynomial. The

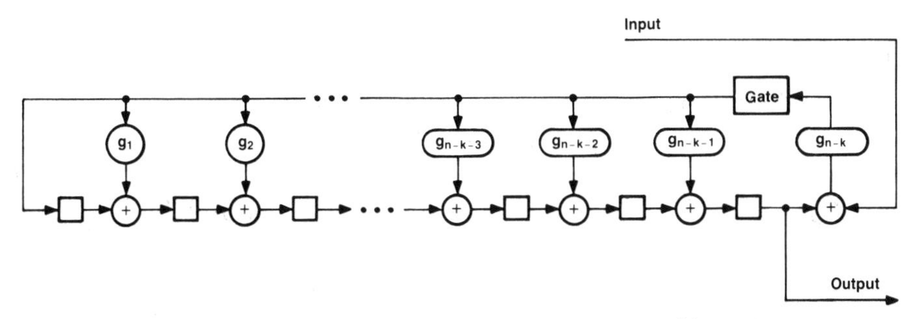

FIGURE 4.2. Encoder for a binary BCH code generated by $g(x)$.

operation of the encoder is as follows:

1. Shift the k information bits into the encoder and simultaneously into the channel. As soon as the k information bits have entered the shift register, the $r = n - k$ bits in the register are the check bits.
2. Disable the feedback circuit.
3. Shift the contents of the register out and into the channel.

As an example, consider an encoder for the length-7 Hamming code generated by $g(x) = x^3 + x + 1$. Using a shift register of the form shown in Fig. 4.2, the feedback tap connections are $g_1 = g_3 = 1$, and $g_2 = 0$. The feedback circuit accomplishes division by $x^3 + x + 1$ in that it sets $x^3 = x + 1$ at each shift of the circuit in step 1. The list of codewords generated by this encoder is shown in Table 4.9. (The correspondence between the table and the encoder circuit is made by taking the rightmost bit in each codeword as the first to be transmitted and the leftmost bit the last.) As an exercise, the reader should verify the encoding procedure for a few codewords in the list. Also, Table 4.9 should be compared with Table 4.5, where the same code was generated in a nonsystematic form.

The encoding circuit that uses an r-stage feedback shift register whose connections are given by the generator polynomial is most convenient for high-rate codes where $k \gg r$. For low-rate codes, a more convenient encoder

TABLE 4.9. The (7,4) Binary Hamming Code Generated from $g(x) = x^3 + x + 1$, Using the Circuit of Fig. 4.2

$i(x)$	$x^3 i(x) \bmod g(x)$	Codeword $p_1 \cdots i_4$
0	0	0000000
1	$1 + x$	1101000
x	$x + x^2$	0110100
$1 + x$	$1 + x^2$	1011100
x^2	$1 + x + x^2$	1110010
$1 + x^2$	x^2	0011010
$x + x^2$	1	1000110
$1 + x + x^2$	x	0101110
x^3	$1 + x^2$	1010001
$1 + x^3$	$x + x^2$	0111001
$x + x^3$	$1 + x$	1100101
$1 + x + x^3$	0	0001101
$x^2 + x^3$	x	0100011
$1 + x^2 + x^3$	1	1001011
$x + x^2 + x^3$	x^2	0010111
$1 + x + x^2 + x^3$	$1 + x + x^2$	1111111

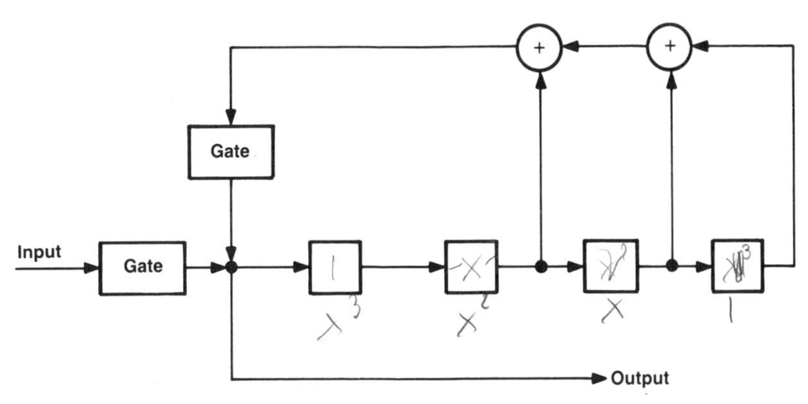

FIGURE 4.3. Encoder for the (7,4) Hamming code generated by $h(x) = x^4 + x^2 + x + 1$.

realization employs a k-stage feedback shift register whose tap connections are given by $h(x) = (x^n - 1)/g(x)$. For the Hamming code considered, we have $h(x) = x^4 + x^2 + x + 1$, and the circuit shown in Fig. 4.3 may be used. The operation of the encoder is as follows:

1. With the feedback circuit disabled, shift the $k = 4$ information bits into the k-stage register and simultaneously into the channel.
2. When the k information bits have entered the encoding register, cycle the register $r = 3$ times with the input disabled. The $r = 3$ bits obtained at the output are the encoded parity bits. The parity bits are shifted into the channel.

4.5. NOTES

In addition to the Peterson tables, extensive tables of irreducible polynomials have been published by Church [26], Green and Taylor [58], Marsh [100], and Watson [176]. Reference 26 gives polynomials over prime fields of order 2, 3, 5, and 7, and reference 58 gives polynomials over prime-power fields up to $GF(16)$. A number of other references can be found in Chapter 4 of MacWilliams and Sloane [109].

Recall from Chapter 1 that Shannon's coding theorem states that there exist codes that can produce arbitrarily small probability of error. That is, if the information rate is held below channel capacity, there exist codes for which the post-decoding error probability can be made to decrease exponentially with increasing block length. This leads us to ask, How good are the codes of any given class as the block lengths are made arbitrarily long? As a criterion of goodness in this regard, one would at least like, for any class of codes, that as block length increases, both the code rate k/n and the ratio of minimum distance to block length, d/n, remain nonzero. Codes having this desirable property are said to be *asymptotically good*. The promise of codes that are asymptotically good is offered by the *Gilbert-Varshamov bound*, which states that if the code rate R is fixed, $0 \leq R \leq 1$, then there exist binary (n,k) distance-d codes with $k/n \geq R$ and $d/n \geq H^{-1}(1 - R)$, where $H^{-1}(y)$ is the *inverse entropy function*, that is, the value of x for which

$$H(x) = -x\log_2 x - (1 - x)\log_2(1 - x)$$

The entropy function was introduced in Section 1.4. (A detailed discussion of bounds on minimum distance for block codes can be found in Peterson and Weldon [131].)

The BCH codes for short to moderate block lengths (say, for block lengths up to about 1023) are reasonably efficient in their use of redundancy, as measured by how close d/n comes to the Gilbert-Varshamov bound for any given code rate k/n. However, the codes are asymptotically weak as the block lengths become arbitrarily long. The asymptotic properties of BCH codes were first published in a paper entitled "Long BCH Codes are Bad" [96]. Later, Berlekamp [8] showed in particular that for a sequence of primitive binary BCH codes of rate $R = k/n$, the minimum distance is given asymptotically by

$$d \doteq 2n\frac{\ln(1/R)}{\log_2 n}, \quad \text{as } n \to \infty$$

where d may be interpreted as the design distance or the true minimum distance. Thus it is seen that at a fixed code rate the ratio d/n goes to zero as the block length goes to infinity, and the BCH codes are said to be *asymptotically bad*.

There are, however, several classes of codes containing codes that are asymptotically good. Justesen was the first to obtain a class of asymptotically good codes that were "constructive" in the strict sense of the term [79]. (That is, code design does not require the use of randomly chosen codes.) However, the Justesen codes cannot be considered practical for real-world application. Perhaps the most important class of codes containing asymptotically good codes is a generalization of the BCH codes, known as *Goppa codes* [55, 56]. In addition to the BCH codes, the Goppa codes include many binary codes of short to moderate block lengths that are better than any BCH codes, though the improvements are often very minor. The Goppa codes are noncyclic (except for the BCH codes). While all BCH codes are asymptotically bad, almost all long Goppa codes are good. In fact, it has been shown that there are Goppa codes that asymptotically satisfy the Gilbert-Varshamov bound. Unfortunately, none of these good codes has yet been explicitly identified. An excellent summary of the properties of Goppa codes is given in a paper by Berlekamp [9].

■ ■ CHAPTER FIVE

Decoding Techniques for Binary BCH Codes

From our earlier discussions we can succinctly describe the problem of decoding a linear block code. For each received word we first determine if the parity-check relationships hold. If all parities check, the word is assumed to be the correct codeword, but if one or more parities fail to check, the word is recognized to contain errors. For error-correction decoding, then, the job of the decoder is to determine what changes have to be made in the received word in order for the parity checks to be satisfied. A brute-force approach would be to change one bit at a time, then two bits at a time in all combinations, and so forth, until the parity checks are finally satisfied. This in fact would accomplish nearest-neighbor decoding but is impractical for any but the simplest of codes. It is out of the question for codes with large block lengths that correct large numbers of errors. Therefore, much work in the coding field has been devoted to finding algorithms for doing error correction with the least amounts of computation. A number of efficient techniques have been devised, some specialized to particular codes and others applicable to wide classes of codes.

In this chapter we describe several of the most important hard-decision decoding techniques in detail. We begin with a discussion of an approach to syndrome decoding in which the correction of an error pattern in a received word is accomplished by solving a set of simultaneous nonlinear equations over a finite field, a discussion that leads to the important Berlekamp iterative algorithm for decoding any BCH code. We then describe the Kasami algorithm, a very efficient procedure for decoding the binary Golay codes. The Kasami algorithm is a special case of a more general class of decoding techniques often called *error trapping*.

In systems where quality measures can be obtained for each received bit, soft-decision decoding techniques can be utilized to increase the power of a code beyond that achievable with algebraic hard-decision decoding. Several such techniques are discussed, ranging from the simplest forms of *erasure filling* to a powerful set of algorithms termed *channel-measurement decoding*.

Another important decoding method, one not covered in this chapter, is *threshold decoding*. While the technique is applicable to block codes, it has found its widest application with convolutional codes, and for that reason it is treated in conjunction with convolutional codes in Chapter 8.

Throughout this chapter we continue to concentrate on binary codes.

5.1. THE PARITY-CHECK MATRIX FOR A BCH CODE

In discussing methods for decoding BCH codes, it is useful to begin by examining the relationship between the generator polynomial and the parity-check matrix, as we did in Chapter 4 for the special case of the cyclic Hamming codes. Since a BCH code is defined by a generator polynomial having roots that are $2t$ consecutive powers of an element in a specified extension field, say $GF(2^m)$, all codeword polynomials in the code have the same sequence of roots, and we can write, for example,

$$c(\alpha^i) = 0, \qquad i = 1, 3, \ldots, 2t - 1$$

where α is a primitive element of $GF(2^m)$. If we evaluate a codeword polynomial at the prescribed roots, we have

$$c_0\alpha^0 + c_1\alpha^1 + c_2\alpha^2 + \cdots + c_{n-1}\alpha^{n-1} = 0$$

$$c_0\alpha^0 + c_1\alpha^3 + c_2\alpha^6 + \cdots + c_{n-1}\alpha^{3(n-1)} = 0$$

$$\vdots$$

$$c_0\alpha^0 + c_1\alpha^{2t-1} + c_2\alpha^{2(2t-1)} + \cdots + c_{n-1}\alpha^{(2t-1)(n-1)} = 0$$

Thus there is a set of simultaneous nonlinear equations in α defining each codeword $c(x)$. The indicated additions and multiplications are performed in $GF(2^m)$. Note that the $c_i = 0$ or 1 are both elements of $GF(2^m)$. We also remind the reader that since $GF(2^m)$ is a field of characteristic 2 (Section 4.1), subtraction is the same operation as addition.

We can compactly represent the above set of simultaneous equations in matrix notation as

$$(c_0 \quad c_1 \quad c_2 \quad c_3 \cdots c_{n-1})[\alpha^0 \quad \alpha^i \quad \alpha^{2i} \quad \alpha^{3i} \cdots \alpha^{(n-1)i}]^T = \mathbf{0}$$

for $i = 1, 3, \ldots, 2t - 1$. We now see that the matrix

$$\mathbf{H} = [\alpha^0 \quad \alpha^i \quad \alpha^{2i} \quad \alpha^{3i} \cdots \alpha^{(n-1)i}], \qquad i = 1, 3, \ldots, 2t - 1$$

is the parity-check matrix for the BCH code having design distance $2t + 1$. As written above, $\mathbf{H}$ is a $t \times n$ matrix of elements in $GF(2^m)$. For example, consider the three-error-correcting length-15 BCH code, the design of which was outlined in Section 4.3.1. Since codewords in this code are specified as having roots α, α^3, and α^5, where α is a primitive element in $GF(2^4)$, we can write the parity-check matrix as

$$\mathbf{H} = \begin{bmatrix} \alpha^0 & \alpha^1 & \alpha^2 & \alpha^3 & \alpha^4 & \alpha^5 & \alpha^6 & \alpha^7 & \alpha^8 & \alpha^9 & \alpha^{10} & \alpha^{11} & \alpha^{12} & \alpha^{13} & \alpha^{14} \\ \alpha^0 & \alpha^3 & \alpha^6 & \alpha^9 & \alpha^{12} & \alpha^0 & \alpha^3 & \alpha^6 & \alpha^9 & \alpha^{12} & \alpha^0 & \alpha^3 & \alpha^6 & \alpha^9 & \alpha^{12} \\ \alpha^0 & \alpha^5 & \alpha^{10} & \alpha^0 & \alpha^5 & \alpha^{10} & \alpha^0 & \alpha^5 & \alpha^{10} & \alpha^0 & \alpha^5 & \alpha^{10} & \alpha^0 & \alpha^5 & \alpha^{10} \end{bmatrix}$$

where we have used $\alpha^{15} = \alpha^0$ to reduce powers of α greater than 14.

It is important to note that while this way of writing the parity-check matrix may seem to be very different from writing it as an $(n - k) \times n$ matrix with binary entries, the two forms are mathematically equivalent as representations of the code. For example, if we were to replace each $GF(2^m)$ field element in the $t \times n$ matrix by its m-bit binary representation, we would always produce an $mt \times n$ matrix of $GF(2)$ elements. However, in many instances, a t-error correcting BCH code has fewer than mt check bits, as we have already seen in Section 4.3, where the design of BCH codes was discussed. In the cases where $n - k$ is less than mt, the $mt \times n$ matrix on $GF(2)$ constructed from the $t \times n$ matrix on $GF(2^m)$ will have only $n - k$ linearly independent rows. The dependent rows would have to be removed from $\mathbf{H}$ (using elementary row operations) if the binary parity-check matrix were to be constructed in this manner. For example, if we were to rewrite the 3×15 parity-check matrix given above for the three-error-correcting length-15 code into its equivalent 12×15 binary form, it would be seen that the last four rows would contain two identical rows and one all-zeros row. (The reader should verify this by using Table 4.3 to express α^0, α^5, and α^{10} as binary 4-tuples.) Therefore, two rows can be struck out of the 12×15 binary form of $\mathbf{H}$, which produces the 10×15 parity-check matrix for the (15,5) distance-7 code.

It is important to point out that while any given binary BCH code with $2t$ consecutive powers of some element as roots in $GF(2^m)$ may or may not have exactly mt binary parity-check equations, the definition of codewords as the polynomials $c(x)$ satisfying the t simultaneous equations $c(\alpha^i) = 0$ is valid in every case. The exact number of parity checks in any given code is, of course, important, but this is automatically taken care of in the design of the generator polynomial $g(x)$. For the purpose of decoding a BCH code, however, we shall see in the following sections that we need not concern ourselves with the exact number of rows in the binary parity-check matrix, but only with the particular sequence of roots in $GF(2^m)$ that defines the code at hand. In fact, it will be seen that it is convenient and useful to consider the parity-check relationships strictly in terms of the polynomial equations $c(\alpha^i) = 0$, $i = 1, 3, \ldots, 2t - 1$.

5.2. THE SYNDROME EQUATIONS

The decoding of a BCH code, as with any linear parity-check code, begins with the calculation of the syndrome of the received word. As we first saw in Chapter 2, where we represented codewords as vectors, calculation of the syndrome is equivalent to a reapplication of the parity-check encoding equations to the received word. Here, where we represent a codeword as a polynomial $c(x)$, we correspondingly represent an error pattern as $e(x)$ and the received word as $r(x)$, where

$$r(x) = c(x) + e(x)$$

To compute the syndrome values $\{S_k\}$ for a binary BCH code, we simply substitute the prescribed roots of the code into the polynomial $r(x)$, that is,

$$S_k = r(\alpha^k) = r_0(\alpha^k)^0 + r_1(\alpha^k)^1 + r_2(\alpha^k)^2 + \cdots + r_{n-1}(\alpha^k)^{n-1}$$

A convenient way to implement the syndrome calculation is to use

$$r(\alpha^k) = \{ \cdots [(r_{n-1}\alpha^k + r_{n-2})\alpha^k + r_{n-3}]\alpha^k + \cdots \}\alpha^k + r_0$$

We now note that

$$\begin{aligned} S_k &= c(\alpha^k) + e(\alpha^k) \\ &= e(\alpha^k), \qquad k = 1, 3, \ldots, 2t - 1 \end{aligned} \tag{5.1}$$

since $c(\alpha^k) = 0$, $k = 1, 3, \ldots, 2t - 1$. That is, each element S_k of the syndrome is simply the error-pattern polynomial $e(x)$ evaluated at $x = \alpha^k$ and thus is some element in the extension field $GF(2^m)$. Let us now assume that there are t errors in the received word, so that $e(x)$ has t nonzero coefficients. (To avoid unduly complicated notation we are letting the actual number of errors be equal to the maximum number correctable by the code, that is, the value t such that the minimum distance of the code is $d = 2t + 1$. For error patterns having fewer than t errors, one can think of the appropriate subset of the assumed t errors having values equal to 0 rather than 1.) If the ith error ($1 \le i \le t$) occurs in received symbol $r_j (0 \le j \le n - 1)$, then we define its *error locator* to be $X_i = \alpha^j$, which is an element of $GF(2^m)$. We thus refer to $GF(2^m)$ as the *locator field*. Since we are considering a binary code, all *error values* are 0 or 1, and we can write for any k

$$e(\alpha^k) = \sum_{i=1}^{t} X_i^k \tag{5.2}$$

where t is the number of errors in the received word.

To make these points clearer let us say, for example, that there are three errors in the received word, in the first, second, and last bit positions. Then the error polynomial evaluated at each root of the code is simply

$$e(\alpha^k) = e_0(\alpha^k)^0 + e_1(\alpha^k)^1 + e_{n-1}(\alpha^k)^{n-1}$$

$$= (\alpha^0)^k + (\alpha^1)^k + (\alpha^{n-1})^k$$

$$= X_1^k + X_2^k + X_3^k$$

where X_1, X_2, and X_3 are the three error locators and the $e_j = \alpha^0 = 1$ are the error values.

From Eqs. (5.1) and (5.2) we see that

$$S_k = \sum_{i=1}^{t} X_i^k, \qquad k = 1, 3, \ldots, 2t-1 \tag{5.3}$$

The decoding problem then is simply to find the error locators X_i from the syndrome values $S_1, \ldots, S_{2t-1}$. Note, however, that Eq. (5.3) represents t nonlinear coupled algebraic equations over the finite field $GF(2^m)$. Direct solution of such equations is generally avoided, and an indirect approach is used instead. To this end we introduce the polynomial

$$\sigma(x) = \prod_{i=1}^{t}(x + X_i) = x^t + \sigma_1 x^{t-1} + \cdots + \sigma_t \tag{5.4}$$

having the error locators as roots, and which we therefore call the *error-locator polynomial.* The coefficients σ_i are seen to be given by the *elementary symmetric functions* of the error locators [131], that is,

$$\sigma_1 = \sum_i X_i$$

$$\sigma_2 = \sum_{i<j} X_i X_j$$

$$\sigma_3 = \sum_{i<j<k} X_i X_j X_k$$

$$\vdots$$

$$\sigma_t = X_1 X_2 X_3 \cdots X_t$$

[*Note*: Some authors define the error-locator polynomial $\sigma(x)$ as a polynomial with factors of the form $(1 + X_i x)$, so that the roots of $\sigma(x)$ are the reciprocals of the error locators. We find the notation used here to be more convenient for purposes of exposition. However, the reciprocal-root formulation of $\sigma(x)$ will be used in later discussions.]

Several approaches can be taken to decoding a BCH code, each having relative advantages and disadvantages that depend largely upon the number of errors the code is designed to correct. Several of the important techniques that are used can be broadly summarized for binary codes as follows:

Step 1. Calculate the syndrome values $S_k = r(\alpha^k)$, $k = 1, 3, \ldots, 2t - 1$.

Step 2. Determine the elementary symmetric functions, that is, the coefficients of the error-locator polynomial $\sigma(x)$, from the syndrome values.

Step 3. Solve for the roots of $\sigma(x)$, which are the error locators.

Step 4. Correct the errors in the positions indicated by the error locators.

In general, the most difficult part of this procedure is step 2, determination of the coefficients of $\sigma(x)$ from the syndrome values, and it is in this step that the most prominent algorithms differ.

5.3. PETERSON'S DIRECT SOLUTION METHOD

We saw in the previous discussion that the syndrome values $S_1, S_3, \ldots, S_{2t-1}$ are the constants in a set of simultaneous nonlinear equations in which the unknowns are the error locators $X_1, X_2, \ldots, X_t$. We now describe a method, due to Peterson [128], for direct solution of these nonlinear equations. In order to describe this method we write the full set of syndrome values $S_1, S_2, \ldots, S_{2t}$ as

$$\begin{aligned} S_k &= r(\alpha^k) \\ &= c(\alpha^k) + e(\alpha^k) \\ &= \sum_{i=1}^{t} X_i^k, \qquad k = 1, 2, \ldots, 2t \end{aligned}$$

which gives the equations

$$\begin{aligned} X_1 + X_2 + \cdots + X_t &= S_1 \\ X_1^2 + X_2^2 + \cdots + X_t^2 &= S_2 \\ &\vdots \\ X_1^{2t} + X_2^{2t} + \cdots + X_t^{2t} &= S_{2t} \end{aligned} \tag{5.5}$$

We call these the *syndrome equations*. The syndrome values $\{S_k\}$ are computed from the received word, and Eq. (5.5) is to be used to obtain the error locators $\{X_i\}$.

Peterson showed that rather than having to solve this set of nonlinear equations in the form written here, it is possible to convert the equations into linear equations that can be solved in conjunction with the error-locator polynomial $\sigma(x)$. This is accomplished by first noting that $\sigma(x)$ evaluated at each error-locator value equals zero. That is,

$$\sigma(X_i) = X_i^t + \sigma_1 X_i^{t-1} + \cdots + \sigma_t = 0, \qquad i = 1, 2, \ldots, t \tag{5.6}$$

Clearly we can multiply Eq. (5.6) through by any power of X_i and the equality is preserved. In particular, let us multiply by X_i^j, so that we have

$$X_i^{t+j} + \sigma_1 X_i^{t+j-1} + \cdots + \sigma_t X_i^j = 0, \qquad i = 1, 2, \ldots, t \tag{5.7}$$

Now, letting j remain general, we sum Eq. (5.7) over $i = 1, 2, \ldots, t$, and using the syndrome equations, Eq. (5.5), we can write

$$S_{t+j} + \sigma_1 S_{t+j-1} + \cdots + \sigma_t S_j = 0 \tag{5.8}$$

The equations defined by Eq. (5.8), with t general, are called *Newton's identities*, which for a binary code can be shown to be equivalent to

$$\begin{aligned} S_1 + \sigma_1 &= 0 \\ S_3 + S_2\sigma_1 + S_1\sigma_2 + \sigma_3 &= 0 \\ S_5 + S_4\sigma_1 + S_3\sigma_2 + S_2\sigma_3 + S_1\sigma_4 + \sigma_5 &= 0 \\ &\vdots \end{aligned} \tag{5.9}$$

In principle, to decode a code of any given minimum distance, we need only truncate Eq. (5.9) in an appropriate manner and solve a set of linear equations for the $\{\sigma_i\}$ in terms of the given syndrome values.

For example, in decoding a single-error-correcting code, there is only one syndrome value, S_1, and the first line of Eq. (5.9) gives

$$S_1 + \sigma_1 = 0$$

so that we have

$$\sigma_1 = S_1$$

For $t = 1$, the error-locator polynomial, Eq. (5.4), is simply $x + \sigma_1$, having the trivial root $x = \sigma_1$, which we have just found to be equal to S_1. Thus for a single-error-correcting BCH code, we have the very simple result that the error locator is equal to the syndrome S_1. This is easily verified by noting that if there is only one error in the received word $r(x)$, at position j, $0 \leq j \leq n - 1$, the syndrome is simply

$$S_1 = r(\alpha) = e(\alpha) = X_1 = \alpha^j$$

For a two-error-correcting code, two syndrome values are computed, S_1 and S_3, and the first two lines of Eq. (5.9) (with $\sigma_3 = 0$) can be written in matrix form as

$$\begin{bmatrix} 1 & 0 \\ S_2 & S_1 \end{bmatrix} \begin{bmatrix} \sigma_1 \\ \sigma_2 \end{bmatrix} = \begin{bmatrix} S_1 \\ S_3 \end{bmatrix}$$

These simultaneous linear equations are solved using the methods of ordinary algebra except that multiplications, divisions, and additions are done using the rules for $GF(2^m)$. We note here that while the S_k's for k even need not be calculated for a binary BCH code, these even-indexed syndrome values appear in the given formulation of the decoding problem. They are readily obtained since it is easy to show that for binary codes, $S_{2k} = S_k^2$ for any k. That is, for elements A and B_i in a field of characteristic 2, if

$$A = \sum_i B_i$$

then the square of A is simply

$$A^2 = \sum_i \sum_j B_i B_j = \sum_i B_i^2$$

so that we have

$$S_1^2 = \sum_{i=1}^{t} X_i^2 = S_2 \tag{5.10}$$

Similarly, $S_4 = S_2^2 = S_1^4$, $S_6 = S_3^2$, and so forth. Thus in solving the simultaneous equations, the solutions can be expressed in terms of only the odd-indexed syndrome values. For the case of two-error correction, we have

$$\sigma_1 = S_1 \qquad \text{and} \qquad \sigma_2 = \frac{S_3 + S_1^3}{S_1} \tag{5.11}$$

As an example let us consider a two-error-correcting BCH code of block length $n = 15$ and reception of a codeword in which errors have occurred at the third and eleventh bit positions. From the definition of error locators at the beginning of Section 5.2, the two errors have locators

$$X_1 = \alpha^2 \qquad \text{and} \qquad X_2 = \alpha^{10}$$

where we assume that the roots of the code are powers of a primitive element. For simplicity we consider a case in which the transmitted codeword was all zeros. Using Table 4.3 to perform arithmetic in $GF(2^4)$, we find the two syndrome values to be

$$S_1 = r(\alpha) = \alpha^2 + \alpha^{10} = \alpha^4$$

$$S_3 = r(\alpha^3) = \alpha^6 + \alpha^0 = \alpha^{13}$$

The coefficients of $\sigma(x)$ are found from Eq. (5.11) to be

$$\sigma_1 = \alpha^4$$

$$\sigma_2 = \frac{\alpha^{13} + \alpha^{12}}{\alpha^4} = \frac{\alpha^1}{\alpha^4} = \alpha^{12}$$

Therefore the error-locator polynomial is

$$\sigma(x) = x^2 + \alpha^4 x + \alpha^{12}$$

The roots of $\sigma(x)$ may be found by direct substitution of elements from $GF(2^4)$; an efficient implementation will be described later. Solution of $\sigma(x) = 0$ yields the roots $x = \alpha^2$ and $x = \alpha^{10}$, which we may verify with the aid of Table 4.3, as follows:

$$(\alpha^2)^2 + \alpha^4\alpha^2 + \alpha^{12} = \alpha^4 + \alpha^6 + \alpha^{12} = \alpha^{12} + \alpha^{12} = 0$$

$$(\alpha^{10})^2 + \alpha^4\alpha^{10} + \alpha^{12} = \alpha^5 + \alpha^{14} + \alpha^{12} = \alpha^{12} + \alpha^{12} = 0$$

Using standard techniques to solve sets of simultaneous linear equations, direct solutions for the coefficients of the error-locator polynomial can be found for any error-correction limit t. The results of such solutions for $t = 3$ through 6 are as follows:

Three-Error Correction

$$\sigma_1 = S_1$$

$$\sigma_2 = \frac{S_1^2 S_3 + S_5}{S_1^3 + S_3}$$

$$\sigma_3 = (S_1^3 + S_3) + S_1\sigma_2$$

Four-Error Correction

$$\sigma_1 = S_1$$

$$\sigma_2 = \frac{S_1(S_7 + S_1^7) + S_3(S_1^5 + S_5)}{S_3(S_1^3 + S_3) + S_1(S_1^5 + S_5)}$$

$$\sigma_3 = (S_1^3 + S_3) + S_1\sigma_2$$

$$\sigma_4 = \frac{(S_5 + S_1^2 S_3) + (S_1^3 + S_3)\sigma_2}{S_1}$$

Five-Error Correction

$$\sigma_1 = S_1$$

$$\begin{aligned}\sigma_2 = {} & \{(S_1^3 + S_3)[(S_1^9 + S_9) + S_1^4(S_5 + S_3S_1^2) + S_3^2(S_1^3 + S_3)] \\ & + [(S_1^5 + S_5)(S_7 + S_1^7) + S_1(S_3^2 + S_1S_5)]\} \\ & \div \{(S_1^3 + S_3)[(S_1^7 + S_7) + S_1S_3(S_1^3 + S_3)] \\ & + [(S_5 + S_1^2S_3)(S_1^5 + S_5)]\}\end{aligned}$$

$$\sigma_3 = (S_3 + S_1^3) + S_1\sigma_2$$

$$\begin{aligned}\sigma_4 = {} & \{[(S_1^9 + S_9) + S_3^2(S_3 + S_1^3) + S_1^4(S_5 + S_3S_1^2)] \\ & + [(S_1^7 + S_7) + S_1S_3(S_1^3 + S_3)]\sigma_2\} \\ & \div \{S_1^5 + S_5\}\end{aligned}$$

$$\sigma_5 = (S_5 + S_3S_1^2) + S_1\sigma_4 + (S_1^3 + S_3)\sigma_2$$

Six-Error Correction

$$\sigma_1 = S_1$$

$$\begin{aligned}\sigma_2 = {} & \{S_3[S_3(S_1^5 + S_5) + S_1(S_1^7 + S_7)] \\ & \times [S_1^2(S_5 + S_1^5) + (S_7 + S_3^2S_1) + S_1(S_9 + S_1^4S_5)] \\ & + [S_3(S_1^3 + S_3) + S_5(S_1^5 + S_5)] \\ & \times [S_1^4S_3(S_5 + S_1^5) + S_1^3(S_3^3 + S_9) + S_1^2(S_7 + S_1^4S_3) + (S_5S_7 + S_1S_{11})]\} \\ & \div \{[S_3(S_1^5 + S_5) + S_1(S_1^7 + S_7)] \\ & \times [S_1(S_1^3 + S_3)^2 + S_3(S_5 + S_1^2S_3) + S_1(S_7 + S_1^4S_3)] \\ & + [S_3(S_1^3 + S_3) + S_1(S_1^5 + S_5)] \\ & \times [(S_5 + S_1^2S_3)^2 + S_1^3(S_7 + S_1^4S_3) + (S_3^3 + S_9)]\}\end{aligned}$$

$$\sigma_3 = (S_1^3 + S_3) + S_1\sigma_2$$

$$\sigma_4 = \Big\{\big[S_1^4S_3(S_1^5 + S_5) + S_1^3(S_3^3 + S_9) + S_1^2(S_7 + S_1^4S_3) + (S_5S_7 + S_1S_{11})\big]$$

$$+\big[(S_5 + S_1^2S_3)^2 + S_1^3(S_7 + S_1^4S_3) + (S_3^3 + S_9)\big]\sigma_2\Big\}$$

$$\div\big\{S_3(S_1^5 + S_5) + S_1(S_1^7 + S_7)\big\}$$

$$\sigma_5 = (S_5 + S_1^2S_3) + S_1\sigma_4 + (S_1^3 + S_3)\sigma_2$$

$$\sigma_6 = \frac{\big[S_1^2(S_1^5 + S_5) + (S_7 + S_3^2S_1)\big] + (S_1^3 + S_3)\sigma_4 + (S_5 + S_1^2S_3)\sigma_2}{S_1}$$

The equations just listed give, for each indicated number of errors, the formulas to be used for calculation of the coefficients of the corresponding error-locator polynomial $\sigma(x)$. In general, however, with use of a t-error-correcting code, any error pattern with fewer than t errors is also correctable, and we do not know at the outset of decoding how many errors there actually are. To use these formulas knowing that the actual number of errors in a received word may be less than t, we start by determining whether the first t lines in Eq. (5.9) can be solved for $\sigma_1, \sigma_2, \ldots, \sigma_t$. This is done by using the determinant test

$$\det\begin{bmatrix} 1 & 0 & 0 & 0 & 0 & \cdots & 0 \\ S_2 & S_1 & 1 & 0 & 0 & \cdots & 0 \\ S_4 & S_3 & S_2 & S_1 & 1 & \cdots & 0 \\ \vdots & \vdots & \vdots & \vdots & \vdots & & \vdots \\ S_{2t-4} & S_{2t-5} & S_{2t-6} & S_{2t-7} & S_{2t-8} & \cdots & S_{t-3} \\ S_{2t-2} & S_{2t-3} & S_{2t-4} & S_{2t-5} & S_{2t-6} & \cdots & S_{t-1} \end{bmatrix} \overset{?}{\neq} 0$$

It can be shown [128] that if there are t or $t-1$ errors in the received word, the determinant will be nonzero. Given this outcome, we proceed with the formulas for t-error correction. If there are actually t errors, the solutions found for $\sigma_1, \sigma_2, \ldots, \sigma_t$ define a degree-t error-locator polynomial. If there are only $t-1$ errors, $\sigma_t = 0$ and thus $\sigma(x)$ has degree $t-1$.

If the determinant shown above is found to be zero, two rows and columns of the matrix are removed, and the determinant of the resulting $(t-2) \times (t-2)$ matrix is tested in the same manner. This procedure is repeated until a nonzero determinant is found and the error-locator polynomial coefficients are determined.

The final steps in decoding a binary BCH code are to find the roots of the error-locator polynomial $\sigma(x)$, which are the error locators, and to correct the errors indicated. A procedure called the *Chien search* [24] accomplishes these two processes without explicitly solving $\sigma(x)$. This can be done with the circuit

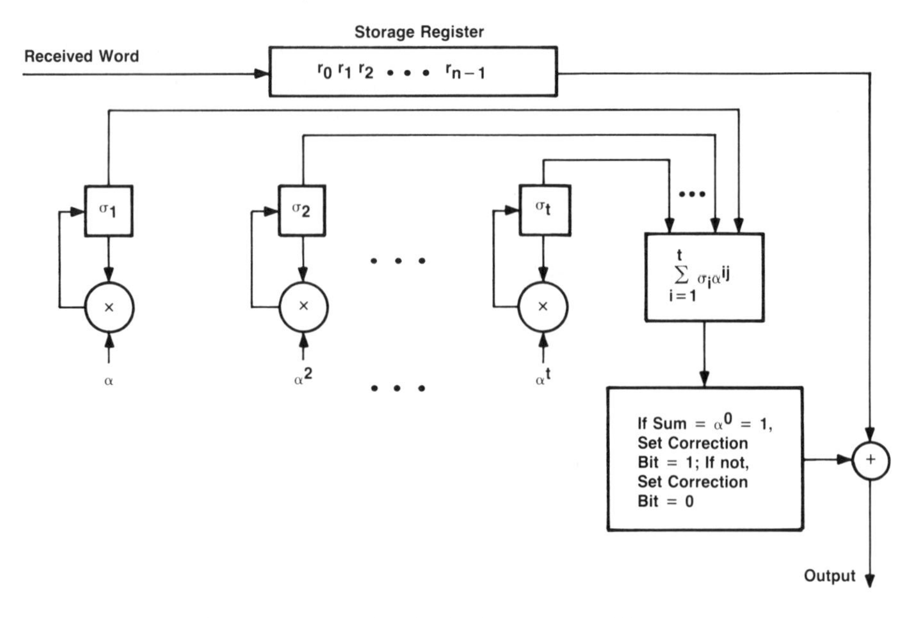

FIGURE 5.1. Procedure for implementing the Chien search.

shown in Fig. 5.1. The circuit steps sequentially through all possible error-locator values and corrects the corresponding bits as the locators are found. To see how the circuit operates, consider the error-locator polynomial $\sigma(x)$ as given by Eq. (5.4) and divide it through by x^t, which gives

$$\frac{\sigma(x)}{x^t} = 1 + \sigma_1 x^{-1} + \sigma_2 x^{-2} + \cdots + \sigma_t x^{-t}$$

The values of x that satisfy $\sigma(x) = 0$ consequently satisfy the equation

$$\sigma_1 x^{-1} + \sigma_2 x^{-2} + \cdots + \sigma_t x^{-t} = 1$$

Assuming the convention of transmitting codewords high-order bits first, it is convenient to apply the root test to locator α^{n-1} first. Note that evaluation of a term x^{-i} at α^{n-1} yields α^{-in+i}, which equals α^i if we are using a full-length BCH code, since we then have $\alpha^n = 1$ and thus $\alpha^{-in} = 1$. Therefore, we see that testing α^{n-1} as a possible root of $\sigma(x)$ is the same as testing for

$$\sigma_1 \alpha + \sigma_2 \alpha^2 + \cdots + \sigma_t \alpha^t \stackrel{?}{=} 1$$

and, in general, testing for α^{n-j} as an error locator is equivalent to finding whether or not α^j satisfies

$$\sum_{i=1}^{t} \sigma_i \alpha^{ij} = \alpha^0 = 1, \qquad j = 0, 1, 2, \ldots, n-1$$

It is seen from Fig. 5.1 that the Chien search performs this sequence of tests by initially loading the σ-registers with the coefficients of $\sigma(x)$ and then clocking the circuit n times, where n is the code block length. After each multiplication, the sum of the current contents of the σ-registers is compared with unity. The storage register is shifted by one bit. If the emerging bit is in error, a one is generated at the output of the comparison circuit and the bit is corrected.

If the received word contains t or fewer errors, where t is the error-correction limit for the code, Peterson's direct solutions will produce the coefficients of the error-locator polynomial and the Chien search, since it is exhaustive, will find the roots of the polynomial and correct the errors in the corresponding locations. However, if the Chien search yields fewer than l roots when given the coefficients of a degree-l error-locator polynomial, this means that the polynomial does not have all its roots in the locator field, and thus the polynomial is not a legitimate l-error-locator polynomial. This type of event is, in fact, the most typical indication of a detected uncorrectable error pattern, a pattern containing more than t errors but not causing a decoding to a wrong codeword. However, in certain cases, patterns with more than t errors escape detection by the Chien search, and for this reason a further check is needed. That is, when the Peterson direct solution indicates the presence of $l < t$ errors and the Chien search in turn makes l bit corrections, the resulting codeword must be checked to verify that the syndrome equations are satisfied. Failure to satisfy one or more of the syndrome equations indicates a detected but uncorrectable error pattern. For the case of t-error correction, all syndrome equations have been used by the decoder and the syndrome of the decoded word need not be checked.

5.4. THE BERLEKAMP ALGORITHM

For correction of more than about six errors in a binary BCH codeword, Peterson's direct method of solving for the coefficients of $\sigma(x)$ from the syndrome values becomes cumbersome and inefficient, since the number of finite field multiplications required increases approximately with the square of the number of errors to be corrected. Instead, it is preferable to use an iterative algorithm developed by Berlekamp [7] for solution of Newton's identities. In contrast with the direct solution method, the *Berlekamp algorithm* has a computational complexity that grows only linearly with the number of errors to be corrected. Another version of this algorithm was given by Massey [104]. The Massey formulation is presented in Section 6.6.2 for the case of nonbinary codes.

In the use of the Berlekamp algorithm, the sequence of calculated syndrome values, $S_1, S_2, \ldots, S_{2t}$ is represented with the polynomial

$$S(z) = S_1 z + S_2 z^2 + \cdots + S_{2t} z^{2t}$$

As a convenience in the algorithm, the error-locator polynomial is replaced with an equivalent polynomial $C(z)$ whose *reciprocal roots* are the error locators X_i, $i = 1, 2, \ldots, t$. That is, $C(z)$ is defined by

$$C(z) = \prod_{i=1}^{t} (1 + X_i z)$$

so that $C(z)$ has roots at $z = Z_i$, where $Z_i = 1/X_i$, $i = 1, 2, \ldots, t$. We call the polynomial $C(z)$ the *reciprocal error-locator polynomial* to distinguish it from the error-locator polynomial $\sigma(x)$. Now writing $C(z)$ in its expanded form we have

$$C(z) = 1 + \sigma_1 z + \sigma_2 z^2 + \cdots + \sigma_t z^t$$

where the coefficients $\{\sigma_i\}$ are again seen to be the elementary symmetric functions of the error locators $X_1, X_2, \ldots, X_t$.

The Berlekamp algorithm is an efficient iterative procedure for finding the minimum-degree reciprocal error-locator polynomial $C(z)$ whose coefficients, taken together with the syndrome values, satisfy all t equations in Newton's identities, Eq. (5.9). The algorithm begins by constructing the polynomial $C^{(1)}(z)$ of least degree satisfying the first line in Eq. (5.9) and then setting $C^{(2)}(z) = C^{(1)}(z)$ and determining whether $C^{(2)}(z)$ satisfies the second line in Eq. (5.9). It is easy to see that these first steps consist in simply letting $C^{(1)}(z) = 1 + S_1 z$, so that $\sigma_1 = S_1$, and then testing (second line) the relationship $S_3 = S_2\sigma_1$, where $S_2 = S_1^2$ from Eq. (5.10). This is equivalent to testing $S_3 = S_1^3$, the relationship that must hold, by Eq. (5.5), if there is only a single error in the received word. If the test of the second line succeeds, then we set $C^{(3)}(z) = C^{(2)}(z)$ and test the third line of Eq. (5.9). If the test of the second line of Eq. (5.9) fails, $C^{(2)}(z)$ is modified by adding a correction term that changes $C^{(2)}(z)$ to a minimum-degree polynomial satisfying the first two lines of Eq. (5.9). With the corrected form of $C^{(2)}(z)$, we let $C^{(3)}(z) = C^{(2)}(z)$ and then test the third line of Eq. (5.9), and so forth. The iteration continues until a reciprocal error-locator polynomial $C^{(l)}(z)$, $l \leq t$, is found that satisfies all t lines in Eq. (5.9). The efficiency of the Berlekamp algorithm is due largely to its provision for constructing the correction term at the ith iteration, if needed, so that the $i - 1$ previous lines in Eq. (5.9) do not have to be retested. It can be shown that if the number of errors in the received word is t or less, the Berlekamp algorithm will end with the correct reciprocal error-locator polynomial. We shall simply provide a brief description of the algorithm here. A detailed discussion of the algorithm and a rigorous proof of its error-correction properties can be found in references 7 and 131. The algorithm as described below is actually a simplification for use with binary BCH codes. There is a more general version of the algorithm applicable to nonbinary codes as well. However, when we discuss the decoding of nonbinary codes (Chapter 6) we shall instead describe a closely related algorithm due to Massey.

The steps in the Berlekamp algorithm are described below. The initialized polynomial $C^{(0)}(z)$ fixes one as the leading term in $C(z)$, while $T^{(0)}(z)$ is the initialized correction polynomial. The quantity $\Delta^{(2k)}$ is the discrepancy found when an interim version of $C(z)$ constructed at one line in Eq. (5.9) fails to satisfy the next line. Superscripts are used to index the steps in the iteration.

The Berlekamp Algorithm for Binary Codes

1. Initialize: $k = 0$, $C^{(0)}(z) = 1$, $T^{(0)}(z) = 1$.
2. If S_{2k+1} is not given, stop. Otherwise, define $\Delta^{(2k)}$ as the coefficient of z^{2k+1} in the product $[1 + S(z)]\, C^{(2k)}(z)$. Let

$$C^{(2k+2)}(z) = C^{(2k)}(z) + \Delta^{(2k)} z T^{(2k)}(z)$$

where

$$T^{(2k+2)} = \begin{cases} z^2 T^{(2k)}(z) & \text{if } \Delta^{(2k)} = 0 \text{ or if } \deg C^{(2k)}(z) > k \\ \dfrac{zC^{(2k)}(z)}{\Delta^{(2k)}} & \text{if } \Delta^{(2k)} \neq 0 \text{ and } \deg C^{(2k)}(z) \leq k \end{cases}$$

3. Set $k = k + 1$ and return to step 2.

Note that the multiplications and additions indicated are all in the locator field $GF(2^m)$.

As an example of the application of the Berlekamp algorithm, let us consider the use of the three-error-correcting (31,16) BCH code, where we suppose that the syndrome values for a received word are $S_1 = \alpha^5$, $S_3 = \alpha^{24}$, and $S_5 = \alpha^8$. Since even- and odd-indexed syndrome values are related by $S_{2k} = (S_k)^2$, we have $S_2 = \alpha^{10}$, $S_4 = \alpha^{20}$, and $S_6 = \alpha^{17}$, and thus we write

$$1 + S(z) = 1 + \alpha^5 z + \alpha^{10} z^2 + \alpha^{24} z^3 + \alpha^{20} z^4 + \alpha^8 z^5 + \alpha^{17} z^6$$

Application of the Berlekamp algorithm, using Table 4.4 for the required multiplications and additions in $GF(32)$, yields the following sequence of calculations:

$2k$	$C^{(2k)}(z)$	$T^{(2k)}(z)$	$\Delta^{(2k)}$
0	1	1	α^5
2	$1 + \alpha^5 z$	$\alpha^{26} z$	α^0
4	$1 + \alpha^5 z + \alpha^{26} z^2$	$z + \alpha^5 z^2$	α^{30}
6	$1 + \alpha^5 z + \alpha^5 z^2 + \alpha^4 z^3$	(STOP)	

Therefore, $C(z) = 1 + \alpha^5 z + \alpha^5 z^2 + \alpha^4 z^3$, and we must find its roots, which are $Z_1 = \alpha^{28}$, $Z_2 = \alpha^{24}$, and $Z_3 = \alpha^6$. The reciprocals of these roots are the error locators $X_1 = \alpha^3$, $X_2 = \alpha^7$, and $X_3 = \alpha^{25}$, corresponding to errors in positions 4, 8, and 26 in the length-31 code block. The reader should verify that this error pattern will indeed produce the syndrome values given at the start of this example.

We note that it is not necessary to know, at the outset of the iterative algorithm, exactly how many errors there are in the received word. For example, consider a case using the same (31,16) code but with syndrome values $S_1 = \alpha^4$, $S_3 = \alpha^{30}$, and $S_5 = \alpha^0$. We now have $1 + S(z) = 1 + \alpha^4 z + \alpha^8 z^2 + \alpha^{30} z^3 + \alpha^{16} z^4 + \alpha^0 z^5 + \alpha^{29} z^6$. Application of the Berlekamp algorithm now yields:

$2k$	$C^{(2k)}(z)$	$T^{(2k)}(z)$	$\Delta^{(2k)}$
0	1	1	α^4
2	$1 + \alpha^4 z$	$\alpha^{27} z$	α^{13}
4	$1 + \alpha^4 z + \alpha^9 z^2$	$\alpha^{18} z + \alpha^{22} z^2$	0
6	$1 + \alpha^4 z + \alpha^9 z^2$	(STOP)	

The resulting polynomial $C(z) = 1 + \alpha^4 z + \alpha^9 z^2$ is found to have roots $Z_1 = \alpha^{29}$ and $Z_2 = \alpha^{24}$, corresponding to error locators $X_1 = \alpha^2$ and $X_2 = \alpha^7$. It is readily verified that $S_1 = \alpha^2 + \alpha^7 = \alpha^4$, $S_3 = \alpha^6 + \alpha^{21} = \alpha^{30}$, and $S_5 = \alpha^{10} + \alpha^4 = \alpha^0$.

Berlekamp shows that if there are t or fewer errors in a received code block, the algorithm will correctly identify the locations of all the errors. If there are more than t errors, a variety of different events can occur. It is possible (though only in rare cases) for the algorithm to terminate with the correct set of error locators, that is, a polynomial $C(z)$ with degree greater than t. Far more likely, however, is that the algorithm will converge to a polynomial $C(z)$ of degree t or less, indicating an apparently correct (though in fact incorrect) solution. The latter events will then divide into two categories when the Chien root search is applied. If the error pattern has produced a received word that lies within the radius-t sphere surrounding some legitimate codeword in the code, the Chien search will produce a decoding to that codeword, this being an incorrect decoding. In many cases, however (except when the code is a perfect code), the Chien search will find the error-locator polynomial to be illegitimate, and a detectable but uncorrectable error pattern will be identified. The illegitimacy of the error-locator polynomial will be discovered by noting that fewer than l distinct roots are found for a degree-l polynomial.

5.5. THE KASAMI ALGORITHM

While the Berlekamp iterative decoding algorithm can be applied to any BCH code, various specialized algorithms have been developed that provide for more efficient implementations in some cases. Here we describe one such algorithm devised by Kasami [80] for decoding the distance-7 (23,12) binary Golay code.

We showed in Section 4.3.3 that the (23,12) Golay code can be described as a nonprimitive BCH code having as roots successive powers of $\beta = \alpha^{89}$, where α is a primitive element of $GF(2^{11})$. Therefore, this code can in principle be decoded using the Berlekamp algorithm, with the necessary finite field arithmetic being done in the locator field $GF(2^{11})$. However, this would be a cumbersome decoding technique, due to the rather large size of the locator field. For example, use of log and antilog tables for multiplication requires 4096 storage locations for the table entries alone. Instead, use of the *Kasami algorithm* is generally preferred for decoding the Golay code.

The Kasami algorithm for decoding the (23,12) Golay code makes effective use of the fact that there are about as many parity bits as information bits in the code, and because of this the general technique might in principle be applied to any code with rate about one-half.

In describing the Kasami algorithm we begin by reviewing the polynomial representation of codewords introduced in Chapter 4. We describe the code of block length n as the set of polynomials $c(x) = i(x)g(x)$, where $g(x)$ is the binary generator polynomial, with degree $r = n - k$, and $i(x)$ is any binary polynomial of degree $k - 1$ or less. In other words, we can define the code by specifying that $c(x) = 0 \bmod g(x)$ for each codeword $c(x)$ in the code. We again represent the received word as

$$r(x) = c(x) + e(x)$$

where $c(x)$ is the transmitted codeword

$$c(x) = c_0 + c_1 x + c_2 x^2 + \cdots + c_{n-1} x^{n-1}$$

and $e(x)$ is the *error polynomial*

$$e(x) = e_0 + e_1 x + e_2 x^2 + \cdots + e_{n-1} x^{n-1}$$

We now define the *syndrome polynomial* $s(x)$ as the remainder after dividing $r(x)$ by $g(x)$, that is,

$$s(x) = r(x) \bmod g(x)$$

Thus, we see that the syndrome can be represented as a polynomial of degree $r - 1$ or less, since $g(x)$ has degree r, and we write

$$s(x) = s_0 + s_1 x + s_2 x^2 + \cdots + s_{r-1} x^{r-1}$$

where the coefficients $\{s_i\}$ are in $GF(2)$.

However, since $r(x) = c(x) + e(x)$, we can write

$$\begin{aligned} s(x) &= [c(x) + e(x)] \bmod g(x) \\ &= e(x) \bmod g(x) \end{aligned}$$

where $c(x) = 0 \bmod g(x)$. This manner of defining the syndrome, as the received error polynomial mod $g(x)$, is completely equivalent to our earlier definition of the syndrome, that is, as the result of applying the encoding parity-check equations to the received word $\mathbf{r} = \mathbf{c} + \mathbf{e}$ or the result of evaluating the received polynomial $r(x)$ at the prescribed roots $x = \alpha^i$ of the generator polynomial. Specifically we have $S_i = s(\alpha^i)$. Here, representing the syndrome as $r(x)$ mod $g(x)$ is convenient for the discussion of the Kasami algorithm.

It should be noted that thc binary degree-$(r-1)$ polynomial $s(x)$ used here to represent the syndrome is different from the representation $S(z)$ used in Section 5.4, which is a degree-$2t$ polynomial with coefficients in the locator field $GF(2^m)$.

Let us write the error polynomial $e(x)$ as the sum of two polynomials $e_p(x)$ and $e_i(x)$, where

$$e_p(x) = e_0 + e_1x + \cdots + e_{r-1}x^{r-1}$$

represents the error pattern in the parity-check positions of the codeword, and

$$e_i(x) = e_rx^r + e_{r+1}x^{r+1} + \cdots + e_{n-1}x^{n-1}$$

represents the error pattern in the information positions of the codeword. Note that this assumes the parity bits are placed in the r lowest-degree positions in the codeword. As we described in Section 4.4, this is done by forming codewords as

$$c(x) = [x^ri(x) \bmod g(x)] + x^ri(x) \tag{5.12}$$

Now, from our earlier discussion we see that the syndrome is

$$\begin{aligned} s(x) &= e(x) \bmod g(x) \\ &= [e_p(x) + e_i(x)] \bmod g(x) \\ &= e_p(x) + e_i(x) \bmod g(x) \end{aligned}$$

Therefore, by referring to Eq. (5.12), we can describe $s(x)$ as the sum of the error pattern in the parity-check positions of the received word plus the parity bits associated with the errors in the information positions of the received word, that is, the parity bits obtained if we treat $e_i(x)$ as a set of information bits to be encoded.

We now observe that if all the errors are in the parity positions of the received word, that is, if $e_i(x) = 0$, the syndrome is simply equal to $e_p(x)$, the error pattern in the parity positions. This observation led Kasami to devise the decoding algorithm we now describe.

We first recall the basic property of cyclic codes, that any cyclic shift of a codeword,

$$x^i c(x) \bmod x^n - 1, \qquad i = 1, \ldots, n-1$$

is again a codeword.

The syndrome polynomial of a cyclic shift of a received word $r(x)$ may be calculated as

$$\begin{aligned} s_i(x) &= \left[x^i r(x) \bmod x^n - 1\right] \bmod g(x) \\ &= x^i r(x) \bmod g(x) \end{aligned}$$

since $g(x)$ divides $x^n - 1$. If after i cyclic shifts of $r(x)$, the errors (assume t or less in number) are located in the first r positions (the parity-check positions) of the shifted error polynomial $[x^i e(x) \bmod x^n - 1]$, then the syndrome $x^i r(x) \bmod g(x)$ duplicates the error pattern of the shifted codeword; that is,

$$x^i e(x) \bmod (x^n - 1) = x^i r(x) \bmod g(x), \qquad i = 1, \ldots, n-1$$

We are interested in the case

$$\mathrm{Wt}\left[x^i r(x) \bmod g(x)\right] \le t$$

where Wt[·] denotes Hamming weight.

Testing the weight of $s_i(x)$ allows us to determine when all the errors have been shifted into the parity positions of the received word. It is easy to show for a code with minimum Hamming distance $2t + 1$ that $\mathrm{Wt}[s_i(x)] > t$ unless all errors are in the parity positions. For example, assume that $t - 1$ errors are in the parity positions and one error is in an information position of the error polynomial. Then we have

$$\mathrm{Wt}\left[e_i(x) \bmod g(x)\right] \ge 2t$$

from the distance structure of the code. Therefore,

$$\mathrm{Wt}\left[s_i(x)\right] \ge \mathrm{Wt}\left[e_i(x) \bmod g(x)\right] - \mathrm{Wt}\left[e_p(x)\right] \ge t + 1$$

and the test fails. When the test on $\mathrm{Wt}\,[s_i(x)]$ succeeds, the correct codeword is given by

$$x^{-i}\left\{\left[x^i r(x) \bmod g(x)\right] + x^i r(x) \bmod x^n - 1\right\} \bmod x^n - 1$$

Multiplication by x^{-i} undoes the cyclic shift that moved the errors into the parity positions. However, this decoding scheme will succeed only if all the errors can be shifted into the first r positions of the received word. This obviously cannot always be done.

For those error polynomials whose errors cannot be shifted entirely into the parity positions, we write the error polynomial as a sum of two polynomials,

$$e(x) = P(x) + Q(x)$$

where $P(x)$ is a member of a set of polynomials such that the ith cyclic shift $x^i P(x) \bmod x^n - 1$ with degree no greater than $r - 1$ coincides with $x^i e(x) \bmod x^n - 1$ in its parity positions. Then $Q(x)$ will be chosen to represent errors that cannot be accounted for with $P(x)$, that is, cannot be shifted into the parity positions. The polynomials $\{Q(x)\}$ are called *covering polynomials*. This representation will include all possible error patterns and their cyclic shifts with finite sets of $P(x)$ and $Q(x)$.

We now proceed as before, except that we calculate the syndrome as

$$s_i(x) = [x^i r(x) + Q(x)] \bmod g(x)$$

and apply the test

$$\mathrm{Wt}[s_i(x)] \leq t, \qquad i = 0, 1, \ldots, n-1$$

If the test is satisfied, then the corrected codeword is given by

$$[x^{-i}(\{[x^i r(x) + Q(x)] \bmod g(x)\} + Q(x)) \bmod x^n - 1] + r(x)$$

where $[x^i r(x) + Q(x)] \bmod g(x)$ is the error pattern in the parity positions and $Q(x)$ is the error pattern in the information positions.

If the test on the weight of $s_i(x)$ fails for all i, then we must try another $Q(x)$ [$\deg Q(x) \geq r$ unless $Q(x) = 0$]. Thus a key problem here is the determination of a minimal set of covering polynomials $Q(x)$ that will allow us to represent all correctable error patterns.

Kasami, in his implementation of the algorithm, actually calculates the syndrome as

$$s_i(x) = x^{r+i} r(x) \bmod g(x), \qquad i = 0, 1, \ldots, n-1$$

where $r = n - k$. Thus the corrected codeword is now given by

$$[x^{-r-i}(\{[x^{r+i} r(x) + Q(x)] \bmod g(x)\} + Q(x)) \bmod x^n - 1] + r(x)$$
$$= [x^{k-i}(\{[x^{r+i} r(x) + Q(x)] \bmod g(x)\} + Q(x)) \bmod x^n - 1] + r(x)$$

The factor x^r is introduced by Kasami for convenience in the implementation of the algorithm.

The Kasami decoding algorithm is summarized in the following steps:

1. Calculate $s_0(x)$ from the received codeword $r(x)$ as $s_0(x) = x^r r(x) \bmod g(x)$.
2. Calculate $s_i(x)$ from $s_{i-1}(x)$ using the relationship $s_i(x) = x s_{i-1}(x) \bmod g(x)$.
3. For each i, calculate $\mathrm{Wt}\{[x^{r+i}r(x) + Q(x)] \bmod g(x)\}$.
4. If, for some $Q(x)$, $\mathrm{Wt}\{[x^{r+i}r(x) + Q(x)] \bmod g(x)\} \leq t$, form the corrected codeword as

$$r(x) + x^{k-i}\big(\{[x^{r+i}r(x) + Q(x)] \bmod g(x)\} + Q(x)\big) \bmod x^n - 1$$

5. If necessary, continue for other choices of $Q(x)$ until the test is satisfied. If the test is never satisfied, announce an uncorrectable error pattern.

5.5.1. Decoding the (23,12) Golay Code

Kasami found that the (23,12) Golay code can be easily decoded by this method using a set of only three covering polynomials $Q(x)$. These three polynomials plus all cyclic shifts of the received word allow us to represent all correctable error patterns as discussed previously.

We now describe in detail the operation of the Kasami decoder for the (23,12) Golay code. The reader will recall (Section 4.3.3) that the Golay code can be constructed with either of two degree-11 generator polynomials. For this discussion we assume the use of $g(x) = x^{11} + x^{10} + x^6 + x^5 + x^4 + x^2 + 1$. The decoding procedure is described as follows, with reference to Fig. 5.2.

1. To start, gates 1, 3, and 5 are opened and gates 2 and 4 are closed (data flows through an open gate). Then $s_0(x)$ is calculated by shifting the entire received word $r(x)$ into the shift register P, which calculates $x^{11}r(x) \bmod g(x)$. At the same time, the received word is stored in the 23-stage shift register SR and the previously stored word that has been corrected is read out through gate 5.

2. Next, gates 1, 4, and 5 are closed and gates 2 and 3 are opened. The contents of the syndrome register P [now $s_0(x)$] are tested for correctable error patterns. The covering polynomial $Q(x)$ is a polynomial of degree 11 or greater chosen so that $Q(x)$ agrees with $e(x)$ in its information positions. Now, $Q(x) \bmod g(x)$ gives the check bits resulting from encoding $Q(x)$ [regarding $Q(x)$ as an information polynomial]. If $Q(x)$ agrees with $x^{11}e(x) \bmod x^{23} - 1$ in its 12 information positions, then $s_0(x) + Q(x) \bmod g(x)$ gives the error polynomial in the check positions [recall the definition of $s(x)$] and $Q(x)$ gives the error polynomial in the information positions. Thus, this gives us the error polynomial or, to be exact,

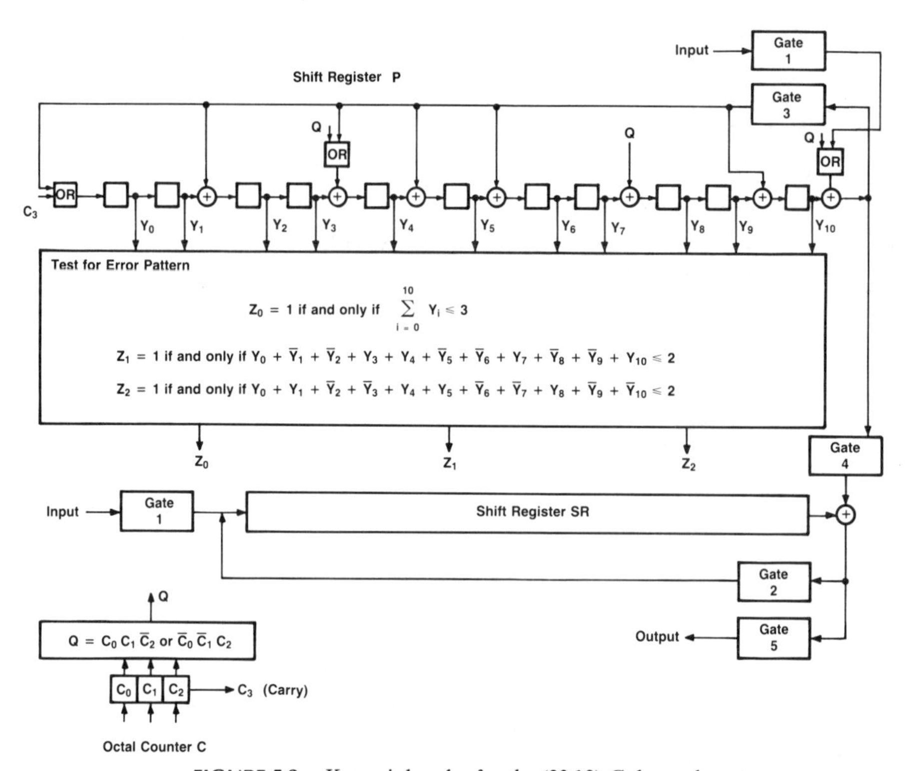

FIGURE 5.2. Kasami decoder for the (23,12) Golay code.

$x^{11}e(x) \bmod x^{23} - 1$, a cyclic shift of $e(x)$. The Kasami decoder for the (23,12) Golay code uses $Q_0(x) = 0$, $Q_1(x) = x^{16}$, and $Q_2(x) = x^{17}$. Agreement of $Q(x)$ with $x^{11}e(x) \bmod x^{23} - 1$ in its information positions is detected by the equation

$$\mathrm{Wt}[s_0(x) + Q(x) \bmod g(x)] + \mathrm{Wt}[Q(x)] \le 3, \qquad \text{for all } Q(x)$$

The test for Z_0 equal to 1 in Fig. 5.2 tells us if $Q_0(x)$ agrees with $x^{11}e(x) \bmod x^{23} - 1$ in its information positions. If $Z_0 = 1$, then $Q_0(x) \bmod g(x) = 0$, and the error polynomial in its check positions is equal to $s_0(x)$. We simply test the contents of the P register [which contains $s_0(x) = x^{11}r(x) \bmod g(x)$] for $\mathrm{Wt}[s_0(x)] \le 3$. If so, the error polynomial in the check positions is in the P register. Because of the 11-bit cyclic shift of $r(x)$, the P register actually contains $e_{12}, e_{13}, \ldots, e_{22}$ of the unshifted error polynomial. We now simply shift out the contents of P and SR one bit at a time. After the 11 bits of P have been shifted out, the received word has been corrected.

The test for Z_1 equal to 1 tells us if $Q_1(x)$ agrees with $x^{11}e(x) \bmod x^{23} - 1$ in its information positions. If it agrees, $Q_1(x) \bmod g(x) = x + x^2 + x^5 + x^6 + x^8 + x^9$. We test the contents of the P register by calculating $s_0(x) + Q_1(x) \bmod g(x)$ and seeing if $\mathrm{Wt}[s_0(x) + Q_1(x) \bmod g(x)] + \mathrm{Wt}[Q_1(x)] \le 3$

or $\mathrm{Wt}[s_0(x) + Q_1(x) \bmod g(x)] \le 2$. This calculation can be performed by testing if $Y_0 + \bar{Y}_1 + \bar{Y}_2 + Y_3 + Y_4 + \bar{Y}_5 + \bar{Y}_6 + Y_7 + \bar{Y}_8 + \bar{Y}_9 + Y_{10} \le 2$. In this calculation $Y_0, Y_1, \ldots, Y_{10}$ are the outputs tapped from shift register P as indicated in Fig. 5.2, the plus signs represent ordinary arithmetic addition, and the overbar represents complementation ($\bar{1} = 0, \bar{0} = 1$). As before, the error polynomial in the check positions of $x^{11}e(x) \bmod x^{23} - 1$ equals $s_0(x) + Q_1(x) \bmod g(x) = s_0(x) + x + x^2 + x^5 + x^6 + x^8 + x^9$. The error correction is performed as follows. The octal counter C is set to 2 and starts counting the clock pulses. Q takes the value 1 if and only if the count in C is 3 or 4. By these two additions (performed mod 2), those positions in $s_0(x)$ corresponding to ones in $Q_1(x) \bmod g(x)$ are complemented. The bit coming out of P, which indicates the error position, is added to the bit coming out of SR. The contents of SR are shifted out in step with P. When C_3, the carry of C_2, takes the value 1, the leftmost flip-flop of P is set to 1 and C stops its count. This corrects the error corresponding to position x^{16}.

The test for Z_2 equal to 1 tells us that if $Q_2(x)$ agrees with $x^{11}e(x) \bmod x^{23} - 1$ in its information positions. If it agrees, $Q_2(x) \bmod g(x) = x^2 + x^3 + x^6 + x^7 + x^9 + x^{10}$. We test the contents of the P register by calculating $s_0(x) + Q_2(x) \bmod g(x)$ and seeing if $\mathrm{Wt}[s_0(x) + Q_2(x) \bmod g(x)] + \mathrm{Wt}[Q_2(x)] \le 3$ or $\mathrm{Wt}[s_0(x) + Q_2(x) \bmod g(x)] \le 2$. This test may be performed by testing for $Y_0 + Y_1 + \bar{Y}_2 + \bar{Y}_3 + Y_4 + Y_5 + \bar{Y}_6 + \bar{Y}_7 + Y_8 + \bar{Y}_9 + \bar{Y}_{10} \le 2$. As before, the error polynomial in the check positions of $x^{11}e(x) \bmod x^{23} - 1$ equals $s_0(x) + Q_2(x) \bmod g(x) = s_0(x) + x^2 + x^3 + x^6 + x^7 + x^9 + x^{10}$. Error correction is performed as follows. The octal counter C is set to 3 and starts counting the clock pulses. Q takes the value 1 if and only if the count in C is 3 or 4. Q is added to Y_3, Y_7, and Y_{10}. By these two additions, those positions in $s_0(x)$ corresponding to ones in $Q_2(x) \bmod g(x)$ are complemented. The bit coming out of P, which indicates the error position, is added to the bit coming out of SR. The contents of SR are shifted out in step with P. When C_3, the carry of C_2, takes the value 1, the leftmost flip-flop of P is set to 1 and C stops its count. This results in the correction of the error position corresponding to x^{17}.

If none of Z_0, Z_1, and Z_2 equals 1, then P and SR are both cyclically shifted once. The P register now contains $s_1(x) = x^{12}r(x) \bmod g(x)$. The preceding tests for Z_0, Z_1, and Z_2 are now repeated. If Z_0, Z_1, or Z_2 now equals 1, we correct the stored codeword as before, but the contents of SR must be recycled to their original position before reading out through gate 5. A maximum of two complete cycles (46 shifts) of SR may be required. One of the Z tests will eventually succeed, since the Golay decoder will never perform error detection without correction (recall that the Golay code is a perfect code).

5.5.2. Decoding the (24,12) Extended Golay Code

The (24,12) code is obtained by attaching an overall parity-check bit to each codeword of the (23,12) Golay code. The additional check bit is simply the modulo-2 sum of the 12 information and 11 check bits of the (23,12) codeword. The bits of a (24,12) codeword thus satisfy the equation $c_0 + c_1 + \cdots + c_{23} = 0$ or $c_0 = c_1 + c_2 + \cdots c_{23}$. The minimum Hamming distance of the (24,12) code is 8, and the code is capable of correcting all error patterns of weight 3 or less and of detecting all error patterns of weight 4. The (24,12) code is not a cyclic code.

A procedure for applying the Kasami decoding algorithm to the (24,12) code is given as follows:

1. The 23 bits of the received word corresponding to a (23,12) codeword are read into the Kasami decoder and are decoded.
2. If the decoder corrects zero, one, or two errors in the 23 bit positions, the word is assumed to be correctly decoded and the overall parity bit is ignored.
3. If the decoder corrects three errors, the overall parity bit is checked. If the overall parity of the decoded codeword checks, the word is assumed to have been correctly decoded. If the overall parity check fails, then a detected error pattern of weight 4 is announced.

Steps 1 through 3 are not specific to the Kasami decoder and may be utilized with any bounded-distance algorithm. In addition, the procedure can be used with any code that has an overall parity-check bit appended. Specifically, if $l < (d - 1)/2$ errors are corrected in step 1, the parity-check bit is not used, since if it does not check, a bounded-distance decoder would conclude that $l + 1$ errors are present in the received word, the last error affecting the overall parity-check bit. Since this is a parity-check position, it need not be corrected. On the other hand, if $t = (d - 1)/2$ errors are corrected in step 1, and if the overall parity check fails, the decoder concludes that $t + 1$ or more errors have occurred, and error detection is announced.

5.6. ERRORS-AND-ERASURES DECODING

Up to this point, we have discussed the decoding problem primarily as one of finding the number and locations of errors in a received word. (When we consider nonbinary codes in chapter 6, we shall also have to be concerned with *error values*.) It has been assumed that at the receiving end of the communication circuit, a definite binary decision is made on each received digit after demodulation and prior to decoding, that is, a hard binary decision. However, as we pointed out in Chapter 1, it is often possible in designing a communication system to provide for quality or confidence estimates for demodulated data. In the simplest example of such schemes, we might test the demodulator output against a preselected magnitude threshold and erase each digit that falls below the threshold. The decoder is then presented with a sequence consisting of definite zeros and ones as well as erasures, and given that sequence, the decoder has the task of deciding which of the valid codewords is most likely to have been transmitted. We call this decoding task one of *errors-and-erasures decoding*.

In Chapter 2 we stated that a block code having minimum Hamming distance d is capable of correcting any pattern of t or fewer errors, where $d = 2t + 1$ or $2t + 2$, for d odd or even, respectively. We now state that a distance-d code is capable of correcting any pattern of l errors and s erasures such that $2l + s < d$, and we show a very simple procedure that demonstrates that this is true. Assuming that we have at our disposal a decoder for correcting up to t errors, where $2t + 1 = d$, we decode for s erasures and an unknown number of errors as follows:

1. Set all s erased bits in the received word equal to 0, and perform error correction of up to t errors. Note the number of errors corrected if decoding can be completed.
2. Next, set all s erased bits equal to 1, and decode the received word again, noting the number of errors corrected if decoding can be completed.
3. If both decodings succeed but produce different codewords, accept the decoding result that required correction of the smaller number of errors.

To explain the procedure, let us assume that the s erased bit positions consist of u positions in which the true (transmitted) bit value was 0 and v positions in which the true bit value was 1, so that $s = u + v$. Let us further assume that there are also l errors in the received word, where $2l + s < d$. In step 1, setting all s erasures equal to 0 results in a total of $l + v$ errors, while in step 2, setting all s erasures equal to 1 results in $l + u$ errors. Now we note that since u and v cannot both be greater than $s/2$, we must have either $(l + u) \le (d - 1)/2$ or $(l + v) \le (d - 1)/2$, and since the code can correct up to $(d - 1)/2$ errors, at least one of the error patterns is correctable. It is possible that only one of the decodings can be completed; for example, if $(l + v) > (d - 1)/2$, the corresponding error pattern might be uncorrectable. On the other hand, the weight-$(l + v)$ error pattern might result in an apparently successful, although in fact incorrect, decoding. However, let us examine this situation more closely. If one of the erasure-filled error patterns, say the one with $l + u$ errors, decodes correctly and the other, with $l + v$ errors, decodes to an incorrect codeword that is distance $d^{(v)}$ away from the transmitted word, then the number of errors (apparently) corrected in decoding the word with the weight-$(l + v)$ error pattern must be $d^{(v)} - (l + v)$. But note that

$$\begin{aligned}(l + u) - \left[d^{(v)} - (l + v)\right] &= 2l + u + v - d^{(v)} \\ &= 2l + s - d^{(v)} \\ &\le d - 1 - d^{(v)} \\ &\le -1\end{aligned}$$

since the distance $d^{(v)}$ to the incorrect codeword must be d or greater. Therefore, we see that if one decoding is correct and the other incorrect, the correct decoding will be identifiable because it will have a smaller number of apparent error corrections. If $l + u$ and $l + v$ are both less than $(d - 1)/2$, both decodings will succeed and will produce the same decoded word, although the number of errors corrected in the two cases may be different (for example, when s is odd.)

Thus, we have a simple procedure for accomplishing errors-and-erasures decoding of a binary block code using two steps of erasure filling with errors-only decoding. As long as the number of errors and erasures satisfies $2l + s < d$, there will be at least one correct decoding, and if there is one incorrect decoding, the correct decoding can be identified.

There are a number of algorithms that use more precise bit quality information than simple erasures; several of these will be discussed next.

5.7. SOFT-DECISION DECODING TECHNIQUES

The errors-and-erasures decoding procedure just described is an example of a general class of algorithms that are usually referred to as *soft-decision decoding* techniques. The simplest of such techniques is Wagner coding, which we mentioned briefly in Section 2.3.2. In this scheme, encoding is done by appending a single overall parity check to a block of k information bits. The decoding procedure can be described as follows. Upon reception of each received digit r_i, the a posteriori probabilities $p(0|r_i)$ and $p(1|r_i)$ are calculated and saved, and a hard bit decision is also made on each of the $k + 1$ digits. Overall parity is checked, and if it is satisfied, the k information bits are accepted as first decoded. If parity fails, the received digit having the smallest difference between its two a posteriori probabilities is inverted before the k information bits are accepted. It is seen that this technique is in fact the simplest application of the errors-and-erasures decoding procedure described in the previous section, where here only a single erasure may be filled but no errors corrected, since the minimum distance of the single-parity-check code is only 2.

A generalization of Wagner coding applicable to any multiple-error-correcting (n,k) code is a scheme called *forced-erasure decoding*. Here we assume that the demodulator, in addition to making a hard binary decision on each received digit, also measures relative reliability; we denote the set of reliability measures by $\rho_1, \rho_2, \ldots, \rho_n$. For many communication channels, the probability of correct bit demodulation is monotonically related to the magnitude of the detected signal, and in such cases the detected signal strength can be taken as a measure of reliability for each bit.

There are several decoding strategies that come under the heading of forced-erasure decoding; we shall describe one such strategy for the purpose of illustration. Let the n bits in the received block be tagged with their reliability measures $\rho_1, \rho_2, \ldots, \rho_n$, where $\rho_i > \rho_j$ implies that the ith bit is judged more likely to be correct than the jth bit. We first erase the bit having lowest reliability and attempt to satisfy all $n - k$ parity-check equations with either zero or one in the erased position. If this is impossible, we erase the two least reliable bits and try all four binary combinations in the erased positions, again seeking to find a combination that causes the overall word to satisfy all $n - k$ parity-check equations. If this fails, we erase the three least reliable bits, and so on. In general, letting s be the number of bit positions erased at each step, we try all 2^s binary patterns in erased positions and, failing to satisfy all parity-check equations, we increase s by one and try again, until at some step one of the following three conditions occurs:

1. One and only one of the 2^s patterns produces a word that satisfies the parity-check equations. The decoding procedure is completed.
2. Two or more of the 2^s erasure-filling patterns produce words that satisfy the parity-check equations. In this case, some procedure must be used to select from among the contending codewords. For example, one can use the erasure-filling pattern that has the fewest disagreements with the original hard bit decisions in its s positions. This is not an optimum strategy; more will be said about this later.
3. The number of bits erased is equal to $n - k$, and no pattern has been found that causes the parity checks to be satisfied. This is a decoding failure, and, depending upon system requirements, the k information bits in this case may be handled in various ways ranging from delivering the bits with their reliability measures and a "decoding failure" flag, to erasing all k bits permanently.

We already know from earlier discussion in this section that if we erase s bits and the remaining $n - s$ bits are demodulated without errors, then one of the 2^s patterns tested will uniquely satisfy the parity-check equations as long as $s \leq d - 1$, where d is the minimum distance of the code. In fact, it is possible to show that as the SNR is increased, the performance of the forced-erasure decoder will approach that of a hard-decision bounded-distance decoder capable of correcting $d - 1 = 2t$ errors in the n-bit block (d is assumed to be odd here). Therefore, loosely speaking, the use of soft-decision information with forced-erasure decoding makes it possible to nearly double the error-correction power of a code.

However, at intermediate levels of SNR, there will, in general, be a significant probability of unknown errors in the $n - s$ bit positions that are not erased; hence the allowance for more than $d - 1$ erasures in the algorithm. When $s > d - 1$, one cannot be sure that only one of the 2^s patterns tested will be found to satisfy the parity-check equations. For some sets of s erased positions, a unique decoding will be found, but for some other sets multiple decodings will be found. Some limited results have been obtained for selected classes of codes concerning the fraction of uniquely reconstructable erasure patterns given that all demodulation errors have occurred in the s erased positions [65]. The use of $s = n - k$ as the maximum number of bit positions erased in the forced-erasure decoding algorithm stems from a well-known result, that no (n,k) linear block code can uniquely reconstruct more than $n - k$ erasures. For $s > n - k$, multiple decodings are sure to be obtained, regardless of which s positions in the code block are erased. It is seen that if the full power of forced-erasure decoding is to be exploited, effective means must be provided to make the proper selection, in each decoding, from among the multiple codewords that can be produced by filling the erasures.

In a paper that unifies much of the theory of soft-decision decoding techniques, Chase [22] provides several useful bounds on probability of code block error achievable with soft-decision decoding techniques. In the cited paper, Chase describes a decoding technique termed *channel-measurement decoding*, which provides a means of making highly effective use of bit reliability information.

To describe this decoding concept, we assume the use of an arbitrary (n,k) binary block code having minimum distance d. Let $\mathbf{r}$ represent the block of hard-decision demodulated bits corresponding to the word that is received, and, for the purpose of an example, let us assume that $\mathbf{r}$ as received can be decoded by correction of t_A errors to a valid codeword $\mathbf{c}_A$, where $t_A \leq t = \lfloor (d-1)/2 \rfloor$, and $\lfloor x \rfloor$ denotes "integer part of x." In other words, $\mathbf{r}$, if decoded by bounded-distance hard-decision decoding, is found to be within the radius-t decoding sphere surrounding $\mathbf{c}_A$, and therefore the decoder produces $\mathbf{c}_A$ as the codeword at the smallest Hamming distance from the received word. The binary error pattern implied by this decoding is

$$\mathbf{e}_A = \mathbf{r} + \mathbf{c}_A \tag{5.13}$$

Now, the central idea behind channel-measurement decoding can be described as follows. While $\mathbf{e}_A$ is estimated to be the most likely error pattern contained in $\mathbf{r}$ by the criterion of smallest Hamming distance to a codeword, it is not necessarily the most likely if channel-measurement information is taken into account. For example, the bits changed (the ones in $\mathbf{e}_A$) in decoding to $\mathbf{c}_A$ might be bits of high reliability, while some of the bits left unchanged might be of relatively low reliability. Therefore, if we had an efficient procedure for probing a region of signal space surrounding the received word $\mathbf{r}$, we might well produce decodings to codewords other than $\mathbf{c}_A$, corresponding to estimated error patterns other than $\mathbf{e}_A$, where one or more decoded words have a higher overall reliability measure than $\mathbf{c}_A$.

The channel-measurement decoding technique provides the desired strategy. We outline the technique as follows. Upon reception of a word $\mathbf{r}$, make hard bit decisions and also quantify the reliability of each received bit using a decision statistic appropriate to the channel being used. Also, rank the received bits in accordance with their reliability measures. Next, erase a specified number (to be detailed below) of the least reliable bits, and create test patterns by filling the erased positions with all possible bit combinations. For each such test pattern, perform bounded-distance decoding of up $t = \lfloor (d-1)/2 \rfloor$ errors. The objective is to produce multiple decodings, each corresponding to a different estimated error pattern, and the decoded word is found that corresponds to the error pattern of least overall reliability. These ideas will be made clearer in the discussion that follows.

We denote a *test pattern* as an n-tuple $\mathbf{T}$ having ones in the erased bit positions to be filled with ones and zeros elsewhere. Thus, adding the test pattern $\mathbf{T}$ to the received hard-decision word $\mathbf{r}$ produces a new word

$$\mathbf{r}' = \mathbf{r} + \mathbf{T}$$

where $\mathbf{r}'$ is simply a perturbed version of $\mathbf{r}$ in which hard bit decisions have been inverted at the positions corresponding to the ones in $\mathbf{T}$. Let us now say that one such test pattern, $\mathbf{T}_i$, produces a word $\mathbf{r}_i'$, which in turn decodes, by correction of a pattern of t or fewer errors, to a codeword $\mathbf{c}_B$, different from $\mathbf{c}_A$, the codeword obtained from $\mathbf{r}$ with strictly hard-decision decoding. The error pattern $\mathbf{e}_B$, relative to $\mathbf{r}$, implied in this decoding is

$$\mathbf{e}_B = \mathbf{r} + \mathbf{c}_B$$

which can be related to the error pattern $\mathbf{e}_A$ by writing [using Eq. (5.13)]

$$\mathbf{e}_B = \mathbf{c}_A + \mathbf{c}_B + \mathbf{e}_A \tag{5.14}$$

Therefore, given that $\mathbf{c}_A$ and $\mathbf{c}_B$ are different codewords, the implied pattern $\mathbf{e}_B$ is seen to be different from the error pattern $\mathbf{e}_A$. A two-dimensional geometric interpretation of the decoding technique is shown in Fig. 5.3. The figure shows

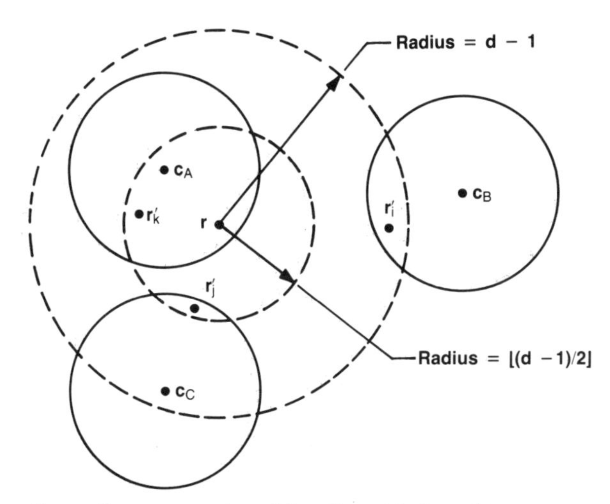

FIGURE 5.3. Geometric representation of decoding with channel-measurement information.

three codewords, $\mathbf{c}_A$, $\mathbf{c}_B$, and $\mathbf{c}_C$, each surrounded by a sphere of radius $t = \lfloor (d-1)/2 \rfloor$. The decoding of the hard-decision received word $\mathbf{r}$ to codeword $\mathbf{c}_A$ is indicated, as is the decoding of $\mathbf{r}_i'$ to codeword $\mathbf{c}_B$. The figure also indicates another perturbed word, $\mathbf{r}_j'$, produced by a test pattern $\mathbf{T}_j$, and its decoding to codeword $\mathbf{c}_C$. We also indicate a perturbed word $\mathbf{r}_k'$, produced by a test pattern $\mathbf{T}_k$, and its decoding to $\mathbf{c}_A$. (An example of this last case is a test pattern $\mathbf{T}_k$ that inverts some of the errors in $\mathbf{r}$ without inserting any new errors.)

Thus we see that perturbing the received word $\mathbf{r}$ with test patterns in a sphere surrounding $\mathbf{r}$ can provide multiple decodings, some of which may yield codewords different from the codeword that is the nearest neighbor, in the Hamming distance sense, of the unperturbed word $\mathbf{r}$. We have also seen that decodings to different codewords imply different error patterns. The channel-measurement decoder will select from among the multiple decodings the codeword output $\mathbf{c}$ that satisfies the minimization

$$\min_m W_\rho(\mathbf{r} + \mathbf{c}_m) \tag{5.15}$$

where $\mathbf{r} + \mathbf{c}_m = \mathbf{e}_m$ is the error pattern implied by the decoding to $\mathbf{c}_m$ and $W_\rho(\mathbf{e}_m)$ is the *analog weight* of $\mathbf{e}_m$, defined as

$$W_\rho(\mathbf{e}_m) = \sum_{i=1}^{n} \rho_i e_{mi} \tag{5.16}$$

where e_{mi} denotes the ith bit in the error pattern $\mathbf{e}_m$ and ρ_i is the reliability measure for the ith received bit. It can be shown that if the minimization in Eq. (5.15) were to be done over all 2^k codewords, with the reliability measures in Eq. (5.16) generated by optimum bit detection, the algorithm would in fact accomplish maximum likelihood decoding [22]. As a practical matter, however, one is interested in having a means for testing subsets of the full set of 2^k codewords, where the subsets are chosen to include the most likely codewords given the received word $\mathbf{r}$. For the class of algorithms to be defined here, the received word $\mathbf{r}$ will be perturbed by various amounts within the sphere of radius $d - 1$ surrounding $\mathbf{r}$. This sphere is indicated in Fig. 5.3.

A flowchart for channel-measurement decoding is shown in Fig. 5.4. Chase defined three algorithms for channel-measurement decoding, that all correspond to Fig. 5.4, but differ in the number of test patterns used. The three algorithms are summarized below.

Summary of Channel-Measurement Decoding Algorithms*

Algorithm 1. Test all error patterns within the sphere of radius $d - 1$ around the received word $\mathbf{r}$.

*Adapted from: Chase, D., *IEEE Trans. Inf. Theory*, *IT-18*, 170–182 (1972).

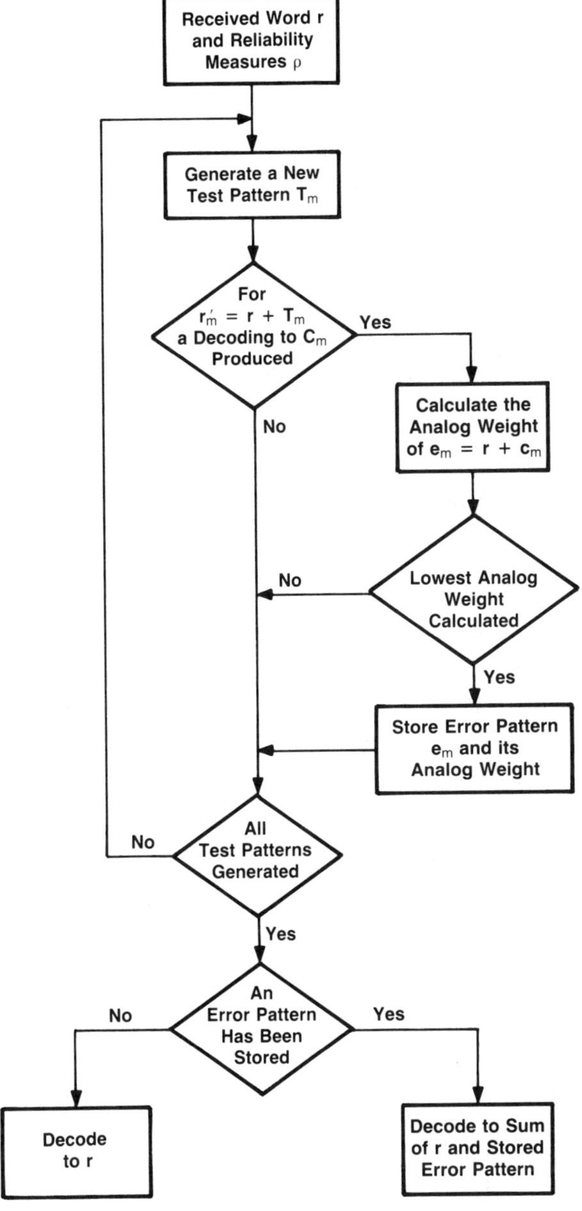

FIGURE 5.4. Flow chart for decoding with channel-measurement information.

Comments. Since the hard-decision decoder can obtain error patterns of weight up to $\lfloor (d-1)/2 \rfloor$, it is sufficient to generate all test patterns having weight $\lfloor d/2 \rfloor$ or less. At most $\binom{n}{\lfloor d/2 \rfloor}$ test patterns need be generated. However, considerable reduction in the number of test patterns can be achieved by eliminating patterns that yield identical error patterns.

Algorithm 2. Test only those error patterns with at most $\lfloor (d-1)/2 \rfloor$ errors located outside the set of bits having the $\lfloor d/2 \rfloor$ lowest reliabilities.

Comments. The error patterns produced have at most $d-1$ errors, but not all patterns of weight $d-1$ or less are generated. It is sufficient to generate the $2^{\lfloor d/2 \rfloor}$ test patterns consisting of all bit combinations in the $\lfloor d/2 \rfloor$ positions with lowest reliability. Considerable reduction in complexity is achieved by eliminating test patterns yielding identical error patterns.

Algorithm 3. Test only those error patterns generated from test patterns having i ones in the i least reliable bit positions, with $i = 1, 3, 5, \ldots, d-1$ for d odd.

Comments. Almost identical to Algorithm 2, except that the number of test patterns is only $\lfloor (d/2) + 1 \rfloor$.

The first of the three algorithms is the most powerful and requires the greatest amount of computation, since it uses as many as $\binom{n}{\lfloor d/2 \rfloor}$ test patterns to generate all possible error patterns in the sphere of radius $d-1$ or less surrounding $\mathbf{r}$. It is shown in reference [22] that for application to the steady-signal Gaussian noise channel and certain fading channels, Algorithm 1 comes very close to achieving the performance of the maximum likelihood decoder. However, because of the computational requirements, the use of Algorithm 1 is unattractive except for codes with relatively small minimum distance and block length. Furthermore, it has been shown that the simpler approaches can be used with relatively little sacrifice in performance.

Algorithm 2 uses considerably fewer test patterns than does Algorithm 1; however, the number of patterns to be generated still grows rapidly with d, and thus its use is, as a practical matter, ruled out for codes of large minimum distance. However, Algorithm 3, though it suffers some performance degradation relative to the other two, can be considered for application even for codes with large minimum distance, since the number of test patterns to be generated grows only linearly with d.

Chase's Algorithm 3 is closely related to a soft-decision decoding technique developed by Forney [37, 38], called *generalized minimum-distance decoding*, often abbreviated as *GMD decoding*. In his original work, Forney considered the application of GMD decoding to nonbinary as well as binary codes. Forney also considered more complex schemes using errors-and-erasures decoding of nonbinary codes [39].

The performance improvements achievable with various forms of erasure-filling and soft-decision decoding, relative to bounded-distance decoding, depend strongly upon the characteristics of the channel on which the coding system is utilized. Therefore, it is not possible to give precise rules as to the gains achievable with a particular decoding strategy for an arbitrary channel. Certain general guidelines can be given, however. A key point to be made here is that the most powerful of the soft-decision decoding techniques, such as Chase's channel-measurement algorithms and Forney's GMD algorithms, approach, with increasing complexity, ideal maximum likelihood decoding. These relationships are brought out in some detail in reference 22, where it is shown that as the SNR is increased, the channel-measurement decoding algorithms outlined above exhibit the same behavior as maximum likelihood decoders. It is further shown that the asymptotic performance of the algorithms closely corresponds to the performance that would be achieved with hard-decision decoding if $d - 1$ errors in each codeword were corrected rather than $\lfloor (d-1)/2 \rfloor$ errors. For the case of the AWGN channel, this leads in a straightforward way to showing that in the limit of very high SNR, the asymptotic improvement for channel-measurement decoding versus conventional bounded-distance error correction is $d/(t+1)$, where t is the error-correction limit of the code. Therefore, for even minimum distance, the asymptotic SNR improvement on the Gaussian channel is exactly 3 dB. However, experience with the application of a variety of soft-decision decoding techniques in practical systems for the Gaussian channel shows that actual improvements of 1 to 2 dB are feasible with algorithms of reasonable complexity.

For channels other than the steady-signal Gaussian noise channel, significantly larger gains in SNR can be realized. For example, Chase shows that in the case of independent Rayleigh fading, the improvement is equivalent to the change in performance of a $(t + 1)$-order diversity receiving system to performance of a d-order diversity system.

5.8. NOTES

It has been pointed out by Wolf [179] that Peterson's method for decoding BCH codes is closely related to a curve-fitting procedure described by Prony in 1795. In the Prony method, a finite sum of exponential terms is made equal, at uniformly spaced sample intervals, to a given set of data samples. The exponents and amplitudes of the terms must be determined, and this requires solving a set of simultaneous equations that are nonlinear in the exponents. Prony's solution provides the exponents and amplitudes in the same manner as Peterson's decoding procedure provides error locators and error values, respectively.

Various forms of error-trapping decoding have been devised for use on burst-error channels. In this type of application, error trapping can provide a very simple approach to decoding, consisting simply in calculating and testing a set of shifted syndrome polynomials without having to use covering polynomials.

Many codes have been designed specifically for correcting errors occurring in dense bursts or clusters. Such codes are called *burst-error-correcting codes*. A burst-error-correcting code will in general correct more errors, occurring in bursts of some maximum length, than a random-error correcting code with the same block length and the same number of parity-check bits. The use of cyclic codes for burst-error correction was first studied by Abramson [1, 2]. Subsequently Fire discovered a large class of burst-error-correcting codes, long known as *Fire codes* [34]. Much other work has followed, including the generation by computer of extensive lists of good burst-error-correcting codes. Detailed discussions of burst-error-correcting codes and methods for decoding them can be found in Peterson and Weldon [131] and Lin and Costello [94].

■ ■ CHAPTER SIX

Nonbinary BCH Codes and Reed-Solomon Codes

This chapter deals with two important classes of cyclic nonbinary linear block codes: nonbinary BCH codes and Reed-Solomon codes, which are defined as a subclass of the BCH codes. The material is covered in a compact presentation, with code structure, encoding, and decoding all being treated. Certain important modifications of the BCH and Reed-Solomon codes are also discussed. The approach to describing the structure of the nonbinary codes closely parallels that used in Chapter 4, and reference is made to that chapter where appropriate.

6.1. ALGEBRA FOR NONBINARY CODES

In our discussion of nonbinary block codes, we again make use of the polynomial representation of codewords, so that a word of length n is written as

$$c(x) = c_0x^0 + c_1x^1 + c_2x^2 + \cdots + c_{n-1}x^{n-1}$$

where now the polynomial coefficients $\{c_i\}$ are taken from a nonbinary alphabet, specifically the finite field $GF(q)$, where q is any prime raised to an integer power. We confine our attention to cyclic codes, so that the various properties of binary cyclic codes, such as their definition in terms of generator polynomials and their roots in extension fields, carry over to the present discussion with the generalization that the field of coefficients is not restricted to be $GF(2)$.

The description of a nonbinary cyclic code follows directly from the binary case. That is, an (n, k) cyclic code on $GF(q)$ can be generated as the set of all polynomials of the form $a(x)g(x)$ where $a(x)$ is any polynomial of degree $k - 1$ or less with coefficients in $GF(q)$ and the generator polynomial $g(x)$ divides $x^n - 1$ and has coefficients in $GF(q)$. As in the binary case, we shall see that the design of a nonbinary code rests upon selection of a generator polynomial having prescribed roots in a field that is an extension of $GF(q)$, say $GF(q^m)$. Therefore, in order to deal with the various mathematical manipulations involved in encoding and decoding, we require a consistent algebra for operations in $GF(q)$ and $GF(q^m)$.

For example, let us construct the field $GF(9)$ using an irreducible polynomial of degree 2 having coefficients in $GF(3)$. This is called a *quadratic extension* of $GF(3)$. We let the elements of $GF(3)$ be 0, 1, and 2. Addition and multiplication are then simply ordinary addition and multiplication modulo 3. The polynomial $f(x) = x^2 + 2x + 2$ is irreducible in $GF(3)$, which is easily verified by seeing that $f(x)$ evaluated at 0, 1, and 2 equals, respectively, 2, 2, and 1 mod 3. We can now generate the eight nonzero elements of $GF(9)$ as the set of polynomials on $GF(3)$ reduced modulo $x^2 + 2x + 2$, where the reduction is equivalent to setting $x^2 = -2x - 2 = x + 1$. The reduction, in turn, is equivalent to solving $x^2 + 2x + 2$ for one of its roots, so that we can represent field elements as functions of a root, say α, which we do in the left-hand and center columns of Table 6.1. The shorthand vector representations of the field elements are given in the right-hand column. It is seen that the table repeats after eight steps. Therefore α is a primitive root and $x^2 + 2x + 2$ is a primitive polynomial. The addition of any two elements in $GF(9)$ is accomplished by adding the corresponding polynomial or vector representations, term by term, using

TABLE 6.1. A Representation of $GF(9)$ Using a Root of $x^2 + 2x + 2$ over $GF(3)$

Zero and Powers of α		Polynomials over $GF(3)$		Vectors over $GF(3)$
0	=	0	=	00
α^0	=	1	=	01
α^1	=	α	=	10
α^2	=	$\alpha + 1$	=	11
α^3	=	$2\alpha + 1$	=	21
α^4	=	2	=	02
α^5	=	2α	=	20
α^6	=	$2\alpha + 2$	=	22
α^7	=	$\alpha + 2$	=	12

modulo-3 addition. Multiplication of nonzero elements may be done using the table by addition of corresponding powers of the primitive element α or equivalently by multiplication of the corresponding polynomials and reduction of the product modulo $\alpha^2 + 2\alpha + 2$.

As an exercise the reader should show that such a table constructed in the same manner using the irreducible polynomial $x^2 + 1$ repeats after four steps; this is due to the fact that $x^2 + 1$ is nonprimitive, its roots all being of order 4.

To generate an extension field $GF(q^m)$ where q is not a prime but rather an integer power of a prime (that is $q = p^s$ with s greater than 1), we use the same procedure as before. That is, we generate the nonzero elements of $GF(q^m)$ as polynomials on $GF(p^s)$ reduced using a primitive polynomial of degree m with coefficients in $GF(p^s)$. The algebraic operations on the polynomials representing elements of $GF(q^m)$ requires addition and multiplication of coefficients in $GF(p^s)$, and these in turn are facilitated by a table generated with a second primitive polynomial, one of degree s with coefficients in $GF(p)$. This will be made clearer with another example.

Consider the polynomial $f(x) = x^2 + x + A$ with coefficients in $GF(4)$, where we denote the elements of $GF(4)$ as $0, 1, A$, and B. The polynomial $f(x)$ is irreducible in $GF(4)$. To show this we need a representation for $GF(4)$, which we obtain from $x^2 + x + 1$, which is irreducible in $GF(2)$. Letting $0 = 0, 1 = x^0$, $A = x$, and $B = x^2 = x + 1$, we readily construct addition and

TABLE 6.2. A Representation of $GF(16)$ Using a Root of $x^2 + x + A$ over $GF(4)$

Zero and Powers of α		Polynomials over $GF(4)$		Vectors over $GF(4)$
0	=	0	=	00
α^0	=	1	=	01
α^1	=	α	=	10
α^2	=	$\alpha + A$	=	$1A$
α^3	=	$B\alpha + A$	=	BA
α^4	=	$\alpha + 1$	=	11
α^5	=	A	=	$0A$
α^6	=	$A\alpha$	=	$A0$
α^7	=	$A\alpha + B$	=	AB
α^8	=	$\alpha + B$	=	$1B$
α^9	=	$A\alpha + A$	=	AA
α^{10}	=	B	=	$0B$
α^{11}	=	$B\alpha$	=	$B0$
α^{12}	=	$B\alpha + 1$	=	$B1$
α^{13}	=	$A\alpha + 1$	=	$A1$
α^{14}	=	$B\alpha + B$	=	BB

multiplication tables for $GF(4)$ as follows:

Addition Table for $GF(4)$

+	0	1	A	B
0	0	1	A	B
1	1	0	B	A
A	A	B	0	1
B	B	A	1	0

Multiplication Table for $GF(4)$

•	0	1	A	B
0	0	0	0	0
1	0	1	A	B
A	0	A	B	1
B	0	B	1	A

We can now verify that $f(x) = x^2 + x + A$ is irreducible in $GF(4)$ by evaluating $f(x)$ at $x = 0$, 1, A, and B, which yields A, A, B, and B, respectively. The nonzero elements of the field $GF(16)$ are now represented as polynomials on $GF(4)$ reduced modulo $x^2 + x + A$, as shown in Table 6.2.

The right-hand column in Table 6.2 gives a representation for each element of $GF(16)$ as a 2-tuple in $GF(4)$, but note that each such 2-tuple can be written

TABLE 6.3. Some Primitive Irreducible Polynomials over Nonbinary Fields

Field	Polynomial
$GF(2)$	$x + 1$
	$x^2 + x + 1$
	$x^3 + x + 1$
	$x^4 + x + 1$
	$x^5 + x^2 + 1$
	$x^6 + x + 1$
$GF(3)$	$x + 1$
	$x^2 + x + 2$
	$x^3 + 2x + 1$
	$x^4 + x + 2$
	$x^5 + 2x + 1$
	$x^6 + x + 2$
$GF(5)$	$x + 1$
	$x^2 + x + 2$
	$x^3 + 3x + 2$
	$x^4 + x^2 + 2x + 2$
	$x^5 + 4x + 2$
$GF(7)$	$x + 1$
	$x^2 + x + 3$
	$x^3 + 3x + 2$
	$x^4 + x^2 + 3x + 5$

as a 4-tuple in $GF(2)$, simply by replacing each element in $GF(4)$ by its representation in $GF(2)$. For example, the element α^3 in Table 6.2, represented by $(BA)_4$, can be rewritten by replacing A with $\gamma = (10)_2$ and B with $\gamma + 1 = (11)_2$, where γ is a primitive element of $GF(4)$, so that α^3 becomes

$$\alpha^3 = (BA)_4 = (1110)_2$$

We make this connection among the representations for $GF(2), GF(4)$, and $GF(16)$ to show a method of representation of $GF(16)$ that is an alternative to using an irreducible polynomial of degree 4 with coefficients in $GF(2)$.

Tables of irreducible polynomials over prime fields $GF(p)$, $p = 2, 3, 5, 7$, and 11 and prime-power fields $GF(q)$, $q = 4, 8, 9$, and 16 are found in Dickson [30], Church [26], and Green and Taylor [58]. A brief list of irreducible polynomials over various nonbinary fields is given in Table 6.3. As was mentioned earlier in the book, an extensive table of irreducible polynomials over $GF(2)$, previously published in Peterson and Weldon [131], is reprinted here in Appendix B.

6.2. MINIMAL POLYNOMIALS OVER $GF(q)$

For each element β in the extension field $GF(q^m)$, we can define the minimal polynomial of β as the monic polynomial of least degree with coefficients in $GF(q)$ having β as a root. We discussed minimal polynomials in some detail for the special case $q = 2$ in Section 4.2.2, and noted that minimal polynomials for specific elements of $GF(2^m)$ are identified in the published tables of irreducible polynomials over $GF(2)$. Such convenient identification of minimal polynomials has typically not been provided in the published lists of irreducible polynomials over nonbinary fields; therefore it is useful to show how minimal polynomials can be constructed.

As an example, let us find the minimal polynomials of the elements of $GF(3^2)$. To do this we use a generalization of property 6 of irreducible polynomials given in Section 4.2.1, which lets us enumerate the roots of an irreducible polynomial when one root is known. Specifically, if $f(x)$ is irreducible over $GF(q)$ with degree m and has a root β in the extension field $GF(q^m)$, then $\beta, \beta^q, \beta^{q^2}, \ldots, \beta^{q^{m-1}}$ are all of the roots of $f(x)$.

We can immediately write down the minimal polynomials of the zero and unity elements of the field $GF(3^2)$, which are, respectively, $m_0(x) = x$ and $m_{\alpha^0}(x) = x - 1 = x + 2$, where $\alpha^0 = 1$ and α is a primitive element of $GF(3^2)$. To find $m_\alpha(x)$, the minimal polynomial of α, we use a generalization of property 4 of minimal polynomials, given in Section 4.2.2, which tells us that a minimal polynomial of a primitive element of an extension field $GF(q^m)$ has degree m and is a primitive polynomial over $GF(q)$. Therefore, in the present example, the minimal polynomial of α must be a primitive polynomial of degree 2 over $GF(3)$. One such polynomial is $x^2 + 2x + 2$, which is used to generate $GF(3^2)$ in Table 6.1.

We can now perform a check on the consistency of Table 6.1, as follows. By generalization of property 6, stated two paragraphs earlier, we see that $m_\alpha(x)$ has α and α^3 as its roots. This is easily verified by writing $m_\alpha(x)$ as

$$m_\alpha(x) = (x - \alpha)(x - \alpha^3)$$

$$= x^2 - (\alpha + \alpha^3)x + \alpha^4$$

To represent the coefficients of $m_\alpha(x)$ in $GF(3)$, we use Table 6.1, which gives us $-(\alpha + \alpha^3) = 2$ and $\alpha^4 = 2$, so that the minimal polynomial of α is $m_\alpha(x) = x^2 + 2x + 2$, which is the primitive polynomial that was used to construct Table 6.1.

We can find the minimal polynomials for other elements of $GF(3^2)$ in a similar manner. The minimal polynomial of α^2 has as its roots α^2 and α^6, so we write, with the aid of Table 6.1,

$$m_{\alpha^2}(x) = (x - \alpha^2)(x - \alpha^6)$$

$$= x^2 - (\alpha^2 + \alpha^6)x + \alpha^8$$

$$= x^2 + 1$$

The minimal polynomial of α^4 has only one root, since $(\alpha^4)^3 = \alpha^{12} = \alpha^4$, so we have

$$m_{\alpha^4}(x) = x - \alpha^4 = x + 1$$

Calculation of $m_{\alpha^5}(x)$, which is left to the reader, completes the set of minimal polynomials of the elements of $GF(9)$, and the results are summarized in Table 6.4.

In a similar manner, we can generate the minimal polynomials of the elements of $GF(16)$ by taking $x^2 + x + A$, used in generating Table 6.2, as the

TABLE 6.4. The Minimal Polynomials of the Elements of *GF*(9) Constructed over *GF*(3)

Elements in $GF(9)$	Minimal Polynomials over $GF(3)$
0	x
1	$x + 2$
α, α^3	$x^2 + 2x + 2$
α^2, α^6	$x^2 + 1$
α^4	$x + 1$
α^5, α^7	$x^2 + x + 2$

TABLE 6.5. The Minimal Polynomials of the Elements of $GF(16)$ Constructed over $GF(4)$

Elements in $GF(16)$	Minimal Polynomials over $GF(4)$
0	x
1	$x + 1$
α, α^4	$x^2 + x + A$
α^2, α^8	$x^2 + x + B$
α^3, α^{12}	$x^2 + Bx + 1$
α^5	$x + A$
α^6, α^9	$x^2 + Ax + 1$
α^7, α^{13}	$x^2 + Ax + A$
α^{10}	$x + B$
α^{11}, α^{14}	$x^2 + Bx + B$

minimal polynomial of α. Then, using Table 6.2, we can calculate the coefficients of all remaining minimal polynomials. The resulting set of minimal polynomials is given in Table 6.5.

6.3. NONBINARY BCH CODES

The binary BCH codes are a special case of a class of cyclic codes that can be constructed for any symbol alphabet defined on a finite field, say $GF(q)$, which can be a prime field or some extension of a prime field. As a generalization of the binary case, a t-error-correcting BCH code on $GF(q)$ is a cyclic code, and all the codewords have roots that include $2t$ consecutive powers of some element β contained in $GF(q^m)$, an extension field of $GF(q)$. It will be convenient to distinguish between the two fields by calling $GF(q)$ the *symbol field* and $GF(q^m)$ the *locator field*. As with the binary codes, BCH codes on $GF(q)$ can be primitive or nonprimitive, depending on whether a primitive or nonprimitive element of $GF(q^m)$ is used to specify the consecutive roots of the codewords. For the present discussion, let us restrict our attention to the case of primitive codes, so that the code is specified to be a set of code polynomials whose roots include the elements $\alpha, \alpha^2, \ldots, \alpha^{2t}$, where α is a primitive element of $GF(q^m)$. The design distance of the code is one greater than the number of consecutive roots, and the true minimum distance can be equal to or greater than the design distance. The generator polynomial of a BCH code on $GF(q)$ is defined as the least common multiple of the minimal polynomials of $\alpha, \alpha^2, \ldots, \alpha^{2t}$, that is,

$$g(x) = \text{LCM}[m_{\alpha^1}(x), m_{\alpha^2}(x), \ldots, m_{\alpha^{2t}}(x)]$$

The block length of the code is the order of the element chosen to prescribe the consecutive roots, and therefore for the primitive codes, where we choose a primitive element of $GF(q^m)$, the block length is $n = q^m - 1$.

As we shall see shortly, a t-error-correcting code may have either odd or even minimum design distance, given by $d = 2t + 1$ or $d = 2t + 2$, respectively. Furthermore, the sequence of powers of α can begin with an arbitrary power, say m_0, so that we can specify the roots as $\alpha^{m_0}, \alpha^{m_0+1}, \ldots, \alpha^{m_0+d-2}$, that is, $d - 1$ consecutive powers of α. Similarly, we can define the generator polynomial as

$$g(x) = \text{LCM}[m_{\alpha^{m_0}}(x), m_{\alpha^{m_0+1}}(x), \ldots, m_{\alpha^{m_0+d-2}}(x)] \tag{6.1}$$

6.3.1. Some Examples of Primitive Codes

We can show with an example how the generator polynomial for a required code is constructed. Let us consider codes defined on the symbol field $GF(3)$ having roots in $GF(3^3)$. If we specify the roots as consecutive powers of a primitive element in $GF(3^3)$, we will have block length $n = 3^3 - 1 = 26$. Now we know that the minimal polynomial of α will have α, α^3, and α^9 as roots; the minimal polynomial of α^2 will have α^2, α^6, and α^{18} as roots, and so on. Therefore we can easily make a list of the roots of the minimal polynomials of elements of $GF(3^3)$ as shown in Table 6.6. (As an aid to readability, the table simply gives the exponents of α.) From this list we can quickly determine the minimal polynomials that must be multiplied together to form the generator polynomial for a code with symbols in $GF(3)$ having block length 26 and a particular error-correction power. If a single-error-correcting code is required, the minimum distance of the code must be 3 or 4, and thus $g(x)$ must have either two or three consecutive powers of α as roots. For example, we might

TABLE 6.6. An Enumeration of the Nonzero Elements of $GF(27)$ According to Their Minimal Polynomials over $GF(3)$

Exponents of Roots of $m(x)$	Degree of $m(x)$
0	1
1, 3, 9	3
2, 6, 18	3
4, 12, 10	3
5, 15, 19	3
7, 21, 11	3
8, 24, 20	3
13	1
14, 16, 22	3
17, 25, 23	3

choose as roots the elements α^0 and α, so that, using Eq. (6.1), the generator polynomial is given by

$$g(x) = m_{\alpha^0}(x)m_{\alpha}(x)$$

Since $m_{\alpha^0}(x)$ has degree 1 and $m_{\alpha}(x)$ has degree 3, the degree of $g(x)$ is 4, yielding a (26,22) distance-3 code. Alternatively, we might choose α and α^2 as the code roots, giving the generator polynomial

$$g(x) = m_{\alpha}(x)m_{\alpha^2}(x)$$

which, from Table 6.6, is a polynomial of degree 6. Note, however, that the roots of $m_{\alpha}(x)$ and $m_{\alpha^2}(x)$ taken together actually include three consecutive powers of α, namely α, α^2, and α^3. Therefore, this choice of roots yields a (26, 20) code of design distance 4 capable of correcting all one-error patterns and detecting all two-error patterns.

Examination of Table 6.6 provides other choices for the code roots similar to the examples just given. For example, the use of α^{13} and α^{14} yields another (26, 22) design-distance-3 code, while the use of α^{14}, α^{15}, and α^{16} provides a (26, 20) design-distance-4 code. As a final example, we might use α^4 and α^5, giving us a (26, 20) design-distance-3 code. In general, with a list such as Table 6.6 one can quickly identify the set of distinct minimal polynomials that has in its aggregate set of roots the required consecutive sequence. It should be evident from the examples just cited that a judicious choice of code roots can minimize the number of parity-check symbols required for a desired design distance.

Once the desired roots and set of minimal polynomials have been identified, the generator polynomial of a code can be obtained as follows. As in Section 6.2, we select an irreducible polynomial on $GF(q)$ and form the extension field $GF(q^m)$. Then, using $GF(q^m)$ arithmetic, we find the minimal polynomials and multiply them together to form the code generator polynomial. Note that the coefficients of the generator polynomial are in $GF(q)$. For example, a block-length-8 distance-5 BCH code on $GF(3)$ can be constructed using Table 6.4. The generator polynomial is $\text{LCM}[m_{\alpha}(x), m_{\alpha^2}(x), m_{\alpha^4}(x)]$ giving

$$g(x) = (x^2 + 2x + 2)(x^2 + 1)(x + 1)$$

$$= x^5 + 2x^3 + 2x^2 + x + 2$$

The degree of $g(x)$ is 5, and therefore we have an (8,3) code on $GF(3)$.

A potential problem with using the stated procedure, though, is the need for the irreducible polynomial on $GF(q)$ to construct the extension field $GF(q^m)$. For large q, the required irreducible polynomial may not be readily available. In that case we may start with a convenient representation for $GF(q^m)$ and find the subfield $GF(q)$. With consistent representations for the symbol and locator fields, it is then a straightforward matter to express the locator field as an extension of the symbol field. As an example we consider a primitive distance-4 code defined on $GF(64)$ with block length 4095.

As we have seen, a nonbinary BCH code with block length $n = q^2 - 1$ may be defined on a q-ary symbol field where, for the current example, $q = 64$. The locator field of the code is $GF(4096)$. Encoding may be implemented with $GF(64)$ arithmetic, and $GF(4096)$ can be represented as a quadratic extension of $GF(64)$.

The coefficients of the codewords are to be in $GF(64)$, a subfield of $GF(4096)$. In order to construct the code, we start by specifying a particular representation for $GF(4096)$. We consider the primitive binary polynomial

$$f(x) = x^{12} + x^6 + x^4 + x + 1$$

Since $f(x)$ is primitive, it has at least one primitive root, α, and we can write

$$\alpha^{12} = \alpha^6 + \alpha^4 + \alpha + 1$$

The field $GF(4096)$ may then be defined as the set of binary polynomials of degree up to 11 with coefficients in $GF(2)$. That is, we can write

$$\begin{aligned}
\alpha^0 &= 1 \\
\alpha &= \alpha \\
&\vdots \\
\alpha^{11} &= \alpha^{11} \\
\alpha^{12} &= \alpha^6 + \alpha^4 + \alpha + 1 \\
\alpha^{13} &= \alpha^7 + \alpha^5 + \alpha^2 + \alpha \\
&\vdots \\
\alpha^{4094} &= \alpha^{11} + \alpha^5 + \alpha^3 + 1 \\
\alpha^{4095} &= 1
\end{aligned}$$

Next, in order to provide a self-consistent algebraic system, we construct the symbol field $GF(64)$ as a subfield of the locator field $GF(4096)$. The subfield of $GF(4096)$ containing 64 elements may be obtained as follows. Let β be a primitive element of the 64-ary subfield. Then

$$\beta = \alpha^j$$

where j is to be determined. We observe that

$$\beta^{63} = 1 = \alpha^{4095}$$

and thus $\beta = \alpha^{65k}$, where k is any integer relatively prime to 4095. A convenient choice is $k = 1$, so that we have

$$\beta = \alpha^{65}$$

Therefore, the 64-ary subfield contains the elements, 0, $\beta^0 = \alpha^0$, and α^{65i}, $1 \le i \le 62$.

For implementation purposes it is more convenient to represent $GF(4096)$ as a quadratic extension of $GF(64)$. Suppose we have a representation for $GF(64)$. Then the polynomial

$$p(x) = x^2 + \beta^i x + \beta^j$$

is required where $p(\alpha) = 0$ and $p(x)$ is irreducible over $GF(64)$. The elements of $GF(4096)$ can then be represented as the polynomials of degree 1 or 0 with coefficients in $GF(64)$.

We next apply again the generalization of property 6 of irreducible polynomials, given for the binary case in Section 4.2.1 and restated for the nonbinary case in Section 6.2. Note that since α is a root of $p(x)$, α^{64} is also a root, and we have

$$\begin{aligned} p(x) &= (x + \alpha)(x + \alpha^{64}) \\ &= x^2 + (\alpha + \alpha^{64})x + \alpha^{65} \end{aligned}$$

As we have already seen, we may choose $\beta = \alpha^{65}$, and using the previous representation for $GF(4096)$ we have $\alpha + \alpha^{64} = \alpha^{1950} = \beta^{30}$. Thus,

$$p(x) = x^2 + \beta^{30}x + \beta$$

and all that remains is to find a representation of $GF(64)$ that is consistent with the representation of $GF(4096)$ that has been assumed.

The first seven elements of the subfield of interest can be listed as binary vectors as follows. [The vectors stand for binary polynomials in α modulo the primitive polynomial $\alpha^{12} + \alpha^6 + \alpha^4 + \alpha + 1$. The coefficient of α^{11} is shown on the left, so that, for example, the last line is to be read $\beta^6 = \alpha^{390} = \alpha^9 + \alpha^8 + \alpha^5 + \alpha^3 + \alpha$.]

$$\begin{aligned} \beta^0 &= \alpha^0 = 000000000001 \\ \beta^1 &= \alpha^{65} = 000110001100 \\ \beta^2 &= \alpha^{130} = 010000101100 \\ \beta^3 &= \alpha^{195} = 010100100011 \\ \beta^4 &= \alpha^{260} = 011001001111 \\ \beta^5 &= \alpha^{325} = 001100101011 \\ \beta^6 &= \alpha^{390} = 001100101010 \end{aligned}$$

What we want to do with this list of vectors is find a set of vectors that sum in a way to define a primitive polynomial, that is, a polynomial having the primitive element β as a root, so that we can construct a representation for $GF(64)$. This is simply a matter of finding a linearly dependent set of vectors in the list. If we had generated the full list of 63 powers of β, there would be a great many such sets to be found and straightforward procedures for finding them, for example, by selecting an arbitrary vector and then choosing others to be added to it so as to cancel the ones in progressively lower-order bit positions. For our purposes, however, the list of seven vectors is sufficient to find a primitive polynomial of degree 6. Specifically, note from the list that

$$\beta^6 + \beta^5 + \beta^0 = 0$$

and therefore the polynomial $x^6 + x^5 + 1$ has β as a root, is primitive, and may be used to generate the desired representation of $GF(64)$. We then have

$$\beta^0 = 1$$

$$\beta^1 = \beta$$

$$\beta^2 = \beta^2$$

$$\vdots$$

$$\beta^6 = \beta^5 + 1$$

$$\vdots$$

$$\beta^{62} = \beta^5 + \beta^4$$

$$\beta^{63} = 1$$

We can now generate $GF(4096)$ using the earlier polynomial $p(x)$ given by

$$p(x) = x^2 + \beta^{30}x + \beta$$

so that

$$x^2 = \beta^{30}x + \beta$$

The circuit shown in Fig. 6.1 can be used to generate all nonzero elements of $GF(64^2)$ as consecutive powers of α.

We now arrive at the central task at hand, finding the generator polynomial for the distance-4 BCH code defined on $GF(64)$. The reader can easily verify

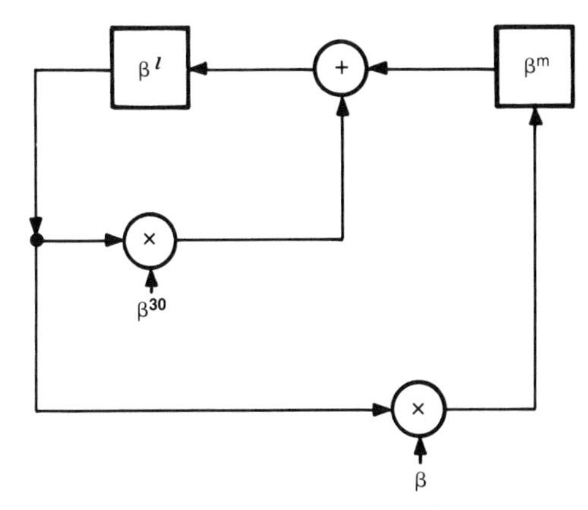

FIGURE 6.1. Circuit for generating the nonzero elements of $GF(4096)$ as consecutive powers of a primitive element.

that a generator polynomial for the code is given by

$$
\begin{aligned}
g(x) &= (x+1)(x+\alpha)(x+\alpha^{64})(x+\alpha^2)(x+\alpha^{128}) \\
&= (x+1)\left[x^2+(\alpha+\alpha^{64})x+\alpha^{65}\right]\left[x^2+(\alpha^2+\alpha^{128})x+\alpha^{130}\right] \\
&= (x+1)(x^2+\beta^{30}x+\beta)(x^2+\beta^{60}x+\beta^2)
\end{aligned}
$$

By using the previously described representation of $GF(64)$, we have

$$g(x) = x^5 + \beta^{25}x^4 + \beta^{35}x^3 + \beta^{16}x^2 + \beta^{54}x + \beta^3$$

and thus we have a (4095,4090) code on $GF(64)$.

6.3.2. Nonprimitive Codes

We now turn our attention to nonprimitive BCH codes on $GF(q)$. If $q^m - 1$ is factorable, a nonprimitive code with design distance d can be formed by specifying its roots to be $d-1$ consecutive powers of β, some nonprimitive element of $GF(q^m)$. The block length n of the code is the order of β, that is, n divides $q^m - 1$. For the moment, let us consider BCH codes on the symbol field $GF(4)$. By specifying the roots of a code as elements in various extension fields, $GF(4^m)$, we can design codes with various block lengths. For example, since $4^3 - 1 = 63 = 3 \times 3 \times 7$, we can obtain nonprimitive codes by specifying roots as consecutive powers of some nonprimitive element of $GF(64)$. If we use the element $\beta = \alpha^7$, where α is a primitive element of $GF(64)$, the block length of the resulting codes will be 9, since the order of β is

9, as seen by

$$\beta^0 = \alpha^0, \qquad \beta = \alpha^7, \qquad \beta^2 = \alpha^{14},$$

$$\beta^3 = \alpha^{21}, \ldots, \qquad \beta^8 = \alpha^{56}, \qquad \beta^9 = \alpha^{63} = \alpha^0$$

An interesting and often very useful feature of nonprimitive BCH codes is that they can at times provide a desired minimum distance with fewer check symbols than are required with a primitive code. A few examples will demonstrate this point. Let us first list the roots of the minimal polynomials over *GF*(4) of elements of *GF*(64), denoted as powers of a primitive element α. This is done in Table 6.7, where, as before, each line in the table gives all the roots of one minimal polynomial over *GF*(4). (Again, to make the table easy to read, we show only the exponents of α.) For brevity, we have written down only a portion of the list, but enumerating the entire list would show that all remaining minimal polynomials are of degree 3 except $m_{\alpha^{21}}(x)$ and $m_{\alpha^{42}}(x)$, both of which have degree 1, since $(\alpha^{21})^4 = \alpha^{84} = \alpha^{21}$ and $(\alpha^{42})^4 = \alpha^{168} = \alpha^{42}$. Given the list in Table 6.7, it is a simple matter to specify the minimal polynomials to be multiplied together to form the generator polynomial for a code of a specified minimum distance. For a primitive code with design distance 6, we can use the table to define a generator polynomial as

$$g(x) = m_{\alpha^0}(x)m_{\alpha^1}(x)m_{\alpha^2}(x)m_{\alpha^3}(x) \tag{6.2}$$

which produces a (63,53) code on *GF*(4). But notice that we can instead specify a nonprimitive code having roots that are consecutive powers of $\beta = \alpha^7$, noting that β has order 9. A shorthand table of roots of the minimal polynomials over

TABLE 6.7. A Partial Enumeration of the Nonzero Elements of *GF*(64) According to Their Minimal Polynomials over *GF*(4)

Exponents of Roots of $m(x)$	Degree of $m(x)$
0	1
1, 4, 16	3
2, 8, 32	3
3, 12, 48	3
5, 20, 17	3
6, 24, 33	3
7, 28, 49	3
9, 36, 18	3

$GF(4)$ of powers of β is constructed as follows:

Roots of $m(x)$	Degree of $m(x)$
$\beta^0 = 1$	1
$\beta^1, \beta^4, \beta^{16} = \beta^7$	3
$\beta^2, \beta^8, \beta^{32} = \beta^5$	3
β^3 (Note: $\beta^{12} = \beta^3$)	1
β^6 (Note: $\beta^{24} = \beta^6$)	1

Note that the list accounts for the minimal polynomials of all the powers of β, since β has order 9. We can now form a design-distance-6 nonprimitive code with the generator polynomial

$$g(x) = m_\beta(x) m_{\beta^2}(x) m_{\beta^3}(x) \tag{6.3}$$

which has degree 7, in contrast with the generator polynomial in Eq. (6.2), which has degree 10. Note, however, that the nonprimitive code generated from Eq. (6.3) has block length 9 in contrast with block length 63 for the primitive code generated from Eq. (6.2).

6.4. REED-SOLOMON CODES

We now consider an important subclass of nonbinary BCH codes, obtained by choosing the locator field to be the same as the symbol field. The resulting codes are called *Reed-Solomon codes* [145] often abbreviated as *RS codes*. Specifically, an RS code on $GF(q)$ with minimum distance d has as roots $d - 1$ consecutive powers of α, a primitive element of $GF(q)$. The minimal polynomial of any element γ is just $x - \gamma$. This means that the generator polynomial $g(x)$ for a design-distance-d RS code is

$$g(x) = (x - \alpha^{m_0})(x - \alpha^{m_0+1}) \cdots (x - \alpha^{m_0+d-2}) \tag{6.4}$$

where m_0 is an arbitrary integer, usually chosen as 0 or 1. Since the order of α is $q - 1$, the block length of an RS code is $q - 1$. For any BCH code the design distance is one greater than the number of consecutive roots in the locator field, and since from Eq. (6.4) the number of check symbols is always equal to the number of prescribed roots, we have for any RS code

$$d = n - k + 1$$

where n is the block length and k is the number of information symbols in each block. An important property of any RS code is that the true minimum distance is always equal to the design distance. It is easy to show that no (n,k)

linear block code can have minimum distance greater than $n - k + 1$. A code for which the minimum distance equals $n - k + 1$ is called a *maximum-distance-separable* (*MDS*) code, or simply a *maximum code* [162]. Therefore, every RS code is an MDS code. Furthermore, the reader can readily verify that shortening the block length of an RS code by omitting information symbols cannot reduce its minimum distance, and therefore we can state that any shortened RS code is also an MDS code.

Another important property of an RS code is the fact that any k positions in the codeword may be used as an information set. That is, given an (n,k) RS code on $GF(q)$, for any k symbol positions there will be one and only one codeword corresponding to each of the q^k assignments in those k positions. An important and very useful consequence of this property is that it enables one to write down the exact weight distribution for any RS code. The weight distribution $\{A_i\}$ for an RS code or any MDS code defined on $GF(q)$ having block length n and minimum distance d is given by

$$A_i = \binom{n}{i}(q-1)\sum_{j=0}^{i-d}(-1)^j\binom{i-1}{j}q^{i-d-j} \tag{6.5}$$

Derivations of this weight distribution formula can be found in Forney [37], Berlekamp [7], Peterson and Weldon [131], and other references. Weight distributions for several RS codes on small alphabets are given in Table 6.8. Unlike binary codes, there are always words of every weight i in the range $d \leq i \leq n$.

If we have formed an RS code with $m_0 = 1$, we can add an overall parity-check symbol, which increases the block length by one while leaving the number of codewords unchanged. The minimum distance of the code is increased by one, regardless of whether the minimum distance of the original

TABLE 6.8. Weight Distributions for Some RS Codes on Small Alphabets

Alphabet q	Length n	Distance $d_{\min}$	A_i for $d_{\min} \leq i \leq n$
3	2	2	2
4	3	2	9, 6
		3	3
5	4	2	24, 48, 52
		3	16, 8
		4	4
7	6	2	90, 600, 2790, 6660, 6666
		3	120, 360, 972, 948
		4	90, 108, 144
		5	36, 12
		6	6

code was odd or even. We call this an *extended RS code*. Note that an RS code initially formed with $m_0 = 0$ cannot be extended in this way, since each of its codewords has $x = \alpha^0$ as a root and therefore already has overall even parity. Since the relationship $d = n - k + 1$ is preserved, an extended RS code is an MDS code, and therefore its weight distribution is given by Eq. (6.5).

A few examples will serve to show how RS codes can be constructed. As the first example, we construct a (2,1) code on $GF(3)$. The code has minimum distance $d = n - k + 1 = 2$, and letting $m_0 = 1$ we have the set of degree-2 polynomials over $GF(3)$ having as a root α, a primitive element of $GF(3)$. If $GF(3)$ is represented by the integers 0, 1, and 2, $\alpha = 2$ is primitive, and we can write the generator polynomial for the code as simply $g(x) = x - 2$, or $g(x) = x + 1$ over $GF(3)$. The code has only three codewords, namely those represented by the all-zeros polynomial and the two polynomials $x + 1$ and $2x + 2$. Therefore, the codewords are simply 00, 11, and 22. It can be seen that this simple code is, in fact, a repetition code on $GF(3)$.

Let us now look at length-3 codes on $GF(4)$, where we represent the symbol field by $0, 1, \alpha$, and α^2. Consider the (3,2) code formed with $m_0 = 1$, which has the generator polynomial $g(x) = x - \alpha$. The 16 words in this code, which are the coefficients of codeword polynomials found by multiplying each of the degree-1 polynomials over $GF(4)$ by $x - \alpha$, are given in Table 6.9. The necessary arithmetic in $GF(4)$ is done using $\alpha^2 = \alpha + 1$ over $GF(2)$. Examination of Table 6.9 shows that this is in fact a single-parity-check code over $GF(4)$.

The (3,1) code on $GF(4)$ formed with $m_0 = 1$ has the generator polynomial

$$\begin{aligned} g(x) &= (x - \alpha)(x - \alpha^2) \\ &= x^2 - (\alpha + \alpha^2)x + \alpha^3 \\ &= x^2 + x + 1 \end{aligned}$$

Therefore the four codewords are 000, 111, $\alpha\alpha\alpha$, and $\alpha^2\alpha^2\alpha^2$. Note that the result is again a repetition code.

We can generate a different (3,1) code on $GF(4)$ by letting $m_0 = 0$, in which case the generator polynomial is

$$\begin{aligned} g(x) &= (x - \alpha^0)(x - \alpha) \\ &= x^2 - (\alpha + 1)x + \alpha \\ &= x^2 + \alpha^2 x + \alpha \end{aligned} \tag{6.6}$$

The resulting set of codewords $\{\mathbf{c}_i\}$ is

$$\{\mathbf{c}_i\} = \{000;\ 1\alpha^2\alpha;\ \alpha 1 \alpha^2;\ \alpha^2\alpha 1\} \tag{6.7}$$

TABLE 6.9. The (3,2) RS Code on $GF(4)$

000	101	$\alpha 0 \alpha$	$\alpha^2 0 \alpha^2$
011	110	$\alpha 1 \alpha^2$	$\alpha^2 1 \alpha$
$0 \alpha \alpha$	$1 \alpha \alpha^2$	$\alpha \alpha 0$	$\alpha^2 \alpha 1$
$0 \alpha^2 \alpha^2$	$1 \alpha^2 \alpha$	$\alpha \alpha^2 1$	$\alpha^2 \alpha^2 0$

In these examples, we could also have specified the codes with a parity-check matrix. For example, the code having the generator polynomial given by Eq. (6.6) has the following parity-check equations:

$$c_2 + c_1 + c_0 = 0$$

$$c_2\alpha^2 + c_1\alpha + c_0\alpha^0 = 0$$

To construct independent parity-check equations from these relationships we simply choose a symbol position, say c_0, as the information position, and solve for c_1 and c_2, yielding

$$c_1 = \alpha c_0, \qquad c_2 = \alpha^2 c_0$$

The reader should verify that these parity-check equations produce the four codewords in Eq. (6.7). It should also be verified that any of the three symbol positions in the code may be treated as the information symbol and the same list of codewords is produced in each case.

As a final example we consider a distance-3 RS code defined on $GF(13)$. Taking $m_0 = 1$ and $\alpha = 2$, we have

$$g(x) = (x - \alpha)(x - \alpha^2) = (x - 2)(x - 4)$$

$$= x^2 + 7x + 8$$

This code has $n = 12$, $r = 2$, and thus is a (12,10) code over $GF(13)$. This code is in fact used by the Office of the Jury Commissioner for Middlesex County, Massachusetts, in an automated juror selection and accounting system. A computer is used to randomly select prospective jurors and each is assigned a 10-digit number for identification. Communication with jurors is done by telephone and mail, and juror responses are entered into the system by manual keypunch. To assure that each response is assigned to the appropriate juror, the (12,10) RS code is used, with two parity-check symbols attached to each 10-digit identification number. Since the parity checks range over the 13-ary code alphabet, three letters (A, C, and E) are used in the two parity positions to represent the code symbols 10, 11, and 12, respectively.

6.5. ENCODING NONBINARY BCH CODES AND RS CODES

The formation of codewords in a BCH code on $GF(q)$ from its generator polynomial $g(x)$ corresponds directly to the binary case treated in Chapter 4. That is, the words in an (n,k) code correspond to the set of all polynomials over $GF(q)$ of degree $n-1$ or less that are divisible by $g(x)$, where the degree of $g(x)$ is $r = n - k$.

The codewords can be generated by multiplying all polynomials over $GF(q)$ having degree $k-1$ or less by $g(x)$. As was seen in the binary case, this will not necessarily produce a systematic code and is generally avoided. As in the case of the binary codes, systematic structure can be provided by forming codewords as

$$c(x) = x^r i(x) \bmod g(x) + x^r i(x)$$

where $i(x)$ denotes the k information symbols on $GF(q)$ to be encoded represented as a polynomial of degree $k-1$ or less.

Encoding can be implemented with a division circuit of the form previously considered for binary BCH codes (see Fig. 4.2). However, multiplications and additions are now done in $GF(q)$, the symbol field of the code. As an example we consider the primitive distance-4 code defined on a 64-ary alphabet with block length 4095 that was described in Section 6.3. The generator polynomial of this code was shown to be

$$g(x) = x^5 + \beta^{25}x^4 + \beta^{35}x^3 + \beta^{16}x^2 + \beta^{54}x + \beta^3$$

An encoder for the (4095,4090) distance-4 code defined on $GF(64)$ is shown in Fig. 6.2. Each stage of the register is a 64-ary storage device, and the feedback lines require multiplication in $GF(64)$. The feedback weights are the coefficients of the generator polynomial. This encoder is, in fact, the nonbinary equivalent of the encoder shown in Fig. 4.2. As with binary cyclic codes, nonbinary codes can also be encoded with a k-stage feedback shift register [131]. The circuit shown in Fig. 6.2 operates as follows:

1. Open (enable) gates G_1, G_2, and G_3. Close (disable) gate G_4. Clock the information symbols to be encoded into the feedback shift register and simultaneously into the codeword buffer.
2. Close gates G_1, G_2, and G_3, and open gate G_4. The five parity symbols are now contained in the five storage elements of the feedback shift register. Clock these five symbols into the buffer to complete formation of the codeword.
3. All stages of the feedback shift register are now reset to zero, and the encoded word is shifted out while the next information set to be encoded is shifted in. Return to step 1.

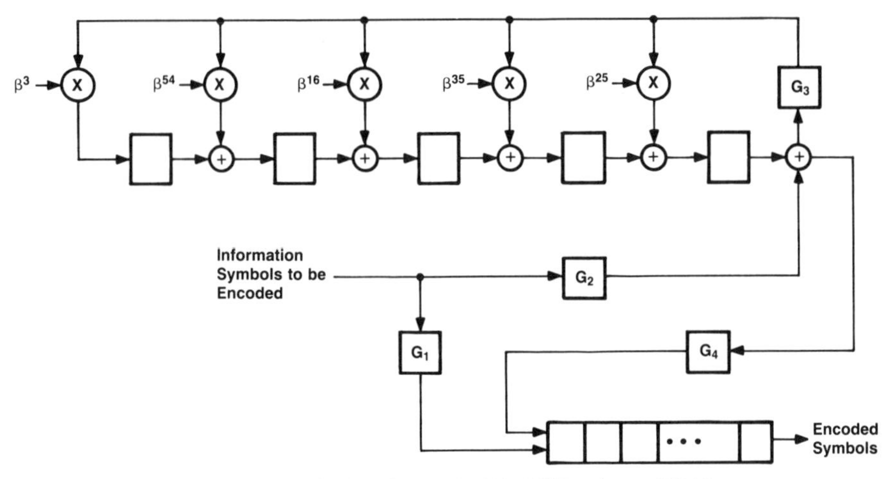

FIGURE 6.2. Encoder for the (4095,4090) BCH code on $GF(64)$.

6.6. DECODING ALGORITHMS FOR BCH AND RS CODES

We now describe algorithms for decoding nonbinary BCH codes and RS codes that are generalizations of the algorithms already presented in Sections 5.2 through 5.4 for the binary BCH codes. We have described these procedures under the heading *syndrome decoding*, which for nonbinary codes we describe as follows. Given the set of syndrome values for the received word, the decoding task is to find the most likely error pattern, within the error-correction limit of the code, that produces the observed syndrome values. Therefore, as in the binary case, decoding is viewed as a problem of solving a set of simultaneous syndrome equations, but one where the set of unknowns now includes the *error values* or *error magnitudes* in addition to the error locators.

We use the notation adopted earlier, letting a transmitted codeword be represented by a polynomial $c(x)$, where here the coefficients of $c(x)$ are elements in $GF(q)$. Similarly, a received error pattern is represented by a polynomial $e(x)$, again with coefficients in $GF(q)$, and the received word is represented by $r(x)$, where

$$r(x) = c(x) + e(x)$$

The syndrome values are obtained by reapplying the encoding rules to the received word, and this in turn is done by evaluating $r(x)$ at the prescribed roots of the generator polynomial, that is,

$$\begin{aligned} S_k &= r(\alpha^k) \\ &= c(\alpha^k) + e(\alpha^k) \\ &= e(\alpha^k), \qquad k = m_0, m_0 + 1, \ldots, m_0 + d - 2 \end{aligned} \tag{6.8}$$

In Eq. (6.8) we let the roots of the code be any arbitrary sequence of consecutive powers of α, although m_0 is usually chosen to be 0 or 1. We restrict our attention to primitive codes, letting α be a primitive element of the locator field $GF(q^m)$, where for BCH codes, m is any integer, and for the special case of RS codes, m equals 1.

The error polynomial $e(x)$ has nonzero terms only in those positions where errors have occurred, so that if there are t errors in the received word, we can write the syndrome values as

$$S_k = \sum_{i=1}^{t} Y_i X_i^k, \qquad k = m_0, m_0 + 1, \ldots, m_0 + d - 2 \tag{6.9}$$

where X_i is the error locator for the ith error and Y_i is its value. Therefore, the decoding task is, given the S's, find the X's and Y's. In a generalization of the procedure outlined in Section 5.2, syndrome decoding of a nonbinary code proceeds as follows:

1. Calculate the syndrome values S_k, $k = m_0, m_0 + 1, \ldots, m_0 + d - 2$.
2. Determine the error-locator polynomial $\sigma(x)$ from the syndrome values.
3. Solve for the roots of $\sigma(x)$, which are the error locators.
4. Given the error locators, calculate the error values.
5. Correct the indicated errors.

The fundamental difference between this sequence of steps and the procedure outlined in Section 5.2 for the binary case is step 4, calculation of the error values. However, once the error locations have been determined, finding the error values is straightforward, since, given the S's and X's, Eq. (6.9) is simply a set of simultaneous linear equations having the error values as unknowns. As in the binary case, the most difficult part of the procedure is usually step 2, determination of the error-locator polynomial $\sigma(x)$ from the syndrome values.

Peterson's direct solution method for finding the coefficients of the error-locator polynomial $\sigma(x)$, described in Section 5.3 for binary codes, generalizes in a straightforward way to the case of nonbinary codes, although there are a few important differences. The set of simultaneous nonlinear (in the X's) syndrome equations, Eq. (6.9), can be converted into a set of linear equations to be solved in conjunction with $\sigma(x)$. To begin, exactly as we did in Eqs. (5.5) through (5.8), we can operate repeatedly on $\sigma(x)$ and invoke the syndrome equations, Eq.(6.9), to establish the relationship

$$S_{t+j} + \sigma_1 S_{t+j-1} + \cdots + \sigma_t S_j = 0, \qquad \text{for all } j \tag{6.10}$$

where the σ's are coefficients of the error-locator polynomial, $\sigma(x)$. That is,

$$\sigma(x) = x^t + \sigma_1 x^{t-1} + \cdots + \sigma_t \tag{6.11}$$

The equations defined by Eq. (6.10) are *Newton's identities*, previously presented as Eq. (5.8).

Let us consider a t-error-correcting nonbinary BCH or RS code, for which we have computed $2t$ syndrome values $S_1, S_2, \ldots, S_{2t}$. From Eq. (6.10) we can construct t simultaneous equations, linear in coefficients of $\sigma(x)$, by letting j range from 1 through t. To illustrate this with an example, we consider the case of a three-error-correcting code so that we have

$$\begin{aligned} S_1\sigma_3 + S_2\sigma_2 + S_3\sigma_1 &= -S_4 \\ S_2\sigma_3 + S_3\sigma_2 + S_4\sigma_1 &= -S_5 \\ S_3\sigma_3 + S_4\sigma_2 + S_5\sigma_1 &= -S_6 \end{aligned} \tag{6.12}$$

The three equations have been written in a form suggesting their use, that is, as a set of three simultaneous linear equations, with coefficients and constants that are the syndrome values. These equations are then solved for the three coefficients of $\sigma(x)$ when three errors are assumed to have occurred.

The reader should compare Eq. (6.12) with Eq. (5.9) for the binary case and note certain differences. First, unlike the binary case, Eq. (6.12) includes equations beginning with the even-indexed syndrome values. This is because the relationships $S_j^2 = S_{2j}$ are specific to the binary case and do not hold for nonbinary codes. Second, the simpler forms of the uppermost lines in Eq. (5.9) are also specific to the binary case and do not apply here. Finally, negative signs are retained when the constants S_4, S_5, and S_6 are moved from the left side in Eq. (6.10) to the right side in Eq. (6.12), since addition and subtraction are identical only when the field is of characteristic 2.

As with binary codes, determining the locations of a given number of errors is done by constructing an appropriate set of simultaneous equations of the form given by Eq. (6.12) and solving the equations for the σ's in terms of the syndrome values $\{S_k\}$. The number of equations to be used is equal to the actual number of errors in the received code block, which must be determined as part of the decoding operation. This is done simply by testing determinants of various sizes corresponding to the various possible numbers of errors. We have already written out the equations for the σ's in the three-error case in Eq. (6.12). We now write the sets of equations for the one-error and two-error cases, in the more compact matrix form, as follows:

$$[S_1][\sigma_1] = [-S_2] \tag{6.13}$$

$$\begin{bmatrix} S_1 & S_2 \\ S_2 & S_3 \end{bmatrix} \begin{bmatrix} \sigma_2 \\ \sigma_1 \end{bmatrix} = \begin{bmatrix} -S_3 \\ -S_4 \end{bmatrix} \tag{6.14}$$

We now define D_2 as the determinant of the coefficient matrix in Eq. (6.14), that is $S_1S_3 - S_2^2$ and D_3 as the determinant of the 3×3 coefficient matrix in

Eq. (6.12), which is

$$S_1S_3S_5 + S_2S_3S_4 + S_2S_3S_4 - S_3^3 - S_1S_4^2 - S_2^2S_5$$

Now, tests of D_2 and D_3 can be used to determine how many errors have occurred, and therefore which set of equations should be used to solve for the σ's. The reader should verify, for example, that if only one error has occurred, D_2 and D_3 will equal zero, and therefore the Eqs. (6.12) and (6.14) will be indeterminate, and Eq. (6.13) is to be used.

Once the σ's have been determined, the error locator polynomial $\sigma(x)$ is formed and its roots obtained. The Chien search, already described in Section 5.3, can be used. The roots of $\sigma(x)$ are the error-locator values, the X's. Once the X's have been determined, they are inserted into the syndrome equations, Eq. (6.9), which are then solved as linear equations for the error values, the Y's. A description of the steps in a direct solution decoding algorithm is presented in the next subsection using an example.

6.6.1. Direct Solution for a Distance-7 RS Code

We now outline the use of Peterson's direct solution method for decoding the (63,57) RS code defined on a 64-ary alphabet. Since the code has distance 7, it can be used to correct up to three errors in a received word or to correct combinations of l errors and s erasures such that $2l + s < 7$. In this discussion, however, we confine our attention to error-correction decoding. Combined errors-and-erasures decoding is treated separately in a later subsection.

Since for an RS code the code symbol field and the locator field are the same, all computations for decoding are done in the field $GF(64)$. Furthermore, since $GF(64)$ is a field of characteristic 2, addition and subtraction are identical operations, which means, for example, that in determining the coefficients of $\sigma(x)$ and calculating error values, the minus signs can be replaced with plus signs. The 64 elements of the field may be represented conveniently as binary 6-tuples. Addition is then implemented with modulo-2 addition, applied bit by bit. For implementation in a processor, finite field multiplication and division are conveniently done with logarithm and antilogarithm tables and table look-up routines, as described in Chapter 4.

An error-correction decoder for the (63,57) RS code can be implemented as follows. Let the polynomial $r(x)$ represent the received word, where the high-order terms correspond to the information symbols and the low-order terms to the check symbols. The steps in the decoding process are

1. Compute the syndrome values S_k, $1 \le k \le 6$, where

$$S_k = r(\alpha^k) = \{ \cdots [(r_{62}\alpha^k + r_{61})\alpha^k + r_{60}]\alpha^k + \cdots \}\alpha^k + r_0, \quad 1 \le k \le 6$$

2. Determine the number of errors in the received word:

a. If $S_k = 0$, $1 \le k \le 6$, the received word is a codeword, and no further processing is necessary.

b. If $D_3 = S_1S_3S_5 + S_1S_4^2 + S_2^2S_5 + S_3^3 \ne 0$, assume three errors are present.

c. If $D_3 = 0$ and $D_2 = S_1S_3 + S_2^2 \ne 0$, assume two errors are present.

d. If $D_2 = D_3 = 0$ and $S_1 \ne 0$, assume one error is present.

3. Compute the coefficients of the error-locator polynomial:

a. If three errors are present, compute

$$\sigma_1 = \frac{1}{D_3}\left[S_1S_3S_6 + S_1S_4S_5 + S_2^2S_6 + S_2S_3S_5 + S_2S_4^2 + S_3^2S_4\right]$$

$$\sigma_2 = \frac{1}{D_3}\left[S_1S_4S_6 + S_1S_5^2 + S_2S_3S_6 + S_2S_4S_5 + S_3^2S_5 + S_3S_4^2\right]$$

$$\sigma_3 = \frac{1}{D_3}\left[S_2S_4S_6 + S_2S_5^2 + S_3^2S_6 + S_4^3\right]$$

b. If two errors are present, compute

$$\sigma_1 = \frac{1}{D_2}\left[S_1S_4 + S_2S_3\right]$$

$$\sigma_2 = \frac{1}{D_2}\left[S_2S_4 + S_3^2\right]$$

c. If one error is present, compute

$$\sigma_1 = X_1 = \frac{S_2}{S_1}$$

4. If three errors are indicated in step 3, find (using the Chien search) the roots of the polynomial $\sigma(x)$, where

$$\sigma(x) = x^3 + \sigma_1x^2 + \sigma_2x + \sigma_3$$

If two errors are indicated, find the roots of

$$\sigma(x) = x^2 + \sigma_1x + \sigma_2$$

Of course, in the case of three errors, three distinct roots of $\sigma(x)$ must be found, and for the two-error case, $\sigma(x)$ must have two distinct roots. If the correct number of roots is not found, error detection is announced.

5. After the error locators are determined, the error values are obtained by solving the syndrome equations.

   One-error case:

$$Y_1 = \frac{S_1^2}{S_2}$$

   Two-error case:

$$Y_1 = \frac{S_1 X_2 + S_2}{X_1 X_2 + X_1^2}$$

$$Y_2 = \frac{S_1 X_1 + S_2}{X_1 X_2 + X_2^2}$$

   Three-error case:

   Let

$$C = X_1 X_2^2 X_3^3 + X_1^3 X_2 X_3^2 + X_1^2 X_2^3 X_3 + X_1^3 X_2^2 X_3 + X_1 X_2^3 X_3^2 + X_1^2 X_2 X_3^3$$

   Then

$$Y_1 = \frac{1}{C}\big[S_1 X_2^2 X_3^3 + S_2 X_2^3 X_3 + S_3 X_2 X_3^2 + S_1 X_2^3 X_3^2 + S_2 X_2 X_3^3 + S_3 X_2^2 X_3\big]$$

$$Y_2 = \frac{1}{C}\big[S_1 X_3^2 X_1^3 + S_2 X_1 X_3^3 + S_3 X_1^2 X_3 + S_1 X_1^2 X_3^3 + S_2 X_1^3 X_3 + S_3 X_1 X_3^2\big]$$

$$Y_3 = \frac{1}{C}\big[S_1 X_1^2 X_2^3 + S_2 X_1^3 X_2 + S_3 X_1 X_2^2 + S_1 X_1^3 X_2^2 + S_2 X_1 X_2^3 + S_3 X_1^2 X_2\big]$$

   It should be noted that when the denominators in the expressions for Y_1, Y_2, and Y_3 are written out, the expressions can be simplified.

6. Correct the received word by adding the computed error values to the symbols received in positions identified as error locations.
7. Compute the syndrome of the corrected word, and if it is not zero, announce error detection.

The correction of both errors and erasures will be discussed in Section 6.6.3. However, we first present an efficient iterative decoding algorithm for correction of errors in nonbinary BCH codes and RS codes.

6.6.2. The Massey-Berlekamp Algorithm

For correction of moderate to large numbers of errors with a nonbinary BCH or RS code, Peterson's direct method of solving for the coefficients of $\sigma(x)$ from the syndrome values becomes cumbersome and inefficient due to the large number of multiplications and divisions that must be performed. Instead, it is preferable to use either of two algorithms developed by Berlekamp [7] and Massey [104] for solution of Newton's identities. The two algorithms are closely related and are often referred to as one procedure, the *Massey-Berlekamp algorithm*. The approach used by Massey in presenting the technique is particularly instructive, and thus we shall follow Massey closely here. Berlekamp's formulation is described in Section 5.4, with simplifications applicable to decoding binary codes. Both Massey's and Berlekamp's versions of the algorithm can be used for binary and nonbinary codes.

We let $m_0 = 1$ and return to the error-locator polynomial $\sigma(x)$ in Eq. (6.11). Then, substituting an error locator X_j for x, we obtain

$$X_j^t + \sigma_1 X_j^{t-1} + \cdots + \sigma_t = 0, \qquad j = 1, 2, \ldots, t \tag{6.15}$$

Multiplying Eq. (6.15) by X_j^k and summing for $j = 1, 2, \ldots, t$, we obtain

$$S_{k+t} + \sigma_1 S_{k+t-1} + \cdots + \sigma_t S_k = 0, \qquad k = 1, 2, \ldots \tag{6.16}$$

which are again Newton's identities. Letting $j = k + t$, we obtain

$$S_j + \sigma_1 S_{j-1} + \cdots + \sigma_t S_{j-t} = 0, \qquad j = t+1,\ t+2, \ldots \tag{6.17}$$

With Newton's identities written in this form, one can recognize that they describe the operation of a linear feedback shift register (FSR) with initial states $S_1, S_2, \ldots, S_t$ and tap connections given by $C_i = \sigma_i$. A diagram of a linear FSR is shown in Fig. 6.3. From the figure it is seen that the FSR

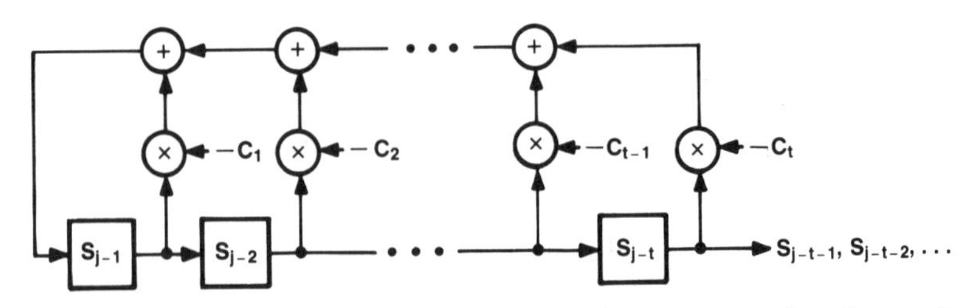

FIGURE 6.3. Linear feedback shift register for generating a sequence of syndrome values.

implements the equations

$$S_j = -C_1 S_{j-1} - C_2 S_{j-2} - \cdots - C_t S_{j-t}, \qquad j = t+1, t+2, \ldots \tag{6.18}$$

or

$$S_j + C_1 S_{j-1} + \cdots + C_t S_{j-t} = 0, \qquad j = t+1, t+2, \ldots \tag{6.19}$$

With $C_i = \sigma_i$, the correspondence with Eq. (6.17) is immediate.

Recognizing the relationship just obtained, Massey established the equivalence between the problem of determining the coefficients of the error-locator polynomial from the syndrome values and that of synthesizing an FSR with minimum length that generates the given sequence of syndromes. We shall provide a rationale for this in the following.

We define the *connection polynomial* as a convenient representation for the coefficients of the syndrome values in Eq. (6.19), that is,

$$C(x) = 1 + C_1 x + C_2 x^2 + \cdots + C_t x^t \tag{6.20}$$

We now state that the problem of determining the error-locator polynomial $\sigma(x)$ is equivalent to that of determining a connection polynomial $C(x)$ for a linear FSR that generates the syndrome values $S_{t+1}, S_{t+2}, \ldots$, given that the FSR is initialized with $S_1, S_2, \ldots, S_t$.

Note that from Eqs. (6.17) through (6.20), as well as in Fig. 6.3, the assumed length of the FSR is t stages, where t is the error-correction limit of the code. However, the iterative algorithm is designed to correct l errors, where $l \leq t$. The number of errors l is not known at the start of decoding and is determined as part of the decoding procedure.

Without delving into the details of the properties of FSRs and the sequences that they generate [54], we simply point out that for a given sequence of syndrome values, there are a determinable number of connection polynomials of various lengths that will generate the syndromes. This corresponds directly to the fact that there are in general a number of error patterns that can account for a given set of syndrome values. However, the task of bounded-distance decoding is to find the lowest-weight error pattern corresponding to the given syndrome. Therefore, in the FSR synthesis problem we seek the lowest degree connection polynomial $C(x)$ that generates the syndrome. In his 1969 paper, Massey described an algorithm that finds the minimal-length FSR. He further showed that given an error pattern of weight $l \leq t$, the algorithm yields the connection polynomial that uniquely corresponds to the correct error-locator polynomial. Massey's algorithm is often called the *FSR synthesis algorithm*.

Before describing the FSR synthesis algorithm in detail, we outline the procedure as follows. The FSR algorithm synthesizes the minimal-length shift register with an iterative routine that begins by postulating the shortest possible shift register and then attempts to generate the entire sequence of given syndrome values in order. The actual syndrome sequence is continually compared with the output of the postulated FSR until either the entire sequence of given syndrome values is reproduced or a discrepancy is encountered. At the first discrepancy, the postulated FSR is modified with a specified rule, and the sequence generation is restarted and continued until all the remaining syndromes are reproduced or another discrepancy is encountered, and so forth. The modification rule is designed to ensure that for a correctable error pattern, the FSR eventually settles into the correct configuration. The FSR synthesis algorithm is described in detail below.

The Massey FSR Synthesis Algorithm

0. *Compute Syndrome Values*
   $S_n, 1 \leq n \leq d - 1$
1. *Initialize Algorithm Variables*
   Let $C(x) = 1 \qquad D(x) = x$
   $L = 0 \qquad n = 1$
2. *Take in New Syndrome Value and Compute Discrepancy*
   $$\delta = S_n + \sum_{i=1}^{L} C_i S_{n-i}$$
3. *Test Discrepancy*
   If $\delta = 0$, go to step 8. Otherwise, go to step 4.
4. *Modify Connection Polynomial*
   Let $C^*(x) = C(x) - \delta D(x)$.
5. *Test Register Length*
   If $2L \geq n$, go to step 7 (i.e., do not extend register). Otherwise, go to step 6.
6. *Change Register Length and Update Correction Term*
   Let $L = n - L$ and $D(x) = c(x)/\delta$.
7. *Update Connection Polynomial*
   Let $C(x) = C^*(x)$.
8. *Update Correction Term*
   Let $D(x) = xD(x)$.
9. *Update Syndrome Counter*
   Let $n = n + 1$.
10. *Test Syndrome Count*
    If $n < d$, go to step 2.
    Otherwise, stop.

In the algorithm, $C(x)$ is the FSR connection polynomial. The algorithm is designed to expediently build up the polynomial $C(x)$ of lowest degree that generates the given sequence of syndromes, $S_1, S_2, \ldots, S_{d-1}$. The connection polynomial is first initialized to its simplest possible form, $C(x) = 1$, and is subsequently modified as needed to correctly reproduce the syndrome values in sequence. The other polynomial formed in the algorithm, $D(x)$, is a correction term that is used to modify $C(x)$ at each iteration in which a discrepancy is encountered between a generated value and the corresponding syndrome value. The syndromes are examined by the algorithm in sequence, one in each iteration. At each iteration the discrepancy δ, the difference between the newly entered syndrome value and the value generated by the FSR in the corresponding sequence position, is computed, using the connection polynomial $C(x)$ as it was structured at the end of the previous iteration. Note that δ is defined in such a way that at the first entry into step 2, it is given the value of the first syndrome S_1, even though there are no previous syndrome values from which S_1 could be generated. At each appearance of a nonzero value for δ, the connection polynomial is modified using the computed value of δ and the correction term (step 4). The formation and use of the correction term is the most important part of the algorithm. One reason is that in addition to zeroing out the encountered discrepancy, the modification of $C(x)$ is such that the new $C(x)$ also correctly generates all the previous syndrome values. This obviates the necessity of having to reexamine previous syndromes each time $C(x)$ is modified, and provides an algorithm in which the number of computations required per decoding is a linear function of l rather than some geometric function. Another important characteristic of the polynomial modification procedure is that it accomplishes the needed sequence modification with the smallest possible increase in the degree of the connection polynomial. The other variable used in the algorithm is L, which is the current length of the FSR. If the algorithm terminates with an FSR connection polynomial of degree greater than t, that is, $2L > d - 1$, then we are not assured that the corresponding error-locator polynomial is correct, and error detection is announced.

We next present examples of the application of the FSR synthesis technique to two different nonbinary codes.

FSR Decoding Example No. 1

Consider the case of the (10,6) RS code defined on the prime field $GF(11)$, having roots α, α^2, α^3, and α^4, where α is a primitive element in $GF(11)$. This is a distance-5 code, capable of correcting up to two errors in each 10-symbol codeword. In practice, for such a small amount of error correction, one would probably not utilize the FSR synthesis procedure but would instead use the direct solution method described earlier. However, this serves as a convenient example, since the reader can easily verify the indicated computations after selection of a primitive element.

From Eq. (6.8) the four syndrome values are calculated as

$$S_k = \sum_{j=0}^{9} r_j \alpha^{jk}, \qquad k = 1, 2, 3, 4$$

where r_j is the $(j + 1)$th symbol in the received word and α is the chosen primitive element of $GF(11)$.

All mathematical operations are done in $GF(11)$, and since 11 is prime, addition and multiplication are simply ordinary addition and multiplication modulo 11.

Let us now consider a simple example in which the word 0300000000 is received. We have $r(x) = 3x$, and therefore the syndrome values (step 0) are calculated as

$$S_1 = r(\alpha) = 3\alpha$$

$$S_2 = r(\alpha^2) = 3\alpha^2$$

$$S_3 = r(\alpha^3) = 3\alpha^3$$

$$S_4 = r(\alpha^4) = 3\alpha^4$$

The FSR algorithm is initialized (step 1) by setting the connection polynomial to $C(x) = 1$, the correction term to $D(x) = x$, the register length to $L = 0$, and the syndrome counter to $n = 1$. The first discrepancy is found (step 2) to be $\delta = S_1 = 3\alpha$, and therefore the connection polynomial is modified (step 4) to $C^*(x) = 1 - 3\alpha x = 1 + 8\alpha x$. The register length test (step 5) finds $2L = 0$, less than $n = 1$, so that L is increased to 1. Step 6 is completed by updating the correction term to $D(x) = (3\alpha)^{-1}$, which equals $4\alpha^{-1}$, since $3 \times 4 = 12 = 1 \bmod 11$. The connection polynomial is replaced with its modified form $C(x) = C^*(x) = 1 + 8\alpha x$ (step 7), the correction term is updated to $(3\alpha)^{-1}x$ (step 8), the syndrome counter is updated to $n = 2$ (step 9), and the algorithm returns to step 2 for an iteration.

At this point, the postulated FSR connection polynomial is $C(x) = 1 + 8\alpha x$, which describes a single-stage register with a feedback line returning the output to the input after multiplication by -8α or 3α (see Fig. 6.3). Therefore, the discrepancy computation (step 2) now produces $\delta = 3\alpha^2 + 8\alpha(3\alpha) = 5\alpha^2$. The connection polynomial is now modified (step 4) to $C^*(x) = 1 + 8\alpha x - 5\alpha^2(4\alpha^{-1})x = 1 - \alpha x$. We need not bother tracing the remaining steps in the synthesis procedure, since we can observe at this point that the connection polynomial $C(x) = C^*(x) = 1 - \alpha x$ describes a one-stage register in which the output is fed back to the input after multiplication by α. Therefore, it is readily seen that if initially loaded with $S_1 = 3\alpha$, the FSR will correctly generate the succeeding syndrome values $3\alpha^2$, $3\alpha^3$, and $3\alpha^4$. Thus the algorithm

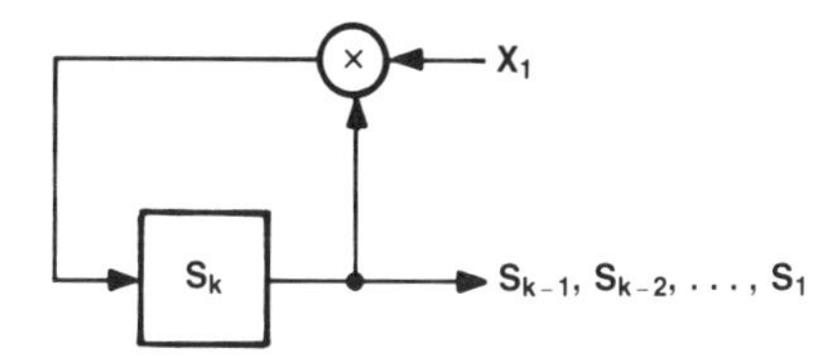

FIGURE 6.4. Minimal-length FSR for the case of a single error.

has found the connection polynomial of lowest degree that generates the given syndrome values. [The only connection polynomial of lower degree is $C(x) = 1$, which generates a sequence of identical symbol values.] We form the error-locator polynomial $\sigma(x)$ by setting $\sigma_1 = C_1 = -\alpha$ and using Eq. (6.11), which gives us

$$\sigma(x) = x - \alpha$$

Therefore, the locator of the apparent single error is $X_1 = \alpha$, which points to received symbol r_1, given our convention that the symbol locations in a codeword are marked by the locators $\alpha^0, \alpha^1, \dots, \alpha^{n-1}$ reading left to right. The error value Y_1 is found from Eq. (6.9) as $Y_1 = 3\alpha\alpha^{-1} = 3$, and subtraction of Y_1 from $r(x)$ in the second symbol position produces the all-zeros codeword. Therefore, if the transmitted codeword was in fact the all-zeros word, single-error correction has been successfully accomplished. The reader may select a particular value for the primitive element of $GF(11)$, of which there are four, and work out the intermediate results in this example in detail.

From this simple example we can readily generalize to any case in which a single error has occurred and anticipate the form of the connection polynomial which the FSR synthesis algorithm must finally produce. For any case of a single error we see from Eq. (6.9) that the computed syndromes have the values $S_1 = Y_1X_1$, $S_2 = Y_1X_1^2, \dots,$ and so forth. We now note that the ratio of each consecutive pair of syndrome values is the same, that is,

$$\frac{S_2}{S_1} = \frac{S_3}{S_2} = \frac{S_4}{S_3} = \cdots = \frac{S_{d-1}}{S_{d-2}} = X_1 \qquad (6.21)$$

where X_1 is the locator of the single error and d denotes the minimum distance of the code. It is readily seen from Eq. (6.21) that the syndromes can be generated with the very simple FSR shown in Fig. 6.4. The one-stage shift register is initially loaded with S_1 and in each clock cycle the current syndrome value is multiplied by X_1 and enters the register as the next succeeding syndrome value. By reference to Fig. 6.3 and Eq. (6.19) we can write the connection polynomial for this FSR as $C(x) = 1 - X_1x$.

FSR Decoding Example No. 2

Consider the case of the (31,25) RS code on $GF(32)$, with $m_0 = 1$, for which codewords all have the six consecutive roots $\alpha, \alpha^2, \dots, \alpha^6$. Let the all-zeros codeword be transmitted, and assume the received word

$$000\alpha^7 00000000\alpha^3 0000000\alpha^{22} 0000000000$$

Thus we have

$$r(x) = \alpha^7 x^3 + \alpha^3 x^{12} + \alpha^{22} x^{20}$$

To represent elements in $GF(32)$, we use the primitive polynomial $p(x) = x^5 + x^2 + 1$, with which Table 4.4 was constructed. The syndrome values are computed as follows:

$$S_1 = r(\alpha) = \alpha^{29}$$

$$S_2 = r(\alpha^2) = \alpha^{28}$$

$$S_3 = r(\alpha^3) = \alpha^9$$

$$S_4 = r(\alpha^4) = \alpha^4$$

$$S_5 = r(\alpha^5) = \alpha^{24}$$

$$S_6 = r(\alpha^6) = \alpha^{19}$$

We next use the FSR synthesis algorithm to find the shortest connection polynomial $C(x)$ that generates the six syndrome values in order. The iterative solution is summarized as follows:

n	S_n	$C(x)$	δ	L
1	α^{29}	1	α^{29}	0
2	α^{28}	$1 + \alpha^{29}x$	α^{14}	1
3	α^9	$1 + \alpha^{30}x$	α^{10}	1
4	α^4	$1 + \alpha^{30}x + \alpha^{12}x^2$	α^{11}	2
5	α^{24}	$1 + \alpha^4 x + \alpha^{23}x^2$	α^{10}	2
6	α^{19}	$1 + \alpha^4 x + \alpha^{12}x^2 + \alpha^{30}x^3$	α^{19}	3
7		$1 + \alpha^6 x + \alpha^{30}x^2 + \alpha^4 x^3$	(STOP)	

Thus the minimal length connection polynomial is found to be

$$C(x) = 1 + \alpha^6 x + \alpha^{30}x^2 + \alpha^4 x^3$$

and with $\sigma_i = C_i$ we can write the error-locator polynomial as

$$\sigma(x) = x^3 + \alpha^6 x^2 + \alpha^{30}x + \alpha^4$$

The three roots of $\sigma(x)$, the error-locator numbers, are found to be

$$X_1 = \alpha^3, \qquad X_2 = \alpha^{12}, \qquad X_3 = \alpha^{20}$$

which point to errors in the fourth, thirteenth, and twenty-first symbol positions.

The error magnitudes, Y_1, Y_2, and Y_3 are now computed using the equations shown for the three-error case in step 5 near the end of Section 6.6.1. The reader should verify that the computations yield

$$Y_1 = \alpha^7, \qquad Y_2 = \alpha^3, \qquad Y_3 = \alpha^{22}$$

Finally, error correction is completed by subtracting the error values from the corresponding received symbols, which yields the all-zeros word as the corrected codeword.

FSR Decoding Example No. 3

As a third example we consider again the (31,25) *RS* code on *GF*(32), assuming transmission of the all-zeros codeword and the received word

$$r(x) = \alpha^6 x^2 + \alpha^{15} x^5 + \alpha^{20} x^6 + \alpha^{23} x^{28}$$

We have four errors, and the error pattern is uncorrectable. The following set of syndromes is obtained:

$$S_1 = \alpha^9, \qquad S_2 = \alpha^{25}, \qquad S_3 = \alpha^{22}$$

$$S_4 = \alpha^{28}, \qquad S_5 = \alpha^6, \qquad S_6 = \alpha$$

Application of the FSR synthesis algorithm yields the following iteration:

n	S_n	$C(x)$	δ	L
1	α^9	1	α^9	0
2	α^{25}	$1 + \alpha^9 x$	α^9	1
3	α^{22}	$1 + \alpha^{16} x$	α^2	1
4	α^{28}	$1 + \alpha^{16} x + \alpha^{14} x^2$	α^0	2
5	α^6	$1 + \alpha^{30} x + \alpha^{18} x^2$	α^{16}	2
6	α	$1 + \alpha^{30} x + \alpha^{24} x^2 + \alpha^{30} x^3$	α^{11}	3
7		$1 + \alpha^5 x + \alpha^{11} x^2 + \alpha^{12} x^3$	(STOP)	

The synthesized connection polynomial has degree 3, which at this point appears to be a legitimate connection polynomial, since the code has distance 7. However, when we search for the roots of the corresponding error-locator polynomial,

$$\sigma(x) = x^3 + \alpha^5 x^2 + \alpha^{11} x + \alpha^{12}$$

we find no roots in $GF(32)$. Therefore, the synthesized connection polynomial is in fact not legitimate, and error detection is announced.

6.6.3. Errors-and-Erasures Decoding

If some procedure is being used to erase unreliable symbols in a received word, then the job of the decoder is to fill in the proper values of the erasures and at the same time locate and correct any unknown errors. We recall from earlier discussions that a code of minimum distance d is capable of correcting any pattern of l errors and s erasures as long as $2l + s < d$. We shall now outline an efficient procedure for simultaneous errors-and-erasures decoding, which was suggested by Forney [36] for nonbinary BCH codes. First we define the *erasure-locator polynomial* $\sigma'(z)$, which is the polynomial of degree s whose roots are the erasure locators. That is,

$$\sigma'(x) = \prod_{i=1}^{s} (z + Z_i)$$

$$= \sigma_0' z^s + \sigma_1' z^{s-1} + \cdots + \sigma_s' \tag{6.22}$$

We have assumed a field of characteristic 2 and omitted the minus signs. It should be noted that $\sigma'(z)$ is written in much the same form as the error-locator polynomial, Eq. (6.11), except that for notational convenience we have given the term of highest degree the coefficient σ_0' even though it always has value 1. Since, by definition, the erasure-location values are known, the coefficients of $\sigma'(z)$ may be computed directly. We also assume use of a primitive code with $m_0 = 1$.

Combined errors-and-erasures decoding begins, as does error-correction decoding, with calculation of the syndrome values, which are

$$S_k = \sum_{i=1}^{n} r_i \alpha^{ik}, \qquad 1 \le k \le d - 1$$

where we denote $d - 1$ rather than $2t$ syndrome values, to allow for both odd and even values of d. The reader may well ask what values should be assigned to the erasures for the syndrome calculation, but it will be seen shortly that these values are immaterial to the decoding procedure. As a practical matter, it is usually advantageous to assign zeros.

To take account of the known erasure-location information in forming the syndromes, Forney introduced a linear transformation on the syndromes,

$$T_i = \sum_{j=0}^{s} \sigma_j' S_{i+s+1-j}, \qquad 0 \le i \le d - s - 2 \tag{6.23}$$

The T's are called the *modified syndromes*. Notice that there are s fewer T's than S's. Thus if one symbol is erased, the $d-1$ original syndromes are transformed by Eq. (6.23) into $d-2$ modified syndromes, and so forth. We shall see how this transformation lets us establish a useful recursion among the T's.

Let us assume the presence of l errors and s erasures. Let the errors be at locations $X_1, X_2, \ldots, X_l$ and have values $Y_1, Y_2, \ldots, Y_l$. Let the known erasure locations be denoted by $Z_1, Z_2, \ldots, Z_s$, and let $D_1, D_2, \ldots, D_s$ designate the erasure-discrepancy values, that is, the difference between the correct symbol values and the values arbitrarily assigned before the syndromes are computed. We can now express the syndromes as

$$S_k = \sum_{m=1}^{l} Y_m X_m^k + \sum_{n=1}^{s} D_n Z_n^k, \qquad 1 \le k \le d-1$$

From Eq. (6.23) we write the modified syndromes as

$$T_i = \sum_{j=0}^{s} \sigma_i' \left[\sum_{m=1}^{l} Y_m X_m^{i+s+1-j} + \sum_{n=1}^{s} D_n Z_n^{i+s+1-j} \right], \qquad 0 \le i \le d-s-2$$

or

$$T_i = \sum_{m=1}^{l} Y_m X_m^{i+1} \sum_{j=0}^{s} \sigma_j' X_m^{s-j} + \sum_{n=1}^{s} D_n Z_n^{i+1} \sum_{j=0}^{s} \sigma_j' Z_n^{s-j} \tag{6.24}$$

However, from Eq. (6.22), we see that the second summation in the last term of Eq. (6.24) is the erasure-locator polynomial evaluated at a root Z_n, which equals zero. Further, we recognize from Eq. (6.22) that the second summation in the first term of the right-hand side of Eq. (6.24) is simply the erasure-locator polynomial evaluated at the error location X_m, which we write as $\sigma'(X_m)$. Therefore, if we define a new quantity E_m as

$$E_m = Y_m X_m \sigma'(X_m) \tag{6.25}$$

we can rewrite Eq. (6.24) as

$$T_i = \sum_{m=1}^{l} E_m X_m^i, \qquad 0 \le i \le d-s-2 \tag{6.26}$$

What is important to note here is that Eq. (6.26) defines the modified syndrome values in a manner essentially the same as that in which the ordinary syndrome values are defined for l-error correction, for example by Eq. (6.9). Thus we see that for the simultaneous decoding of l errors and s erasures, the transformation in Eq. (6.23) has the effect of folding the known erasure locators into the original syndromes in such a way that we preserve the form of the syndrome equations in terms of the error locators. Now, by starting with Eq. (6.26) as the formulation of a new decoding problem, where l error locators X_m are to be determined, we can perform decoding with much the same overall procedure as is done for the case of ordinary error correction. That is, we first find the l error locators from the T's and then compute the values of code symbols in the $l + s$ error and erasure locations.

If Peterson's direct solution method is used, error-locator polynomial coefficients are computed from the T's in the same way as they are computed from the S's in the earlier discussion of errors-only decoding. After solving for the roots of $\sigma(x)$, any $l + s$ of the syndrome equations can be used as a set of simultaneous linear equations to solve for the Y's and D's.

Alternatively, the Massey FSR synthesis technique may be applied in almost the same way as for ordinary error correction. That is, using Eq. (6.26) instead of Eq. (6.9), we treat the relationship of the T's to the σ's in a manner that exactly parallels the discussion in Eqs. (6.15) to (6.20), developing along the way a recursion relationship for the T's equivalent to Eq. (6.17), that is,

$$T_j + \sigma_1 T_{j-1} + \cdots + \sigma_l T_{j-l} = 0, \qquad j = l, l+1, \ldots$$

Thus the problem of finding the coefficients of the error-locator polynomial can be formulated again as an FSR synthesis problem, where the FSR must now be synthesized to generate a given sequence of modified syndrome values $\{T_j\}$ rather than original syndrome values $\{S_k\}$.

Once the error-locator polynomial is obtained, the roots are efficiently found using the Chien search. The l error locators, taken together with the s known erasure locators, in effect constitute $l + s$ erasures whose values are to be computed from the original syndromes. This can be done by solving $l + s$ syndrome equations, as in the direct method. However, a more efficient method of determining the erasure values has also been given by Forney [36]. Although we mention this method as part of the errors-and-erasures decoding procedure, it is also applicable in the case of errors-only decoding, since it is applied at the point in decoding where all of the unknown errors have been located. The suggested erasure-filling procedure is now described.

Let us denote the given erasure locators and computed error locators together by $Z_1, Z_2, \ldots, Z_{l+s}$. Now consider deleting Z_1 from the set of $l + s$ erasure locators, and forming the erasure-locator polynomial ${}_1\sigma(z)$ which has as roots the remaining $l + s - 1$ locators. Next, we calculate the coefficients of

$${}_1\sigma(z) = {}_1\sigma_0 z^{l+s-1} + {}_1\sigma_1 z^{l+s-2} + \cdots + {}_1\sigma_{l+s-1}$$

Then the erasure correction value, to be subtracted from the received or assigned value in the location Z_1, is given by

$$D_1 = \frac{\sum_{k=0}^{l+s-1} {}_1\sigma_k S_{l+s-k}}{\sum_{j=0}^{l+s-1} {}_1\sigma_j Z_1^{l+s-j}}$$

or in general by

$$D_i = \frac{\sum_{k=0}^{l+s-1} {}_i\sigma_k S_{l+s-k}}{\sum_{j=0}^{l+s-1} {}_i\sigma_j Z_i^{l+s-j}}, \qquad 1 \le i \le l+s \tag{6.27}$$

By deleting one erasure at a time, all erasure values are calculated in turn by Eq. (6.27). Another procedure requiring even fewer computations can also be used. If after computing D_1 the syndrome values are modified using

$$S'_k = S_k + D_1 Z_1^k$$

it is only necessary to form an $(l + s - 2)$-order erasure-locator polynomial, with coefficients ${}_2\sigma_1, {}_2\sigma_2, \ldots, {}_2\sigma_{l+s-2}$, in order to find D_2 from Eq. (6.27), and so forth.

We now summarize the procedure for errors-and-erasures decoding, assuming the use of the FSR synthesis algorithm, as follows:

1. Inspect the received word for erasures and compute the syndrome.
   a. If $s > d - 1$, declare the word undecodable.
   b. Otherwise, compute the syndrome values $S_1, S_2, \ldots, S_{d-1}$. If all syndromes are zero, the received word is a valid codeword, and no further processing is to be done.
2. If no symbols have been erased ($s = 0$), follow the procedure for errors-only decoding.
3. Compute the modified syndrome (if necessary).
   a. If $s = d - 1$, go to step 6.
   b. If $0 < s < d - 1$, compute the modified syndrome values using Eq. (6.23).
4. Determine the number of errors in the received word.
   a. If all $T_i = 0$, $0 \le i \le d - s - 2$, assume no errors are present and go to step 6.
   b. If some $T_i \ne 0$, use the FSR synthesis algorithm to find $\sigma_1, \sigma_2, \ldots, \sigma_l$, the coefficients of the error-locator polynomial $\sigma(x)$.

5. Determine the error locators, the roots of $\sigma(x)$, using the Chien search. Put the l computed error locators together with the given s erasure locators to make up the new set of erasure locators $Z_1, Z_2, \ldots, Z_{l+s}$.
6. Compute the $l + s$ erasure magnitudes, using Eq. (6.27) or the more efficient procedure discussed immediately following Eq. (6.27).

6.7. FOURIER TRANSFORM TECHNIQUES FOR RS CODES

There is an alternative decoding procedure for BCH codes that, depending upon code parameters, can be computationally more efficient than the procedures considered thus far. The technique is based on the close relationship that exists between the calculation of syndrome values for a received word and the computation of a Fourier transform in a finite field. The transform approach can be advantageous when the code parameters permit use of well-known *fast-transform algorithms*. The transform technique can also be used for encoding in the case of nonsystematic codes. In addition to providing possible savings in decoding complexity, the Fourier transform formulation represents an alternative way of describing the structure of BCH codes, one that may be appealing to readers familiar with the theory and methods of digital signal processing. We consider the case of decoding RS codes.

6.7.1. The Finite Field Fourier Transform

Let f_i, $i = 0, 1, 2, \ldots, N-1$, be a sequence of numbers, either real or complex. The *discrete Fourier transform* (*DFT*) of the given sequence is a sequence of N spectral values defined as

$$F_k = \sum_{i=0}^{N-1} f_i e^{-j(2\pi/N)ik}, \qquad k = 0, 1, \ldots, N-1 \tag{6.28}$$

where $j = \sqrt{-1}$. The *inverse DFT* of the sequence of spectral values, which reconstructs the original sequence, is defined as

$$f_i = \frac{1}{N} \sum_{k=0}^{N-1} F_k e^{j(2\pi/N)ik}, \qquad i = 0, 1, \ldots, N-1$$

We shall call F_k, $k = 0, 1, 2, \ldots, N-1$ the *spectral coefficients* of the original sequence. The DFT plays a role in the analysis of sampled-data signals that directly parallels that of the classical Fourier transform in the analysis of continuous signals. A thorough treatment of the DFT may be found in Oppenheim and Schafer [123]. A number of journal papers on the subject can be found in two IEEE volumes of reprints on digital signal processing [124,140]. (*Note*: The literature contains several variations in the notation used to define the DFT, some involving a multiplicative factor $1/N$ or $1/\sqrt{N}$. The various forms are equivalent, each simply requiring an appropriate compensation in the form of the inverse transform. Here we use a form closely resembling, although not identical to, the one recommended by the IEEE committee on digital signal processing [124]. Our variation consists in using a subscript notation rather than a function notation for the sample and spectral values.)

We observe that the exponential factor $e^{-j(2\pi/N)}$ is an *Nth root of unity* in the field of complex numbers. Recalling the introductory discussion of finite fields in Section 4.1, we can say that a field element α of order N is also an Nth root of unity. Thus, drawing upon this analogy between $e^{-j(2\pi/N)}$ and α, we can define a *finite field Fourier transform* directly paralleling Eq. (6.28) as follows. Let $\mathbf{r} = r_0, r_1, \ldots, r_{N-1}$ be a vector of finite field elements in $GF(q)$, where $N = q - 1$, the order of a primitive element α of $GF(q)$. The finite field Fourier transform of $\mathbf{r}$ is another vector of N elements in $GF(q)$, which we denote as $\mathbf{R} = \{R_k\}$, $k = 0, 1, \ldots, N - 1$, where the elements of $\mathbf{R}$ are given by

$$R_k = \sum_{i=0}^{N-1} r_i \alpha^{ik}, \qquad k = 0, 1, \ldots, N - 1 \tag{6.29}$$

The correspondence to the DFT, Eq. (6.28), is immediately seen. Similarly, directly paralleling the inverse DFT, we can recover the original sequence $\mathbf{r}$ with the *inverse finite field Fourier transform* given by

$$r_i = \frac{1}{N \bmod p} \sum_{k=0}^{N-1} R_k \alpha^{-ik}, \qquad i = 0, 1, \ldots, N - 1 \tag{6.30}$$

where p is the characteristic of the field $GF(q)$. (Recall that if the size of the field, q, is prime, the characteristic is simply q, while if q is a prime power, $q = p^m$, the characteristic is p.) The preceding discussion provides a somewhat restricted definition of the finite field Fourier transform, one that is adequate for our purposes in treating RS codes. More general and rigorous treatments of the finite field Fourier transform and the many parallels with the classical Fourier transform can be found in a number of references, including Pollard [135] and Blahut [13, 14].

For our present purposes it is useful to write Eqs. (6.29) and (6.30) in the form of matrix equations as

$$\mathbf{R} = \mathbf{T}\mathbf{r}$$

and

$$\mathbf{r} = \mathbf{T}'\mathbf{R}$$

where $\mathbf{T}$ and $\mathbf{T}'$ compactly represent the forward and inverse transformations, respectively. The (m,n)th elements in $\mathbf{T}$ and $\mathbf{T}'$ are $\alpha^{(m-1)(n-1)}$ and $\alpha^{-(m-1)(n-1)}$ respectively. Further, we note that since

$$\mathbf{r} = \mathbf{T}'\mathbf{R} = \mathbf{T}'\mathbf{T}\mathbf{r}$$

the inverse transform matrix $\mathbf{T}'$ is mathematically equivalent to $\mathbf{T}^{-1}$, the inverse of $\mathbf{T}$.

6.7.2. Transform Decoding for Errors Only

Referring to our earlier discussion of syndrome decoding of a distance-d RS code on $GF(q)$ (Section 6.6), we recall that the syndrome values for a received word $\mathbf{r}$ of length n are given by

$$S_k = \sum_{i=0}^{n-1} r_i \alpha^{ik}, \qquad k = m_0, m_0 + 1, \ldots, m_0 + d - 2$$

where α is a primitive element in $GF(q)$. We also recall that for an RS code, the error locators and the code symbols are both drawn from $GF(q)$. Therefore we see by comparison with Eq. (6.29) that with $m_0 = 0$, the $d - 1$ syndromes of $\mathbf{r}$ are simply the first $d - 1$ spectral coefficients of $\mathbf{Tr}$, the finite field Fourier transform of $\mathbf{r}$. We note in passing that this provides an alternative way of defining an RS code. That is, we can define a distance-d RS code on $GF(q)$ as the set of vectors on $GF(q)$, with length $n = q - 1$, whose first $d - 1$ spectral coefficients are equal to zero. In fact, it is possible to show that for the particular case of RS codes, a code may be defined by setting any $d - 1$ of its spectral coefficients to zero. This is directly connected to the property of RS codes that any $n - k$ of its symbols may be taken as parity-check symbols. The paper by Blahut [13] includes an extensive discussion of the frequency domain formulation of the broad class of cyclic codes, which includes RS codes as a special case.

To examine the use of transform techniques for decoding, we return to the familiar representation of a received word as $\mathbf{r} = \mathbf{c} + \mathbf{e}$, where $\mathbf{c}$ is the transmitted codeword and $\mathbf{e}$ the additive error pattern. Since the transformation is linear, we can write

$$\mathbf{Tr} = \mathbf{Tc} + \mathbf{Te} \tag{6.31}$$

Thus we see that if it were possible to compute $\mathbf{Te}$, the transform of the error pattern, we could determine the transform of the codeword $\mathbf{c}$ by computing

$$\mathbf{Tc} = \mathbf{Tr} - \mathbf{Te}$$

and then reconstruct the transmitted word by taking the inverse transform. Gore [57], in fact, showed that $\mathbf{Te}$ can be found from $\mathbf{Tr}$, provided that the number of received errors, l, is within the error-correction capability of the code ($2l < d$). The complete decoding procedure, which we term *transform decoding*, will be outlined below. Before describing the procedure, we point out with reference to Eq. (6.31) that since the first $d - 1$ elements in **Tc** are zero, the first $d - 1$ spectral coefficients in **Te** are equal to the first $d - 1$ elements in **Tr**, which in turn are the syndrome values $S_0, S_1, S_2, \ldots, S_{d-2}$. [Note that in order to be consistent with the notation in Eqs. (6.28) and (6.29) we have indexed the syndrome values from 0 to $d - 2$, although earlier in the book we used index numbers ranging over 1 to $d - 1$. This corresponds to the choice of $m_0 = 0$ in defining the roots of the code rather than $m_0 = 1$ and simplifies the presentation. However, the choice of notation is immaterial to the basic decoding approach.] Let us denote the transform of the error pattern as **E**, given by **E** = **Te**, and the jth component of **E** by E_j.

Returning to Newton's identities [see Eq. (6.17)] for RS codes, we have

$$S_j + \sigma_1 S_{j-1} + \cdots + \sigma_t S_{j-t} = 0, \qquad j = t, t + 1, \ldots$$

or

$$E_j + \sigma_1 E_{j-1} + \cdots + \sigma_t E_{j-t} = 0, \qquad j = t, t + 1, \ldots$$

where the σ's are the coefficients of the error-locator polynomial. The key observation made by Gore is that the recursion given above is true for all $j \geq t$, and therefore, once the error-locator polynomial is determined, the complete spectrum of the error pattern can be obtained. In fact, **E** = **Te** is precisely the sequence of digits produced by the minimal-length FSR found by the Massey algorithm when it is initialized with the syndrome sequence. The transform decoding procedure can therefore be described as follows:

1. Compute **Tr**, the transform of the received word. Extract the syndromes, which are the first $d - 1$ spectral coefficients in the transform.
2. Using the Massey FSR synthesis algorithm, find the minimal length FSR that generates the syndromes $S_0, S_1, \ldots, S_{d-2}$.
3. Use the synthesized FSR to recursively generate the remaining elements of **E**, namely, $E_{d-1}, E_d, \ldots, E_{n-1}$. In simple terms, one can think of this as simply clocking the FSR an additional $n - d + 1$ times immediately after the successful reproduction of syndrome values in the Massey FSR synthesis algorithm. This completes the computation of **E** = **Te**.
4. Compute **Tr** − **E**, and then find the transmitted codeword as

$$\mathbf{c} = \mathbf{T}'[\mathbf{Tr} - \mathbf{E}]$$

To compare the transform decoder with the general procedure outlined earlier, we summarize the general procedure as follows:

1. Calculate the syndrome values for the received word.
2. Calculate the coefficients of the error-locator polynomial, using the Massey FSR algorithm.

3. Determine the error locations, using the Chien search.
4. Compute the error values and correct the indicated errors.

Note that the use of the Massey algorithm (step 2) is common to the two decoding procedures. Furthermore, it can be shown that step 3 in either procedure requires about the same number of arithmetic operations. Therefore, the transform decoder will be computationally more efficient in cases where the transform computations can be done more efficiently than the syndrome computations plus the error-value calculations. In general, improved efficiency is found for the transform decoder when the code block length is highly composite (factorable), so that *fast Fourier transform* (FFT) techniques are applicable.

More specifically, there are two categories of cases in which the transform decoder considered here may offer an improvement in efficiency as measured by the number of decoding computations. First, since the transform complexity is independent of the code rate, the transform decoder may be advantageous for decoding low-rate codes (codes for which a large amount of error correction is to be done), which entails a relatively large amount of computation for steps 1 and 4 in the standard decoder. Second, even for moderate- to high-rate codes, the transform decoder can improve efficiency if the block length is long and highly composite.

6.7.3. Errors-Only Decoding with Frequency-Domain Encoding

The discussion of transform decoding in Section 6.7.2 assumes conventional, systematic *time-domain encoding*, in which a transmitted codeword is composed of k source-generated information symbols and $n - k$ calculated parity-check symbols. Decoding starts with use of the forward transform of $\mathbf{r}$ and finishes with computation of the inverse transform of $\mathbf{Tr} - \mathbf{Te}$. An interesting and sometimes useful variation of the transform procedure consists in transmitting a *frequency-domain codeword* instead, with the result that only one transform operation, the inverse $\mathbf{T}'$, is needed in decoding. We describe the technique as follows.

Let the k source information symbols be represented by the polynomial $i(x)$, where

$$i(x) = i_0 + i_1 x + \cdots + i_{k-1} x^{k-1}$$

We now encode $i(x)$ into the code polynomial $F(f)$, where

$$F(f) = F_0 + F_1 f + \cdots + F_{n-1} f^{n-1}$$

by the rule

$$F_i = i(x)|_{x=\alpha^i}, \qquad 0 \leq i \leq n-1$$

Alternatively, we can write the information polynomial as an n-element vector by appending $n - k$ zeros, and in matrix notation we have

$$[F_0 \quad F_1 \quad \cdots \quad F_{n-1}] = \begin{bmatrix} 1 & 1 & 1 & \cdots & 1 \\ 1 & \alpha & \alpha^2 & \cdots & \alpha^{1(n-1)} \\ 1 & \alpha^2 & \alpha^4 & \cdots & \alpha^{2(n-1)} \\ \vdots & \vdots & \vdots & & \vdots \\ 1 & \alpha^{(n-1)} & \alpha^{2(n-1)} & \cdots & \alpha^{(n-1)(n-1)} \end{bmatrix} \begin{bmatrix} i_0 \\ i_1 \\ \vdots \\ i_{k-1} \\ 0 \\ \vdots \\ 0 \end{bmatrix}$$

and recognize that the codeword $F(f)$ is the n-point finite field transform of the information vector. Note that the codeword is nonsystematic, since the information symbols are not preserved directly.

Using Eq. (6.30), the inverse transform relationships are

$$\frac{1}{n \bmod p} F(1) = i_0$$

$$\frac{1}{n \bmod p} F(\alpha^{-1}) = i_1$$

$$\frac{1}{n \bmod p} F(\alpha^{-2}) = i_2$$

$$\vdots$$

$$\frac{1}{n \bmod p} F(\alpha^{-(k-1)}) = i_{k-1}$$

and

$$\frac{1}{n \bmod p} F(\alpha^{-(k-1)-j}) = 0, \qquad 1 \le j \le n - k$$

or, since $\alpha^n = 1$,

$$F(\alpha^i) = 0, \qquad 1 \le i \le n - k = d - 1 \tag{6.32}$$

For a more compact notation, we let $\mathbf{C}$ represent the transmitted frequency domain codeword, $\mathbf{C} = \{F_i\} = \mathbf{Ti}$ and also write the error pattern and the received word as frequency domain vectors $\mathbf{E}$ and $\mathbf{R}$, respectively, giving

$$\mathbf{R} = \mathbf{C} + \mathbf{E}$$

Let us now start the decoding procedure by applying an inverse transformation to $\mathbf{R}$, which yields

$$\begin{aligned}\mathbf{T'R} &= \mathbf{T'C} + \mathbf{T'E} \\ &= \mathbf{T'Ti} + \mathbf{T'E} \\ &= \mathbf{i} + \mathbf{e}\end{aligned} \tag{6.33}$$

From Eq. (6.32) we see that for error-free reception, the last $d - 1$ elements of $\mathbf{T'R}$ are all zero, and thus nonzero values for those terms constitute a syndrome that can be interpreted in the same way as in the application of a forward transformation to a time-domain word $\mathbf{r}$. Thus decoding proceeds in the same manner as before. That is, the Massey algorithm is used to synthesize an FSR that generates the last $d - 1$ values of $\mathbf{T'R}$, and the recursion is extended to form $\mathbf{T'E}$. But now we see from Eq. (6.33) that since $\mathbf{T'E} = \mathbf{e}$, we simply have $\mathbf{i} = \mathbf{T'R} - \mathbf{T'E}$, and no further transform operations are needed. Of course, for this technique the encoder is more complex, since a transformation is required. However, the frequency-domain encoder may be simpler than the time-domain encoder if the block length is composite, since a fast-transform routine may require fewer operations than the more usual encoding routine, polynomial division.

6.7.4. Transform Decoding for Errors and Erasures

It was shown by Gore [57] and Michelson [111] that transform techniques can be used to simultaneously correct l unknown errors and fill s erasures provided that $2l + s < d$. We now outline the decoding procedure for the case of time-domain encoding. We begin, as in the case of errors-only decoding, by computing $\mathbf{Tr}$, the transform of the received word, with erased symbol values arbitrarily set to zero for computational convenience. The first $d - 1$ spectral coefficients are taken as the syndrome values $S_0, S_1, \ldots, S_{d-2}$. Then exactly as was described in Section 6.6.3, the erasure-locator polynomial is formed and its coefficients are used, with Eq. (6.23), to form the $d - s - 1$ modified syndromes. Next, the Massey algorithm is used to find the minimal length FSR that generates the modified syndrome, and the FSR is then used recursively to produce the complete transform of the error pattern, that is T_i, $0 \le i \le n - 1$.

However, in order to complete the decoding procedure in this case, we require an estimate of the transform of the error-plus-erasure pattern, that is, S_i, $0 \le i \le n - 1$. This can be obtained by inverting the linear transformation of the syndrome that was used initially to produce the modified syndrome, Eq. (6.23). Notice that Eq. (6.23) describes the operation of an s-stage linear feedforward shift register whose tap connections are given by the coefficients of the erasure-locator polynomial. The inverse transformation required can thus be implemented with an s-stage linear feedback shift register whose tap connections are again the coefficients of the erasure-locator polynomial. The transform of the codeword is then found by subtracting the transform of the error-and-erasure pattern from **Tr**. Finally, the estimate of the transmitted codeword is formed by taking the inverse finite field Fourier transform.

6.7.5. An Example: Fast-Transform Decoding in *GF*(64)

Let us consider the specific case of decoding an RS code on $GF(64)$ with block length $n = 63$. Since 63 is composite, a fast-transform decomposition exists for the 63-point transform in $GF(64)$. In fact, the 63×63 transform matrix **T** can be put in the form **U** shown below by using a row rearrangement procedure [50].

$$\mathbf{U} = \begin{bmatrix} \mathbf{C} & \mathbf{C} & \mathbf{C} & \mathbf{C} & \mathbf{C} & \mathbf{C} & \mathbf{C} \\ \mathbf{C}\,\mathbf{D}_1 & \mathbf{C}\,\mathbf{D}_2 & \mathbf{C}\,\mathbf{D}_3 & \mathbf{C}\,\mathbf{D}_4 & \mathbf{C}\,\mathbf{D}_5 & \mathbf{C}\,\mathbf{D}_6 & \mathbf{C}\,\mathbf{D}_7 \\ \mathbf{C}\,\mathbf{D}_1^2 & \mathbf{C}\,\mathbf{D}_2^2 & \mathbf{C}\,\mathbf{D}_3^2 & \mathbf{C}\,\mathbf{D}_4^2 & \mathbf{C}\,\mathbf{D}_5^2 & \mathbf{C}\,\mathbf{D}_6^2 & \mathbf{C}\,\mathbf{D}_7^2 \\ \mathbf{C}\,\mathbf{D}_1^3 & \mathbf{C}\,\mathbf{D}_2^3 & \mathbf{C}\,\mathbf{D}_3^3 & \mathbf{C}\,\mathbf{D}_4^3 & \mathbf{C}\,\mathbf{D}_5^3 & \mathbf{C}\,\mathbf{D}_6^3 & \mathbf{C}\,\mathbf{D}_7^3 \\ \mathbf{C}\,\mathbf{D}_1^4 & \mathbf{C}\,\mathbf{D}_2^4 & \mathbf{C}\,\mathbf{D}_3^4 & \mathbf{C}\,\mathbf{D}_4^4 & \mathbf{C}\,\mathbf{D}_5^4 & \mathbf{C}\,\mathbf{D}_6^4 & \mathbf{C}\,\mathbf{D}_7^4 \\ \mathbf{C}\,\mathbf{D}_1^5 & \mathbf{C}\,\mathbf{D}_2^5 & \mathbf{C}\,\mathbf{D}_3^5 & \mathbf{C}\,\mathbf{D}_4^5 & \mathbf{C}\,\mathbf{D}_5^5 & \mathbf{C}\,\mathbf{D}_6^5 & \mathbf{C}\,\mathbf{D}_7^5 \\ \mathbf{C}\,\mathbf{D}_1^6 & \mathbf{C}\,\mathbf{D}_2^6 & \mathbf{C}\,\mathbf{D}_3^6 & \mathbf{C}\,\mathbf{D}_4^6 & \mathbf{C}\,\mathbf{D}_5^6 & \mathbf{C}\,\mathbf{D}_6^6 & \mathbf{C}\,\mathbf{D}_7^6 \end{bmatrix}$$

The matrix **C** is shown below where for typographical convenience we write only exponents of α, a primitive element of $GF(64)$, as

$$\mathbf{C} = \begin{bmatrix} 0 & 0 & 0 & 0 & 0 & 0 & 0 & 0 & 0 \\ 0 & 7 & 14 & 21 & 28 & 35 & 42 & 49 & 56 \\ 0 & 14 & 28 & 42 & 56 & 7 & 21 & 35 & 49 \\ 0 & 21 & 42 & 0 & 21 & 42 & 0 & 21 & 42 \\ 0 & 28 & 56 & 21 & 49 & 14 & 42 & 7 & 35 \\ 0 & 35 & 7 & 42 & 14 & 49 & 21 & 56 & 28 \\ 0 & 42 & 21 & 0 & 42 & 21 & 0 & 42 & 21 \\ 0 & 49 & 35 & 21 & 7 & 56 & 42 & 28 & 14 \\ 0 & 56 & 49 & 42 & 35 & 28 & 21 & 14 & 7 \end{bmatrix}$$

and $\mathbf{D}_1$ is the diagonal matrix

$$\mathbf{D}_1 = \begin{bmatrix} \alpha^0 & & & & & & & & \\ & \alpha^1 & & & & & & & \\ & & \alpha^2 & & & & & & \\ & & & \alpha^3 & & & & & \\ & & & & \alpha^4 & & & & \\ & & & & & \alpha^5 & & & \\ & & & & & & \alpha^6 & & \\ & & & & & & & \alpha^7 & \\ & & & & & & & & \alpha^8 \end{bmatrix}$$

and

$$\mathbf{D}_2 = \alpha^9\mathbf{D}_1, \quad \mathbf{D}_3 = \alpha^{18}\mathbf{D}_1 \quad, \ldots, \quad \mathbf{D}_7 = \alpha^{54}\mathbf{D}_1$$

The computational advantage that is achieved by rearranging the rows of **T** to form **U** can now be described. Our intention is of course to compute **Tr** where **r** is the received word. Equivalently we could compute $\mathbf{Ur}'$ where $\mathbf{r}'$ is simply **r** with its elements rearranged in the same manner as the rows of **T** are rearranged to form **U**. Notice that $\mathbf{r}'$ is a 63-element vector that may be segmented into seven 9-element vectors $\mathbf{r}'_i$. That is, we let

$$\mathbf{r}' = (\mathbf{r}'_1, \mathbf{r}'_2, \mathbf{r}'_3, \mathbf{r}'_4, \mathbf{r}'_5, \mathbf{r}'_6, \mathbf{r}'_7)$$

Then, to compute the first nine elements of $\mathbf{Ur}'$ we may form

$$\mathbf{Cr}'_1 + \mathbf{Cr}'_2 + \mathbf{Cr}'_3 + \mathbf{Cr}'_4 + \mathbf{Cr}'_5 + \mathbf{Cr}'_6 + \mathbf{Cr}'_7$$

which requires $9 \times 9 \times 7 = 567$ multiplications. However, we could also compute these nine spectral coefficients by forming

$$\mathbf{C}(\mathbf{r}'_1 + \mathbf{r}'_2 + \mathbf{r}'_3 + \mathbf{r}'_4 + \mathbf{r}'_5 + \mathbf{r}'_6 + \mathbf{r}'_7)$$

which requires 81 multiplications, a significant saving.

A similar reduction in the number of multiplications can be achieved for other segments of $\mathbf{Ur}'$. For example, for the second nine-element segment we form

$$\mathbf{C}\sum_{i=1}^{7} \mathbf{D}_i\mathbf{r}'_i$$

Note that since $\mathbf{D}_i$ is a diagonal matrix, only nine multiplications are required in the product $\mathbf{D}_i\mathbf{r}'_i$.

The computational advantage of using the matrix **U** rather than **T** to compute the transform of **r** can be improved further. Specifically, the 9×9 matrix **C** can be decomposed to $\mathbf{C}'$, where

$$\mathbf{C}' = \begin{bmatrix} \mathbf{B} & \mathbf{B} & \mathbf{B} \\ \mathbf{BD}'_1 & \mathbf{BD}'_2 & \mathbf{BD}'_3 \\ \mathbf{BD}'^2_1 & \mathbf{BD}'^2_2 & \mathbf{BD}'^2_3 \end{bmatrix}$$

with

$$\mathbf{B} = \begin{bmatrix} \alpha^0 & \alpha^0 & \alpha^0 \\ \alpha^0 & \alpha^{21} & \alpha^{42} \\ \alpha^0 & \alpha^{42} & \alpha^{21} \end{bmatrix}$$

and

$$\mathbf{D}_1' = \begin{bmatrix} \alpha^0 & 0 & 0 \\ 0 & \alpha^7 & 0 \\ 0 & 0 & \alpha^{14} \end{bmatrix}$$

$$\mathbf{D}_2' = \alpha^{21}\mathbf{D}_1'$$

$$\mathbf{D}_3' = \alpha^{42}\mathbf{D}_1'$$

We now compare decoding complexity for RS codes on $GF(64)$ given transform decoding on the one hand and the standard decoding procedure on the other. We assume that in the standard decoder the syndrome is computed as a discrete transform. That is,

$$S_k = \left[\cdots \left\{ \left(r_{62}\alpha^k + r_{61} \right) \alpha^k + r_{60} \right\} \alpha^k \cdots \right] \alpha^k + r_0, \qquad 1 \le k \le d-1$$

Since there are $d - 1$ syndrome values to compute, the syndrome calculation requires $62 \times (d - 1)$ multiplications. Further, it has been shown [36] that the error-value calculation can be implemented with approximately t^2 multiplications when t values are computed. Therefore, the syndrome calculation plus the error-value calculation requires approximately $62 \times (d - 1) + t^2$ multiplications. However, the transform operation can be implemented with about 500 multiplications in $GF(64)$ if we omit multiplications by unity. We should therefore consider using the fast-transform decoder for nonsystematic codes when

$$500 < 62(d - 1) + t^2$$

and for systematic codes when

$$1000 < 62(d - 1) + t^2$$

For example, the (63,33) RS code on $GF(64)$ has distance 31 and can correct 15 errors. We have $d = 31$ and $t = 15$, and for nonsystematic codes we have

$$500 < 62(30) + 15^2 = 2085$$

Thus the transform decoder requires fewer multiplications. For lower rate codes the relative advantage of the fast-transform technique increases.

6.8. MODIFICATIONS OF BCH AND RS CODES

Applications often arise where nonbinary codes designed by strict adherence to the definitions presented in this chapter have parameters that are in some way inconvenient for the user's needs. In particular, the code block length or the alphabet size may be close to but not exactly equal to those required. It is desirable, therefore, to have means to modify the strict-definition code designs. We shall next describe several useful modification procedures.

6.8.1. Simple Code Shortening

As was pointed out in Section 4.3.4 for binary codes, it is a simple matter to shorten any linear block code by setting some selected number of the information symbols to zero as part of the encoding procedure. The characteristics of a nonbinary code formed in this way are much the same as the shortened binary codes. We simply summarize the important points as follows:

1. The minimum distance of the shortened code is at least as great as that of the unshortened code. It is, in general, not a simple matter to predict what the actual minimum distance of a shortened code will be. (An exception is the case of RS codes, for which shortening leaves the minimum distance unchanged since d always equals $n - k + 1$.)
2. A shortened code is not necessarily cyclic. In fact, it is rarely cyclic.
3. As with the binary codes, there is no general theory to give guidance as to which symbols are best to omit for a required amount of shortening. In some cases, certain choices of symbol positions to be omitted can result in greater minimum distance than other choices. (Again, RS codes are an exception.) However, shortening is usually done in the most convenient manner, setting the string of highest order consecutive information symbols equal to zero.

6.8.2. Adding Information Symbols to an RS Code

While simple shortening of any nonbinary BCH or RS code is readily accomplished, there is no corresponding all-purpose rule for lengthening a code, that is, increasing the number of information symbols while leaving the number of check symbols and the minimum distance of the code unchanged. There are cases, however, where a modest amount of lengthening can be done. Wolf [180] has shown that RS codes and certain other nonbinary BCH codes can be lengthened by up to two information symbols without changing the minimum distance. The resulting lengthened codes are linear, though not cyclic, and can be decoded with an algorithm that makes use of a standard decoder. Because the lengthened codes are not cyclic, we discuss them in terms of parity-check matrices rather than generator polynomials. We confine our attention to RS codes, since the lengthening can be applied to any RS code, without restrictions.

We begin by constructing the parity-check matrix for the original RS code on $GF(q)$ from the specification of the roots of the generator polynomial as

$d-1$ consecutive powers of some primitive element α of $GF(q)$. We do this in a manner directly analogous to that outlined in Section 4.2.3 for binary cyclic codes. That is, letting $c(x)$ be a polynomial representing a codeword, we describe the distance-d RS code on $GF(q)$ as the set of polynomials $c(x)$, with coefficients in $GF(q)$, such that

$$c(x) = 0, \qquad x = \alpha^{m_0}, \alpha^{m_0+1}, \ldots, \alpha^{m_0+d-2}$$

Expanding the polynomials, evaluated at the prescribed values of x, and letting $m_0 = 0$ for simplicity, we have

$$\begin{array}{llllll}
c_0\alpha^0 + c_1\alpha^0 & + c_2\alpha^0 & + c_3\alpha^0 & + \cdots & + c_{n-1}\alpha^0 & = 0 \\
c_0\alpha^0 + c_1\alpha^1 & + c_2\alpha^2 & + c_3\alpha^3 & + \cdots & + c_{n-1}\alpha^{n-1} & = 0 \\
\vdots & \vdots & \vdots & \cdots & \vdots & \vdots \\
c_0\alpha^0 + c_1\alpha^{(d-2)} & + c_2\alpha^{2(d-2)} & + c_3\alpha^{3(d-2)} & + \cdots & + c_{n-1}\alpha^{(n-1)(d-2)} & = 0
\end{array}$$

We now write this set of $d-1$ simultaneous equations in matrix form, $\mathbf{Hc} = \mathbf{0}$, as

$$\begin{bmatrix} \alpha^0 & \alpha^0 & \alpha^0 & \alpha^0 & \cdots & \alpha^0 \\ \alpha^0 & \alpha^1 & \alpha^2 & \alpha^3 & \cdots & \alpha^{n-1} \\ \alpha^0 & \alpha^2 & \alpha^4 & \alpha^6 & \cdots & \alpha^{2(n-1)} \\ \vdots & \vdots & \vdots & \vdots & \vdots & \vdots \\ \alpha^0 & \alpha^{d-2} & \alpha^{2(d-2)} & \alpha^{3(d-2)} & \cdots & \alpha^{(n-1)(d-2)} \end{bmatrix} \begin{bmatrix} c_0 \\ c_1 \\ c_2 \\ c_3 \\ \vdots \\ c_{n-1} \end{bmatrix} = \mathbf{0} \qquad (6.34)$$

Thus we see that the $\mathbf{H}$ matrix for the RS code can be constructed in a straightforward manner by evaluating the appropriate powers of a primitive element of $GF(q)$. We now append two columns to the check matrix, where each new column has only a single nonzero entry, $\alpha^0 = 1$, producing a new check matrix, $\mathbf{H}'$, given by

$$\mathbf{H}' = \begin{bmatrix} 1 & 0 & 1 & 1 & 1 & 1 & \cdots & 1 \\ 0 & 0 & 1 & \alpha^1 & \alpha^2 & \alpha^3 & \cdots & \alpha^{n'-3} \\ 0 & 0 & 1 & \alpha^2 & \alpha^4 & \alpha^6 & \cdots & \alpha^{2(n'-3)} \\ \vdots & \vdots & \vdots & \vdots & \vdots & \vdots & \vdots & \vdots \\ 0 & 1 & 1 & \alpha^{d-2} & \alpha^{2(d-2)} & \alpha^{3(d-2)} & \cdots & \alpha^{(n'-3)(d-2)} \end{bmatrix} \qquad (6.35)$$

This modification of the parity-check matrix has lengthened the original code block length by two symbols, from n to $n' = n + 2$. Note that we have not increased the number of check symbols, which is equal to the number of rows in the check matrix.

We now want to show that the new code defined by $\mathbf{H}'$ has the same minimum distance, and hence error correction power, as the original code. To do this we make use of an interpretation of minimum distance first introduced in Chapter 3, namely, that if a code has minimum distance d, then any $d - 1$ columns of its parity-check matrix are linearly independent over the field on which the matrix elements are defined, $GF(q)$.

Since the original code, defined by Eq. (6.34), is designed to have minimum distance d, we are assured that any $d - 1$ or fewer columns in the $n' - 2$ rightmost columns of the matrix in Eq. (6.35) are linearly independent. We do not bother to show this formally. Proofs can be found in a number of texts, for example, Forney [37]. Given linear independence of the $n' - 2$ columns of $\mathbf{H}'$ taken from the $\mathbf{H}$ matrix of the original code, it is a simple matter to show that any $d - 1$ columns of $\mathbf{H}'$ are likewise linearly independent. We shall do this using well-known properties of determinants. Namely, linear independence of any m columns in a matrix having m rows is demonstrated by showing that the determinant for any such set of m columns is nonzero. We now consider taking all possible sets of $d - 1$ columns of $\mathbf{H}'$ (note that there are $d - 1$ rows in $\mathbf{H}'$) and classifying them into three groups, as follows:

Group 1. All $d - 1$ columns are in the $n' - 2$ rightmost columns of $\mathbf{H}'$.

Group 2. One column is either the first or second column (from left) and all remaining $d - 2$ columns are in the $n' - 2$ rightmost columns of $\mathbf{H}'$.

Group 3. Two of the columns are the first and second columns of $\mathbf{H}'$, and the remaining $d - 3$ columns are in the $n' - 2$ rightmost columns of $\mathbf{H}'$.

The various arrangements of columns of $\mathbf{H}'$, of course, correspond directly to classes of error patterns, which we want to show are correctable. It is readily seen that all arrangements of columns in group 1 are linearly independent, since this case is equivalent to restricting columns to the original parity-check matrix $\mathbf{H}$. In examining groups 2 and 3, we make use of the standard expansion of a determinant by cofactors (see Appendix A). That is, the expansion of an $m \times m$ determinant is the sum of m $(m - 1) \times (m - 1)$ determinants, where the expansion can be taken about the elements of any one row or column of the $m \times m$ matrix. The coefficient in each term of the expansion is one of the elements in the chosen row or column (with an appropriate sign), and the corresponding $(m - 1) \times (m - 1)$ determinant is the determinant of the $m - 1$ rows and $m - 1$ columns left by striking the row and column of the particular element.

In arrangements belonging to group 2, we want to expand about the one column containing a single one and all remaining elements zero. This will correspond to evaluating each $(d - 1) \times (d - 1)$ determinant as one $(d - 2) \times (d - 2)$ determinant, where all $d - 2$ columns are taken from the $n' - 2$

rightmost columns of $\mathbf{H}'$. We know as a result that each such determinant must be nonzero.

For arrangements in group 3, we begin by expanding about one of the first two columns of $\mathbf{H}'$, which again yields only one $(d-2) \times (d-2)$ determinant. However, each such $(d-2) \times (d-2)$ determinant includes a column having only a single nonzero element, and thus the $(d-2) \times (d-2)$ determinant is conveniently expanded about that column, yielding only a single $(d-3) \times (d-3)$ determinant with all $d-3$ columns in the $n'-2$ rightmost columns of $\mathbf{H}'$. Therefore we know that every determinant in group 3 will also be nonzero, and we conclude that any $d-1$ columns of $\mathbf{H}'$ are linearly independent. This in turn means that this method of lengthening the original code by two symbols preserves the minimum distance of the original code, and since the number of check symbols is not increased, the additional two symbols are available as new information symbols. A very useful aspect of the use of this technique with an RS code is the following: Since the minimum distance d and the number of parity checks, $n-k$, are both left unchanged by the lengthening, the relationship $d = n - k + 1$ is unchanged. It will be recalled from earlier discussion in Section 6.4 that the relationship $d = n - k + 1$ is true only for maximum-distance separable (MDS) codes, and the codeword weight distribution formula given in Eq. (6.5) applies to any MDS code. This means that the lengthened RS codes are MDS and the weight distributions may be readily computed. Finally, we note that the preservation of d and $n-k$ applies equally if the code is lengthened by only one symbol, only column arrangement groups 1 and 2 being applicable in such a case.

Wolf states that while it is possible in certain cases to append more than two columns to the original parity-check matrix of a nonbinary BCH code, there are few general results for such cases. A detailed discussion of various properties of the lengthened codes may be found in reference 180.

Letting $\mathbf{r}$ be the received word, decoding of the lengthened code proceeds as follows:

1. Compute the syndrome,

$$\mathbf{S} = \mathbf{r}(\mathbf{H}')^T = S_0, S_1, S_2, \ldots, S_{d-2}$$

2. Ignore S_0 and S_{d-2}, and attempt up to $(d-3)/2$ error corrections on the symbol positions associated with the original code before lengthening. (Note that the first extended symbol affects only S_0, and the second affects only S_{d-2}.)
3. If $(d-5)/2$ or fewer errors are found and corrected in step 2, recalculate S_0 and S_{d-2} (if necessary), and check the two symbols in the extended positions using the possibly new values for S_0 and S_{d-2}. If S_0 or S_{d-2} or both do not check, the appropriate error value(s) may be computed, since S_0 and S_{d-2} each now check a single distinct symbol.

4. If, in step 2, $(d-3)/2$ errors are found and corrected, recalculate S_0 and S_{d-2}. Using the extended symbols, check S_0 and S_{d-2}.
   a. If S_0 and S_{d-2} check, stop; decoding is completed.
   b. If either S_0 or S_{d-2} checks, but not both, compute the appropriate error value for the single extended symbol in error.
   c. If neither S_0 nor S_{d-2} checks, assume that the extended symbol position values are correct as received. Return to the original received word and its syndrome, modify S_0 and S_{d-2} using the assumption that the extended symbol position values are correct, and attempt $(d-1)/2$-error correction on the symbols in positions corresponding to the original code, before lengthening.
5. If, in step 2, $(d-3)/2$-error correction fails, that is, the event more than $(d-3)/2$ errors is indicated, assume that the two extended symbols are correct and proceed as in step 4c.

6.8.3. Designing Codes for Non-Field Alphabets

There occasionally occur applications where it is desirable to have a code design in which all code symbols, both information and parity, are confined to an alphabet that does not correspond to a finite field. Examples might be systems in which all symbols must be confined to a 26-ary (English letters) or 36-ary (letters plus numbers) alphabet. A procedure first described by Solomon [164], and apparently not widely known, provides a means of accomplishing this by modifying an RS code defined on a finite field alphabet larger than the required non-field alphabet. The technique, which involves *nonlinear parity checks*, is described by presentation of a specific example.

Suppose that it is desired to have a coded system in which all transmitted symbols are to be confined to English letters, numbers, standard punctuation marks, and a small set of control characters. Let us say that an alphabet size in the vicinity of 55 would serve our purposes well. We now show how we can design a three-error-correcting code with block length 63, having all symbols confined to a 55-ary alphabet. We start with a (63,57) RS code on $GF(64)$, which has minimum distance 7.

If one were to use this code and merely constrain the 57 information symbols to an alphabet size smaller than 64, the parity characters would in general still have variability 64. Therefore, we need a technique for reducing the variability of the parity characters as well. The technique described involves reducing the usable information set by one character. A 55-ary (63,56) nonlinear subcode of the 64-ary code is obtained.

To describe how the variability of the check set of the RS code can be reduced to 55-ary, we begin by considering how the alphabet size of the check set can be reduced to 63 symbols by eliminating the possibility of the occurrence of the symbol zero. We first form the information set of the desired

$\mathbf{i} = [\, i_{56} \mid i_{55} \mid i_{54} \mid \cdots \mid i_1 \mid 0 \,]$

FIGURE 6.5. The 57-character information set to be encoded with the (63,57) encoder.

$\mathbf{C} = [\, i_{56} \mid i_{55} \mid i_{54} \mid \cdots \mid i_1 \mid 0 \mid P_1 \mid P_2 \mid P_3 \mid P_4 \mid P_5 \mid P_6 \,]$

FIGURE 6.6. The (63,57) RS codeword associated with the information set shown in Fig. 6.5.

$\mathbf{G} = [\, 0 \mid 0 \mid 0 \mid \cdots \mid 1 \mid P_1' \mid P_2' \mid P_3' \mid P_4' \mid P_5' \mid P_6' \,]$

FIGURE 6.7. The RS codeword associated with the code generator polynomial.

(63,56) codeword as shown in Fig. 6.5 and append a zero at the end of the information set. The 57 characters shown in Fig. 6.5 are then used to form a (63,57) RS codeword, which is shown in Fig. 6.6 and denoted by $\mathbf{C}$. That is, the P's indicated are the RS parity symbols associated with the information set shown in Fig. 6.5. In Fig. 6.7 the codeword associated with the generator polynomial of the (63,57) RS code is shown and denoted by $\mathbf{G}$. Notice that if we can find a nonzero constant A such that $\mathbf{C} + A\mathbf{G}$ contains no zeros in the check set, we have defined a (63,56) 63-ary subcode of the 64-ary RS code. We simply form and transmit $\mathbf{C} + A\mathbf{G}$.

To see that A can always be found, consider the six ratios

$$B_i = \frac{P_i}{P_i'}, \qquad 1 \le i \le 6$$

Clearly A cannot be equal to any of the B's, or a zero symbol would result in the parity characters. However, since the additive and multiplicative inverses of a field element are unique, any of at least $63 - 6 = 57$ possible values for A can be used. If two or more of the B's are identical, more than 57 values of A can be used.

It is straightforward to see that the (63,56) code is a *nonlinear subcode* of the RS code, since, first, the (63,56) codewords are RS codewords and, second, linear combinations of (63,56) codewords are not necessarily (63,56) codewords. That is, they may contain zeros. For example, if $\mathbf{C}$ is a (63,56) codeword, $\mathbf{C} + \mathbf{C} = \mathbf{0}$ is an RS codeword but not a word in the (63,56) subcode.

To reduce the alphabet size of the (63,56) code to 55-ary, additional constraints on A are required. For example, to eliminate zero and $\alpha^0, \alpha^1, \ldots, \alpha^7$, where α is a primitive element of $GF(64)$, we require

$$P_i - AP_i' \neq 0$$

$$P_i - AP_i' \neq 1$$

$$P_i - AP_i' \neq \alpha$$

$$P_i - AP_i' \neq \alpha^2$$

$$P_i - AP_i' \neq \alpha^3 \qquad 1 \leq i \leq 6$$

$$P_i - AP_i' \neq \alpha^4$$

$$P_i - AP_i' \neq \alpha^5$$

$$P_i - AP_i' \neq \alpha^6$$

$$P_i - AP_i' \neq \alpha^7$$

To see that A can always be found, note that each of the nine constraints specifies up to six values out of 55 that cannot be used, but since $6 \times 9 = 54 < 55$, at least one value for A must exist. It is also seen that the symbol variability cannot be reduced to less than 55-ary by this technique without appropriating more than one information symbol as nonlinear parity checks.

An encoder for the (63,56) nonlinear subcode may be implemented using the (63,57) RS encoder as follows:

1. Form a 57-character set by appending a zero to the 56 information characters to be encoded.
2. Encode the associated (63,57) RS codeword to form $C_0(x)$.
3. Find the smallest i such that $C_0(x) + \alpha^i g(x)$ contains only symbols in the 55-ary alphabet.
4. The (63,56) codeword is then $C_0(x) + \alpha^i g(x)$.

Decoding the (63,56) code is also straightforward. Since the encoding procedure described produces RS codewords in the (63,57) code, any suitable decoder for the (63,57) code is first applied. (Received symbols outside the selected 55-ary alphabet are treated as erasures.) It is then necessary only to verify that the decoded word is a word in the (63,56) code. This requires both of the following:

1. All decoded symbols must be in the set of permissible 55-ary characters.
2. The smallest value of the exponent of α must be used to reduce the variability of the check set.

6.9. NOTES

The RS codes have been described here as a special case of nonbinary BCH codes, that is, BCH codes in which the symbol field and the locator field are the same. However, it is also possible to derive a BCH code from an RS code. That is, if we start with an RS code over a field $GF(q^m)$, then a $GF(q)$ *subfield subcode* of the RS code can be formed by collecting all the codewords that contain only symbols in $GF(q)$. The resulting subfield subcode is linear over the field $GF(q)$, since all linear combinations of codewords in $GF(q)$ must be in the subfield subcode. It can, in fact, be shown that every (n,k) BCH code with design distance d, whether binary or nonbinary, is a subfield subcode of an (n,k') RS code of distance d, with $k \leq k'$. The dimension k of the BCH code will usually be smaller than the dimension k' of the RS code. Detailed discussions of the properties of subfield subcodes can be found in a number of texts, including references 7, 14, 109, and 131. Blahut [14] describes subfield subcodes having the useful property that they are transform-decodable. The book by Blahut [14] is unique among available texts on coding in that it makes extensive use of transform techniques in treating both the structure of codes and the techniques for encoding and decoding.

Moderate- to high-rate RS codes can yield very efficient communication when used with orthogonal signaling on steady-signal AWGN channels. This is dealt with in some detail in Chapter 11. While RS codes are naturally suited for application with nonbinary modulation alphabets, they can sometimes be used very effectively with binary modulation. For example, Gore [57] has shown that RS codes on $GF(2^m)$, implemented by simply treating successive m-bit blocks as 2^m-ary symbols, outperform binary codes with the same rate and block length at low output error rates. The reason for the superiority of the RS codes at low error rates is that even though a single bit error results in an RS character error, the distance properties of the RS codes are much better than binary BCH codes.

The effectiveness of RS codes for correction of multiple binary errors can be used to advantage in computer memory systems, where errors are likely to occur in bursts. One prominent example is the application of RS codes in a photo-digital storage system [122].

The application of the distance-3 RS code on $GF(13)$ to the juror selection system for Middlesex County, Massachusetts, came about after an earlier unsuccessful attempt to use a single-character parity-check scheme. With the earlier scheme, it was found that an unexpectedly high number of incorrect identification codes were escaping detection. It was eventually learned that keypunch operators were defeating the system by simply substituting one check character after another until parity was satisfied rather than reentering the entire codeword. With the adoption of two check characters, operators do not attempt to use the trial-and-error scheme, since reentering the codeword is simpler than trying to guess two check characters. The distance-3 RS coding procedure was implemented by J. Romanow, then of the Office of the Jury

Commissioner, Middlesex County. The automated juror selection system is currently being adopted statewide.

There are a number of error-detecting and error-correcting codes that use ordinary arithmetic. These codes are called *arithmetic codes*. They can be used for error control in data transmission, with encoding and decoding operations conveniently performed in a general purpose computer, or they can be used to check various mathematical operations within a computer system. Some of the most powerful of these codes bear a close resemblance to nonbinary cyclic codes designed on finite fields. The book by Peterson and Weldon [131] has a chapter devoted to arithmetic codes, and there is a book by Rao [143] devoted entirely to coding for arithmetic processors.

■ ■ CHAPTER SEVEN

The Performance of Linear Block Codes with Bounded-Distance Decoding

In Chapters 4 through 6 we described several classes of block codes with good distance properties and presented efficient encoding and decoding algorithms. In order to apply coding in a practical communication system, it is also necessary to determine the performance improvements that can be expected with various candidate techniques. As is always the case in trying to achieve a cost-effective design, the complexity and cost of the alternative techniques must be evaluated and compared with the benefits that are obtained.

It is important to note that the costs associated with an error-control scheme involve system-level functions in addition to the encoders and decoders themselves. Many items must be considered when one decides to introduce coding into a system that might otherwise be configured to operate without coding. These items include more complex signaling techniques that utilize increased bandwidth to maintain the data transmission rate, more sophisticated equipment to acquire bit synchronization and codeword framing at low channel symbol SNRs, the time delay and data buffering associated with encoding and decoding, and perhaps interleavers and more complex multiplexers as well. The system issues vary a great deal from application to application. In this chapter we are concerned primarily with showing how to determine the performance gains that can be achieved with some of the more important block coding techniques, in order to provide a key element in design trade-offs.

Throughout this chapter we assume a discrete memoryless channel; for binary codes we use the binary symmetric channel. That is, we assume that for any transmitted bit, either a one or a zero, there is a probability $1 - p$ that the bit is received correctly and a probability p that it is received in error. Furthermore, bit errors are assumed to be independent of one another. The channel error rate p can in turn be related to the channel symbol SNR, E_s/N_0, for a particular signaling scheme. We use the discrete channel model here for both convenience and generality.

7.1. BINARY BLOCK CODES USED FOR ERROR DETECTION

A well-known and the most widely employed error-control technique makes use of block codes only for error detection. The technique, called *automatic repeat request* (*ARQ*), can be utilized when a feedback communication channel is available from the receiver to the transmitter to convey requests for retransmission of data when errors are detected in received words. Many specific ARQ strategies have been devised and analyzed. Several will be described in Chapter 11, where it will be seen that, when feasible, ARQ provides a highly efficient and robust error-control technique. In this section, we consider the calculation of the probability of correct and incorrect decoding for linear block codes used only for error detection.

The performance calculations in this chapter will be simplified by assuming that the transmitted codeword is the all-zeros word. On the discrete memoryless channel, if we restrict ourselves to linear codes we lose no generality, since, as we have seen before, if $\mathbf{r}$ is the received word, $\mathbf{c}$ the transmitted codeword, and $\mathbf{e}$ an additive error pattern, we have

$$\mathbf{r} = \mathbf{c} + \mathbf{e} \tag{7.1}$$

and the syndrome $\mathbf{S}$ is given by

$$\mathbf{S} = (\mathbf{c} + \mathbf{e})\mathbf{H}^T = \mathbf{e}\mathbf{H}^T$$

That is, $\mathbf{S}$ depends only on the error pattern and is unchanged by the addition of any codeword to both sides of Eq. (7.1). Thus, an error pattern that causes incorrect decoding or decoding failure with the transmission of an arbitrary codeword would produce the same result for transmission of the all-zeros codeword.

The calculation of the probabilities of correct and incorrect decoding with an error-detection decoder is straightforward. Specifically, correct decoding occurs if and only if all the bits in a received word are correct, so that the probability of correct decoding is given by

$$P_{\text{CD}} = (1 - p)^n$$

where n is the code block length. To calculate the probability of incorrect decoding, we note that incorrect decoding occurs only when the error pattern $\mathbf{e}$ is identical to a codeword, since error detection is announced when the syndrome is not all zeros. Since codewords can assume weights ranging from d to n, we can write

$$P_{\text{ICD}} = \sum_{i=d}^{n} P(i)P(\text{ICD}|i) \tag{7.2}$$

where $P(i)$ is the probability of i errors in the n bit positions, and $P(\text{ICD}|i)$ is the probability of incorrect decoding given i errors in the received word. On the binary symmetric channel we have

$$P(i) = \binom{n}{i} p^i (1-p)^{n-i}$$

$P(\text{ICD}|i)$ depends on the codeword weight distribution $\{A_i\}$, $i = 0, d, d+1, \ldots, n$, and is given by the ratio of the number of codewords of weight i to the total number of possible received words of weight i, that is,

$$P(\text{ICD}|i) = \frac{A_i}{\binom{n}{i}}$$

Thus we have

$$P_{\text{ICD}} = \sum_{i=d}^{n} A_i p^i (1-p)^{n-i} \tag{7.3}$$

The probability of error detection (or decoding failure) is then given by

$$\begin{aligned} P_{\text{ED}} &= 1 - (P_{\text{CD}} + P_{\text{ICD}}) \\ &= 1 - \left[(1-p)^n + \sum_{i=d}^{n} A_i p^i (1-p)^{n-i} \right] \end{aligned}$$

The reader may notice that Eq. (7.3) could have been written by inspection, since we are simply enumerating all error patterns of weight i that result in incorrect decoding, multiplying by the probability of occurrence of each, and summing over all codeword weights greater than zero. This presentation is instructive, however, and is used again in Section 7.2.

It is also interesting to note that when $p = \frac{1}{2}$ (i.e., $E_s/N_0 = 0$), we have

$$P_{\text{ICD}} = \sum_{i=d}^{n} \left(\frac{1}{2}\right)^n A_i = \frac{2^k - 1}{2^n} = 2^{-r} - 2^{-n}$$

where r is the number of parity-check bits, $n - k$. This result states that the chances of randomly generating an incorrect codeword, a set of k information bits, and the particular set of r check bits associated with those k information bits is bounded by 2^{-r}. Furthermore, since 2^{-n} is typically much less than 2^{-r}, the bound is tight. It is then tempting to conjecture that 2^{-r} provides an upper bound on the incorrect-decoding probability for all $p < 1/2$. One intuitively expects that as the quality of a channel improves, performance should improve, so that like modem performance curves, code performance curves should exhibit monotonic behavior. Although true in many cases, the reader is cautioned that, in general, the conjecture is false. A few codes have, in fact, been found that have $P_{\mathrm{ICD}}(p) > 2^{-r}$ for some $p < 1/2$. See, for example, references 88, 92, 125, and 182.

We have just seen that the performance calculations for a linear binary block code used for error detection can be carried out exactly, provided the weight distribution of the code is known. Often, the weight distribution of a chosen code is not readily determined, but if the weight distribution of the dual of the code is known, exact performance results can also be obtained. Letting $A(z)$ be the *weight-enumerator polynomial* of the code, and $B(z)$ the weight-enumerator polynomial of the dual, it can be shown that

$$B(z) = 2^{-k}(1 + z)^n A\left(\frac{1 - z}{1 + z}\right)$$

This expression was obtained originally by MacWilliams and is known as the *MacWilliams identity* [108]. The weight enumerator polynomial is again the polynomial whose coefficients specify the codeword weight distribution, that is,

$$A(z) = \sum_{i=0}^{n} A_i z^i$$

Returning to Eq. (7.3) we write

$$\begin{aligned} P_{\mathrm{ICD}} &= \sum_{i=d}^{n} A_i p^i (1 - p)^{n-i} \\ &= \sum_{i=0}^{n} A_i \left(\frac{p}{1 - p}\right)^i (1 - p)^n - (1 - p)^n \\ &= (1 - p)^n \left[A\left(\frac{p}{1 - p}\right) - 1\right] \end{aligned}$$

But, from the MacWilliams identity, we have

$$B(1 - 2p) = 2^{(n-k)}(1 - p)^n A\left(\frac{p}{1 - p}\right)$$

or

$$A\left(\frac{p}{1-p}\right) = 2^{-(n-k)}(1-p)^{-n}B(1-2p)$$

and

$$P_{\text{ICD}} = 2^{-(n-k)}B(1-2p) - (1-p)^n \tag{7.4}$$

If neither the weight distribution of a code nor its dual is known, performance results may be obtained by using approximations to either $A(z)$ or $B(z)$. An approximation that is often used is based on the assumption that codeword weights follow a binomial distribution over their nonzero range. We know that for an (n,k) linear binary code with distance d for which the all-ones word is a codeword,

$$A_0 = A_n = 1$$

and

$$A_i = 0, \qquad 1 \le i \le d-1 \quad \text{and} \quad n-d+1 \le i \le n-1$$

The binomial approximation then gives

$$A_i \simeq \binom{n}{i}\frac{2^k - 2}{\sum_{j=d}^{n-d}\binom{n}{j}} \approx 2^{-r}\binom{n}{i}, \qquad d \le i \le n-d$$

Generally speaking, this approximation has greatest accuracy for high-rate codes, except possibly for weights in the tails of the distribution, that is, weights near d and $n-d$. Therefore, the binomial approximation can be used with Eq. (7.3) to estimate the performance of a high-rate code and with Eq. (7.4) to determine the performance of a low-rate code.

There is a problem in using the binomial approximation, however, in that the numbers of very low- and high-weight codewords can be underestimated, and for some bit-error rates the incorrect-decoding probability can be underestimated. In addition, it can be shown by induction that assuming a binomial codeword weight distribution necessarily results in a monotonic error-probability estimate. For many codes, however, one can accurately estimate the number of low-weight codewords experimentally. At some point the measured distribution begins to show satisfactory agreement with the binomial approximation, and one can safely use an appropriately scaled binomial distribution from that point on.

For example, suppose an (n,k) code can be used to correct a maximum of t errors per codeword. Clearly, some weight-$(i-t)$ error patterns would be decoded into weight-i error patterns with a t-error-correcting decoder. We let the fraction of such error patterns that are at distance-t from a weight-i word be denoted as F_i and observe that

$$F_i\binom{n}{i-t} = A_i\binom{i}{i-t} \tag{7.5}$$

TABLE 7.1. Weight Distribution and Values of the Fractions F_i for the (31,15) Distance-8 Binary BCH Code

i	A_i	F_i
8	465	0.153
12	8,680	0.0947
16	18,259	0.0496
20	5,208	0.0224
24	155	0.0071

For many codes, the fractions $\{F_i\}$ can be measured experimentally by use of a t-error-correcting decoder, and Eq. (7.5) then provides an estimate of A_i for small i. In Tables 7.1 through 7.3, the fractions $\{F_i\}$ are given for some binary BCH codes whose weight distributions are known. The weight distributions for these codes were found by Peterson [130]. Note that the $\{F_i\}$ shown are sufficiently large to be measured in simulation experiments without requiring exorbitant numbers of decoding trials.

As an example of this procedure we consider the (1024,993) distance-8 binary BCH code. This code is formed by appending an overall parity-check

TABLE 7.2. A Portion of the Weight Distribution and Values of the Fractions F_i for the (63,45) Distance-8 Binary BCH code

i	A_i	F_i
8	2.3877×10^4	0.190
10	4.2336×10^5	0.092
12	1.0350×10^7	0.096
14	1.4255×10^8	0.084
16	1.3972×10^9	0.0747
18	9.8751×10^9	0.0660
.		
.		
.		
30	3.2832×10^{12}	0.0272
32	3.4959×10^{12}	0.0228
34	2.8969×10^{12}	0.0189

TABLE 7.3. A Portion of the Weight Distribution and Values of the Fractions F_i for the Distance-7 (127,105) Binary BCH code

i	A_i	F_i
7	4.8387×10^4	0.164
8	7.2581×10^5	0.160
9	8.2499×10^6	0.134
10	9.7349×10^7	0.131
11	1.0652×10^9	0.131
12	1.0297×10^{10}	0.128
13	9.0631×10^{10}	0.124

bit to the codewords of the primitive (1023,993) distance-7 code. The (1024,993) code contains no words of odd weight, and the number of words of even weight i is the number of words of weight i plus the number of weight $i - 1$ in the (1023,993) code.

Exact results for the (1024,993) code can be obtained, since the weight distribution of its dual is known [109]. The probability of incorrect decoding obtained with Eq. (7.4) is shown in Fig. 7.1, and the weight distribution of the dual is given in Table 7.4. In addition we recall that the MacWilliams identity

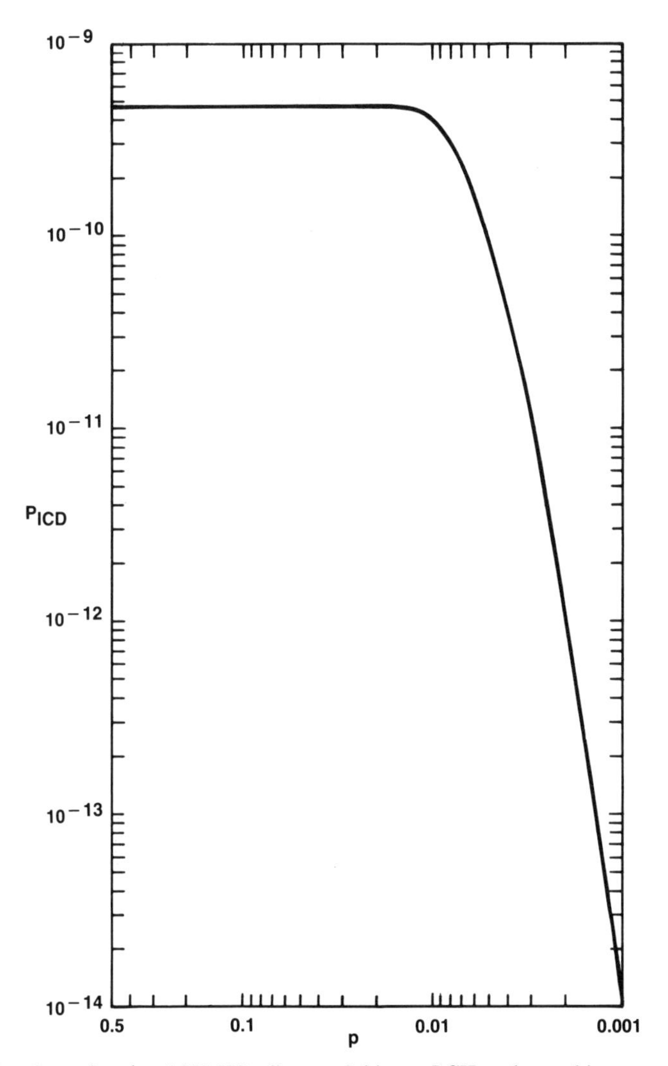

FIGURE 7.1. P_{ICD} for the (1024,993) distance-8 binary BCH code vs. bit-error rate in the channel. Code used for error detection only. From Wolf et al. [181], © 1981 IEEE, reprinted with permission.

TABLE 7.4. Weight Distribution of the Dual of the (1024,993) Distance-8 Binary BCH Code

Codeword Weight, i	Number of Words of Weight i, A_i
0,1024	1
448,576	1,113,024
480,544	156,434,432
496,528	430,194,688
512	971,999,358

TABLE 7.5. Portion of Weight Distribution of the (1024,993) Code, Weight Distribution Based on a Binomial Assumption, and an Experimentally Determined Weight Distribution

i	A_i	A_i^B	A_i^E
8	2.795×10^{10}	2.717×10^{10}	2.796×10^{10}
10	3.112×10^{14}	3.113×10^{14}	3.124×10^{14}
12	2.423×10^{18}	2.422×10^{18}	2.427×10^{18}
14	1.362×10^{22}	1.362×10^{22}	1.361×10^{22}
16	5.783×10^{25}	5.783×10^{25}	5.802×10^{25}
18	1.918×10^{29}	1.918×10^{29}	1.918×10^{29}

Source: J. K. Wolf et al. [181], © 1981 IEEE, reprinted with permission.

can be used to obtain the weight distribution of a code in terms of its dual,

$$A(z) = 2^{(k-n)}(1+z)^n B\left(\frac{1-z}{1+z}\right) = \sum_{i=0}^{n} A_i z^i$$

Thus we may compute the weight distribution of the (1024,993) distance-8 code.

A portion of the weight distribution of the (1024,993) code is shown in Table 7.5 as well as the weight distribution obtained by assuming binomial codeword weights $\{A_i^B\}$. Also indicated is the weight distribution $\{A_i^E\}$ based on estimates of the $\{F_i\}$ that were measured in a series of simulation experiments. We note that in this case the binomial approximation underestimates the number of minimum weight words by about 3 percent, but for larger weights the estimates are quite accurate. In all cases, the experimental estimates are good.

7.2. BINARY BLOCK CODES USED FOR ERROR DETECTION AND CORRECTION

The approach employed in the previous section to compute performance results can be extended to include the case of error-detection-and-correction decoding. Again we use the binary symmetric channel model and note that if an (n,k) code with distance $d = 2t + 1$ is used to correct up to l errors where

$l \leq t$, correct decoding will result if the received word lies within a radius-l sphere centered on the intended word. Thus we have

$$P_{\mathrm{CD}} = \sum_{i=0}^{l} \binom{n}{i} p^i (1-p)^{n-i} \tag{7.6}$$

since $\binom{n}{i}$ is the number of weight-i error patterns that can occur in n bit positions, and $p^i(1-p)^{n-i}$ is the probability of each individual weight-i error pattern.

When, however, more than l errors are present in a received word, decoding will result in either error detection or incorrect decoding. If l' errors are present where $l + 1 \leq l' \leq d - l - 1$, error detection will certainly occur. We may thus obtain bounds on the probabilities of error detection and incorrect decoding if we assume that all error events that contain $d - l$ or more errors always yield incorrect decoding. We have then

$$P_{\mathrm{ICD}} \leq \sum_{i=d-l}^{n} \binom{n}{i} p^i (1-p)^{n-i}$$

$$P_{\mathrm{ED}} \geq \sum_{i=l+1}^{d-l-1} \binom{n}{i} p^i (1-p)^{n-i}$$

These expressions are clearly bounds, because some channel error events that contain more than $d - l - 1$ errors will result in error detection. In addition, we may obtain an upper bound on the post-decoding bit-error probability by assuming that all incorrect decoding events produce $i + l$ post-decoding errors. If i errors are present in a received word, the decoder can insert at most l additional errors. We have then

$$P_b \leq \frac{1}{n} \sum_{i=d-l}^{n} (i + l) \binom{n}{i} p^i (1-p)^{n-i}$$

where P_b is the post-decoding bit-error probability. It is important to note that here and throughout the remainder of this chapter, we are ignoring post-decoding errors associated with error-detection events.

For closely packed codes, the bounds just presented are tight. Of course, for perfect codes with $l = t$, the expression for P_{ICD} holds with equality, and the probability of error detection is zero. For most codes, however, when $d - l$ or more errors are present, the probability of error detection is several orders of magnitude larger than the probability of incorrect decoding, and the bounds are not tight. When exact results are required for an arbitrary code, a more detailed analysis is necessary.

Specifically, one may compute the probability of incorrect decoding as

$$P_{\mathrm{ICD}} = \sum_{i=d-l}^{n} P(i) P(\mathrm{ICD} \mid i) \tag{7.7}$$

where for operation on the binary symmetric channel

$$P(i) = \binom{n}{i} p^i (1-p)^{n-i}$$

The calculation of $P(\text{ICD}|i)$ can proceed by careful enumeration of all possible post-decoding error events. As in the case of decoding for error detection only, knowledge of the codeword weight distribution is required. We assume again that the transmitted codeword is the all-zeros word, since this simplifies the discussion and results in no loss of generality for linear codes. As an example, we consider the (23,12) Golay code, whose weight distribution is given in Table 7.6.

As we pointed out in Chapter 3, the Golay code is a perfect distance-7 code that can be used to correct up to three errors per codeword. We begin by considering the occurrence of four errors in a received word. Clearly an l-error-correcting decoder can add or delete up to l errors in a received word, and thus one might expect an ($l = 3$)-error-correcting decoder for the Golay code to produce a word of weight 1 to 7 given a weight-4 error pattern. However, in Table 7.6, we see that there are no words with weights 1 through 6. Therefore, if a received word contains four errors, incorrect decoding can only yield a weight-7 codeword. Furthermore, a weight-4 error pattern will be decoded into a particular weight-7 codeword only if the four errors have occurred in four of the seven nonzero bit positions of that weight-7 codeword. Since there are $\binom{7}{4}$ ways that a weight-4 error pattern can coincide with four of the seven nonzero positions in a weight-7 word, and since there are 253 weight-7 codewords, incorrect decoding can occur in 253 $\binom{7}{4} = 8855$ possible ways. However, there are $\binom{23}{4} = 8855$ weight-4 error patterns, and thus we are reminded that the (23,12) Golay code is indeed a perfect code.

When i is larger than 4, that is, when more than four errors occur in a received word, several post-decoding error events are possible. For example, with $i = 5$, weight-5 error patterns may be decoded into either weight-7 or weight-8 codewords. A weight-5 error pattern will be decoded into a weight-7

TABLE 7.6. Weight Distribution of the (23,12) Golay Code

Code Word Weights, i	Number of Words of Weight i, A_i
0	1
7	253
8	506
11	1,288
12	1,288
15	506
16	253
23	1
all other i	0

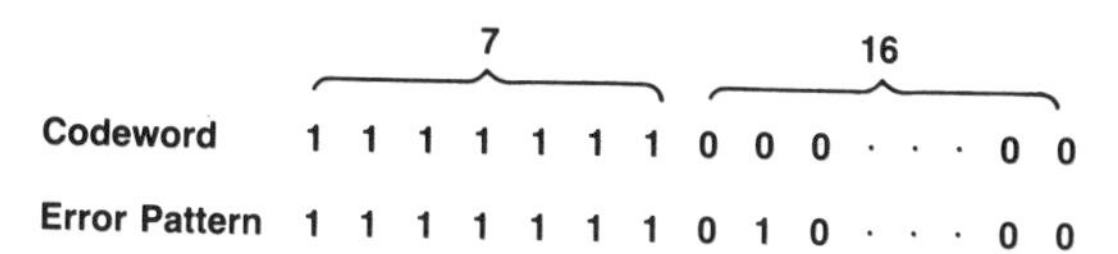

FIGURE 7.2. Representation of a weight-7 codeword and a weight-8 error pattern.

codeword by double-error correction if the five errors coincide with five nonzero positions of a weight-7 codeword, which can occur $\binom{7}{5}$ ways for each weight-7 codeword. Also a weight-5 error pattern can be decoded into a weight-8 codeword if the five errors line up with five ones in a weight-8 codeword, which can occur $\binom{8}{5}$ ways for each weight-8 word. Then, since there are 253 weight-7 words and 506 weight-8 words, we have accounted for

$$\binom{7}{5}253 + \binom{8}{5}506 = 33{,}649 = \binom{23}{5}$$

weight-5 error patterns, which is all of them.

When $i = 8$, additional post-decoding error events must be considered. First we note that a weight-8 error pattern may be decoded to a weight-7, -8, or -11 error pattern; we consider the weight-7 post-decoding error patterns first. In Fig. 7.2 we show a representation of a weight-7 codeword and a particular weight-8 error pattern. (In this and subsequent figures we actually show a permutation of a codeword, in which the nonzero symbols are left-justified. This representation simplifies the bookkeeping and does not affect the analysis.) Note that the seven errors indicated line up with the seven ones in the codeword and that the distance between the error pattern and the codeword is 1. Furthermore, the one in the error pattern appearing in a codeword position that contains a zero can occur in any of 16 bit positions without changing the distance between the codeword and the error pattern. Thus, since there are 253 weight-7 words, we have

$$8 \overset{d}{\rightarrow} 7 \text{ via 1-EC } (16)(253) = 4048 \text{ ways}$$

where the notation above is read, "weight-8 error patterns can be decoded into weight-7 codewords by single-error correction in a total of (16)(253) = 4048 ways."

Another possibility is shown in Fig. 7.3. Again we have a weight-7 codeword and a weight-8 error pattern, but this time only six errors are aligned with

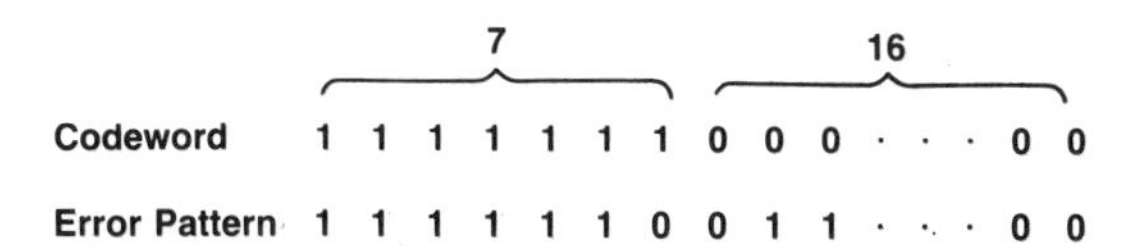

FIGURE 7.3. Representation of a weight-7 codeword and another weight-8 error pattern.

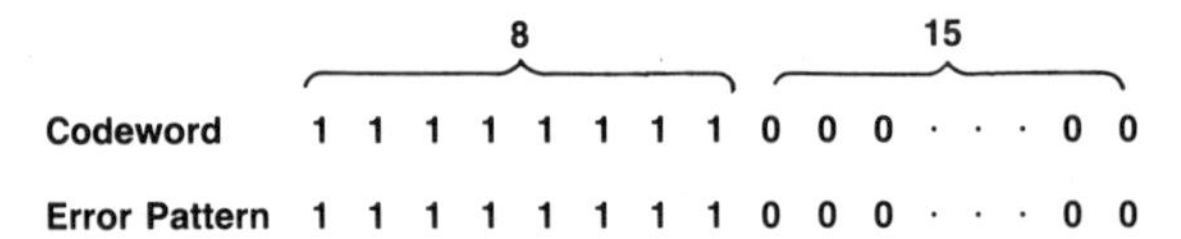

FIGURE 7.4. Representation of a weight-8 codeword and a weight-8 error pattern.

nonzero codeword bit positions and the two additional errors can be placed anywhere in the 16 remaining positions. Since the distance between the error pattern and the codeword shown is 3, we have

$$8 \overset{d}{\rightarrow} 7 \text{ via 3-EC } \binom{7}{6}\binom{16}{2}253 = 212{,}520 \text{ ways}$$

With the aid of Figs. 7.4 and 7.5 we can also see that weight-8 error patterns may be decoded into weight-8 words by zero- or double-error correction, and we have

$$8 \overset{d}{\rightarrow} 8 \text{ via 0-EC } (1)(506) = 506 \text{ ways}$$

$$8 \overset{d}{\rightarrow} 8 \text{ via 2-EC } \binom{8}{7}(15)(506) = 60{,}720 \text{ ways}$$

Finally, it is easy to see that weight-8 error patterns may be decoded into weight-11 words by triple-error correction, and thus we have

$$8 \overset{d}{\rightarrow} 11 \text{ via 3-EC } \binom{11}{8}(1288) = 212{,}520 \text{ ways}$$

By continuing in this manner it is possible to enumerate all the incorrect-decoding events for the Golay code, and in Table 7.7 we show these events for $i \leq 11$. The events for $i > 11$ can be obtained in a similar fashion.

It should now be apparent that if $l = 3$, $P(\text{ICD}|i) = 1$ when $i \geq 4$. However, if $l = 2$, all triple-error correction events included in Table 7.7 become

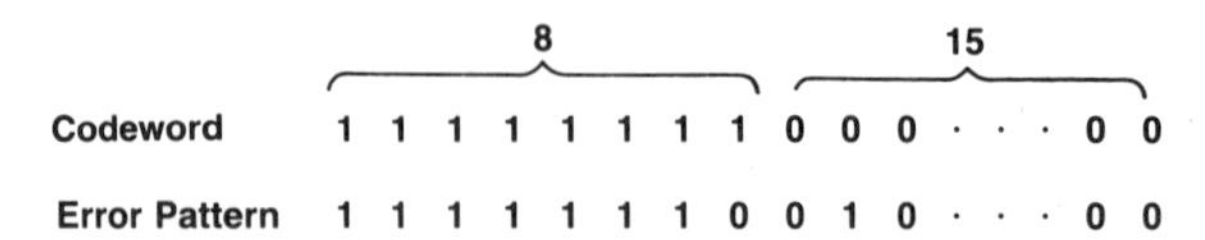

FIGURE 7.5. Representation of a weight-8 codeword and another weight-8 error pattern.

TABLE 7.7. Enumeration of the Incorrect Decoding Events for the Golay Code for up to 11 Channel Errors

$4 \xrightarrow{d} 7$	$\binom{7}{4}253 = 8{,}855$ ways
$5 \xrightarrow{d} 7$	$\binom{7}{5}253 = 5{,}313$ ways
$5 \xrightarrow{d} 8$	$\binom{8}{5}506 = 28{,}336$ ways
$6 \xrightarrow{d} 7$	$\left[\binom{7}{6} + \binom{7}{5}16\right] 253 = 86{,}779$ ways
$6 \xrightarrow{d} 8$	$\binom{8}{6} 506 = 14{,}168$ ways
$7 \xrightarrow{d} 7$	$\left[1 + \binom{7}{6}16\right] 253 = 28{,}589$ ways
$7 \xrightarrow{d} 8$	$\left[\binom{8}{7} + 15\binom{8}{6}\right]506 = 216{,}568$ ways
$8 \xrightarrow{d} 8$	$\left[1 + 15\binom{8}{7}\right]506 = 61{,}\ 226$ ways
$8 \xrightarrow{d} 7$	$\left[16 + \binom{7}{6}\binom{16}{2}\right]253 = 216{,}568$ ways
$8 \xrightarrow{d} 11$	$\binom{11}{3}1{,}288 = 212{,}520$ ways
$9 \xrightarrow{d} 7$	$\binom{16}{2} 253 = 30{,}360$ ways
$9 \xrightarrow{d} 8$	$\left[15 + \binom{8}{7}\binom{15}{2}\right]506 = 432{,}630$ ways
$9 \xrightarrow{d} 11$	$\binom{11}{9} 1{,}288 = 70{,}840$ ways
$9 \xrightarrow{d} 12$	$\binom{12}{9} 1{,}288 = 283{,}360$ ways
$10 \xrightarrow{d} 7$	$\binom{16}{3}253 = 141{,}680$ ways
$10 \xrightarrow{d} 8$	$\binom{15}{2} 506 = 53{,}130$ ways
$10 \xrightarrow{d} 11$	$\left[\binom{11}{10} + \binom{11}{9} 12\right] 1{,}288 = 864{,}248$ ways
$10 \xrightarrow{d} 12$	$\binom{12}{10} 1{,}288 = 85{,}008$ ways
$11 \xrightarrow{d} 8$	$\binom{15}{3} 506 = 230{,}230$ ways
$11 \xrightarrow{d} 11$	$\left[1 + \binom{11}{10} 12\right] 1{,}288 = 171{,}304$ ways
$11 \xrightarrow{d} 12$	$\left[\binom{12}{11} + \binom{12}{10}\binom{11}{1}\right]1{,}288 = 950{,}544$ ways

error-detection events and we have, for example,

$$P(\mathrm{ICD}|4) = 0$$

$$P(\mathrm{ICD}|5) = \frac{\binom{7}{5}253}{\binom{23}{5}} = 0.1579$$

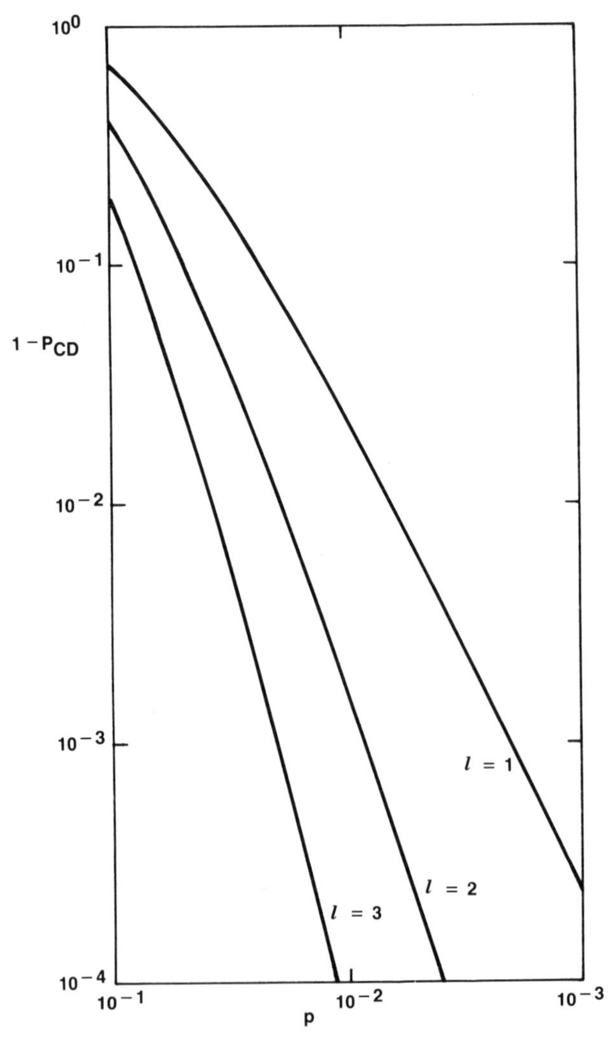

FIGURE 7.6. $1 - P_{\mathrm{CD}}$ vs. bit-error rate in the channel for the (23,12) Golay code with $l = 1$, 2, and 3.

Thus the data contained in Table 7.7 provide all the information necessary to use Eq. (7.7) to determine P_{ICD} for the case in which $l = 2$. Of course, if $l = 1$, we eliminate all double- and triple-error-correction events to compute P_{ICD}.

This analysis can be extended to determine the post-decoding bit-error probability. Note that $P_{\text{ICD}}(h|i)$, the probability of incorrect decoding to a weight-h codeword conditioned on i received errors, is given by the ratio of the number of ways that a weight-i error pattern may be decoded into a weight-h codeword divided by the total number of possible received weight-i error

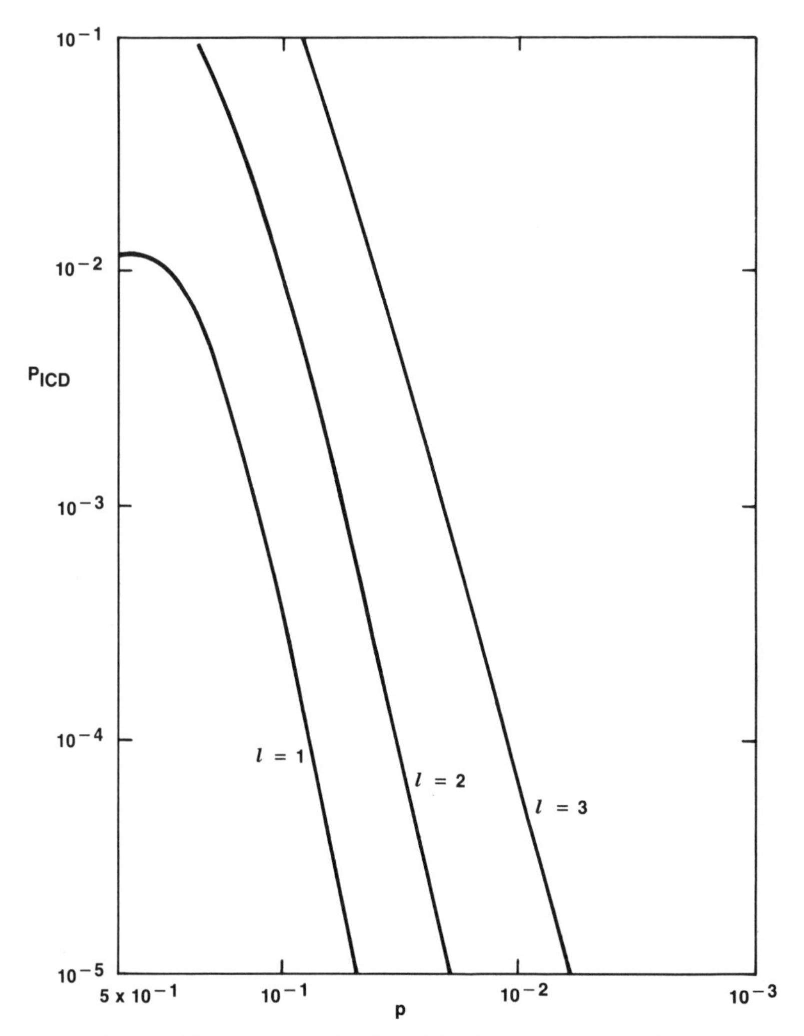

FIGURE 7.7. P_{ICD} vs. bit-error rate in the channel for the (23,12) Golay code with $l = 1$, 2, and 3. From Michelson [112], © 1976 IEEE, reprinted with permission.

patterns. Therefore, for example, we have

$$P_{\text{ICD}}(8|7) = \frac{216{,}568}{\binom{23}{7}} = 0.8834$$

Then the probability of incorrect decoding to a weight-h codeword, $P_{\text{ICD}}(h)$,

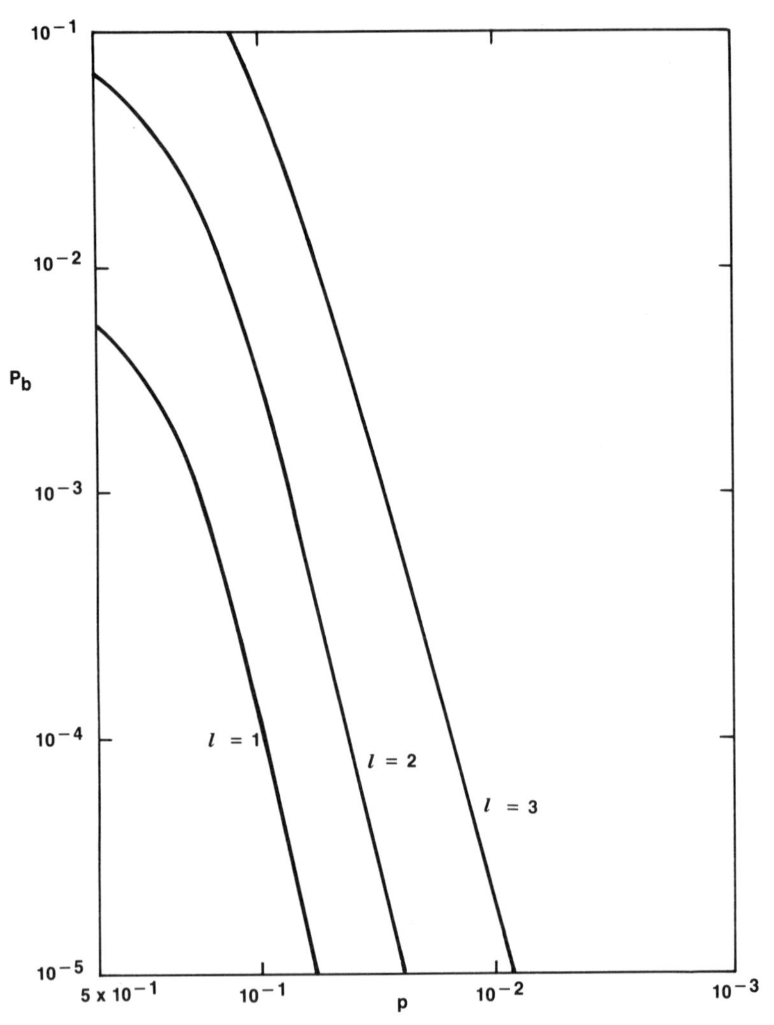

FIGURE 7.8. Post-decoding bit-error rate vs. bit-error rate in the channel for the (23,12) Golay code with l = 1, 2, and 3. From Michelson [112], © 1976 IEEE, reprinted with permission.

can be found by computing

$$P_{\text{ICD}}(h) = \sum_{i=h-l}^{h+l} P_{\text{ICD}}(h|i)P(i)$$

Finally, the post-decoding bit-error probability P_b is given by

$$P_b = \frac{1}{n} \sum_{h=d}^{n} hP_{\text{ICD}}(h)$$

As examples of these calculations we show $1 - P_{\text{CD}}, P_{\text{ICD}}$, and P_b for the (23,12) Golay code as a function of p with $l = 1$, 2, and 3 in Figs. 7.6, 7.7, and 7.8, respectively.

7.3. GENERALIZATION TO NONBINARY CODES

The procedure outlined in the previous section can be generalized to obtain results for linear nonbinary codes used with bounded-distance decoding. In this section we derive analytical expressions for the probability of correct decoding, incorrect decoding, and error detection, as well as the probability of post-decoding character error. The codes considered are defined on q-ary alphabets, and expressions for binary codes will be obtained by letting $q = 2$. We further assume that the communication channel can be modeled as a q-ary symmetric independent-error channel. That is, we assume that if a symbol s_i is transmitted, there is a probability p that s_j is received $(i \neq j)$ and a probability $1 - (q - 1)p$ that s_i is received correctly. The channel model is shown in Fig. 7.9. Note that $(q - 1)p$ is the character-error probability and is denoted P_{CE}.

The calculation of the probability of correct decoding for a nonbinary code with bounded-distance decoding is similar to the calculation for the binary

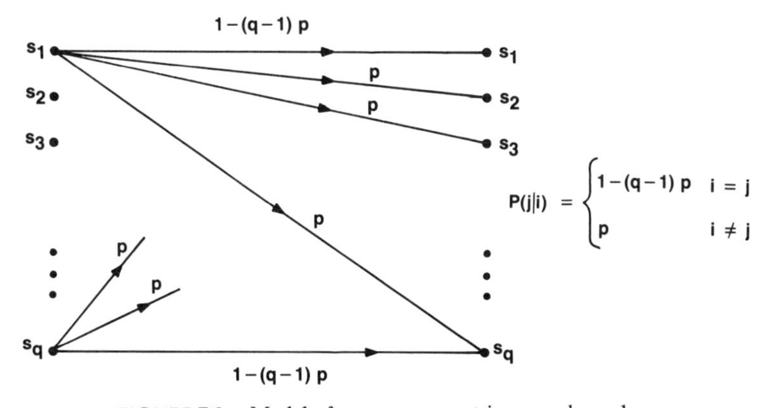

FIGURE 7.9. Model of a q-ary symmetric error channel.

case. For an (n,k) code defined on a q-ary alphabet that is used for l-error correction ($l \leq t$), a received word will be decoded correctly if it contains l or fewer errors. That is, if the received word is contained within a radius-l sphere surrounding the transmitted codeword, decoding will be successful, and thus we have

$$P_{\text{CD}} = \sum_{i=0}^{l} \binom{n}{i}(q-1)^i p^i (1-(q-1)p)^{n-i}$$

$$= \sum_{i=0}^{l} \binom{n}{i} P_{\text{CE}}^i (1-P_{\text{CE}})^{n-i}$$

since there are $\binom{n}{i}(q-1)^i$ weight-i error patterns and each occurs with probability $p^i(1-(q-1)p)^{n-i}$. The factor $(q-1)^i$ is necessary since a weight-i error pattern can assume any one of $q-1$ values in each of the i error locations.

Once again, if more than l errors are present in a received word, the decoder will either decode incorrectly or recognize the presence of an uncorrectable error pattern and announce error detection. For linear codes on the symmetric error channel, we may again assume without loss of generality that the all-zeros word is transmitted and that incorrect decoding will occur only when the received word (or error pattern) is contained within a radius-l sphere centered on a nonzero codeword. To compute the probability of incorrect decoding, we first compute $P_{\text{ICD}}(h)$, the probability of incorrect decoding to a weight-h codeword, and use

$$P_{\text{ICD}} = \sum_{h=d}^{n} P_{\text{ICD}}(h) \tag{7.8}$$

The post-decoding character-error probability is then given by

$$P'_{\text{CE}} = \frac{1}{n} \sum_{h=d}^{n} h P_{\text{ICD}}(h) \tag{7.9}$$

In the analysis that follows, it will be necessary to enumerate somewhat more involved error patterns and codeword geometries than we have considered thus far. For example, for a nonbinary code, the ith symbol of a codeword and the ith symbol of an error pattern can be either equal or not equal and zero or nonzero. That is, in each symbol position we may have

1. $c_i = e_i = 0$
2. $c_i = e_i \neq 0$
3. $c_i = 0,\ e_i \neq 0$
4. $c_i \neq 0,\ e_i = 0$
5. $c_i \neq 0,\ e_i \neq 0,\ c_i \neq e_i$

	7							n-7					
Codeword	c_1	c_2	c_3	c_4	c_5	c_6	c_7	0	0	0	· · ·	0	0
Error Pattern	e_1	e_2	e_3	e_4	0	0	0	0	0	0	· · ·	0	0

$c_1 = e_1, c_2 \neq e_2, c_3 \neq e_3, c_4 \neq e_4$

FIGURE 7.10. Weight-7 codeword and a weight-4 error pattern.

where c_i is the ith codeword symbol and e_i is the error value in the ith symbol position. These possibilities are illustrated in Figs. 7.10 and 7.11, where we show a particular weight-7 codeword and a weight-4 and weight-5 error pattern. The reader should verify that the Hamming distance between the codeword and the error pattern shown is 6 in Fig. 7.10 and 8 in Fig. 7.11. Again, the Hamming distance between two words is the number of symbol positions in which the words differ.

In order to determine the probability of incorrect decoding to a weight-h codeword, we begin by defining $N_{k,s}(h)$, the number of weight-k error patterns that are at distance s from a particular codeword of weight h. When $s \leq l$ and $h \neq 0$, each such error pattern will be decoded incorrectly to the weight-h codeword. Further, $P(k)$, the probability of occurrence of a particular weight-k error pattern, is given by

$$P(k) = p^k[1-(q-1)p]^{n-k} = \frac{P_{\mathrm{CE}}^k}{(q-1)^k}(1-P_{\mathrm{CE}})^{n-k}$$

Since there are A_h weight-h codewords, we have

$$P_{\mathrm{ICD}}(h) = A_h \sum_{s=0}^{l} \sum_{k=h-s}^{h+s} N_{k,s}(h)P(k), \quad d \leq h \leq n \tag{7.10}$$

In Eq. (7.10) we have simply enumerated the set of events that lead to incorrect decoding, multiplied by the probability of occurrence, and summed exhaustively over the set of events of interest.

	7							n−7					
Codeword	c_1	c_2	c_3	c_4	c_5	c_6	c_7	0	0	0	· · ·	0	0
Error Pattern	e_1	e_2	e_3	0	0	0	0	e_4	0	e_5	· · ·	0	0

$c_1 = e_1, c_2 \neq e_2, c_3 \neq e_3, e_4 \neq 0, e_5 \neq 0$

FIGURE 7.11. Weight-7 codeword and a weight-5 error pattern.

Thus, if the weight distribution of a code is known, the probability of incorrect decoding can be determined once an expression for $N_{k,s}(h)$ is obtained. To find $N_{k,s}(h)$, we observe that by definition

1. k is the number of symbol positions in which $e_i \neq 0$.
2. h is the number of symbol positions in which $c_i \neq 0$.

and we let

3. m be the number of positions in which $c_i = e_i \neq 0$.
4. j be the number of positions in which $c_i \neq e_i$ and $e_i \neq 0$, $c_i \neq 0$.
5. v be the number of positions in which $c_i \neq 0$ and $e_i = 0$.
6. r be the number of positions in which $c_i = 0$ and $e_i \neq 0$.

We then have

$$k = m + j + r \tag{7.11}$$

$$h = m + j + v \tag{7.12}$$

$$s = j + v + r \tag{7.13}$$

As an example, consider Fig. 7.10, where $k = 4$, $h = 7$, $s = 6$, and

$$m = 1, \qquad j = 3, \qquad v = 3, \qquad r = 0$$

Also in Fig. 7.11 we have $k = 5$, $h = 7$, $s = 8$, and

$$m = 1, \qquad j = 2, \qquad v = 4, \qquad r = 2$$

From Fig. 7.11 we may also determine the total number of weight-5 error patterns that are at distance 8 from a weight-7 codeword with $m = 1$, $j = 2$, $v = 4$, and $r = 2$. Clearly there are

1. $\binom{7}{1}$ ways that a single nonzero error value can be equal to one of the seven nonzero code symbols.
2. $\binom{6}{2}$ ways that the two nonzero error symbols that are not equal to the code symbols can be arranged in the six remaining nonzero codeword positions.
3. $\binom{n-7}{2}$ ways that the two remaining errors can be placed in the $n - 7$ zero positions of the codeword.

Furthermore, we note that each error symbol identified in item 2 can assume any one of $q-2$ values ($e_i \neq 0$, $e_i \neq c_i$) and each error symbol identified in item 3 can assume any one of $q-1$ values ($e_i \neq 0$). Thus we can write

$$N_{5,8}(h=7|m=1,\ j=2,\ v=4,\ r=2)$$

$$=\binom{7}{1}\binom{6}{2}(q-2)^2\binom{n-7}{2}(q-1)^2$$

where $N_{k,s}(h|m,j,v,r)$ represents the number of weight-k error patterns that are at distance s from a weight-h codeword given values specified for m,j,v, and r. In general, we have

$$N_{k,s}(h|m,j,v,r)=\binom{h}{m}\binom{h-m}{j}(q-2)^j\binom{n-h}{r}(q-1)^r$$

But Eqs. (7.11), (7.12), and (7.13) may be solved for m and j in terms of h, k, s, and r, and we have

$$m=h-s+r$$
$$j=k-h+s-2r \tag{7.14}$$

Therefore,

$$N_{k,s}(h|r)=\binom{h}{h-s+r}\binom{s-r}{k-h+s-2r}$$
$$\times\binom{n-h}{r}(q-2)^{k-h+s-2r}(q-1)^r \tag{7.15}$$

The values that r can assume clearly depend on h and k. For example, if the weight h of the codeword is greater than k, then r can be zero (consider the case $h=n$). When $h<k$, the smallest value that r can assume is $k-h$. Thus we have

$$r \geq 0 \qquad \text{if } h \geq k$$
$$r \geq k-h \qquad \text{if } k > h$$

and the minimum value that r can assume is given by

$$r_1=\max\{0,k-h\}$$

It can also be seen that the maximum value that r can assume is

$$r_2=\left\lfloor\frac{k-h+s}{2}\right\rfloor$$

where $\lfloor a \rfloor$ means the largest integer less than or equal to a. Therefore we have

$$N_{k,s}(h) = \sum_{r=r_1}^{r_2} N_{k,s}(h|r)$$

$$= \sum_{r=r_1}^{r_2} \binom{h}{h-s+r}\binom{s-r}{k-h+s-2r}\binom{n-h}{r}(q-2)^{k-h+s-2r}(q-1)^r \tag{7.16}$$

Combining Eqs. (7.8), (7.10), and (7.16), we may write

$$P_{\text{ICD}} = \sum_{h=d}^{n} P_{\text{ICD}}(h)$$

$$= \sum_{h=d}^{n}\sum_{s=0}^{l}\sum_{k=h-s}^{h+s}\sum_{r=r_1}^{r_2} \binom{h}{h-s+r}\binom{s-r}{k-h+s-2r}$$

$$\times \binom{n-h}{r}(q-2)^{k-h+s-2r}(q-1)^r P(k) A_h$$

The probability of error detection is simply

$$P_{\text{ED}} = 1 - P_{\text{CD}} - P_{\text{ICD}} = 1 - \sum_{i=0}^{l}\binom{n}{i}P_{\text{CE}}^i(1-P_{\text{CE}})^{n-i} - P_{\text{ICD}}$$

For binary codes, j, the number of codeword positions in which $c_i \neq e_i$, $c_i \neq 0$, and $e_i \neq 0$, is zero. Thus from Eq. (7.14) we have

$$r = \frac{k-h+s}{2}$$

and r must assume a single value when k, h, and s are specified. Substituting in Eq. (7.16), we have

$$N_{k,s}(h) = \binom{h}{\frac{h+k-s}{2}}\binom{n-h}{\frac{k-h+s}{2}} \tag{7.17}$$

and using Eqs. (7.8), (7.10), and (7.17) we have for the binary case

$$P_{\text{ICD}} = \sum_{h=d}^{n}\sum_{s=0}^{l}\sum_{k=h-s}^{h+s}\binom{h}{\frac{h+k-s}{2}}\binom{n-h}{\frac{k-h+s}{2}}A_h P(k)$$

where $P(k) = p^k(1-p)^{n-k}$.

The reader should verify that when $(h + k - s)/2$ or $(k - h + s)/2$ is not an integer, the error pattern type specified will not decode for the assumed value of s, and hence we use the convention $\binom{a}{b} = 0$ when b is not an integer.

7.4. SELECTED PERFORMANCE RESULTS

In this section we present examples of the performance calculations that have been described for linear block codes with bounded-distance decoding. In particular, $1 - P_{CD}$ is shown as a function of channel bit-error rate p for a selection of primitive binary BCH codes with block lengths $n = 31, 63, 127, 255$, and 511 in Figs. 7.12 to 7.16. In Figs. 7.17 to 7.20, the probability of incorrect decoding and the post-decoding bit-error rate for selected binary BCH codes are given as a function of p, where the assumption is made that the codeword weight distributions are binomial. In all cases we have used $l = t$, that is, it is assumed that the codes are used with an error-correction limit set to the maximum value. The performance curves shown are sometimes called *waterfall curves*, a term suggested by their shape.

Similar results are also given for some RS codes in Figs. 7.21 to 7.30. Again, $1 - P_{CD}$ is shown as a function of character-error rate in the channel, P_{CE}, for codes with block lengths $n = 31, 63, 127$, and 255. Then, recognizing that the weight distributions of the RS codes are known (see Section 6.4), we note that the probability of incorrect decoding and the post-decoding character-error rate can be computed exactly. Selected results are given for RS codes with block lengths $n = 31, 63$, and 127.

The reader must be cautioned in the interpretation of the results presented. First we have assumed use of an independent-error channel. Furthermore, the post-decoding bit- and character-error rate results include only the errors associated with incorrect-decoding events. That is, the assumption is made that received words that contain detected but uncorrected errors can be ignored. Although this can be safely assumed for many communication systems, it does not always apply. When the system requirements make the assumption inappropriate, the errors associated with error-detection events must also be counted. Since the probability of error detection can be much larger than the probability of incorrect decoding, post-decoding error rates can be dominated by the error-detection events. More will be said on this point in Chapter 11.

As a further caution, we point out that the results presented in this chapter indicate that the performance of block-coded systems with bounded-distance decoding improves continually as one reduces the rate of the code employed. For example, when we consider $1 - P_{CD}$ or post-decoding error probability as a function of the error rate in the channel (or equivalently, E_s/N_0), performance always seems to improve as we reduce the code rate. However, as we pointed out in Chapters 1 and 2, this way of looking at performance can be misleading, since the issue of communication efficiency is not addressed.

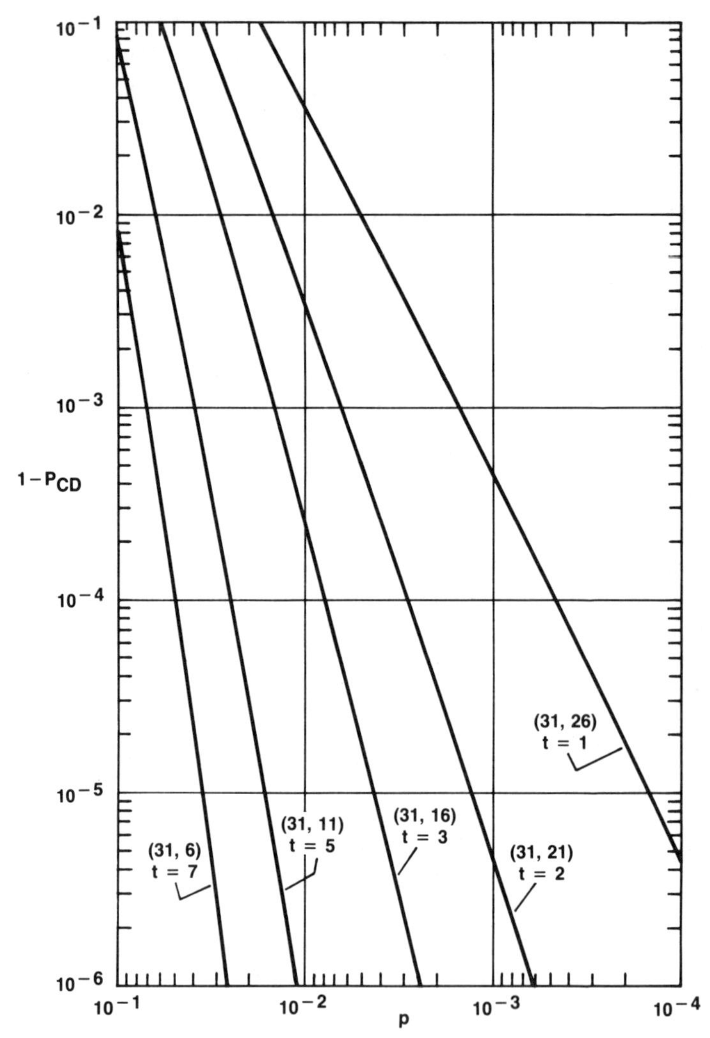

FIGURE 7.12. $1 - P_{CD}$ vs. bit-error rate in the channel for some binary BCH codes with block length 31.

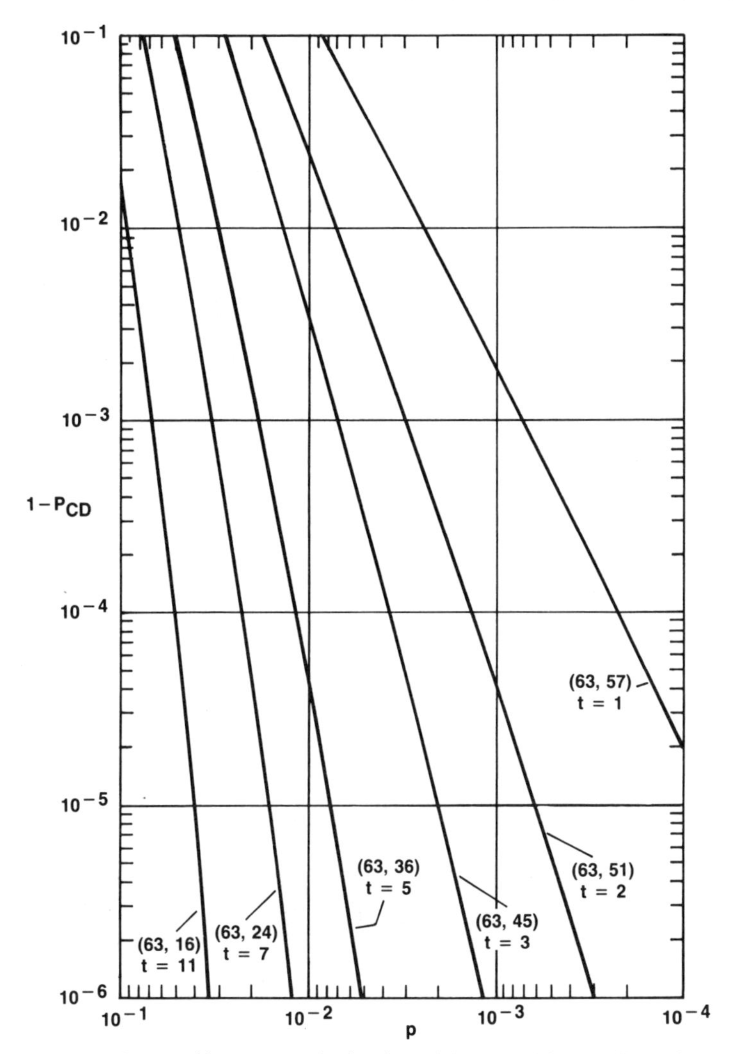

FIGURE 7.13. $1 - P_{CD}$ vs. bit-error rate in the channel for some binary BCH codes with block length 63.

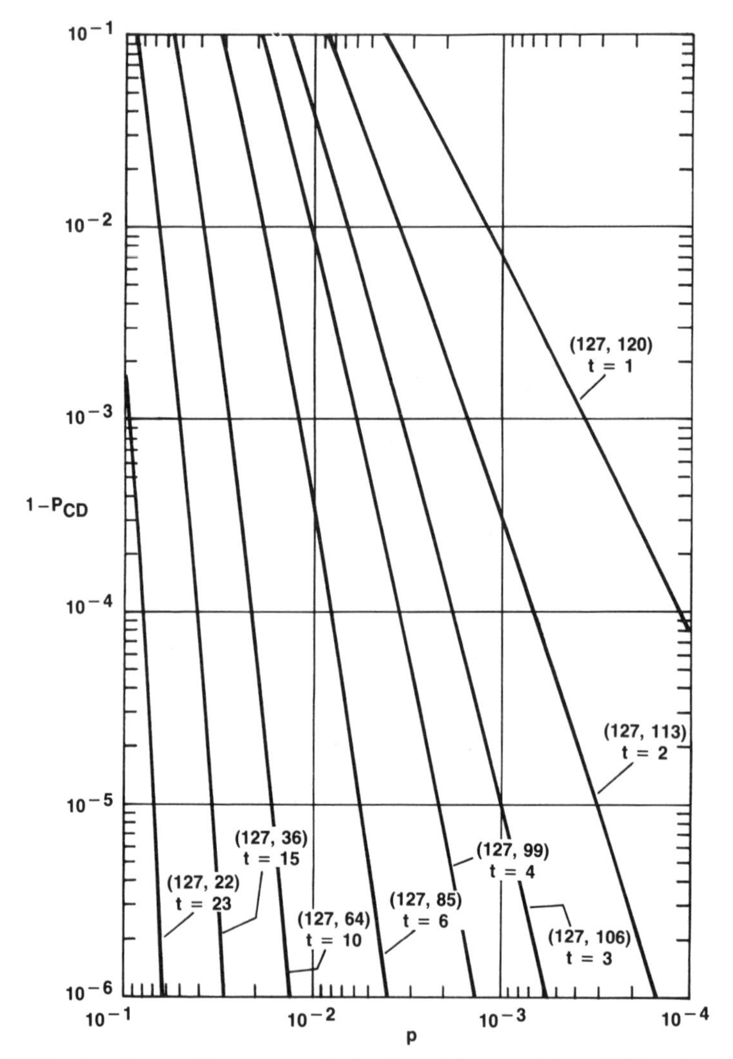

FIGURE 7.14. $1 - P_{\text{CD}}$ vs. bit-error rate in the channel for some binary BCH codes with block length 127.

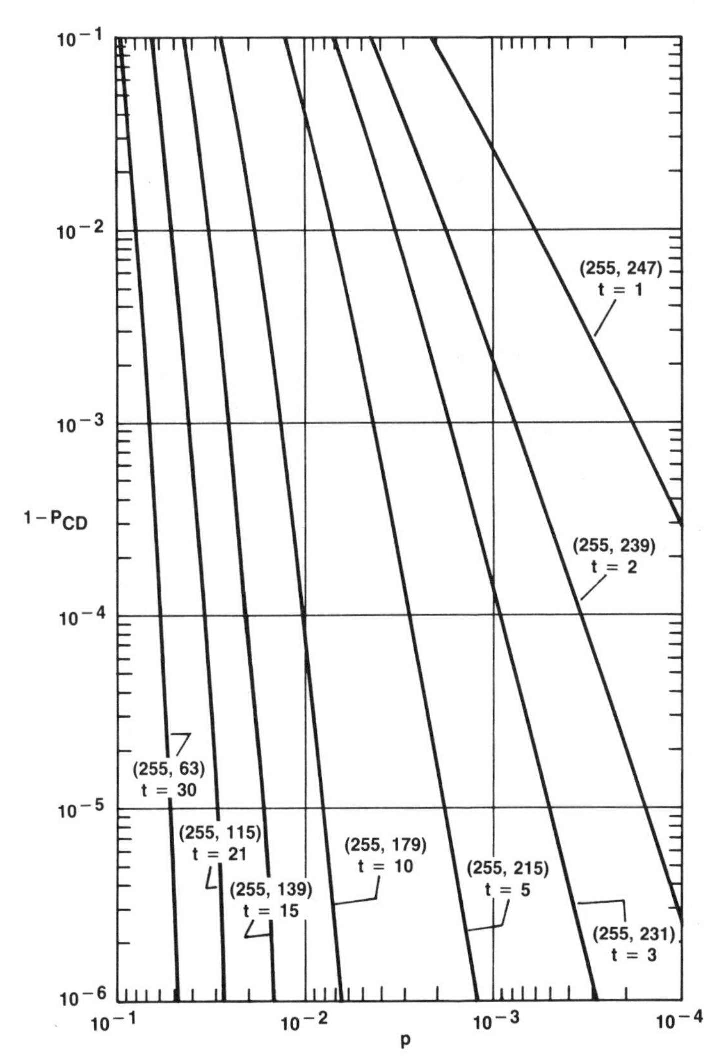

FIGURE 7.15. $1 - P_{CD}$ vs. bit-error rate in the channel for some binary BCH codes with block length 255.

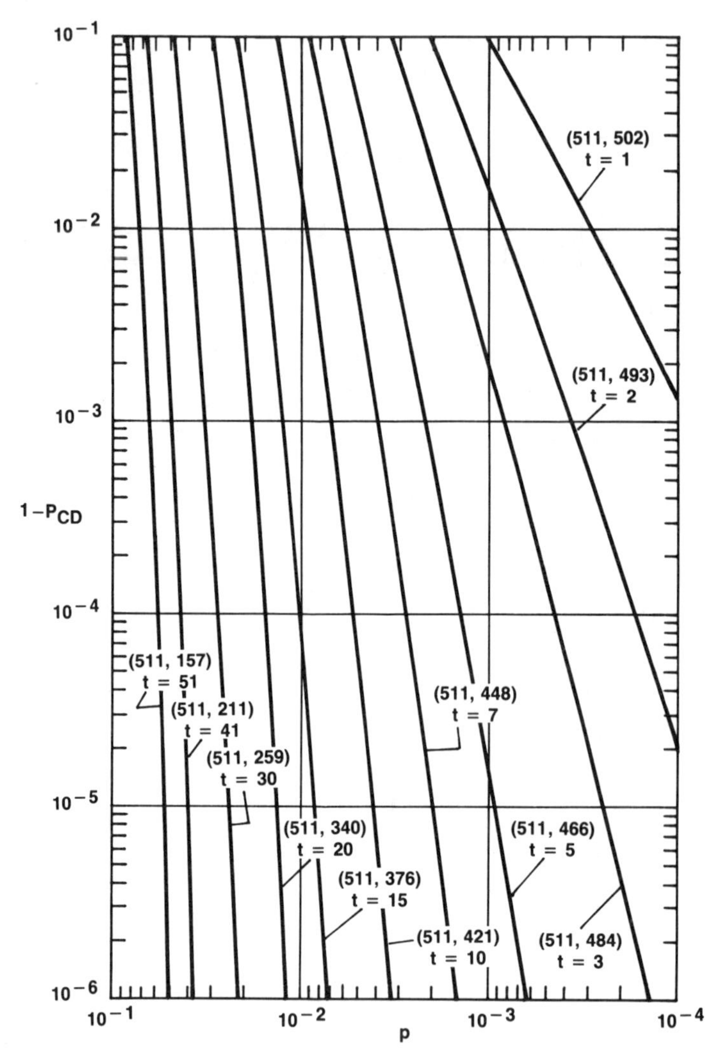

FIGURE 7.16. $1 - P_{CD}$ vs. bit-error rate in the channel for some binary BCH codes with block length 511.

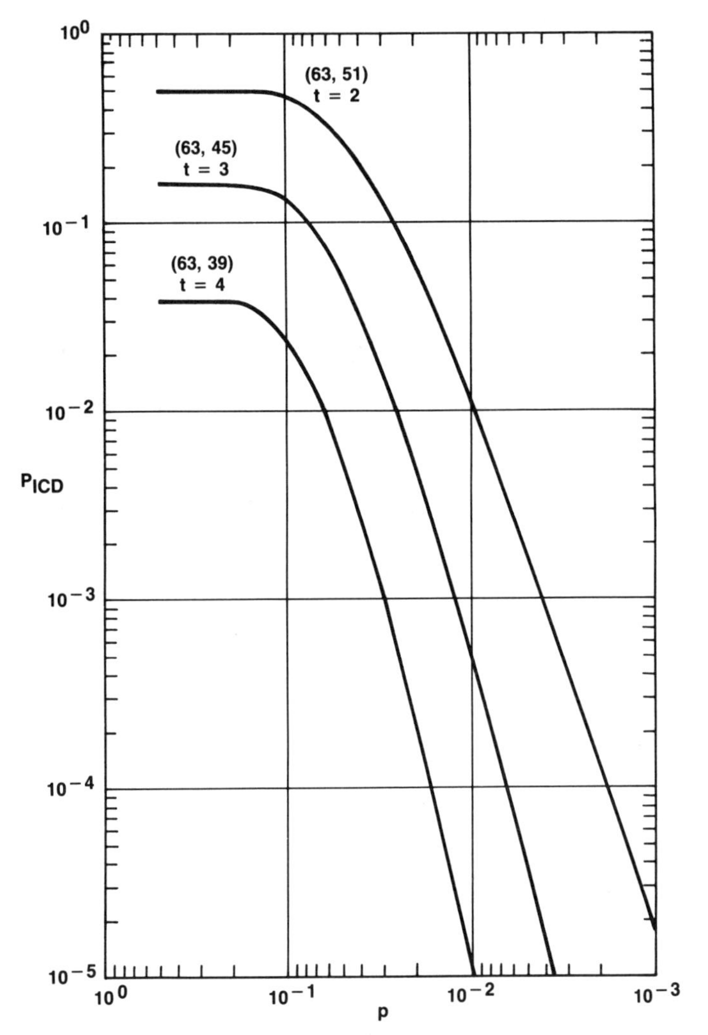

FIGURE 7.17. P_{ICD} vs. bit-error rate in the channel for some binary BCH codes with block length 63. Binomial codeword weights assumed.

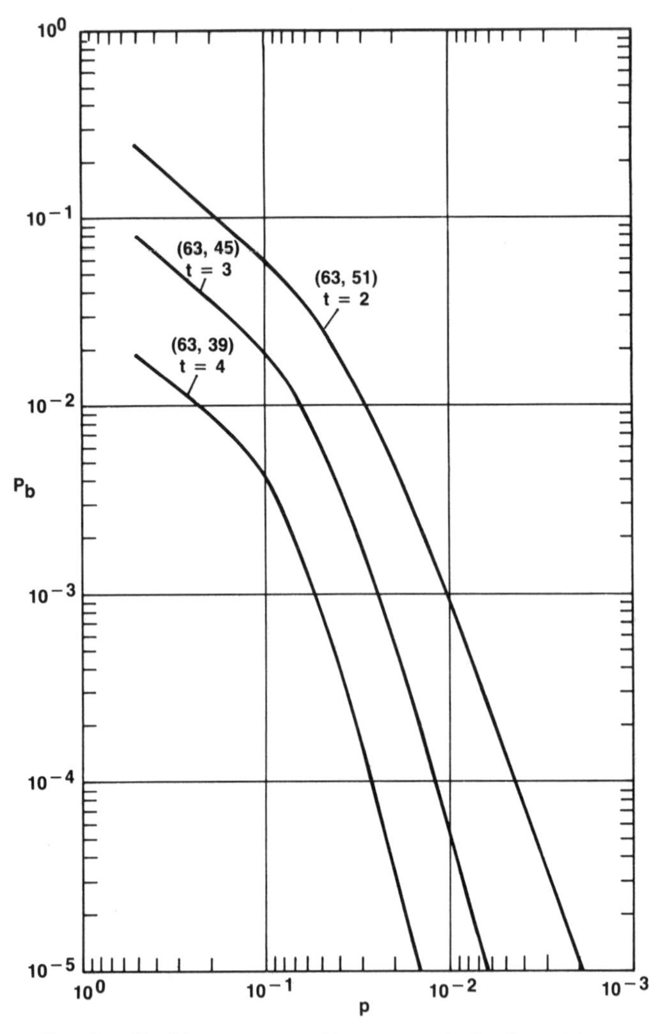

FIGURE 7.18. Post-decoding bit-error rate vs. bit-error rate in the channel for some binary BCH codes with block length 63. Binomial codeword weights assumed.

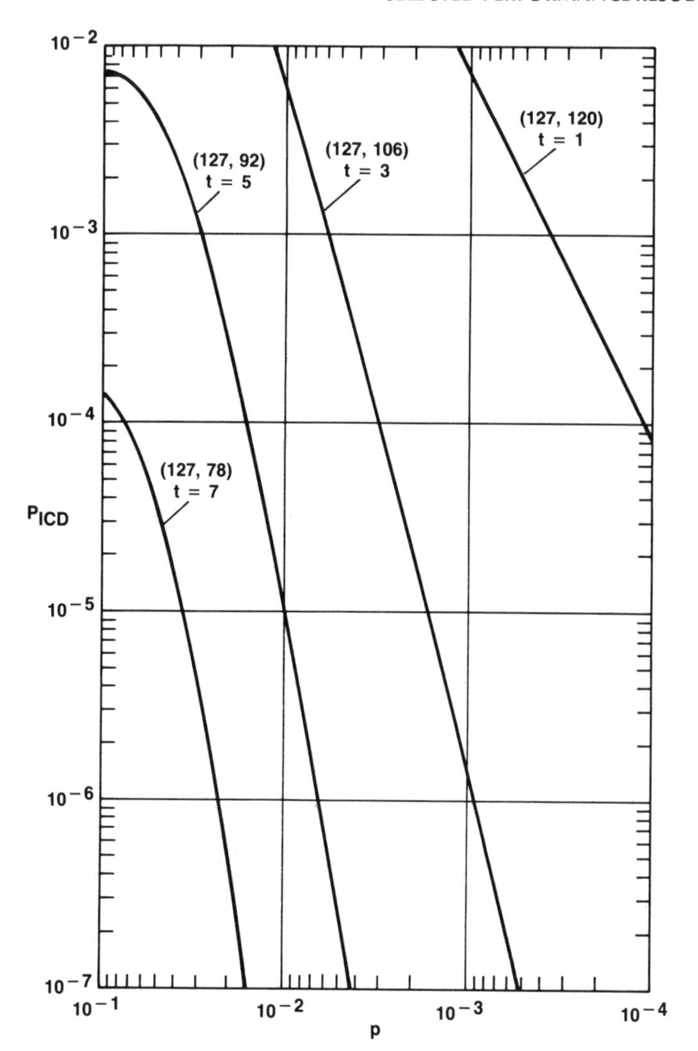

FIGURE 7.19. P_{ICD} vs. bit-error rate in the channel for some binary BCH codes with block length 127. Binomial codeword weights assumed.

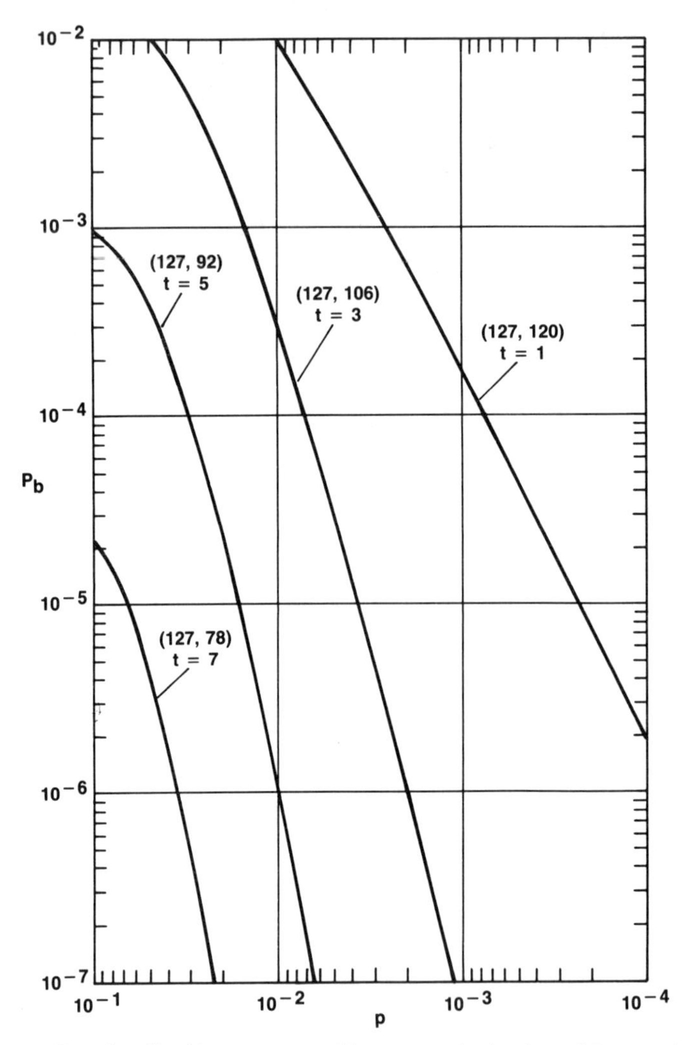

FIGURE 7.20. Post-decoding bit-error rate vs. bit-error rate in the channel for some binary BCH codes with block length 127. Binomial codeword weights assumed.

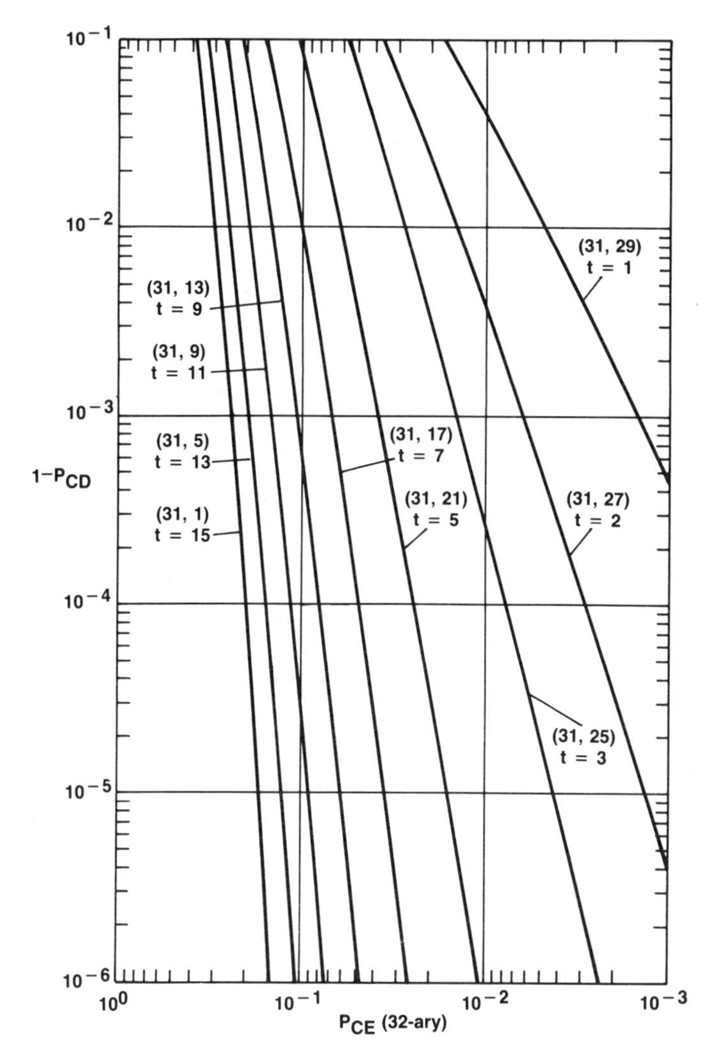

FIGURE 7.21. $1 - P_{CD}$ vs. character-error rate in the channel for some RS codes defined on a 32-ary alphabet.

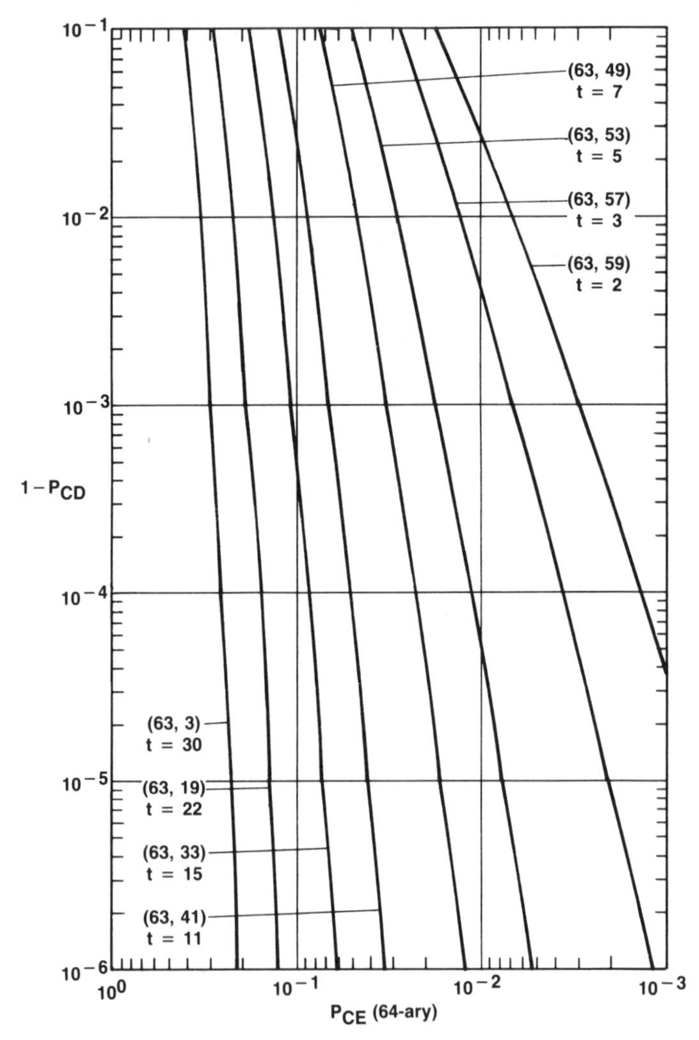

FIGURE 7.22. $1 - P_{CD}$ vs. character-error rate in the channel for some RS codes defined on a 64-ary alphabet.

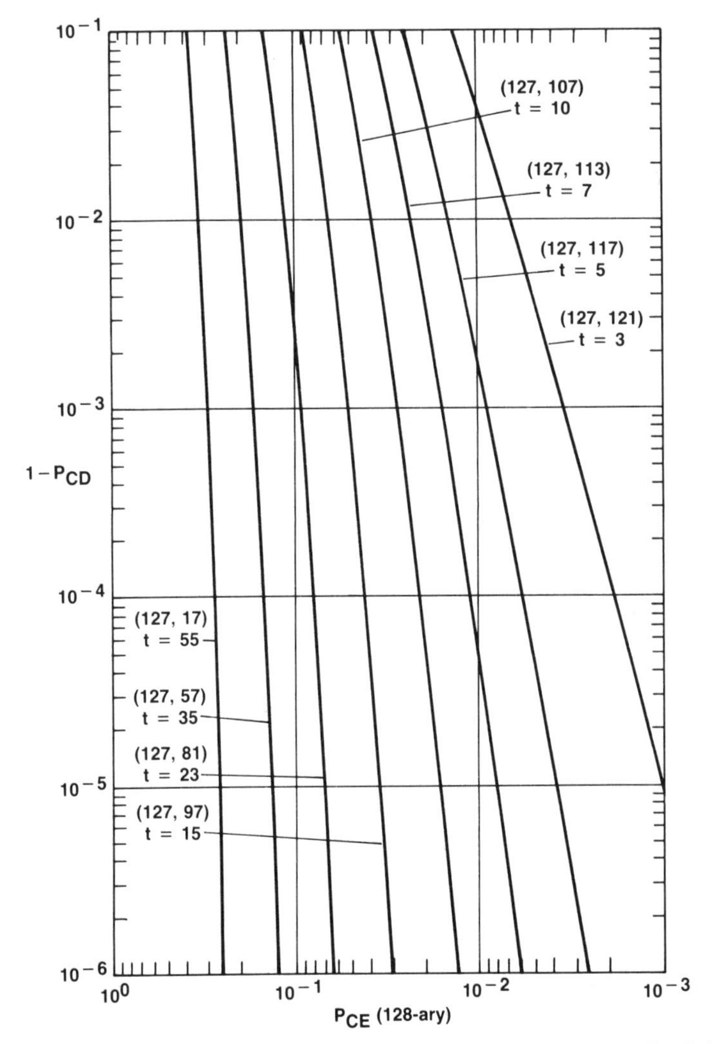

FIGURE 7.23. $1 - P_{CD}$ vs. character-error rate in the channel for some RS codes defined on a 128-ary alphabet.

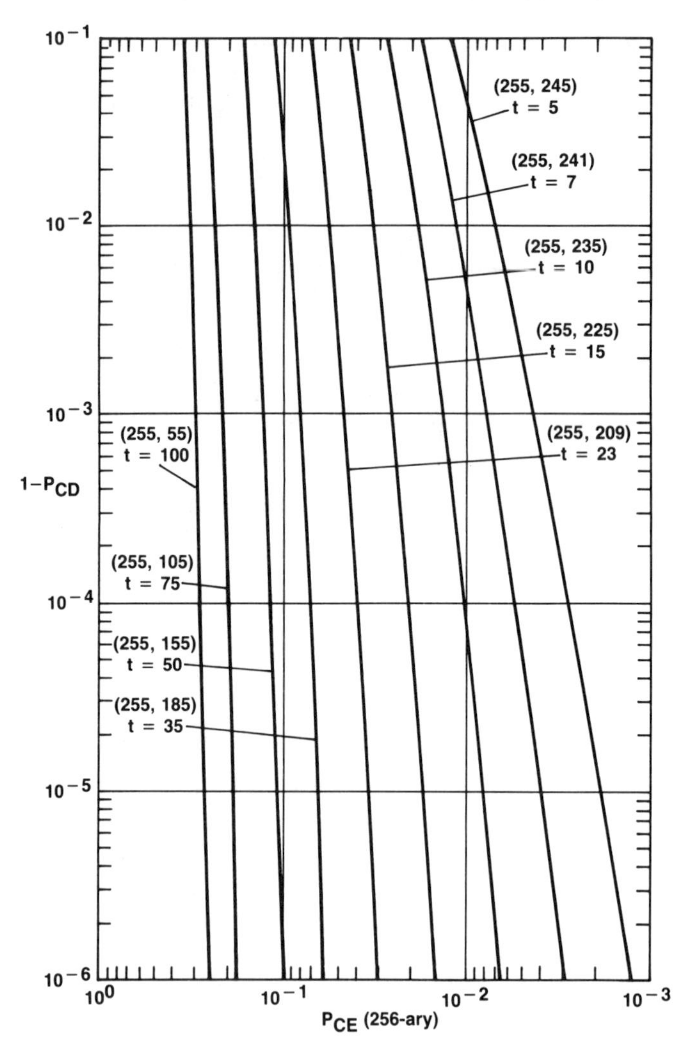

FIGURE 7.24. $1 - P_{CD}$ vs. character-error rate in the channel for some RS codes defined on a 256-ary alphabet.

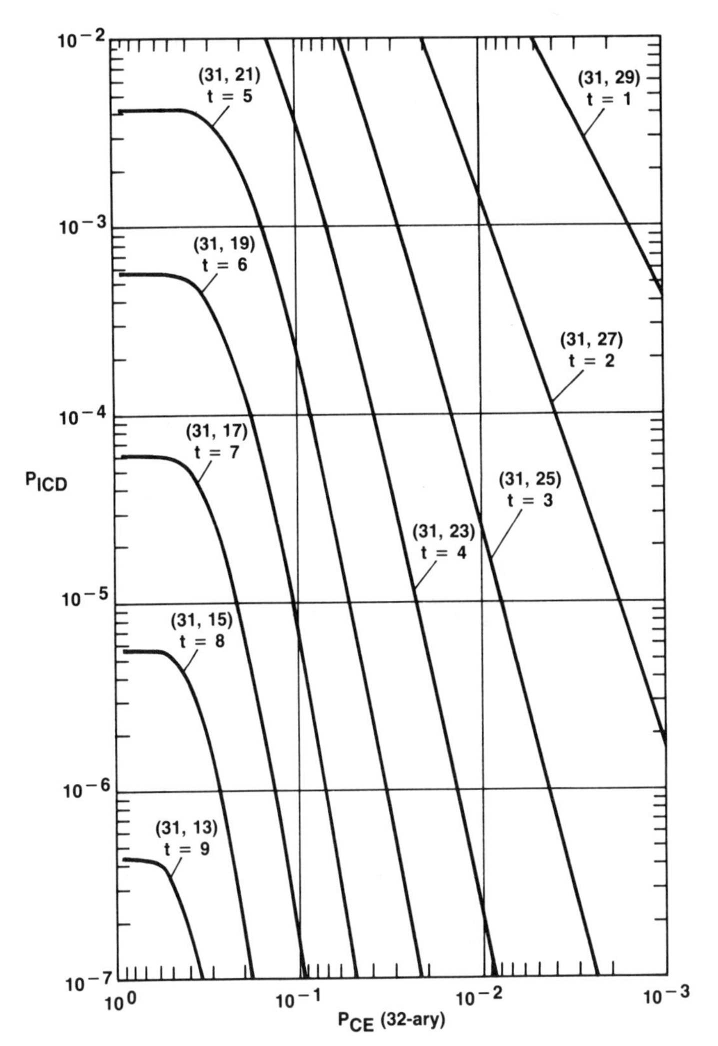

FIGURE 7.25. P_{ICD} vs. character-error rate in the channel for some RS codes defined on a 32-ary alphabet.

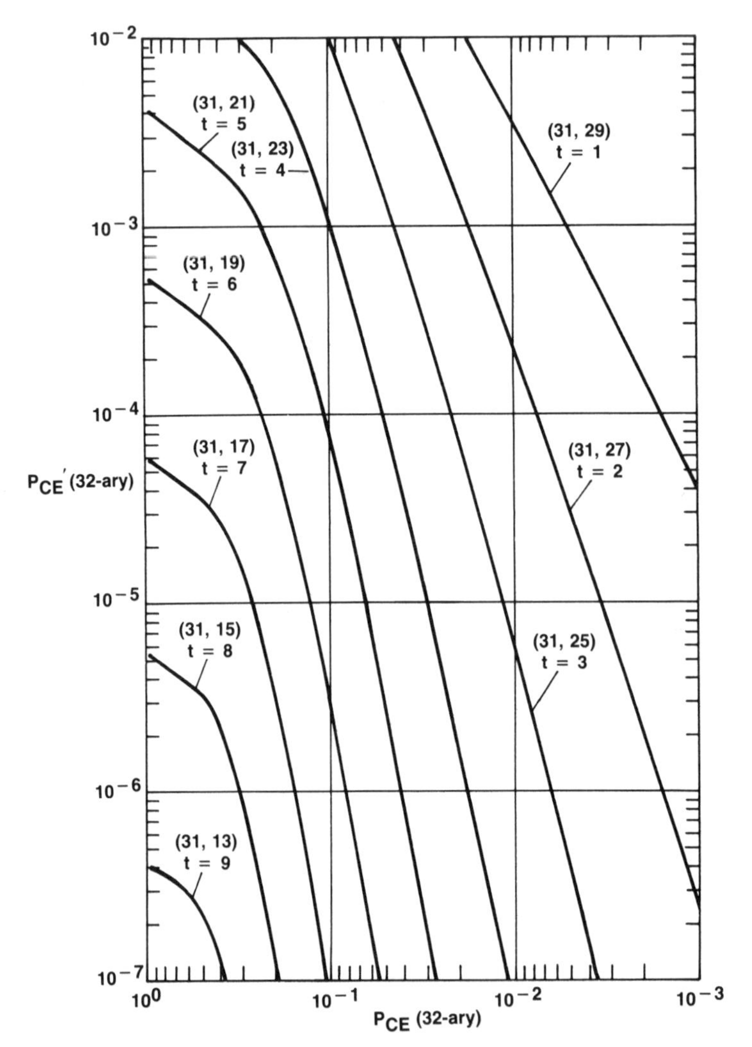

FIGURE 7.26. Post-decoding character-error rate vs. P_{CE} for some RS codes defined on a 32-ary alphabet.

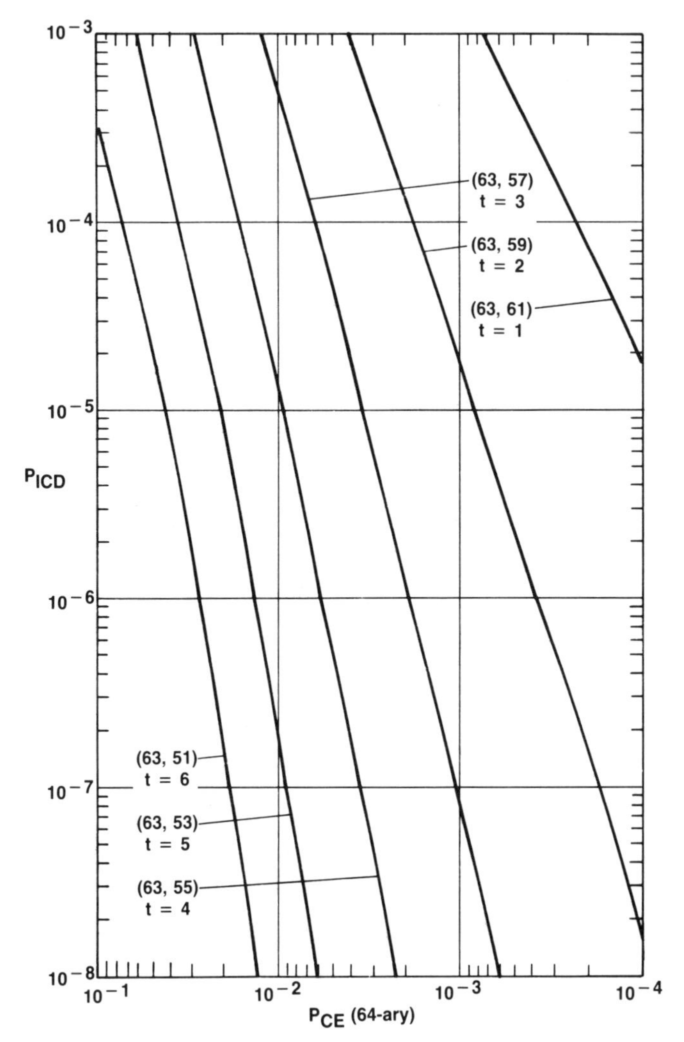

FIGURE 7.27. P_{ICD} vs. character-error rate in the channel for some RS codes defined on a 64-ary alphabet.

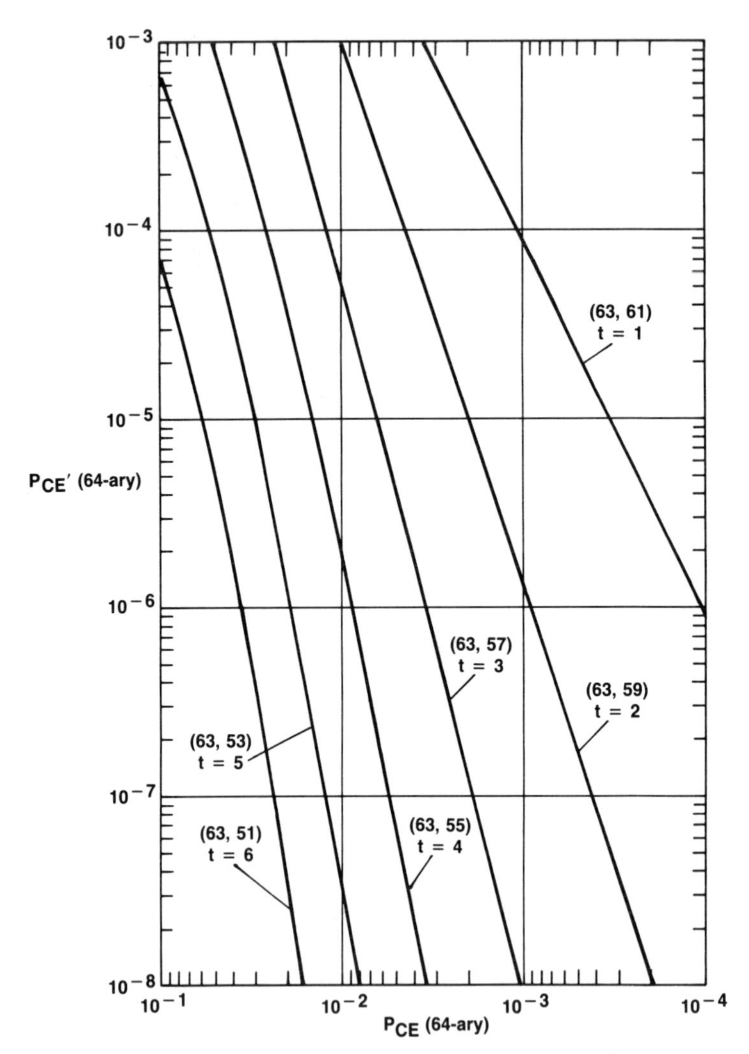

FIGURE 7.28. Post-decoding character-error rate vs. P_{CE} for some RS codes defined on a 64-ary alphabet.

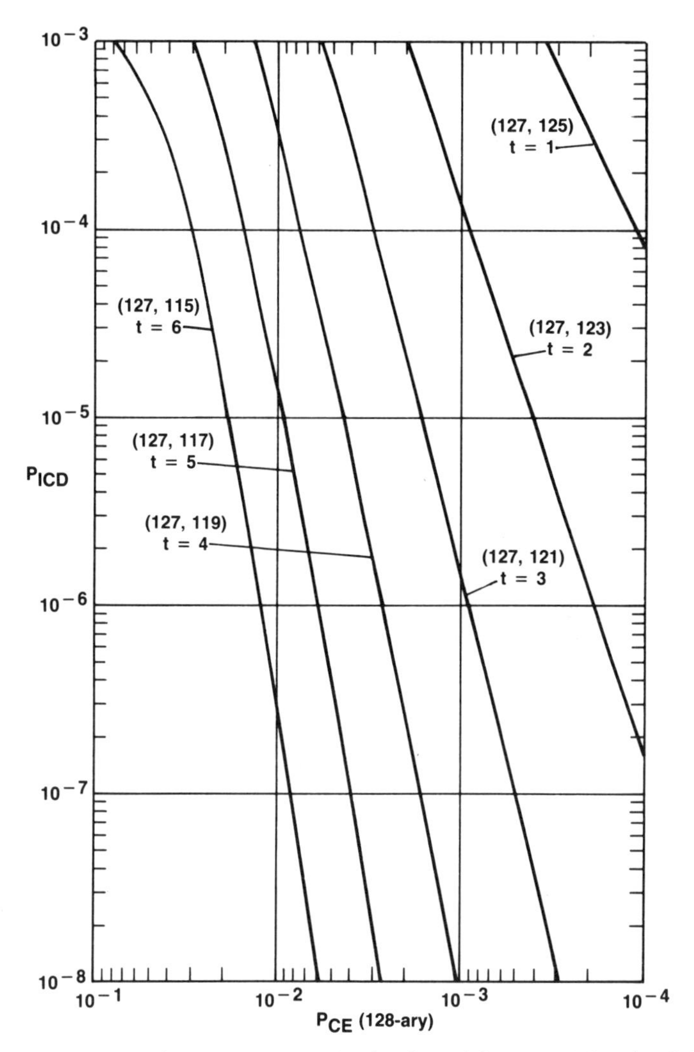

FIGURE 7.29. P_{ICD} vs. character-error rate in the channel for some RS codes defined on a 128-ary alphabet.

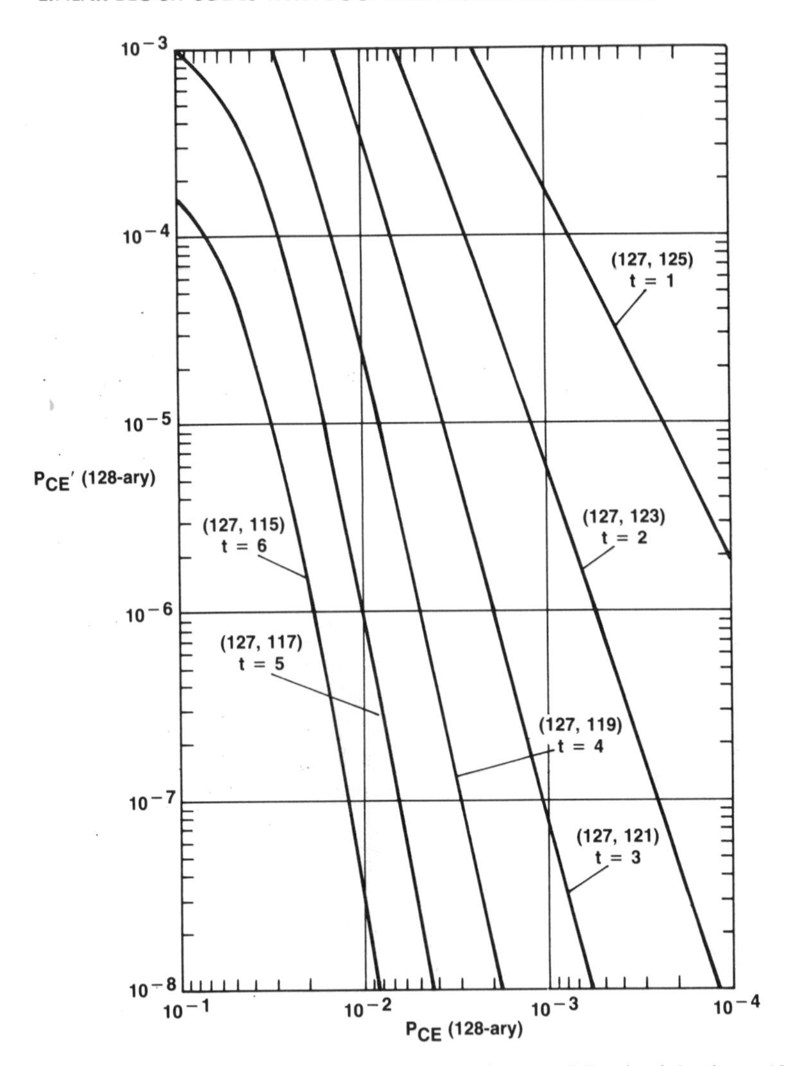

FIGURE 7.30. Post-decoding character-error rate vs. P_{CE} for some RS codes defined on a 128-ary alphabet.

A more meaningful way to compare the performance of alternative designs is on the basis of SNR per information bit, $E_b/N_0 = (1/R)E_s/N_0$. For the binary case, if we assume coherent PSK signaling, one finds that an optimum code rate exists for block codes and bounded-distance decoding, which is typically in the vicinity of $R = 1/2$. See, for example, Wozencraft and Jacobs [185], Section 6.5. For RS codes used with orthogonal signaling and noncoherent detection, an optimum rate also exists, typically in the approximate range $R \approx 0.6$ to 0.7. For higher and lower code rates, performance degrades, although in general the required level of E_b/N_0 increases rather gradually for a substantial range of code rates about the optimum. In subsequent chapters we shall see that for the coherent channel, more complex error-control schemes exist that provide highly efficient communication with very low code rates.

7.5. NOTES

The presentation used in this chapter follows references 69, 112, and 181 closely. Additional results on the 2^{-r} error probability "bound" have been obtained by Leung-Yan-Cheong [93], and Leung [91]. It has been shown that product codes do not obey the bound but that unshortened distance-3 and -5 binary BCH codes do. However, shortened distance-3 and -5 binary BCH codes may not satisfy the bound. Other related results for binary codes have been given by Padovani and Wolf [125] and Kasami [82]. Kasami and Lin [81] have shown that RS codes have a monotonic error-probability characteristic when used either for error detection only or for error detection and correction.

■ ■ CHAPTER EIGHT

Introduction to Convolutional Codes

In the previous chapters we have seen that a systematic (n,k) block code can be used for error control in a digital communication system by first partitioning the source data stream into segments containing k symbols each. Then, using an encoding rule to associate a set of $r = n - k$ parity symbols with each k-symbol information set, a sequence of n-symbol codewords is formed for transmission on the communication channel. A key point to note is that each r-symbol check set is formed independently of all others and is a function only of the information symbols in the same codeword. However, with *convolutional codes,* which were originally called *recurrent codes*, the encoded data does not have this simple block structure. Rather, the encoder for a convolutional code operates on the source data stream using a "sliding window" and produces a continuous stream of encoded symbols. Each information symbol in turn can affect a finite number of consecutive symbols in the output stream.

Many of the concepts presented earlier, such as Hamming distance, minimum distance, parity checks, linearity, and the syndrome, carry over to convolutional codes, although the definitions must be modified somewhat to account for the lack of simple block structure. As is the case with block codes, convolutional codes can be systematic or nonsystematic, where for a systematic code the unmodified information stream is contained in the encoded data sequence. Convolutional codes can also be used for error detection or error detection and correction. However, as we have already seen, there are many good high-rate block codes that can be used effectively for error detection with simple encoders and decoders. Thus error detection without correction is almost always provided for with block codes.

In this chapter we present several descriptions of convolutional codes, corresponding to different ways in which the code space can be viewed. Each description will prove useful in the sequel. Two important yet simple decoding algorithms are described, and the issue of "good" code design is introduced. The first good systematic convolutional codes that were found are presented. In the chapters that follow, more complex decoders for convolutional codes are considered.

The reader will no doubt be struck in this and subsequent chapters by the lack of a detailed mathematical framework underlying the presentation of convolutional codes and the major decoding algorithms. This is in stark contrast with our treatment of block coding techniques in earlier chapters. Forney [40] has given a rigorous theoretical treatment of convolutional codes, but mathematical rigor is not essential to an understanding of the design and application of the most powerful techniques that have been developed to date. As we shall see, the more powerful decoding algorithms for convolutional codes do not make use of algebraic structure and can in principle be used to decode any convolutional code. Nonetheless, the performance gains achievable, particularly with the most sophisticated techniques, are quite dramatic.

8.1. SYSTEMATIC RATE-1/2 CODES AND THE TREE DIAGRAM

Convolutional codes are defined as subsets of arbitrarily long sequences of symbols that obey a linear encoding rule of a particular form. Initially we consider systematic rate-1/2 codes. That is, for each information symbol to be transmitted we associate two channel symbols, where one is the information symbol itself. Let i_j represent an information symbol to be transmitted. Then the associated parity symbol p_j is given by

$$p_j = \sum_{l=0}^{k-1} G_l i_{j-l}, \qquad j = 1, 2, 3, \ldots$$

That is, a particular parity symbol in the encoded sequence is obtained by forming a linear combination of the k most recent information symbols. The encoded data stream consists of the sequences $\{i_j\}$ and $\{p_j\}$ interlaced. We see that a particular code is completely defined by the set of coefficients G_l, $l = 0, 1, \ldots, k-1$, which is called the *generator* of the code. The generator of a convolutional code is analogous to the generator polynomial $g(x)$ for a cyclic block code. Code generators will be discussed more fully later in this chapter.

In general, the information and parity symbols are taken from a finite field with q elements, and $GF(q)$ arithmetic is used to compute the parity symbols. In Chapter 11 we shall see an example of a convolutional code defined on $GF(32)$. However, in most applications, the information and parity symbols are

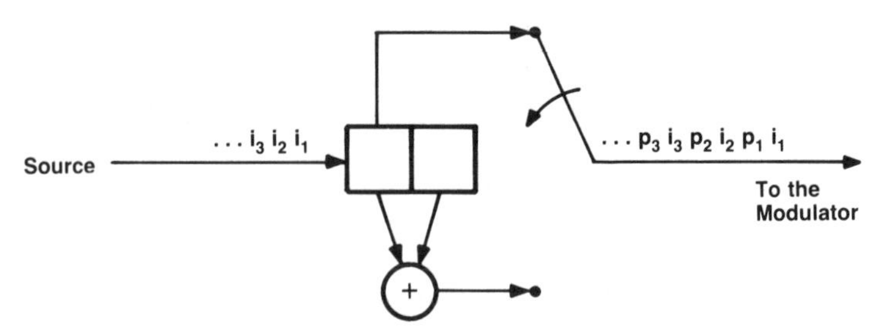

FIGURE 8.1. Encoder for a simple convolutional code.

binary, and modulo-2 arithmetic is employed. For convolutional codes defined on a binary alphabet, the coefficients $\{G_l\}$ are zeros and ones, and therefore each parity bit is simply the modulo-2 sum of a subset of the k most recent information bits. We restrict our attention here to binary convolutional codes.

Convolutional codes are also called *tree codes*, since the set of codewords in the code space can be visualized as a tree structure. Convolutional codes are actually the subset of tree codes that can be encoded with linear feedforward shift registers. Consider, for example, the systematic rate-1/2 code defined by the parity relationship

$$p_j = i_j + i_{j-1}, \qquad j = 1, 2, 3, \ldots$$

The encoder for this code is shown in Fig. 8.1.

The encoder is a two-stage binary feedforward shift register with three taps and one modulo-2 adder. We assume that the contents of the register are initially cleared, that is, the register initially contains two zeros. The source data to be encoded enters the register from the left one bit at a time. Output bits are interlaced by use of the commutator shown. To encode an information bit we proceed as follows. The bits in the register are shifted one stage to the right, and the new information bit to be encoded is inserted in the leftmost stage. Then two output bits are computed by forming linear combinations of the contents of the shift register, the two most recent information bits. Therefore, two output bits are produced for each information bit, and we have a rate-1/2 code. The first output bit is simply the current information bit, while the second is the sum of the current bit and its immediate predecessor. Thus we have a systematic code structure. The reader may wish to verify that the information sequence $i_1 = 1,\ i_2 = 1,\ i_3 = 0,\ i_4 = 1, \ldots,$ produces the parity sequence $p_1 = 1,\ p_2 = 0,\ p_3 = 1,\ p_4 = 1, \ldots,$ that is, the information sequence 1101... is encoded into the channel sequence 11 10 01 11.... Note that the information and parity symbols in Fig. 8.1 are written in a way to suggest the order of their movement into and out of the encoder.

This simple convolutional code may be represented pictorially in several ways. First we consider showing the encoding rule and the set of codewords in the code space as a branching structure called the *code tree*. This is done in

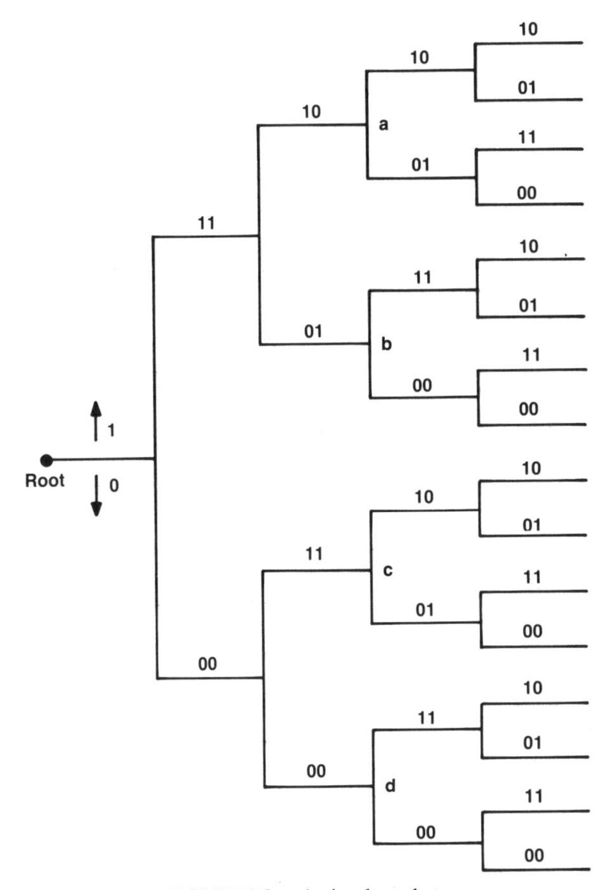

FIGURE 8.2. A simple code tree.

Fig. 8.2, where the encoding rule has simply been restated. To obtain a sequence of channel bits from the source data bits by using the code tree, one begins at the root of the tree and follows the diagram upward if the first data bit is a one or downward if it is a zero, reading the corresponding two channel bits on the line segment or *branch* that is reached. For example, if the first data bit is a one, the associated two channel bits are 11. This process continues from the first branch reached, so that if the second data bit is a zero, one follows the lower branch and reads the two channel bits 01. Subsequent data bits are treated in the same manner, and thus Fig. 8.2 can be used to trace the channel sequences corresponding to any initial four information bits.

As we have seen, convolutional codes are formally defined as arbitrarily long sequences of encoded data. Of course, in any application, our interest is in transmitting finite-length messages that are sometimes segmented into blocks. The procedure for termination or segmentation of a convolutional code sequence is called *truncation*, and a question arises concerning how best to truncate the encoding and transmission processes. Note that if one simply stops with the last information bit in the message, that is, shifts the last data bit into the first stage of the shift register and concludes the transmission with the two associated channel bits, the last information bit affects only the bits on one code branch. All other information bits affect the channel bits on two successive code branches. Therefore, if one wants to provide the same number of estimates of the last data bit as for all the others, an additional known bit should be appended to the information sequence and follow in the encoding process. This last bit encoded and the associated channel bits are called *tail bits*.

For example, suppose we wish to use this simple convolutional code to transmit message blocks that contain four information bits each. A known fifth bit, by convention a zero, is appended to the information sequence, and a fifth level is added to the code tree as shown in Fig. 8.3. Note that there are 16

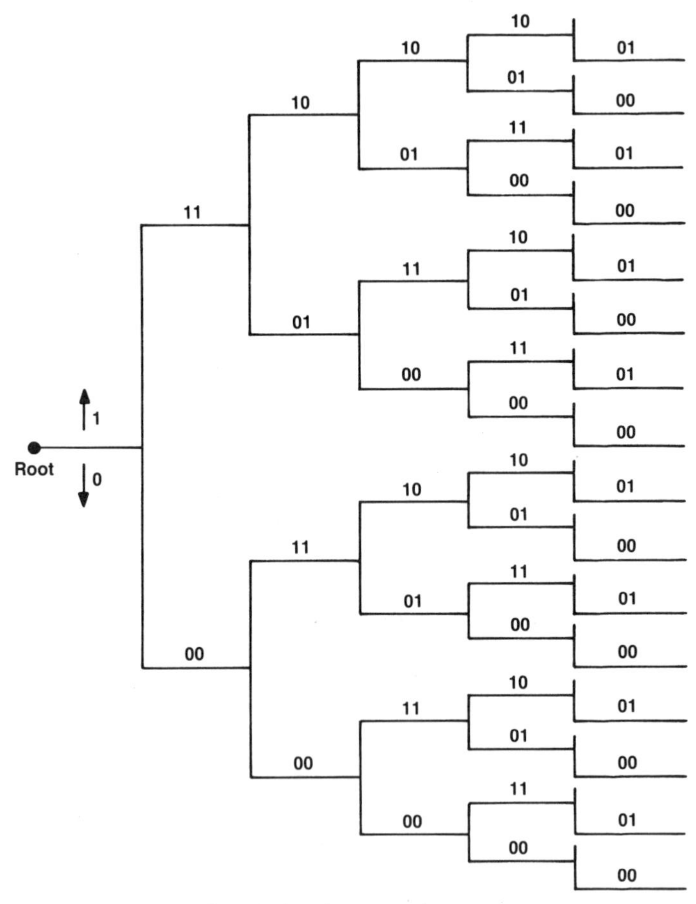

FIGURE 8.3. A truncated tree code.

paths through the tree representing the channel sequences for all 16 possible codewords. However, now there are 10 channel bits associated with each four-bit information set and the code rate is reduced to 0.4. Clearly, for long messages the code rate is close to 1/2.

8.2. THE TRELLIS AND THE STATE DIAGRAM

In this section we consider more compact ways to visualize the encoded sequences in a convolutional code. We begin by pointing out an important property of convolutional codes that can be observed from the tree diagram shown in Fig. 8.2. Note that nodes **a** and **c** at the third level of the tree are closely related. Specifically, if one should stand at either node, one would see identical trees growing to the right. The same can be said about nodes **b** and **d**. Thus, without neglecting any encoded sequences, node **a** may be merged with node **c** and node **b** merged with node **d** to form a more compact representation of the code. The resulting diagram takes the form of a *trellis*, which is shown in Fig. 8.4. We have adopted the convention that the parity bits associated with the information bit value zero are read by following the solid lines. This modification of the code tree not only leads to a more compact representation of the code space, but also, as we shall see in the next chapter, results in a considerable simplification of maximum likelihood decoding.

There is yet another representation of the encoder and code space that proves useful in deriving bounds on code performance. This representation is borrowed from the field of finite state automata [84]. Specifically, we note that a *linear finite-state machine* is defined as any l-input, m-output sequential circuit that is constructed only with $GF(q)$ storage elements, $GF(q)$ adders, and $GF(q)$ multipliers. The input and output symbols are, of course, constrained to be elements of $GF(q)$. The *state* of the machine at any instant of time is defined to be the contents of the storage elements. Thus we see that the encoder for a convolutional code is a linear finite-state machine.

For binary codes the input and output symbols are elements of $GF(2)$, and $GF(2)$ arithmetic is used. The storage elements hold binary values, and the

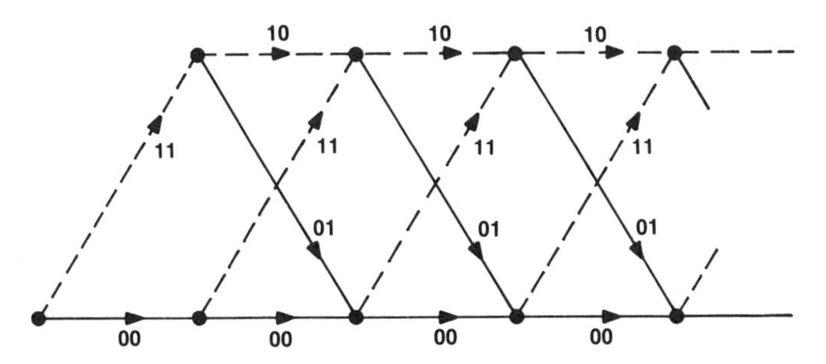

FIGURE 8.4. Code trellis diagram.

state of the encoder is defined as the contents of the encoding shift register. It is important to note that a finite-state machine can be completely described with a *state diagram*. The state diagram is a figure that shows the state transitions and the associated outputs for all possible inputs.

The configuration of the encoder shown in Fig. 8.1 is perhaps misleading, since it appears that a four-state machine model is required for this simple convolutional code. This is not the case, and in fact a two-state model will suffice, as will be seen next.

In general, a finite-state machine can be defined in two similar but distinct ways. With the first approach one specifies the output $y(t)$ at time t to be a function of the present state of the machine $S(t)$. In addition, the next machine state, $S(t+1)$, is given as a function of the state at time t and the current input $x(t)$. This is called a *Moore machine* model and can be summarized as

$$y(t) = \lambda[S(t)]$$

$$S(t+1) = \delta[S(t), x(t)]$$

where λ is called the *output function* and δ is called the *state transition function*. The encoder shown in Fig. 8.1 is a Moore machine.

A straightforward generalization of the Moore machine model permits specification of the output as a function of the present state and the current input. This is called a *Mealy machine* and is summarized by

$$y(t) = \lambda[S(t), x(t)]$$

$$S(t+1) = \delta[S(t), x(t)]$$

The key point here is that for any Moore machine there is an equivalent Mealy machine that has a smaller number of states.

It can be seen that the encoder shown in Fig. 8.1 is equivalent to the Mealy machine shown in Fig. 8.5. That is, both encoders produce the same output

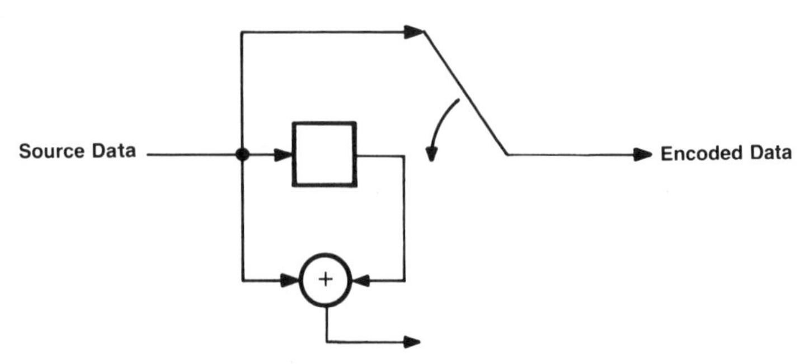

FIGURE 8.5. Two-state equivalent encoder.

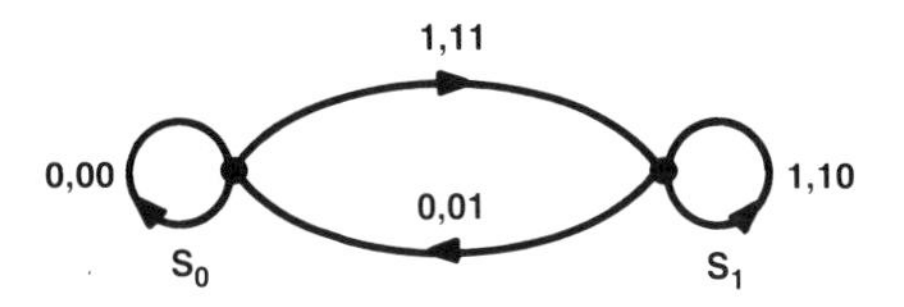

FIGURE 8.6. Code state diagram.

sequence when the input sequence is the same. Note that the Mealy machine shown in Fig. 8.5 has only two states. Because the encoder for this code is equivalent to a two-state machine, the trellis diagram has only two horizontal line sections and the tree diagram has just two distinguishable nodes from the second level on.

The state diagram for the two-state encoder is shown in Fig. 8.6. The two machine states are represented as S_0 and S_1, where S_0 corresponds to the contents of the single-stage register being zero and S_1 to the contents being one. It is conventional and has been assumed that the encoding shift register is cleared prior to encoding, which means starting at state S_0. The state transition line segments shown are labeled x, yz, where x represents the current source data bit to be encoded and yz the two associated channel bits computed by the encoder. We now have a very compact representation for the code.

8.3. RATE-b/V CODES AND A VIEW OF ENCODING AS LINEAR FILTERING

In general, a binary convolutional code is encoded with a linear feedforward shift register that contains bk binary stages. For each cycle of the encoder, the contents of the register are shifted once in b-bit groups. Information bits are shifted into the encoder b bits at a time, and for each b-bit shift V output bits are produced. The output bits are linear combinations of the bk most recent information bits. The rate R of the code is

$$R = \frac{b}{V} \text{ information bits per channel bit}$$

if we ignore the effect of the tail. In the previous sections we considered a rate-1/2 code with $b = 1$ and $k = 2$. We call k the *constraint length* of the code. However, there are two other commonly used definitions of constraint length. Specifically, some authors define constraint length as the number of binary stages contained in the encoder (bk), while others use bkV. We shall consistently define the constraint length as k.

An encoder for an arbitrary convolutional code is shown in Fig. 8.7. We are following the convention of showing the machine output as a function of only the state of the encoder, since this practice is common in the coding literature.

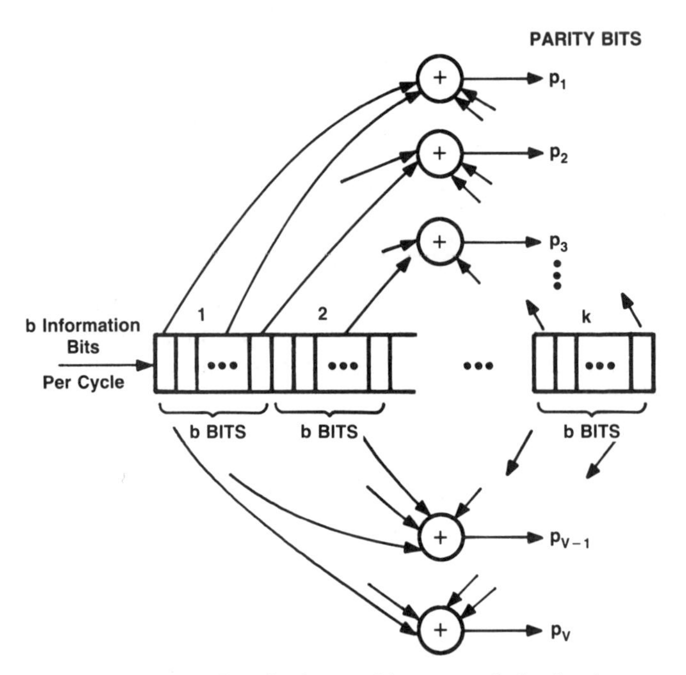

FIGURE 8.7. Encoder for an arbitrary convolutional code.

The reader should keep in mind, however, that the encoder is equivalent to a $2^{b(k-1)}$-state machine. Furthermore, a set of b information bits is seen to affect k consecutive groups of V parity bits. In Fig. 8.8 we show an encoder for a nonsystematic rate-1/3 convolutional code with $b = 1$ and $k = 6$. The output parity sequences $\{p_{1j}\}$, $\{p_{2j}\}$, and $\{p_{3j}\}$ are given by

$$p_{1j} = i_j + i_{j-1} + i_{j-5}$$

$$p_{2j} = i_{j-1} + i_{j-3} + i_{j-4}$$

$$p_{3j} = i_j + i_{j-1} + i_{j-2} + i_{j-3} + i_{j-5}$$

where p_{ij} is the value of the parity bit in the ith output sequence at time j. In Fig. 8.9 an encoder for a rate-2/3 code with $b = 2$ and $k = 3$ is shown.

The set of codewords in any convolutional code may be represented as a tree or trellis, and the encoder can be described with a state diagram. In general, since information bits are encoded in b-bit groups, there are 2^b branches leaving each node in the code tree or trellis. Furthermore, it should be apparent that the tree becomes repetitive after k branches, the trellis has $2^{b(k-1)}$ horizontal line sections, and the state diagram contains $2^{b(k-1)}$ states. Finally, in order to provide the same number of estimates for the last b information bits of a finite-length message, a tail of $b(k - 1)$ zeros is appended

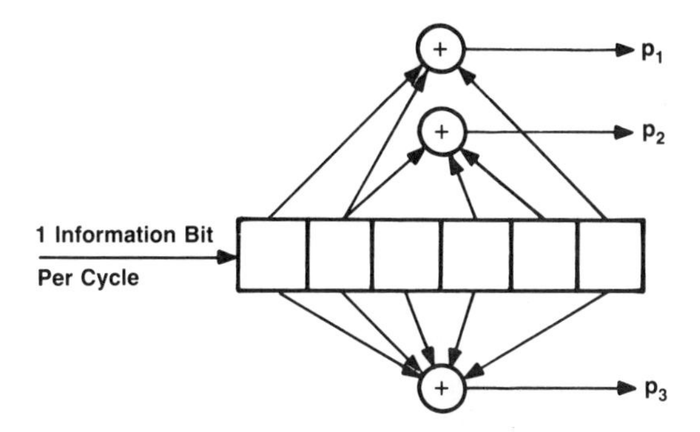

FIGURE 8.8. Encoder for a rate-1/3 code with $k = 6$, $b = 1$, and $V = 3$.

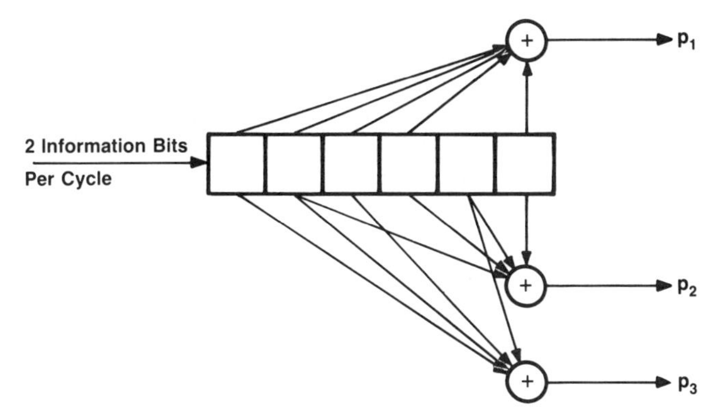

FIGURE 8.9. Encoder for a rate-2/3 code with $k = 3$, $b = 2$, and $V = 3$.

to the information sequence and encoded, and the associated $V(k - 1)$ tail bits are transmitted and used for decoding.

The encoders for the more complex codes indicated in Figs. 8.7 to 8.9 suggest an alternative way to visualize convolutional codes: The encoders can be viewed as linear filters that operate on the input information sequence to produce the output sequences. Actually, the encoder comprises V feedforward filters that compute linear transformations of the input sequence, where the transformations are convolutions of the information sequence with the *impulse responses* of the V parity networks. The impulse response of an encoding filter is defined as the channel sequence generated when the input sequence is a single one followed by an infinite string of zeros. The encoded sequence generated by an arbitrary information stream may then be determined by adding (modulo 2, bit by bit) replicas of the impulse responses that have been appropriately shifted in time to correspond with the positions of ones in the information sequence. This is a consequence of the linearity of the encoder, which is constructed with only modulo-2 adders and storage elements.

For example, the impulse response of the encoder shown in Fig. 8.1 is 11 01 00..., and thus the codeword associated with the information sequence 11000... is seen to be

$$
\begin{array}{cccccc}
 & 11 & 01 & 00 & 00 & \cdots \\
+ & 00 & 11 & 01 & 00 & \cdots \\
\hline
 & 11 & 10 & 01 & 00 & \cdots
\end{array}
$$

It is evident that convolutional codes are linear and that the code sequences form a group under this place-by-place modulo-2 addition rule.

The description of encoding as a linear filtering operation can be made precise. Specifically, convolution in the time domain can be represented as polynomial multiplication in a transform domain. First, we consider the $b = 1$ rate-$1/V$ codes and let the V parity connections be represented as binary polynomials in the delay operator D. The presence of a one in a polynomial position means that the corresponding stage of the encoding shift register is connected to the modulo-2 adder, and a zero indicates that it is not. The position D^0 is associated with the first information bit, the position D^1 with the second, and so on. Then, letting $I(D)$ be the D-transform [84] of the information sequence and $P_i(D)$ the D-transform of the ith encoded parity sequence, that is,

$$I(D) = i_0 + i_1 D + i_2 D^2 + \cdots$$

$$P_i(D) = p_{i0} + p_{i1} D + p_{i2} D^2 + \cdots, \qquad 1 \leq i \leq V$$

we have

$$P_i(D) = G_i(D) I(D), \qquad 1 \leq i \leq V \tag{8.1}$$

where $G_i(D)$ is the polynomial representation of the tap connections for the ith parity network [103]. The $\{G_i(D)\}$ are called the *code generators* or *generator polynomials*.

In Eq. (8.1), modulo-2 addition is assumed after like powers of D are collected. For example, for the encoder shown in Fig. 8.1, we have

$$G_1(D) = 1$$

$$G_2(D) = 1 + D$$

and the response to the all-ones information sequence is

$$P_1(D) = (1) \sum_{i=0}^{\infty} D^i = \frac{1}{1 + D}$$

$$P_2(D) = (1 + D)\left(\frac{1}{1 + D}\right) = 1$$

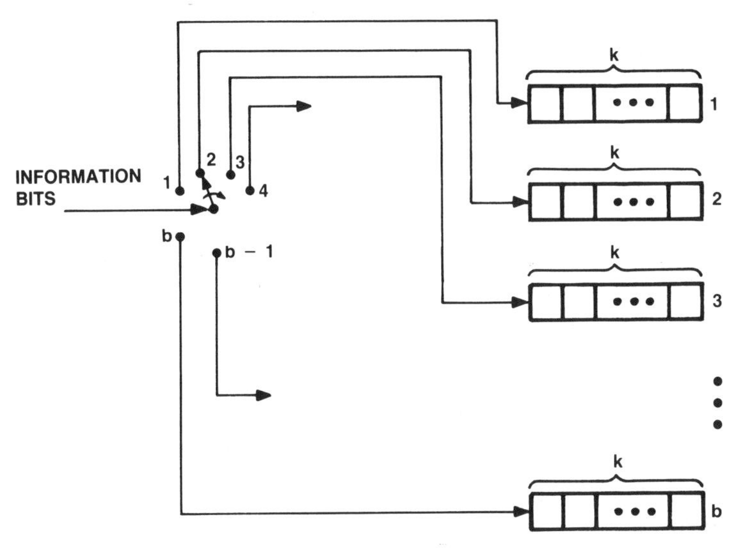

FIGURE 8.10. Formation of b information subsequences for encoding.

The reader should verify that if the input to the encoder in Fig. 8.1 is all ones, the output sequence $\{p_1\}$ is all ones and $\{p_2\}$ is a single one followed by all zeros.

For rate-b/V codes, it is useful to imagine separating the information sequence into b subsequences and encoding with b shift registers containing k stages each rather than one long register that contains bk stages. This is indicated in Fig. 8.10, where a commutator is used to form the information subsequences. Then, for each parity-bit sequence, a generator polynomial of degree up to $k-1$ is associated with each k-bit binary shift register, and a particular parity-bit sequence is obtained by forming sums of convolutions of the generator polynomials with the contents of the associated k-bit shift registers. In the transform domain, the D-transform of the ith parity sequence is given by

$$P_i(D) = \sum_{j=1}^{b} G_{ij}(D) I_j(D), \qquad 1 \le i \le V \tag{8.2}$$

where $I_j(D)$ is the D-transform of the jth information subsequence and $G_{ij}(D)$ is the generator polynomial associated with the ith parity bit sequence and the jth encoding shift register. The arithmetic operations with the polynomial coefficients indicated in Eq. (8.2) are modulo-2 addition and multiplication. Addition is applied after terms with like powers of D are collected together. The reader can verify that the generator polynomials for the code indicated in Fig. 8.9 are

$$G_{11}(D) = 1 + D$$

$$G_{12}(D) = 1 + D + D^2$$

$$G_{21}(D) = D^2$$

$$G_{22}(D) = 1 + D + D^2$$

$$G_{31}(D) = 1 + D + D^2$$

$$G_{32}(D) = 1$$

We have adopted the D-transform notation in this section because it is used extensively in the coding literature. The reader should note that this is equivalent to the z-transform formulation found in circuit and control theory and digital signal processing [123].

8.4. MINIMUM DISTANCE, DECODING DISTANCE, AND MINIMUM FREE DISTANCE

The reader might conjecture that the impulse response of an encoder is the codeword that contains the smallest number of ones in the code space. However, this is not the case for block codes, since an information set that contains a single one is not necessarily mapped into a minimum-weight codeword. With the aid of a simple example, it is easy to see that this is also not necessarily true for convolutional codes. The encoder shown in Fig. 8.11 has the impulse response 11 01 01 01 00 ..., which has weight 5, but the codeword corresponding to an information sequence with two leading ones followed by all zeros is 11 10 00 00 01 00 ..., which has weight 4.

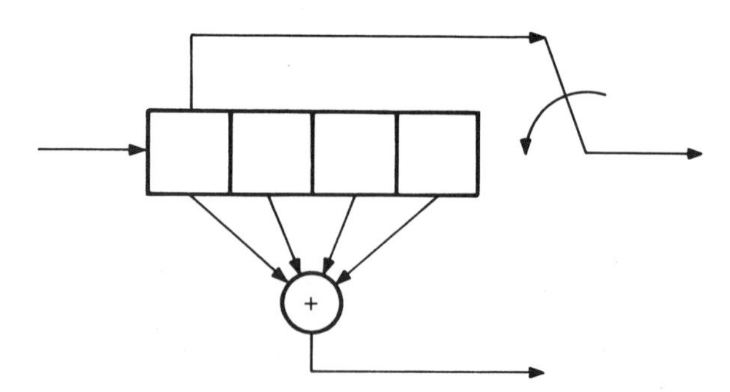

FIGURE 8.11. Encoder for a rate-1/2, $b = 1$, $k = 4$ code.

This example raises an interesting question about the meaning of the distance between encoded sequences or codewords and of the minimum distance of a convolutional code. Since convolutional codes are linear, the minimum codeword weight is the minimum distance of the code, but over how many groups of consecutive encoded bits should the distance measure be computed? In fact, three definitions of the distance between code sequences have been used.

The *minimum distance* of a convolutional code is defined as the smallest Hamming distance between all pairs of encoded sequences taken over a relatively small number of consecutive bits. Since each information bit can affect k consecutive groups of V parity bits, the smallest number of consecutive encoded bits that can be considered is kV. Therefore, we let

1. C_i and C_j be two codeword paths that originate at the same node of the code tree, each containing kV encoded bits.
2. The first information bit associated with C_i be the complement of the first information bit associated with C_j.

We can now define $d_{\min}$, the minimum distance of the convolutional code, as

$$d_{\min} \triangleq \text{minimum number of bit positions in which } C_i \text{ and } C_j \text{ differ for any pair of code sequences satisfying conditions 1 and 2 given above.}$$

Note that the number of bit positions in which C_i and C_j differ is the Hamming distance between C_i and C_j taken over kV consecutive bits.

Clearly we need not be constrained to consider only kV consecutive encoded bits in defining a distance measure for code sequences. For example, a parameter L can be chosen with $L > k$ and the definition above modified to include LV consecutive encoded bits. L is called the *decoding constraint length* or *decoding depth*, and the resulting distance d_L the *decoding minimum distance*. If we now let L go to infinity, d_L is called the *minimum free distance* of the code and is denoted as d_f. We have

$$d_f \geq d_L \geq d_{\min}$$

Decoding algorithms for convolutional codes also need not be constrained to operate on only a small number of received bits prior to making a decision, and one can anticipate that as the decoding constraint length is increased, performance will improve. Ideally a decoder would inspect the entire received message before making a decision. The most powerful decoding algorithms for convolutional codes do, in fact, utilize a long decoding constraint length, which accounts for a large part of the performance advantage achieved. However, as we shall see, the more complex decoders also readily accommodate soft-decision information, which provides an additional important enhancement.

8.5. FEEDBACK DECODING

Thus far in this chapter we have introduced convolutional codes and described several ways in which the code space can be viewed. Measures of the distance between encoded sequences have also been defined. As with block coding techniques, the key to the successful application of convolutional codes has been the development of efficient decoding algorithms that provide significant coding gains. Much of the literature on convolutional codes is devoted to this topic.

In this section we consider a straightforward technique for decoding convolutional codes called *feedback decoding*. Feedback decoding is a simple technique that provides a small but, for some applications, useful coding gain. In the next two chapters we will consider more complex techniques that provide more dramatic enhancements.

Feedback decoding can be used with either systematic or nonsystematic rate-b/V codes, but to simplify the presentation we restrict our attention to systematic rate-1/2 codes with $b = 1$. A feedback decoder makes output bit decisions based on a relatively small number of consecutive received channel bits. For each cycle of the decoder, an estimate of a single information bit is produced. The input to the decoder for each decoding cycle is the current pair of received channel bits, and the output bit decision applies to a previous information bit. The approach is algebraic in the sense that the Hamming distances between the sequence of received bits and the possible transmitted code sequences are compared and an effort is made to decode to an information bit contained in the closest code sequence.

The number of consecutive received channel bits on which a decoding decision is based is LV, where L is the decoding depth and $V = 2$ for the code considered. A feedback decoder, then, after reception of the first $2L$ bits in the message, makes a hard irreversible decision on the first transmitted information bit. The decision can be made by comparing the first $2L$ received bits with all the $2L$-bit code sequences originating at the root of the code tree and finding the path through the tree to the Lth node that has the fewest discrepancies. The decoder chooses the first information bit on the path closest to the received sequence as its first output bit decision. In the case of ties, two or more paths equally distant from the received sequence, the decoder decides in favor of the first bit on any of the closest paths.

After reception of the next two channel bits, the first bits associated with the $(L + 1)$th information bit, the decoder steps forward one node in the code tree and makes a decision on the second information bit transmitted. This decision is made in the same way except that the distance between the received sequence and the encoded sequences that begin with the node associated with the first bit decision are compared. That is, the decoder treats the node associated with the first information bit as the root of a new code tree. This process continues until the end of the message is reached. While decoding through the tail, knowledge of the tail bit values is used. That is, only the set of paths corresponding to the known tail bits are considered. This procedure is sometimes called *minimum-distance feedback decoding*, since the favored paths are those closest in Hamming distance to the received sequence.

It is apparent that each time a bit decision is made by the decoder and a new tree root chosen, the decision affects future bit decisions. At each step in the process, the decoder restricts its attention to only a portion of the code tree, and at each step a new portion of the tree is identified for exploration. Current bit decisions are thus seen to affect future decisions, a process that is implemented with feedback.

We have described the decision-making procedure in feedback decoding as the calculation and comparison of Hamming distances between the received sequence and the possible encoded sequences. As a practical matter, there are more efficient ways to make output bit decisions. In the following sections we consider two alternatives. The first makes use of a syndrome, and the second a majority-logic circuit.

8.5.1. Syndrome Feedback Decoding of Systematic Codes

Clearly, with a systematic convolutional code we can encode the received sequence of information bits and compare the computed parity bits with those actually received. The resulting sequence of discrepancies defines a *syndrome for a convolutional code*. In this section we consider a simple example of a feedback decoder that uses the syndrome.

Consider a *syndrome feedback decoder* for the systematic $k = 3$, $b = 1$ rate-1/2 code for which the parity-bit sequence is given by

$$p_j = i_j + i_{j-1} + i_{j-2}$$

In Fig. 8.12 we show an encoder for this code, a model for an additive error channel, and a syndrome feedback decoder. We assume transmission on the binary symmetric error channel. The sequence of information bits transmitted is represented as $\{i_j\}$, and the associated parity-bit sequence as $\{p_j\}$. The channel error sequences that affect the transmitted information and parity bits are represented as $\{e_j\}$ and $\{e'_j\}$, respectively.

Letting $\hat{\imath}_j$ and $\hat{p}_j$ be the jth received information and parity bits, respectively, the syndrome sequence $\{S_j\}$ is computed as

$$S_j = \sum_{k=0}^{2} \hat{\imath}_{j-k} + \hat{p}_j$$

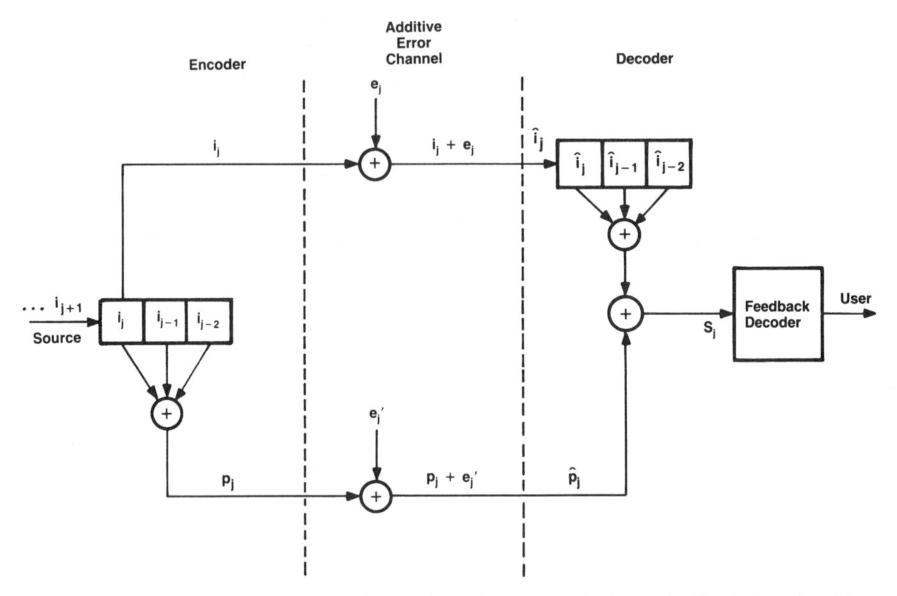

FIGURE 8.12. Encoder, channel model, and syndrome circuit for a feedback decoder. From Heller [63], © 1975 Academic Press, reprinted with permission.

But for the additive error channel we have

$$S_j = \sum_{k=0}^{2} \left(i_{j-k} + e_{j-k}\right) + p_j + e_j' = \sum_{k=0}^{2} e_{j-k} + e_j' \tag{8.3}$$

We can now point out a number of similarities between syndrome decoding of block and convolutional codes. For example, in order to produce an estimate of the transmitted information bits, a syndrome decoder for a convolutional code attempts to find the most likely channel-error sequence that is associated with the observed syndrome. From Eq. (8.3) we note that the syndrome for a convolutional code depends only on the channel-error sequence and not on the transmitted codeword. When no channel errors occur, the syndrome is all zeros. However, the all-zeros syndrome is also produced when the channel-error sequence is a nonzero code sequence. In addition, a nonzero syndrome can be associated with many possible channel-error sequences, each corresponding to the sum of the actual error pattern and a nonzero code sequence. The job of a syndrome decoder for a convolutional code is to find the most likely channel-error pattern, that is, the one that contains the smallest number of errors.

It should be noted that errors affecting information bits are potentially more troublesome than those affecting parity bits. Referring to Fig. 8.12, we see that if $e_j = 1$, meaning that the jth information bit is received in error, the next three syndrome bits are affected. However, if $e_j' = 1$, only S_j is affected. In general, up to k consecutive syndrome bits may be nonzero when an error occurs in an information bit position, but only one syndrome bit will be nonzero when an error occurs in a parity bit position. A similar effect is observed with block codes, since if an error occurs in an information position of a received code block, its affect is spread over many parity bit positions when the syndrome is computed by reencoding.

At this point some differences between syndrome decoders for block and convolutional codes can be identified. A syndrome feedback decoder for a convolutional code inspects a block of L consecutive syndrome bits and at time $j + L - 1$ makes a decision on information bit i_j. This is shown in Fig. 8.13. The received information bits are stored in an L-stage binary shift register, as are the L syndrome bits relevant to the decoding decision on information bit i_j. When $\hat{e}_j$, the estimate of the channel error bit, is zero, the information bit $\hat{\imath}_j$ is decoded as received, and the contents of the syndrome register remain unchanged. When $\hat{e}_j = 1$, however, $\hat{\imath}_j$ and the last three syndrome bits are complemented. This eliminates the effect of the channel error from the syndrome through the use of feedback.

The mapping from the syndrome sequence to the estimate of the channel-error bit can be implemented with a table-look-up procedure. The L syndrome bits are used as a pointer to a 2^L-entry table in which binary values are stored

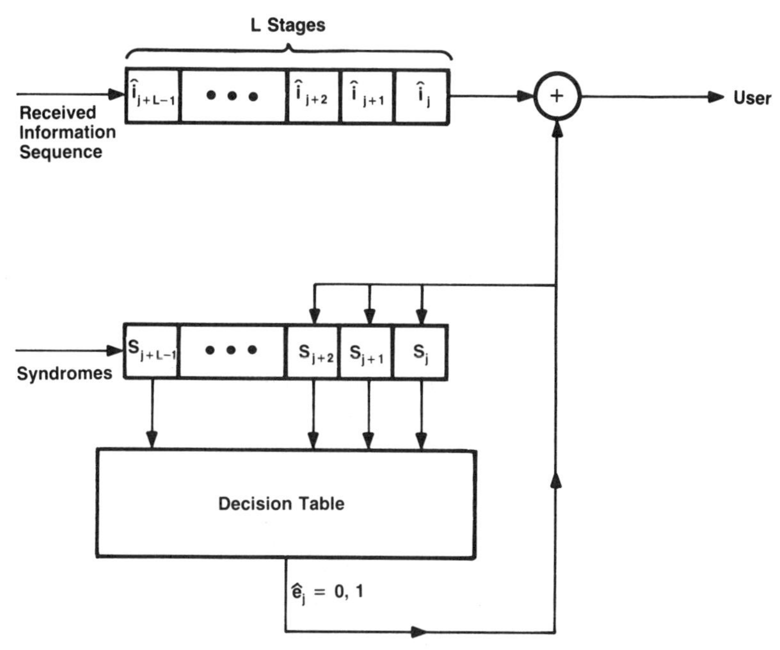

FIGURE 8.13. Syndrome feedback decoder. From Heller [63], © 1975 Academic Press, reprinted with permission.

(0 to indicate no error and 1 to indicate $\hat{e}_j = 1$). The decision table indicated in Fig. 8.13 is constructed under the assumption that all previous decoding bit decisions are correct and hence that the L current syndrome bits depend only on the $2L$ most recent channel-error bits. For any particular code, the mapping from syndrome bits to the most likely error pattern can be determined by computer search. The channel-error sequence containing the smallest number of errors is sought.

It is important to note that with a feedback decoder the effect of a decoding error event may be more severe than a single post-decoding bit error. Since decoding decisions are fed back and used in future bit decisions, it is possible for decoding error events to span many adjacent bits. This effect is compounded by the fact that the 2^L-entry decision table is constructed under the assumption that all previous bit decisions are correct, which, of course, is invalid after a decoding error. It is even possible for a feedback decoder to never recover and to continue making incorrect decoding decisions long after channel errors have ceased. This effect, called *infinite error propagation*, can be avoided if L is sufficiently large and the decision table is constructed as described previously. For systematic codes, $L > k$ is sufficient.

Two serious limitations of syndrome feedback decoding can now be stated. First, the complexity of the decoder grows quite rapidly as the code constraint length and the decoding depth are increased. Since $L > k$, the size of the decision table and hence decoder complexity grow exponentially with constraint length. Thus, syndrome feedback decoding is suitable only for relatively short-constraint-length codes. Since a long constraint length may be needed to provide a required level of performance, this is a serious limitation. Second, syndrome decoding is a hard-decision technique, which imposes a further limitation on achievable decoder performance. However, syndrome feedback decoders have been designed and built and provide modest performance gains (1 to 2 dB) on the independent error channel [63].

8.5.2. A Feedback Decoder That Uses a Majority-Logic Circuit and Threshold Decoding

Feedback decoders for some convolutional codes can also be constructed using a simple majority-logic circuit to make the decoding decisions. As an example, we consider the systematic $b = 1$, $k = 2$ rate-1/2 code that was described in Section 8.1. Again, each transmitted parity bit is simply the modulo-2 sum of the most recent two information bits. The encoder for this code, an additive error channel, and a feedback decoder are shown in Fig. 8.14. The notation is the same as that used in the previous section.

We note that the decoder consists of two modulo-2 adders, two delay elements, and a majority-logic circuit with three inputs. Each input to the majority-logic circuit is an estimate of the information bit to be decoded. At time $j + 1$ indicated, information bit i_j is to be decoded. Three estimates of i_j are derived as follows: at input (**A**), from a received information bit; at (**B**),

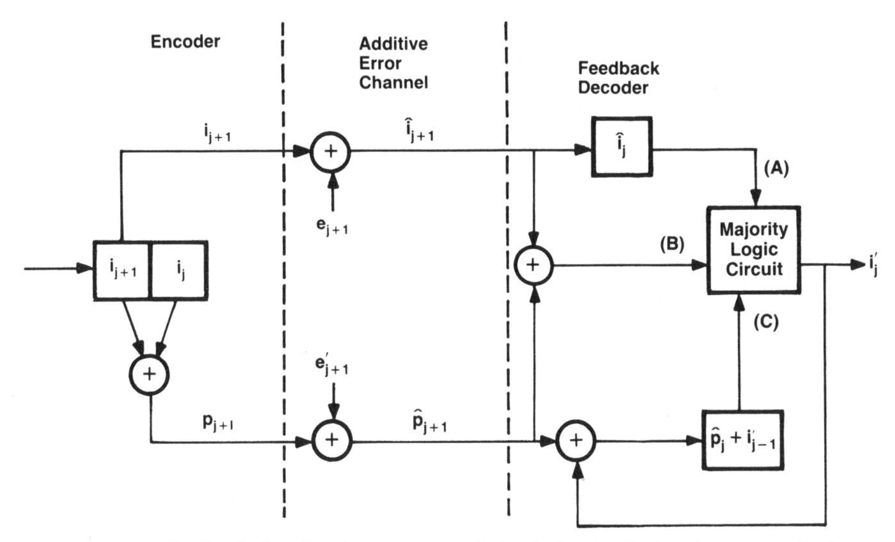

FIGURE 8.14. Feedback decoder that uses a majority-logic circuit for the systematic $k = 2$, rate-1/2 code.

from a received information and parity bit; and at **(C)** from a received parity bit and a previous information bit decision that is obtained through feedback. We let i'_{j-1} be the decoded value of the previous information bit. That three estimates of the current information bit are produced by the circuit can be seen from

$$\text{(A)} \qquad \hat{\imath}_j = i_j + e_j \tag{8.4}$$

$$\begin{aligned} \text{(B)}\ \hat{\imath}_{j+1} + \hat{p}_{j+1} &= i_{j+1} + e_{j+1} + i_{j+1} + i_j + e'_{j+1} \\ &= i_j + e_{j+1} + e'_{j+1} \end{aligned} \tag{8.5}$$

$$\begin{aligned} \text{(C)}\quad \hat{p}_j + i'_{j-1} &= i_j + i_{j-1} + e'_j + i'_{j-1} \\ &= i_j + e'_j \qquad \text{when } i'_{j-1} = i_{j-1} \end{aligned} \tag{8.6}$$

Note that we have three independent estimates of i_j. Therefore, the output bit decision is set equal to the value assumed by two or three of the inputs to the majority-logic circuit.

An interesting property of this decoder can be seen with the aid of the code tree shown in Fig. 8.2. Inspection of the tree shows that the minimum distance of this simple code is 3. Furthermore, since the decoder uses only the received information and parity bits on two successive code branches to make a decoding decision, the decoding depth is 2. Therefore, one might expect that only a single error among four consecutive received bits could be corrected. Indeed from Eqs. (8.4) through (8.6), we see that if there is only one channel error, that is, either e_j, e_j', e_{j+1} or $e_{j+1}' = 1$, single-error correction will be successful when $i_{j-1}' = i_{j-1}$. However, if $e_{j+1} = e_{j+1}' = 1$, i_j will be decoded correctly as long as the previous decoding decision is correct and either e_j or e_j' is zero. Thus, in this case, three channel errors on two consecutive code branches result in correct decoding of the current information bit. This amounts to triple-error correction! If three channel errors have occurred and i_j is decoded correctly, i_{j+1} will necessarily be decoded incorrectly, but it is remarkable nonetheless that a correct decoding decision is made when three errors occur among the four received bits used in the decoding decision.

The capability to provide for correction of more than $(d_{\min} - 1)/2$ errors in a block of kV received bits is a property of a class of convolutional codes studied originally by Massey [103]. These codes are called *majority-logic-decodable codes*, because the decoding decisions can be implemented with a majority-logic circuit. In addition, the decoder can be readily configured to accept soft-decision information from the demodulator, and the decoding decisions can be made with a soft-decision threshold test. Thus these codes are also called *threshold-decodable codes*.

Majority-logic decoding is based on the notion of *orthogonal parity checks*. A set of parity-check equations for a convolutional code can be constructed by forming linear combinations of the syndrome sequence. As we have seen, the syndrome is independent of the transmitted code sequence and depends only on the channel-error bits. A parity-check equation is said to check the channel-error bit e_j if e_j is included in the equation. A set of J parity-check equations is said to be orthogonal on e_j if e_j is checked by each equation but no other error bit appears in more than one equation.

A *majority-logic decoder* for a convolutional code produces an estimate of a single channel-error bit during each cycle of the decoder. Assuming that we have J orthogonal parity checks on the error bit e_j, the decoding rule is given as follows: Set the estimate of e_j equal to 1 when more than $J/2$ parity-check equations fail (give value 1) and set $e_j = 0$ otherwise. After making a decision on e_j, the decoder steps forward to the next channel-error bit and proceeds in the same manner. In this way, estimates of the sequence of channel-error bits are obtained and are used in turn to reconstruct the transmitted information sequence. When the decoder is configured to implement a soft-decision threshold test, the procedure is called *threshold decoding*.

With majority-logic or threshold decoding, current decoding decisions may or may not be used in future bit decisions. When a majority-logic or threshold decoder does not employ feedback, the procedure is called *definite decoding*. Threshold decoding can also be applied to block codes [103]. As is the case with threshold decoding of convolutional codes, a set of orthogonal parity-check equations is required.

We see that threshold decoding of convolutional codes has two clear advantages over syndrome feedback decoding: The complexity of the decoder grows slowly with constraint length, and soft-decision information can be readily accommodated. However, the necessity of starting with a set of orthogonal parity-check equations is a severe restriction, and consequently threshold decoding cannot be used with an arbitrary convolutional code. Furthermore, the codes that can be used do not have good distance properties, that is, for a given code rate and constraint length, threshold-decodable codes have smaller minimum distance than other convolutional codes. Nonetheless, threshold decoders have been designed and built and provide 1 to 3 dB of coding gain on the additive white Gaussian noise channel. More detailed treatments of feedback, majority-logic, and threshold decoding can be found in references 27, 94, and 103.

8.6. THE DESIGN OF CONVOLUTIONAL CODES

The development of convolutional codes has progressed in a very different manner from that of block codes. As we have seen, large classes of block codes have been discovered that have rich algebraic structure and known, relatively large minimum distances. The problem with the block codes was finding efficient decoding algorithms once the codes had been discovered. In contrast, the decoding algorithms for convolutional codes were generally developed first, and their development in turn stimulated the search for good codes with parameters suitable for a particular decoder. Furthermore, the most powerful decoding algorithms for convolutional codes do not make use of algebraic structure.

In this section we consider the design of good convolutional codes. First we show how to avoid a serious problem associated with some convolutional codes, and then we present a portion of the early work in the design of systematic codes.

8.6.1. Infinite Error Propagation and Code Design

We mentioned previously that infinite error propagation can occur with feedback decoding if the decoder is not designed properly. However, error propagation can also result from a poor code design. For example, consider the nonsystematic code generated by the encoder shown in Fig. 8.15. The two parity-bit sequences for this code are given by

$$p_{1j} = i_j + i_{j-1}$$

$$p_{2j} = i_j + i_{j-2}$$

It is easy to verify that for this code the all-ones information set is encoded into the code sequence

$$\mathbf{c} = 11\ 01\ 00\ 00\ \ldots$$

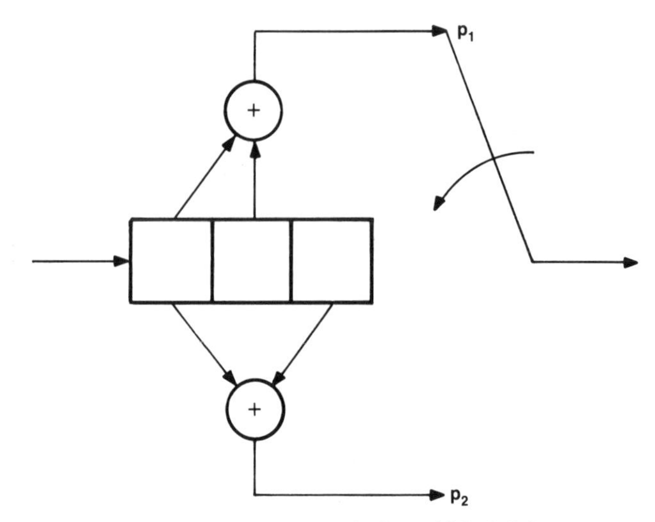

FIGURE 8.15. Rate-1/2, $k = 3$, $b = 1$ code that exhibits infinite error propagation.

This codeword has weight 3 and is quite close to the all-zeros sequence that is the codeword associated with the all-zeros information set. Furthermore, note what happens if the following channel error sequence occurs:

$$\mathbf{e} = 11\ 01\ 00\ 00\ \ldots = \mathbf{c}$$

In this event the received sequence is all zeros, and any sensible decoding algorithm would decide in favor of the all-zeros information sequence. Thus an error event of finite weight and duration produces an unlimited number of decoding errors; that is, infinite error propagation occurs.

In general, error propagation can be due to either a poor code or decoder design. In the example just cited, the problem clearly lies with the code. Massey and Sain have determined necessary and sufficient conditions for the design of convolutional codes so that this type of error propagation does not occur [107]. Simply stated, if an encoder has an inverse that is free of feedback, error propagation due to code design will not occur.

The *inverse* of a linear finite-state machine $\mathbf{M}$ is another linear finite-state machine $\mathbf{M}^{-1}$ that undoes the transformation performed by $\mathbf{M}$ [84]. For example, the inverse of an encoder $\mathbf{E}$ for a convolutional code can be used to compute the information sequence, given knowledge of the parity-bit sequence and the initial state of $\mathbf{E}$ as shown in Fig. 8.16. The output of $\mathbf{E}^{-1}$ is the input information sequence perhaps shifted by a constant delay. Note that if $\mathbf{E}^{-1}$ is a linear feedforward circuit, its state at any instant of time depends only on a finite sum of some number (say l) of the most recent input parity bits. As a consequence, the state of $\mathbf{E}^{-1}$ must be the all-zeros state after l or more consecutive parity bits equal to zero are used as input. From that point on, if

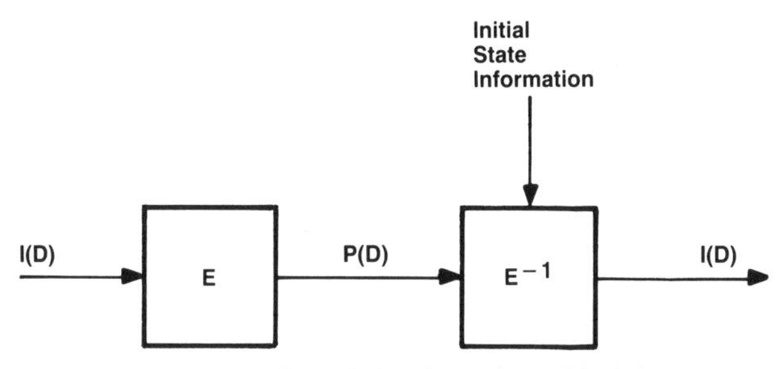

FIGURE 8.16. Convolutional encoder and its inverse.

consecutive zeros continue to occur, $\mathbf{E}^{-1}$ must decide in favor of the all-zeros information set.

The previous argument shows that the encoder depicted in Fig. 8.15 cannot have a feedforward inverse. In fact, if any encoder produces a finite number of nonzero output parity bits when the input contains an infinite number of ones, no feedforward inverse can exist, since a feedforward inverse with memory span l must begin producing the all-zeros output sequence after reception of l consecutive zeros.

We now state without proof the conditions on the generators of a convolutional code that assure the existence of a feedforward inverse, that is, generators that do not produce error propagation. Let $G_i(D)$ be the code generator polynomial for the ith parity connection, $1 \leq i \leq V$. For the rate-$1/V$ codes, we state that error propagation due to code design will not occur if and only if the greatest common divisor of the set of polynomials $\{G_i(D)\}$ is D^m, where $m \geq 0$. That is,

$$\text{GCD}[G_1(D), G_2(D), \ldots, G_V(D)] = D^m, \qquad m \geq 0$$

We see immediately that for systematic codes $G_1(D) = 1$, and the condition is always satisfied with $m = 0$. Therefore, the encoders for all systematic rate-$1/V$ codes possess a feedforward inverse, and error propagation due to code design does not occur.

For more general rate-b/V convolutional codes, the conditions for the existence of a feedforward inverse are slightly more involved. First we must define the generator matrix $\mathbf{G}$ whose rows are the generators of the code. That is, Eq. (8.2) is written in vector matrix notation, and we define

$$\mathbf{G} = \begin{bmatrix} G_{11}(D) & G_{12}(D) & \cdots & G_{1b}(D) \\ G_{21}(D) & G_{22}(D) & \cdots & G_{2b}(D) \\ \vdots & & & \vdots \\ G_{V1}(D) & G_{V2}(D) & \cdots & G_{Vb}(D) \end{bmatrix}$$

G is a $V \times b$ matrix whose rows are the generator polynomials described previously. Note that there are $\binom{V}{b}$ distinct ways in which the rows of **G** may be arranged in $b \times b$ submatrices. We state that error propagation due to code design does not occur with a rate-b/V code if and only if

$$\mathrm{GCD}\left[\Delta_i(D), \qquad i = 1, 2, 3, \ldots, \binom{V}{b}\right] = D^m, \qquad m \geq 0$$

where $\Delta_i(D)$ represents the determinant of the ith $b \times b$ submatrix of **G**.

As an example of a nonsystematic rate-b/V code, we consider the code produced by the encoder shown in Fig. 8.17. This is a $b = 2$, $V = 3$ rate-2/3 code with $k = 2$. The reader can easily verify that

$$\mathbf{G} = \begin{bmatrix} 1 + D & D \\ D & 1 \\ 1 + D & 1 \end{bmatrix}$$

and

$$\Delta_1 = 1 + D + D^2, \qquad \Delta_2 = 1 + D^2, \qquad \Delta_3 = 1$$

Then, since GCD $[\Delta_1, \Delta_2, \Delta_3] = 1$, this code does not exhibit infinite error propagation.

For systematic rate-b/V codes we have

$$G_{ij} = 1 \qquad \text{for } i = j \leq b$$

$$G_{ij} = 0 \qquad \text{for } i \neq j, i \leq b, j \leq b$$

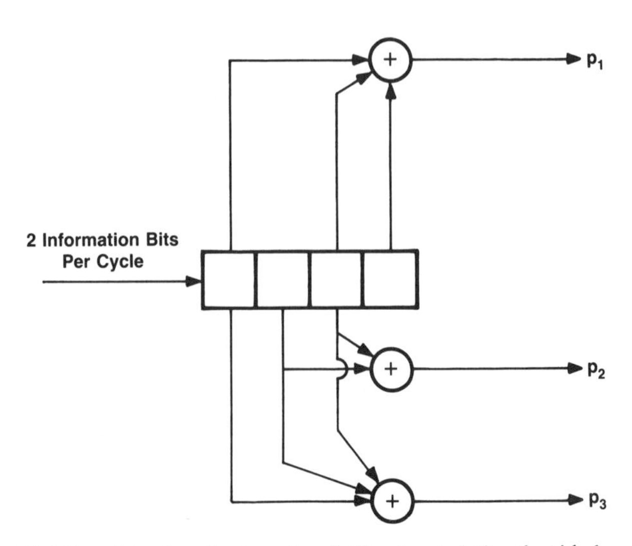

FIGURE 8.17. Encoder for a $b = 2$, $V = 3$, rate-2/3 code with $k = 2$.

and therefore a $b \times b$ submatrix exists that is the identity matrix with determinant equal to unity. Thus we have the result that the systematic rate-b/V codes do not produce error propagation.

8.6.2. Code Generators for Some Systematic Codes

Several authors have addressed the issue of finding good code generators for systematic short- to moderate-constraint-length convolutional codes. The results have been obtained with clever computer searches for codes having the largest minimum distance for a given constraint length. It does not follow that the code with the largest minimum distance necessarily provides the best performance of all the codes with the same rate and constraint length, since

TABLE 8.1. Generators G_2 Found by Bussgang for Good Rate-1/2 Systematic Convolutional Codes with $k \leq 16$

Decoding Depth, L	Constraint Length, k	d_L	n_{d_L}	G_2
2	2	3	2	11
3	2	3	1	11
	3			111
4	4	4	3	1101
	3			111
5	4	4	1	1101
	5			11101
6	6	5	5	110101
	6			111011
7	6	5	2	110101
	6			111011
8	6	6	11	110111
	7			1110011
9	6	6	5	110111
	9			111001101
10	9	6	1	110111001
	10			1110011001
11	9	7	12	110111001
	10			1110011001
12	12	7	5	110101001111
	12			110111000011
	12			110111001001
	11			11011110101
	12			111001001101
	10			1110011001
	10			1110011011
	11			11101100011
13	13	8	29	1101110010011
	13			1110011001001
14	13	8	12	1101110010011
	14			11100110010011
15	13	8	5	1100101111011
	15			110111000010111
	13			1101110000111
	15			111001101100111
	14			11100110110111
	15			111101101000001
16	16	9	42	1101010110010111
	13			1110110101001

Source: J. J. Bussgang [21], © 1965 IEEE, reprinted with permission.

post-decoding error probability depends on the entire codeword weight distribution. However, simulations have shown that the minimum distance or minimum decoding distance is usually the single most important determinant of achievable performance. Here we consider some of the early work of Bussgang [21].

Bussgang searched for the best rate-1/2 systematic codes with constraint length and decoding depth up to 16. The search was exhaustive, and codes with the largest minimum distance or largest decoding distance were found. When there were several codes that had the same d_L, the code containing the smallest number of words at the minimum distance, n_{d_L}, was chosen. When several codes had both the same d_L and the same n_{d_L}, the code that had the smallest number of words at distance $d_L + 1$ was selected.

The results of the search are shown in Table 8.1, where $G_2(D)$ is expressed as a binary sequence instead of a polynomial. The tap connections are read in ascending order from the left so that, for example, the fourth table entry 1101 represents the generator

$$G_2(D) = 1 + D + D^3$$

Bussgang also found the best (as defined above) rate-1/3 systematic codes with $k \leq 7$. These generators are shown in Table 8.2.

Other authors have considered the design of good convolutional codes. For example, Lin and Lyne used another search procedure to find more codes with good distance properties [95]. Their procedure was not exhaustive but employed a technique for adding tap connections to good short-constraint-length codes so that large minimum distances resulted for the longer constraint lengths. They found good rate-$1/V$ codes with $2 \leq V \leq 9$ as well as some good

TABLE 8.2. Generators G_1 and G_2 Found by Bussgang for Good Rate-1/3 Systematic Convolutional Codes with $k \leq 7$

Decoding Depth, L	Constraint Length, k	d_L	n_{d_L}	G_1	G_2
2	2	4	1	10	11
3	3	5	1	101	110
	3			101	111
	3			110	111
4	4	6	1	1011	1100
	4			1011	1110
	4			1101	1111
5	4	7	2	1011	1110
	5			11011	11110
6	6	8	3	101101	111000
	6			101111	110010
	6			101111	110011
	5			11011	11110
	6			110110	111101
7	7	9	5	1011010	1110001
	7			1101100	1111001

Source: J. J. Bussgang [21], © 1965 IEEE, reprinted with permission.

rate-2/3 and rate-2/5 codes. We shall return to the issue of designing good convolutional codes in subsequent chapters.

8.7. PERFORMANCE RESULTS FOR SYNDROME FEEDBACK DECODING

Syndrome feedback decoders for short-constraint-length convolutional codes have been available commercially for a number of years. Performance results have been obtained experimentally with actual decoders, and in Fig. 8.18 we show the performance measured with feedback decoders for rate-1/2, rate-2/3,

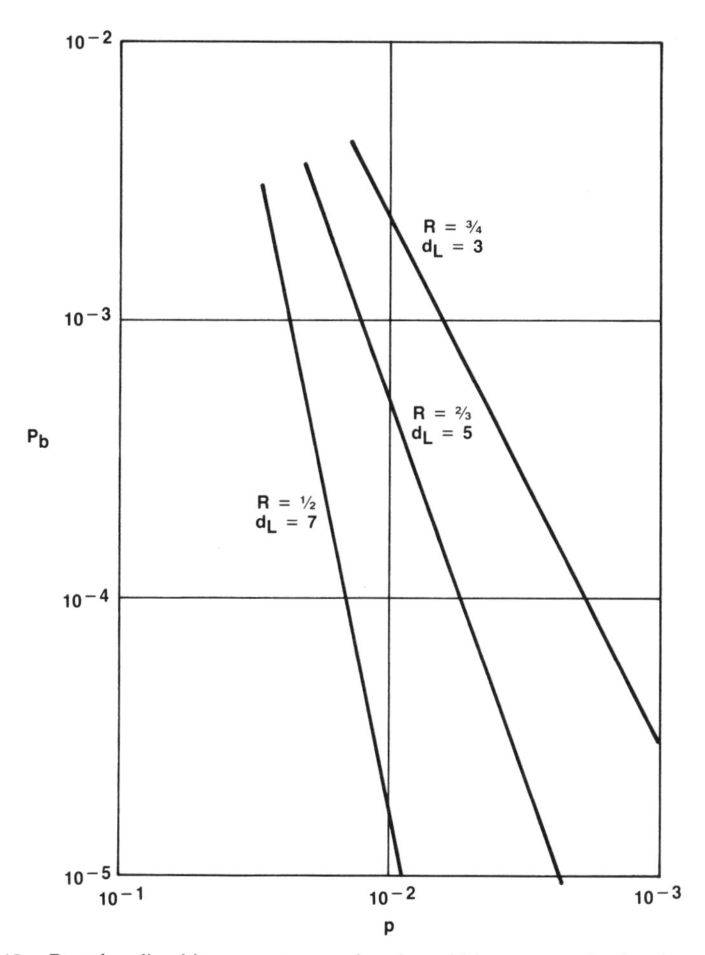

FIGURE 8.18. Post-decoding bit-error rate as a function of bit-error rate in the channel for a syndrome feedback decoder; rate-1/2, -2/3, and -3/4 convolutional codes. From Heller [63], © 1975 Academic Press, reprinted with permission.

and rate-3/4 convolutional codes on the binary symmetric error channel [63]. The reader should verify that with $p_b = 10^{-5}$, the feedback decoder for the rate-1/2 code provides a gain of 2 dB over uncoded operation with binary PSK signaling and coherent detection.

8.8. NOTES

In addition to the references cited in the text, course notes prepared by Professors J. K. Wolf and J. L. Massey were used for presentation of many of the basic concepts in this chapter. The description of feedback decoding closely follows that given originally by Heller [63].

The topic of threshold decoding was treated briefly in this chapter for illustrative purposes. However, threshold-decodable codes and decoding techniques have been active areas of research in coding theory for many years. Threshold decoding was proposed for use with block codes by Reed [144] for a class of codes described by Muller [116]. Massey [103], Hartmann and Rudolph [62], and Weldon [177] subsequently described soft-decision threshold-decoding algorithms for block codes. Good convolutional codes for use with threshold decoding were obtained by Robinson and Bernstein [150], Klieber [83], and Massey [103].

As a final note, we point out that more general distance measures than were considered in this chapter have been used to design convolutional codes. For example, Johannesson used and compared several approaches and found good systematic and nonsystematic rate-1/2 codes [76]. In a later paper, good systematic rate-1/3 and rate-1/4 codes were listed [77].

CHAPTER NINE

Maximum Likelihood Decoding of Convolutional Codes

We have seen in our earlier discussions of block and convolutional codes that the function of an error-control decoder is to find the codeword in the set of all possible transmitted words that most closely resembles the received data. Maximum likelihood decoding involves searching the entire code space and generally is impractical because of the large associated computational burden. However, a decoding algorithm due to Viterbi provides a *maximum likelihood decoding* procedure that is practical for use with short-constraint-length convolutional codes [173]. The *Viterbi algorithm* makes use of the highly repetitive structure of the code tree to dramatically reduce the number of computations required to search the entire code space. This chapter is devoted to a presentation of the Viterbi decoding algorithm.

Although the fundamental observation on which the Viterbi algorithm is based is straightforward, the development of a practical Viterbi decoder required extensive experimental work. Computer simulations played a central role in the development of efficient hardware designs, and computer searches were used to find the short-constraint-length codes with the best distance properties for use with Viterbi decoding. Examples of experimental work will therefore be included in our discussion.

We begin with a description of the Viterbi algorithm for a rate-1/2 constraint-length-3 code applied to the binary symmetric channel. Generalization to other code rates and the AWGN channel follow. Performance bounds are also presented, and the design of a Viterbi decoder for a constraint-length-4 code is described in some detail. Finally, a list of good short-constraint-length codes for use with Viterbi decoding is given as well as performance results for the range of code parameters currently of interest.

9.1. THE VITERBI DECODING ALGORITHM—HARD-DECISION DECODING

A discussion of maximum likelihood decoding of convolutional codes is facilitated by use of a particular example. We consider the rate-1/2, constraint-length-3, $b = 1$, nonsystematic convolutional code whose encoder is shown in Fig. 9.1. As was pointed out in Chapter 8, the encoder for a $k = 3$, $b = 1$ code is equivalent to the four-state machine shown in Fig. 9.2. We call the four machine states **a**, **b**, **c**, and **d**, where

$$\text{State } \mathbf{a} = 00$$

$$\text{State } \mathbf{b} = 10$$

$$\text{State } \mathbf{c} = 01$$

$$\text{State } \mathbf{d} = 11$$

That is, we say that the encoder is in state **a** when the contents of the two-stage register indicated in Fig. 9.2 are the digits 00, the encoder is in state **b** when the digits are 10, and so on. The first five stages of the code tree appear in Fig. 9.3, where the output parity bits and the current state of the encoder are shown for each code branch.

Initially we consider transmission on the binary symmetric channel and hard-decision decoding. That is, for each bit transmitted there is a probability $1 - p$ that the bit is received correctly and probability p that it is received in error. The probability that a bit is in error or is correct is assumed independent from bit to bit.

At first glance it might seem that maximum likelihood decoding of even this relatively simple convolutional code would not be practical. On the binary

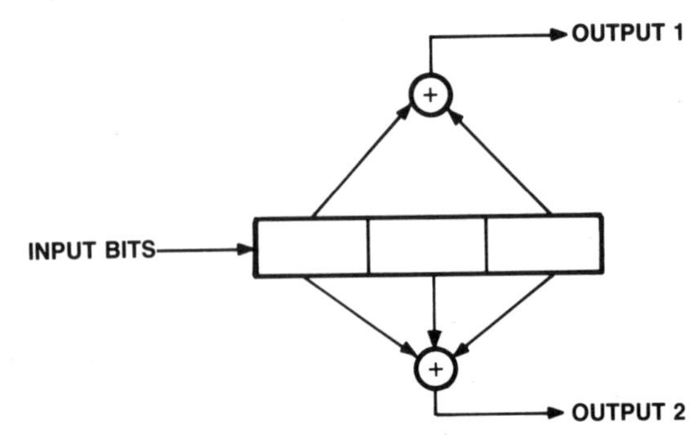

FIGURE 9.1. Conventional encoding circuit for a $k = 3$, $b = 1$, rate-1/2 convolutional code.

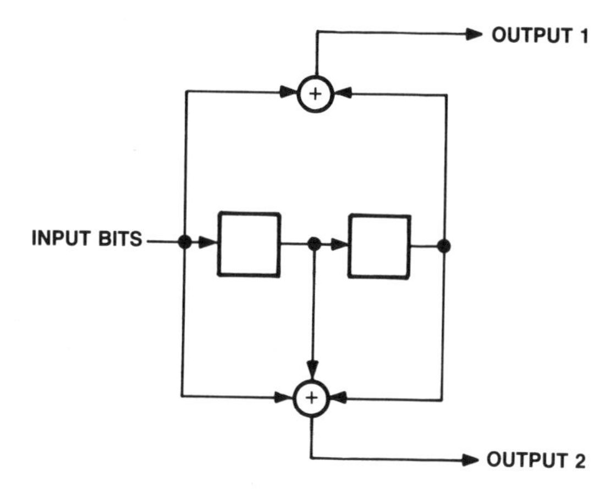

FIGURE 9.2. Four-state equivalent encoder for the $k = 3$, $b = 1$, rate-1/2 code.

symmetric channel, decoding to the maximum likelihood codeword involves comparing all possible transmitted codewords with the received data and choosing the codeword that differs in the smallest number of bit positions. Since the code tree represents the space of all possible codewords, maximum likelihood decoding can be viewed as computing the Hamming distance between the received data sequence and all paths through the code tree and then choosing the path that is closest to the received word. Having found the maximum likelihood path through the code tree, the intended information sequence can be reconstructed by use of the convention that an upward move in the tree corresponds to the transmission of an information bit equal to one, and a downward move corresponds to the transmission of a zero.

The problem with implementing this brute-force maximum likelihood decoder is, of course, that the number of codewords to be examined, that is, the number of distinct paths through the code tree, grows exponentially with the number of information bits in the message. For a message that represents M information bits, there would be 2^M paths through the code tree to compare with the received data, and M need not be very large for this to be totally impractical.

However, using the procedure suggested by Viterbi, an effective maximum likelihood decoder can be implemented, at least for short-constraint-length codes. The technique makes use of the repetitive structure of the code tree to reduce the required number of computations to a manageable level. For example, suppose the first six parity bits received are 11 01 01, and consider the two paths through the code tree that end at the points **1** and **2** (Fig. 9.3). The sequence of parity bits observed along the first path are 11 10 01, and those on the second are 00 11 10. Thus the first path is at Hamming distance 2 from the received sequence, and the second is at distance 5. But note that in both cases the encoder arrives at state **d**, and thus if we imagine ourselves standing at point **1** in the code tree and then at point **2**, we see identical trees

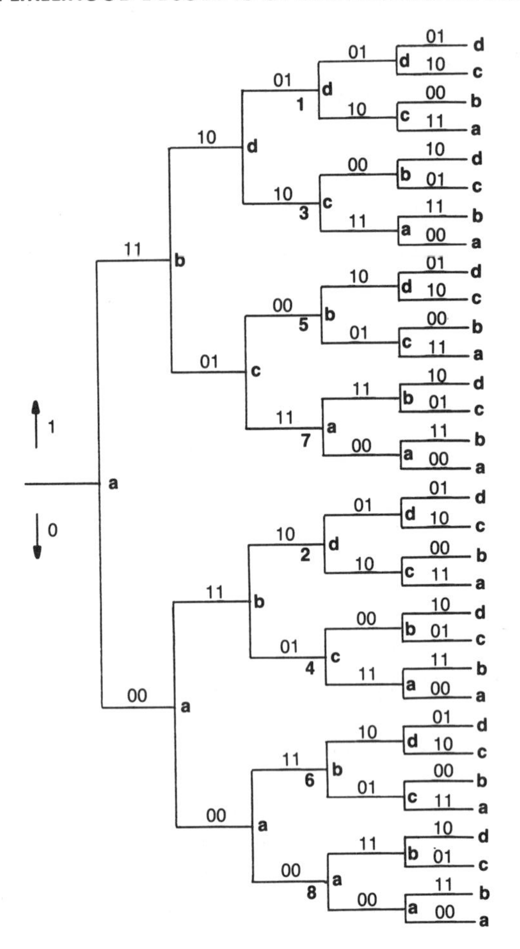

FIGURE 9.3. Code tree for the $k = 3$, $b = 1$, rate-1/2 code.

growing off to the right. As a consequence, the parity bits yet to be received, those associated with the fourth and subsequent information bits, will match and disagree with the paths leaving node **1** in the same way as they will with the paths leaving node **2**. Therefore, since the first path is closer to the received data than the second, no path extended from node **2** could ever catch up, that is, be closer in Hamming distance to the received word than some path leaving node **1**. Therefore we might as well delete all path extensions from node **2** from further consideration.

The same argument can be made for other pairs of nodes at the third level of the code tree that correspond to the same encoder state. For the example considered, the reader can easily verify that for state **c**, node **4** should be favored over node **3**; for state **b**, node **5** should be favored over node **6**; and for state **a**, node **7** should be favored over node **8**. The paths that are favored in this process are called the *survivors*, and we see that at the third level of the code tree there are just four surviving paths.

Clearly, this process can be repeated at the fourth and subsequent levels of the code tree. The four surviving nodes can be extended, producing eight paths, which can again be reduced to four using the simple procedure of retaining only the most likely path to each possible state of the encoder. In the case of ties, that is, two paths to the same state with the same Hamming distance from the received data, one could choose either (flip a fair coin).

The key observation on which the Viterbi algorithm is based is the following. Since the encoder is a finite-state machine, if one computes the maximum likelihood path to each state of the encoder at each level of the code tree, then the maximum likelihood path through the entire code tree must be in that set regardless of how far into the tree decoding has progressed. Thus the number of paths to be extended and retained remains constant and is equal to the number of states of the encoder.

We now see that the brute-force maximum likelihood decoder considered initially would do a large amount of unnecessary work. However, although the number of computations need not grow exponentially with the message length, the number of states of a convolutional encoder does grow exponentially with the constraint length. Thus Viterbi decoding is practical only for relatively short-constraint-length codes.

Thus far we have considered a code with $b = 1$, that is, a code for which information bits are encoded one bit at a time. These codes have rate $1/V$, where V is the number of output bits produced for each cycle of the encoder and the highest rate that can be achieved is $1/2$. Viterbi decoders can also be constructed for rate-b/V codes although the structure is somewhat different.

For a general rate-b/V convolutional code, information bits are encoded b bits at a time, and the encoder contains k b-bit sections and has $2^{b(k-1)}$ states. A Viterbi decoder can be constructed for a rate-b/V code if $b(k-1)$ is sufficiently small, but when a node in the code tree is extended, 2^b new branches are formed, and the structure of the decoder is somewhat more complex than that described for the case $b = 1$. Now at each node in the code tree there will be 2^b paths to extend rather than just two. There will also be $2^{b(k-1)}$ survivors at each step of decoding.

9.2. VITERBI DECODING FOR THE AWGN CHANNEL

Viterbi decoding of short-constraint-length convolutional codes has gained wide acceptance as an error-control technique, principally because the decoder is not unduly complex and provides sufficient coding gain in many applications to justify its cost. Part of the coding gain that can be achieved is due to the fact that a Viterbi decoder can readily accommodate soft-decision information and thus need not suffer the degradation associated with hard-decision decoding. The arguments presented in the previous section could be repeated here using a more general distance measure. For example, rather than computing the Hamming distance between the received data and various paths through the code tree, one can compute the likelihoods of the paths if one knows the probability distribution at the input to the decoder. The process of finding the most likely path to each state of the encoder then involves computing and comparing path likelihoods rather than Hamming distances.

We consider binary antipodal signaling on the AWGN channel with coherent detection. N_0 represents the single-sided noise power density and $\sqrt{E_s}$ the mean of the matched-filter output. Then, after normalization of the matched-filter output by $\sqrt{2/N_0}$, the probability density function observed at the input to the decoder given that a zero was transmitted is

$$p_S(s|0) = \frac{1}{\sqrt{2\pi}} e^{-(s-a)^2/2}$$

where $a = \sqrt{2E_s/N_0}$. For transmission of a one we have

$$p_S(s|1) = \frac{1}{\sqrt{2\pi}} e^{-(s+a)^2/2}$$

That is, we assume that transmission of a positive channel symbol is associated with the transmission of an encoded bit value of zero, and a negative channel symbol with a bit value of one. The SNR per information bit, E_b/N_0, is, of course, given by $E_b/N_0 = (1/R)\ E_s/N_0$, where R is the code rate. The likelihood of a received bit being a zero is the density function $p_S(s|0)$ evaluated at the particular matched-filter output value observed, and the likelihood of one is $p_S(s|1)$ evaluated at the same value. The likelihood of a branch associated with the transmission of a zero or one information bit is the product of the associated bit likelihoods on that branch, and the path likelihood function is the product of the branch likelihoods.

For soft-decision decoding on the AWGN channel it is useful to represent the transmitted code symbols as a sequence of $\pm 1s$ rather than zeros and ones, which is the convenient representation for the case of hard-decision decoding. Clearly, the code tree may be redrawn with 0 replaced with 1 and 1 replaced with -1. We let the kth code symbol on the jth branch be x_{jk}, where $x_{jk} = 1$ for transmission of a logical 0 and $x_{jk} = -1$ for transmission of a logical 1. The received symbols at the input to the decoder are then Gaussian random variables with mean $x_{jk}a = x_{jk}\sqrt{2E_s/N_0}$ and unit variance.

To compute the likelihood of the kth received channel bit on the jth branch, we assume that the matched-filter output, scaled by $\sqrt{2/N_0}$, is s_{jk}. The likelihood L of the kth bit on the jth branch is then given by

$$L = p_S(s_{jk}|0) = \frac{1}{\sqrt{2\pi}} e^{-(s_{jk}-x_{jk}a)^2/2}$$

and we have

$$\begin{aligned}\ln L &= -\tfrac{1}{2}(s_{jk} - x_{jk}a)^2 - \ln\sqrt{2\pi} \\ &= s_{jk}x_{jk}a - \tfrac{1}{2}\left(s_{jk}^2 + a^2\right) - \ln\sqrt{2\pi} \\ &= As_{jk}x_{jk} - B\end{aligned}$$

where A and B are constants that do not depend on the structure of the code. In the expression above we have used the fact that $x_{jk}^2 = 1$.

Of course, to compute the logarithm of the likelihood of a code branch, we add the log likelihoods of all the bits on the branch, and to obtain a path log likelihood we add the branch log likelihoods. Thus we see that computing the likelihood of a path through the code tree is equivalent to forming the inner product between the received sequence $\{s_{jk}\}$ and the code sequence on that path, $\{x_{jk}\}$.

The measure of the "goodness" of a received bit, branch, or codeword that is used for decoding is called the *decoding metric*. Therefore, for the AWGN channel, we will assume that the decoding metric is the inner product of the received sequence and the code sequences. For hard-decision decoding, we use the Hamming metric.

In selecting a particular survivor node on the AWGN channel, a Viterbi decoder computes the branch metrics associated with the state transitions to the node, and then the branch metrics are added to previous path metrics. The contending path metrics are compared, and the path with the largest metric is selected as the survivor. Thus, the key arithmetic and logical operation in Viterbi decoding is called the *add-compare-select function*.

The continuous (unquantized) AWGN channel model that has been considered in this section is useful in deriving performance bounds for Viterbi decoding. These bounds are most accurate in the region of high SNR. We shall see shortly, however, that use of a relatively simple decoder quantizer results in only a small performance degradation relative to unquantized operation for Viterbi decoding.

9.3. THE GENERATING FUNCTION OF A CONVOLUTIONAL CODE

Bounds on the performance afforded by a Viterbi decoder can be obtained if one has knowledge of the distribution of codeword weights for the code employed [174]. We shall define the *generating function* or *transfer function*, whose expansion provides the necessary distance information.

We shall continue to use the example presented earlier, the code produced by the encoder shown in Fig. 9.1. The state diagram for the code is shown in Fig. 9.4. The state transition line segments are labeled ij/k, where i and j represent the output parity bits associated with the input information bit k.

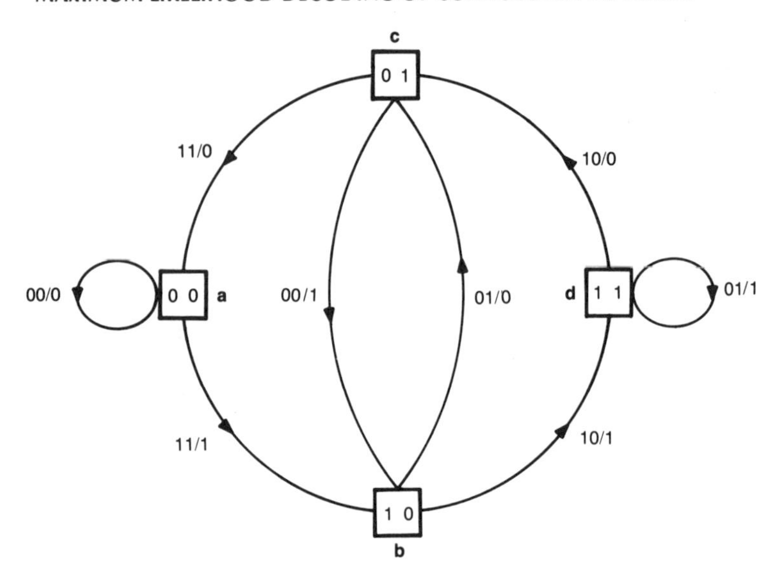

FIGURE 9.4. State diagram for the $k = 3$, $b = 1$, rate-1/2 code.

Convolutional codes are linear, and thus the distance properties do not depend on which particular code sequence is considered. Therefore, we assume transmission of the all-zeros sequence, and incorrect decoding events will correspond to decoding to nonzero information sequences. That is, we are interested in determining the probability that the path favored by a Viterbi decoder departs from the all-zeros path for some period of time before eventually returning to the correct path. It is useful to redraw the state diagram with the zero state, state **a**, split into two states corresponding to the initial and final states of interest, $\mathbf{a}_i$ and $\mathbf{a}_f$, as is seen in Fig. 9.5. In addition we label the branches of the new diagram with 1, D, or D^2, where the power of D indicates the number of output ones associated with each particular state transition.

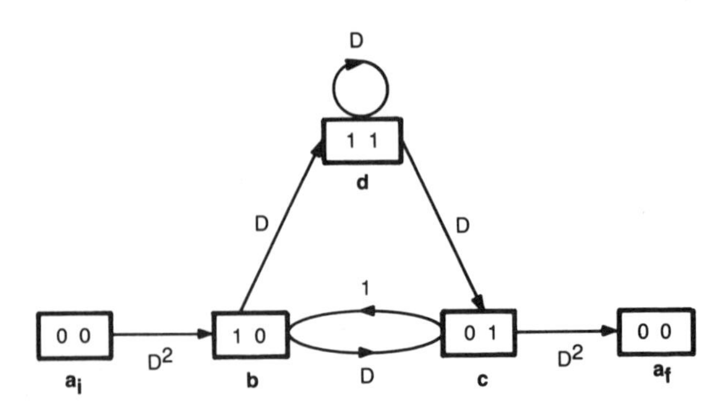

FIGURE 9.5. Modified state diagram for the $k = 3$, $b = 1$, rate-1/2 code.

For reasonably good channels, the most likely error events are those that contain small numbers of output errors, that is, those paths that differ from the all-zeros path in the fewest places. For the code considered here we may obtain the minimum weight of a nonzero path, the code's minimum free distance, by inspection of Fig. 9.5. By definition, as one moves from state to state in the figure, the weight of the path pursued increases by the power of D indicated on the branch connecting the states. Thus, to find the path corresponding to the minimum weight error event, we simply look for the sequence of states starting at $\mathbf{a}_i$ and ending at $\mathbf{a}_f$ such that the accumulated exponent of D is minimum. A quick glance at Fig. 9.5 shows that the state sequence $\mathbf{a}_i\mathbf{bca}_f$ produces a weight-5 path and no other sequence of state transitions provides another path of weight 5 or any path of weight less than 5. For example, the sequences $\mathbf{a}_i\mathbf{bdca}_f$ and $\mathbf{a}_i\mathbf{bcbca}_f$ each produce a weight-6 path. Thus the minimum free distance of the code is 5, and there is just one weight-5 excursion from the all-zeros path. Careful examination of the diagram will also show that there are only two weight-6 excursions.

In order to enumerate the weights of all the paths that originate at state $\mathbf{a}_i$ and terminate at $\mathbf{a}_f$, we proceed as follows. Let $x_{\mathbf{j}}(t)$ be a variable representing the accumulated weight of each path that enters state $\mathbf{j}$ at time t. The state diagram then provides the following recursive relationships:

$$x_{\mathbf{b}}(t) = D^2 x_{\mathbf{a}_i}(t-1) + x_{\mathbf{c}}(t-1)$$

$$x_{\mathbf{c}}(t) = D x_{\mathbf{b}}(t-1) + D x_{\mathbf{d}}(t-1)$$

$$x_{\mathbf{d}}(t) = D x_{\mathbf{b}}(t-1) + D x_{\mathbf{d}}(t-1)$$

$$x_{\mathbf{a}_f}(t) = D^2 x_{\mathbf{c}}(t-1)$$

Taking z-transforms [123], we have

$$X_{\mathbf{b}}(z) = D^2 z^{-1} X_{\mathbf{a}_i}(z) + z^{-1} X_{\mathbf{c}}(z)$$

$$X_{\mathbf{c}}(z) = D z^{-1} X_{\mathbf{b}}(z) + D z^{-1} X_{\mathbf{d}}(z)$$

$$X_{\mathbf{d}}(z) = D z^{-1} X_{\mathbf{b}}(z) + D z^{-1} X_{\mathbf{d}}(z)$$

$$X_{\mathbf{a}_f}(z) = D^2 z^{-1} X_{\mathbf{c}}(z)$$

The transfer function associated with all transitions from state $\mathbf{a}_i$ to $\mathbf{a}_f$ then provides the required enumeration of path weights. That is, we consider

$$T(D,z) = \frac{X_{\mathbf{a}_f}(z)}{X_{\mathbf{a}_i}(z)}$$

$T(D,z)$ is obtained by eliminating the intermediate variables $X_b(z)$, $X_c(z)$, and $X_d(z)$ in the four transform equations. After some algebraic manipulations we have

$$T(D,z) = \frac{D^5 z^{-3}}{1 - Dz^{-1} - Dz^{-2}}$$

$$= D^5 z^{-3} + D^6 z^{-4} + D^6 z^{-5} + D^7 z^{-5} + 2D^7 z^{-6} + \cdots$$

In this expansion of the transfer function, each term represents one or more sequences of state transitions from $\mathbf{a}_i$ to $\mathbf{a}_f$ through the state diagram. The exponent of D gives the path weight, and the exponent of z gives the path length, that is, the number of state transitions associated with the path. The coefficient of each term represents the number of paths with the given weight and length. The first three terms in the expansion of $T(D,z)$ show that there is one path of weight 5 and two of weight 6 with lengths 3, 4, and 5, respectively. Furthermore, the fifth term shows that there are two paths with weight 7 and length 6.

The generating function of a convolutional code [174] is defined as the transfer function evaluated at $z = 1$. For the example considered we have

$$T(D) = T(D,z)\Big|_{z=1} = \frac{D^5}{1 - 2D}$$

$$= D^5 + 2D^6 + 4D^7 + \cdots + 2^{k-5}D^k + \cdots$$

Note that the coefficient of each term in $T(D)$ specifies the total number of paths of a given weight. That is, there are 2^{k-5} paths of weight k for $k \geq 5$. In general, there are a_k weight-k paths, and we have

$$T(D) = \sum_{i=d_f}^{\infty} a_k D^k \tag{9.1}$$

where d_f is the minimum free distance of the code.

There is an equivalent and sometimes more convenient procedure to obtain the transfer function or generating function of a convolutional code. Specifically, the state diagram can be viewed as a signal flow graph and the transfer function evaluated using straightforward graph-reduction procedures [102]. Each state in the state diagram is associated with a node in the flow graph. The signal represented at each node is the accumulated weight of all the paths that terminate at the node. Furthermore, even more detailed information on the structure of a convolutional code can be obtained with a more general transfer function. As an example we consider the modified state diagram indicated in Fig. 9.6 interpreted as a signal flow graph.

The more general state diagram of Fig. 9.6 also represents the operation of the encoder shown in Fig. 9.1. The factor N has been included on those

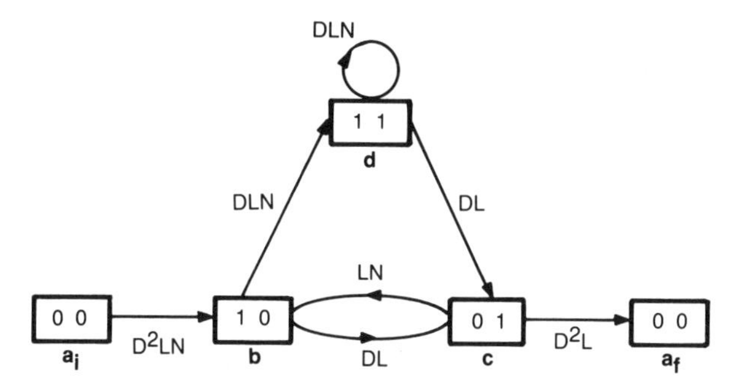

FIGURE 9.6. Another modified state diagram for the $k = 3$, $b = 1$, rate-1/2 code.

branches that correspond to a state transition caused by an input equal to 1, and the factor L has been included on each branch. Note that now as we trace paths through the directed graph, the exponent of L gives the length of the path and the exponent of N indicates the number of input bits on the path that differ from the all-zero information sequence. With this formulation, L plays the role of z^{-1}.

Proceeding as before, we could now write four node equations and eliminate the intermediate variables to obtain the transfer function from node $\mathbf{a}_i$ to $\mathbf{a}_f$. Alternatively, the transfer function can be obtained by inspection using the rule given by Mason [102]. We then have

$$T(D,L,N) = \frac{D^5L^3N}{1 - DLN(1 + L)}$$

$$= D^5L^3N + D^6L^4(1 + L)N^2 + D^7L^5(1 + L)^2N^3 + \cdots$$

$$+ D^{k+5}L^{k+3}(1 + L)^kN^{k+1} + \cdots$$

From this expression we can see the following: (1) There is one nonzero path of length 3 and weight 5 that differs from the all-zeros information sequence by one bit; (2) there is one path of length 4 and one of length 5, each having weight 6 differing from the all-zeros information sequence in two bit positions; and so on. Thus it is apparent that the distance structure of a convolutional code is revealed by the code's generating or transfer function.

9.4. PERFORMANCE BOUNDS FOR VITERBI DECODING

The generating function for a convolutional code will now be used to obtain bounds on the performance that can be achieved with maximum likelihood decoding. We follow Viterbi [174] and Proakis [138] and consider first hard-decision decoding on the binary symmetric channel. Results are then obtained for the unquantized AWGN channel.

9.4.1. The Binary Symmetric Channel

As usual we consider the case in which the all-zeros information sequence is transmitted. The example of the $k = 3$, $b = 1$ code whose transfer function was obtained in the previous section will be continued.

The minimum free distance of this code was observed to be 5, and the sequence of state transitions that results in a weight-5 excursion from the all-zeros path was shown to be $\mathbf{a}_i\mathbf{bca}_f$. The associated sequence of parity bits for this sequence of state transitions is as follows:

Parity bit:	⋯	1	1	0	1	1	1	⋯
Time:	⋯	i	$i+1$	$i+2$	$i+3$	$i+4$	$i+5$	⋯

Note that should the channel produce an error pattern that is closer to this parity-bit sequence than the all-zeros sequence, a maximum likelihood decoder may favor the incorrect path. The most likely way in which this can happen is for three errors to occur among the five bits transmitted at times i, $i + 1$, $i + 3$, $i + 4$, and $i + 5$. But four or five errors in these bit positions could also produce incorrect decoding. The total probability of three, four, or five errors in five particular bit positions is given by

$$P_5 = \sum_{i=3}^{5} \binom{5}{i} p^i(1 - p)^{5-i}$$

Note that whether the bit at time $i + 2$ is received correctly or not is immaterial, since it affects the distance between the correct and incorrect paths in the same way for either event.

The occurrence of one of these error patterns, however, does not assure that a Viterbi decoder will necessarily decode incorrectly. A weight-5 path will be favored only when at least three errors occur and the necessary state transitions are survivors in the state sequence traced by the decoder. Thus P_5 provides an upper bound on the probability that a maximum likelihood decoder produces a weight-5 excursion from the correct path at a particular step of the decoding process. Since this path corresponds to an information sequence containing a single one, a single post-decoding bit error will result.

This procedure can be extended to larger-weight error patterns. We have observed that, for the code considered, there are two weight-6 paths, and a weight-4, -5, or -6 error pattern can produce an incorrect decoding event to a weight-6 path in the same way that the previously considered weight-3, -4, or -5 error patterns produced a weight-5 path excursion. Now, however, a weight-3 error pattern can result in a tie between a weight-6 path and the all-zeros path, which we assume is resolved by a coin flip. We then have

$$P_6 = \frac{1}{2}\binom{6}{3}p^3(1-p)^3 + \sum_{i=4}^{6}\binom{6}{i}p^i(1-p)^{6-i}$$

or in general

$$P_k = \begin{cases} \displaystyle\sum_{i=(k+1)/2}^{k}\binom{k}{i}p^i(1-p)^{k-i}, & k \text{ odd} \\ \displaystyle\frac{1}{2}\binom{k}{k/2}p^{k/2}(1-p)^{k/2} + \sum_{i=(k/2)+1}^{k}\binom{k}{i}p^i(1-p)^{k-i}, & k \text{ even} \end{cases}$$

These expressions provide an upper bound on the probability that a Viterbi decoder will favor a particular weight-k path at a given step in the decoding process. If there are a_k weight-k paths, the total probability of any weight-k path excursion is $a_k P_k$, and we have

$$P_{E_1} < \sum_{k=d_f}^{\infty} a_k P_k$$

where P_{E_1} is the probability that a Viterbi decoder will depart from the all-zeros information sequence at a particular node, follow some nonzero sequence for a period of time, and then return to the all-zeros sequence. Recall that the a_k are given by the code transfer function, Eq. (9.1).

P_{E_1} is called the *first-event error probability* and can be used to bound the probability of message error with Viterbi decoding. Note that a Viterbi decoder will make an incorrect decoding decision if the all-zeros path is deleted from the list of surviving paths. But the probability that this occurs at a particular node is simply P_{E_1}, and after the first error event there is some path, diverging from and then returning to the zero path, whose metric is larger than the metric of the correct path. From that point on, the probability that another portion of the decoded message diverges from the all-zeros sequence at some other node is again bounded by P_{E_1}. Thus we may use a union bound to obtain an upper bound on the probability of message error, P_E. For an M-bit message we have

$$P_E < MP_{E_1} < M\sum_{k=d_f}^{\infty} a_k P_k \tag{9.2}$$

Viterbi showed [173] that the probability of a weight-k path excursion on the binary symmetric channel is bounded as

$$P_k < \left[2\sqrt{p(1-p)}\right]^k \tag{9.3}$$

Using Eq. (9.1) we have

$$P_E < MT(D)\Big|_{D=2\sqrt{p(1-p)}} = M\left(\frac{D^5}{1-2D}\right)\Big|_{D=2\sqrt{p(1-p)}} \tag{9.4}$$

where $T(D)$ is the transfer function of the code. Equation (9.4) is a somewhat weaker result but simpler to compute than Eq. (9.2).

We can find an upper bound on the post-decoding bit-error rate for Viterbi decoding using a similar approach. First consider the transfer function $T(D,N)$, where

$$T(D,N) = T(D,N,L)\,\big|_{L=1}$$

and for our example

$$T(D,N) = D^5N + 2D^6N^2 + \cdots = \sum_{k=5}^{\infty} 2^{k-5}D^kN^{k-4}$$

Note that each term in this expression represents the total number of paths of weight k corresponding to an information sequence that differs from the all-zeros sequence in $k-4$ bit positions. To compute the post-decoding bit-error rate we must weight the various possible incorrect decoding events by the numbers of output errors that are produced. To accomplish this weighting analytically, we differentiate with respect to N and set $N = 1$, with the result

$$\frac{d}{dN}T(D,N)\Big|_{N=1} = \sum_{k=5}^{\infty}(k-4)2^{k-5}D^k \triangleq \sum_{k=d_f}^{\infty} b_kD^k$$

Note that b_k defined above is the product of the number of paths of weight k times the number of corresponding output bit errors.

The expected number of bit errors associated with the first-event error probability is then bounded by

$$E(\text{bit errors for } E_1) < \sum_{k=d_f}^{\infty} b_kP_k \tag{9.5}$$

and the post-decoding bit-error rate P_b for an M-bit message can be bounded by

$$P_b < \frac{1}{M}\left(M\sum_{k=d_f}^{\infty} b_kP_k\right) = \sum_{k=d_f}^{\infty} b_kP_k \tag{9.6}$$

Using Eq. (9.2) we also have

$$P_b < \frac{d}{dN} T(D,N)\bigg|_{\substack{N=1 \\ D=2\sqrt{p(1-p)}}} \tag{9.7}$$

Note that Eq. (9.6) provides a tighter bound than Eq. (9.7).

For rate-b/V codes with $b > 1$, the bounds on P_b must be divided by b to normalize the output bit-error rate.

9.4.2. The AWGN Channel

Performance bounds similar to those just presented for the binary symmetric channel can be obtained for more general channels as well. We next outline the derivation of the results for binary antipodal signaling on the AWGN channel. We assume that the decoder accepts unquantized matched-filter outputs scaled by $\sqrt{2/N_0}$ and uses the inner product between the received data and the code sequences as the decoding metric. Again, transmission of a positive channel symbol is associated with transmission of a logical 0, and a negative channel symbol with 1. The intended sequence of channel symbols are Gaussian random variables with mean $a = \sqrt{2E_s/N_0}$ and unit variance.

Again a bound on the first-event error probability is given by

$$P_{E_1} < \sum_{k=d_f}^{\infty} a_k P_k$$

where the a_k are the coefficients of the code transfer function $T(D)$, and P_k is the probability that a weight-k path is favored over the correct all-zeros path. But a weight-k path will have a larger metric than the transmitted sequence only if the sum of the scaled matched filter outputs for k particular channel bits is negative. Thus P_k is the probability that the sum of k independent Gaussian random variables with mean $a = \sqrt{2E_s/N_0}$ and unit variance is less than 0, that is,

$$P_k = \frac{1}{\sqrt{2\pi k}} \int_{-\infty}^{0} e^{-(z-ka)^2/2k}\, dz$$

$$= \frac{1}{\sqrt{\pi}} \int_{a\sqrt{k/2}}^{\infty} e^{-y^2}\, dy = \frac{1}{2}\operatorname{erfc}\sqrt{kE_s/N_0}$$

A somewhat looser bound on the first-event error probability can be obtained by using

$$\operatorname{erfc}\sqrt{x+y} \le e^{-y} \operatorname{erfc}\sqrt{x} \tag{9.8}$$

when $x \geq 0$ and $y \geq 0$. Letting $l = k - d_f$, we have

$$P_k = \frac{1}{2}\operatorname{erfc}\sqrt{(d_f + l)E_s/N_0}$$

$$\leq \frac{1}{2}e^{-lE_s/N_0}\operatorname{erfc}\sqrt{d_f E_s/N_0}$$

But

$$P_{E_1} < \sum_{k=d_f}^{\infty} a_k P_k \leq \sum_{k=d_f}^{\infty} a_k \left(\frac{1}{2}\right) e^{-(k-d_f)E_s/N_0}\operatorname{erfc}\sqrt{d_f E_s/N_0}$$

and thus

$$P_{E_1} < \frac{1}{2}\operatorname{erfc}\sqrt{d_f E_s/N_0}\, e^{d_f E_s/N_0} T(D)\Big|_{D=e^{-E_s/N_0}}$$

An upper bound on the post-decoding bit-error probability can be obtained in a similar fashion as follows:

$$P_b < \sum_{k=d_f}^{\infty} b_k P_k$$

where

$$\frac{d}{dN}T(D,N)\Big|_{N=1} \triangleq \sum_{k=d_f}^{\infty} b_k D^k$$

Thus the post-decoding bit-error probability is bounded by

$$P_b < \frac{1}{2}\sum_{k=d_f}^{\infty} b_k \operatorname{erfc}\sqrt{kE_s/N_0} \tag{9.9}$$

or, using Eq. (9.8), more loosely by

$$P_b < \frac{1}{2}\left[\operatorname{erfc}\sqrt{d_f E_s/N_0}\right] e^{d_f E_s/N_0} \frac{d}{dN} T(D,N)\Big|_{\substack{N=1 \\ D=e^{-E_s/N_0}}} \tag{9.10}$$

Again for codes with $b > 1$, the bounds on P_b must be divided by b for normalization.

9.5. SOME PRACTICAL DESIGN CONSIDERATIONS

We next consider the design of a Viterbi decoder for a particular convolutional code in some detail. As we mentioned earlier, computer simulations have proved to be very useful in the design of cost-effective

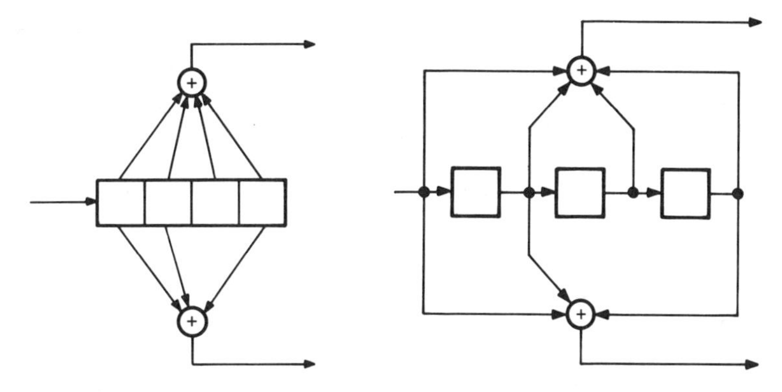

FIGURE 9.7. Equivalent encoders for the $k = 4$, $b = 1$, rate-1/2 code.

decoders. In fact, final designs for Viterbi decoders have typically been obtained by first implementing the decoding algorithm on a computer and executing a series of simulation experiments in which many elements of the decoder design are varied. Those portions of the decoder whose complexity can be reduced without appreciably sacrificing performance have thereby been determined.

Two equivalent encoders for the particular convolutional code to be considered in this section are shown in Fig. 9.7. The code has $k = 4$, $b = 1$, and rate 1/2. This is also a relatively simple code for which we could easily draw the state diagram. However, for more complex codes the state diagram becomes cumbersome and a state table provides the more convenient description.

The *state table* for a finite-state machine is a figure that shows the input-output and state transition relationships that define the machine. The state table associated with the encoder considered is given in Fig. 9.8. Each row of the table shows the present state, next state, and outputs associated with the possible inputs and present states of the encoder. The first row of the table

Present State	Next State		Output Parity Bits	
	Input = 0	Input = 1	Input = 0	Input = 1
0 0 0	0 0 0	1 0 0	0 0	1 1
0 0 1	0 0 0	1 0 0	1 1	0 0
0 1 0	0 0 1	1 0 1	1 0	0 1
0 1 1	0 0 1	1 0 1	0 1	1 0
1 0 0	0 1 0	1 1 0	1 1	0 0
1 0 1	0 1 0	1 1 0	0 0	1 1
1 1 0	0 1 1	1 1 1	0 1	1 0
1 1 1	0 1 1	1 1 1	1 0	0 1

FIGURE 9.8. State table for the $k = 4$, $b = 1$, rate-1/2 code.

indicates that if the encoder is in state 000 and the input bit is a 0, the next state is 000 and the output bits are 00. If the input is a 1, the next state is 100 and the output bits are 11.

The state table provides a convenient way to summarize the information that is necessary to define the operations involved in Viterbi decoding. A Viterbi decoder continually computes the most likely path through the code tree to each of the possible states of the encoder. Suppose we are computing the most likely path to state 101. First we must know from what states we could have come to reach state 101, and a glance at the state table shows that the possible predecessors are 010 and 011. In addition we see that the transition from state 010 to 101 requires the input equal to 1 and produces the output bits 01. Similarly, the 011 to 101 transition also requires the input equal to 1 and produces the output bits 10. In this way the state table can be used to determine the predecessor states to each state of the encoder as well as the corresponding input and output bits. The state table is particularly useful for codes with $b > 1$.

9.5.1. Path-History Storage

There are a number of issues associated with the design of a practical Viterbi decoder that we have not yet addressed. For example, how and when are output bit decisions to be made? We have tacitly assumed for simplicity of presentation that a Viterbi decoder may postpone making an output decision until an entire message has been received and processed. At that point the path with the largest metric would be selected and the bit decisions associated with the entire path would be delivered to the user. If this could actually be done, the complete code space would have been searched, and a true maximum likelihood decision made on the received message. However, even for moderately long messages, a prohibitively large amount of storage would be required.

First we clearly need a strategy for storing the individual path histories. Viterbi decoding has been described as the process of determining the most likely sequence of state transitions to each of the states of the encoder. Of course, for each state transition there is an associated input information bit and we might choose to store the bit sequences directly in an auxiliary memory reserved for the path histories. There would be $2^{b(k-1)}$ information-bit sequences stored, one for each state of the encoder. Alternatively, we could store the state transitions, that is, the most likely sequence of predecessor states, to each state of the encoder. With this approach, it is necessary to track back through the memory in order to reconstruct a particular path, and the bit decisions must be determined from the state transitions found in the memory. The latter approach is generally preferred for processor implementation.

Next we must determine how long the individual path histories should be, that is, how many consecutive bit or state decisions must be stored prior to making an output decision. We are assuming that the decoder will choose the oldest bit on some path as the output decision and release one information bit after processing each code branch. It can be expected that as this storage is increased from some small value, performance will improve to the point where, with some large but finite memory, essentially maximum likelihood behavior is observed. We shall assume first that unquantized matched-filter outputs are used to compute branch and path metrics. In addition, the following output bit decision rule will be employed: After processing each pair of received parity bits and updating path metrics and histories, the largest path metric is found. Then the oldest bit on the path with the largest metric will be chosen as the output bit decision. This decision strategy is called the *maximum likelihood output rule*. Note that the k most recent bits on a particular path are defined by the state associated with that path, and thus, when path histories are stored as bit sequences, that storage may begin with the bit decision associated with the $(k + 1)$th previous node.

The simulation results to be presented in this chapter are based on use of binary antipodal signaling, coherent detection, and additive white Gaussian noise. We consider first the unquantized channel described in Section 9.2. In Fig. 9.9 we show post-Viterbi-decoding bit-error rates that were measured in simulation experiments run for the $k = 4$, $b = 1$, rate-1/2 code with various amounts of path-history storage. The error rates are plotted as a function of SNR per information bit, E_b/N_0. In particular, results for 4, 8, 12, and 16 bits of path-history storage per node are indicated, that is k, $2k$, $3k$, and $4k$ bits per path. For path histories larger than $4k$, results essentially identical to those shown for $4k$ bits were obtained, indicating that no further performance gains can be realized for the unquantized channel and the maximum likelihood output rule. Note also that the performance improvement in going from $3k$ to $4k$ bits of path storage is small, about 0.1 dB, but should too little memory be considered (say, k bits) the performance loss can be appreciable, about 1 dB.

Simulation results are most easily obtained for low SNRs, the region of relatively high output bit-error rates. However, the analytical bound that was obtained in Section 9.4.2 for the unquantized channel is tight for high SNRs, and use of the bound in conjunction with simulation results provides an excellent estimate of decoder performance for all operating regions. For the $k = 4$, $b = 1$, rate-1/2 code, the reader can verify that the transfer function is given by

$$T(N,D) = \frac{N^2D^6 + ND^7 - N^2D^8}{1 - 2ND - ND^3}$$

The bound obtained with Eq. (9.9) is also indicated in Fig. 9.9. Note that for $E_b/N_0 > 4$ dB and sufficiently large path-history storage, the bound is indeed tight.

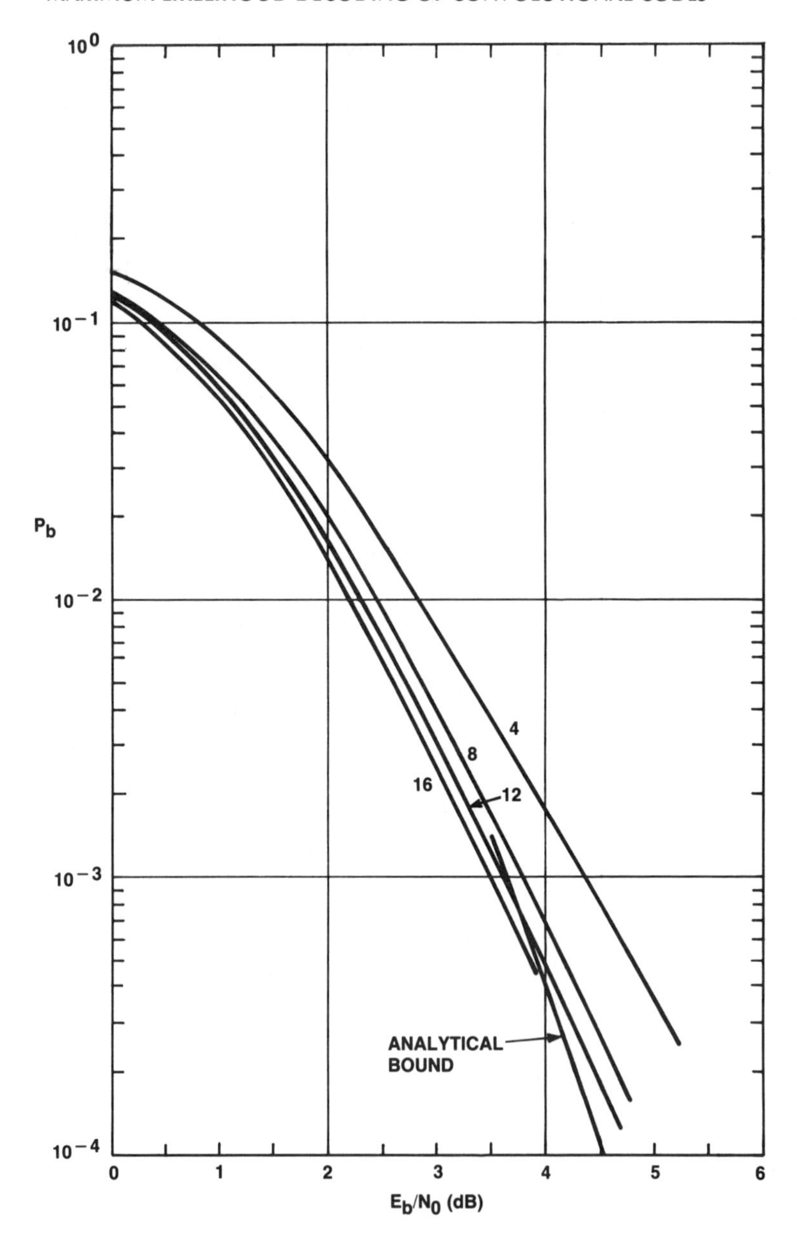

FIGURE 9.9. Post-decoding bit error rate vs. E_b/N_0 for the $k = 4$, rate-1/2 code with 4, 8, 12, and 16 bits of path history per path. Analytical bound for the unquantized channel also shown.

9.5.2. Quantization and Metrics

Thus far in the discussion of Viterbi decoding we have considered only the unquantized AWGN channel and the binary symmetric channel. However, the unquantized channel is not convenient for implementation, and the use of the binary symmetric channel results in the significant performance penalty associated with hard-decision decoding. In this section we shall see that a relatively simple quantizer can be used with Viterbi decoding and the associated degradation from operation on the unquantized channel is small. We have also considered only two decoding metrics, Hamming distance and the inner product between code sequences and the received data. We shall also see that a decoding metric only slightly more complex than Hamming distance provides performance quite close to the unquantized inner product.

Specifically, we consider the performance that can be obtained with two, four, and eight uniformly spaced quantization intervals. The three quantizers considered are shown in Fig. 9.10, and in the next chapter we shall see that the quantization intervals indicated are close to optimum. The horizontal axes labeled s in the figure represent the matched-filter output values scaled by $\sqrt{2/N_0}$. Again we use the convention that a positive channel symbol is associated with the transmission of a logical 0. Thus, a matched-filter output observed in a quantization interval on the far right provides a strong indication that the intended code bit is a zero, and those on the far left indicate that a one was transmitted.

The decoding metrics conditioned on the transmission of a zero code bit are also indicated in Fig. 9.10. For the three-bit quantizer we use the integers 0 to

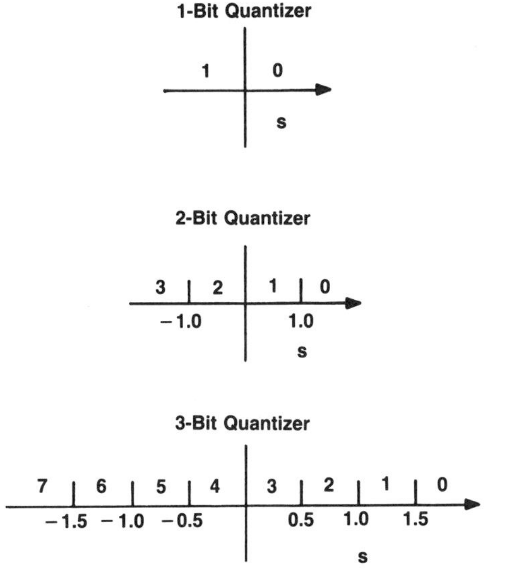

FIGURE 9.10. The two-, four-, and eight-level quantizers to be considered for Viterbi decoding.

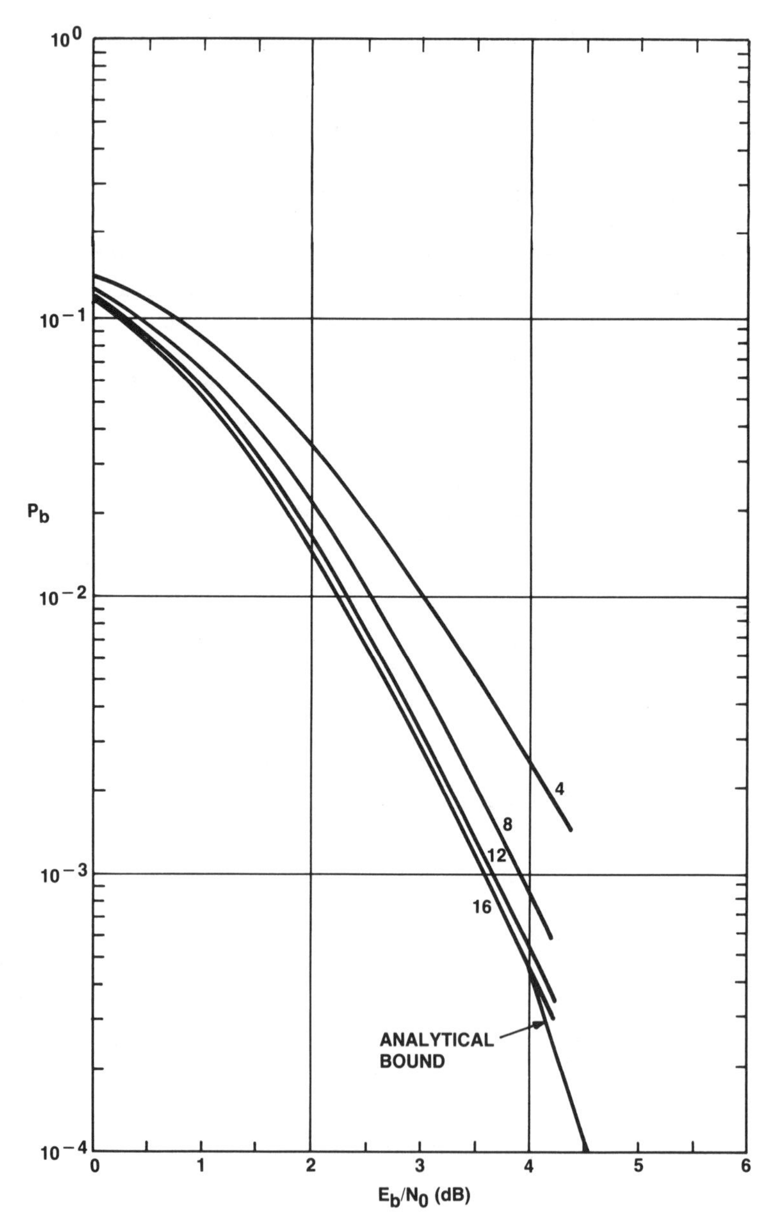

FIGURE 9.11. Post-decoding bit-error rate vs. E_b/N_0 for the $k = 4$, rate-$1/2$ code with $4, 8, 12$, and 16 bits of path history per path; eight-level quantization and three-bit metric range.

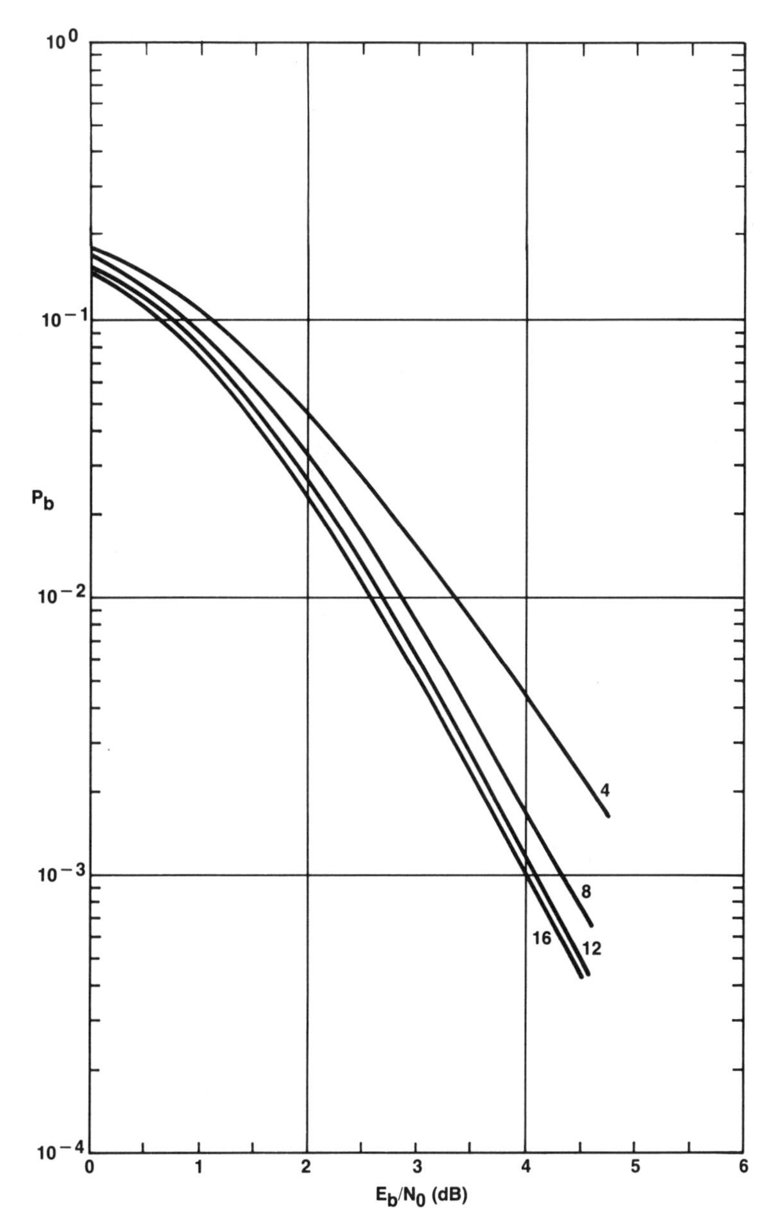

FIGURE 9.12. Post-decoding bit-error rate vs. E_b/N_0 for the $k = 4$, rate-1/2 code with 4, 8, 12, and 16 bits of path history per path; four-level quantization and two-bit metric range.

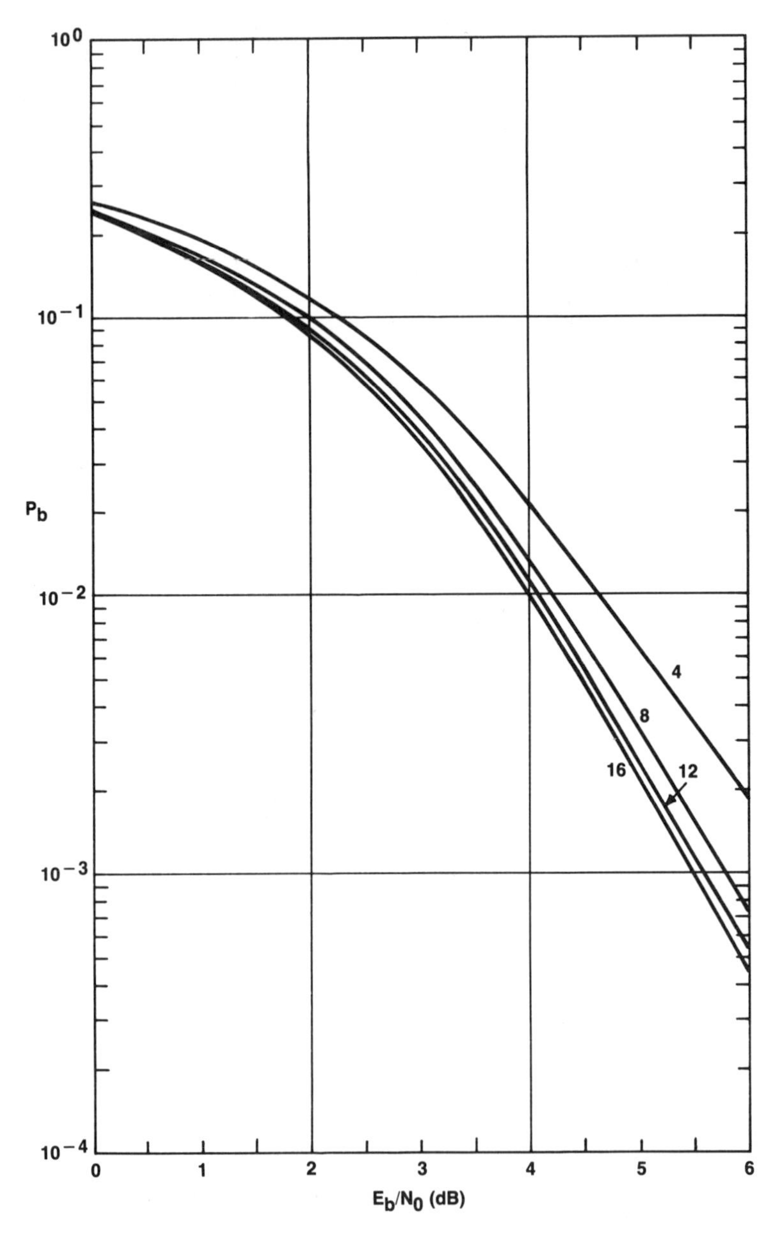

FIGURE 9.13. Post-decoding bit-error rate vs. E_b/N_0 for the $k = 4$, rate-1/2 code with 4, 8, 12, and 16 bits of path history per path; two-level hard-decision quantization and one-bit metric range.

7, for the two-bit quantizer we use 0 to 3, and for the 2-level hard-decision quantizer, we uşe 0 and 1, which is the Hamming metric. The metric values are ordered so that a strong indication that a transmitted bit is a zero is associated with the small metric numbers, and large numbers indicate that the likelihood of a zero is small. The metrics conditioned on the transmission of a logical 1 are simply the reverse of the values shown; for example, 0 is replaced with 7, 1 with 6, and so on.

To compute the metric for a particular code branch with, for example, the 8-level quantizer, we proceed as follows. Suppose that the parity bits on the branch given the transmission of a zero information bit are 01 and for a one they are 10. Let the corresponding scaled matched-filter outputs be $s = 1.4$ and 0.2, respectively. We have then

$$L_0 = 1 + 4 = 5$$

$$L_1 = 6 + 3 = 9$$

where L_0 is the branch metric associated with transmission of a zero information bit and L_1 is the branch metric associated with a transmitted one. For this branch, the transmission of a zero is more likely than a one.

The results of simulation experiments executed with the $k = 4$, $b = 1$, rate-1/2 code and the three quantizers and metric sets considered are shown in Figs. 9.11 through 9.13. The path-history memory provided is again varied from k to $4k$ bits per path, and the output decision rule remains the same, maximum likelihood. We note first that the effects of varying the path-history memory are generally the same for the quantized channels as those observed for the unquantized channel. In addition, with the eight-level quantizer, very little degradation in performance from the continuous case is observed (about 0.1 dB). For the four-level quantizer, the degradation is about 0.5 dB, and for two-level quantization, the hard-decision channel, 2 dB of degradation is observed. Therefore, the choice of eight-level quantization and three-bit metrics is advisable for Viterbi decoding.

9.5.3. Other Design Issues

The results just presented for the $k = 4$, rate-1/2 code also apply for longer constraint length codes and for codes with $b > 1$. Namely, $4k$ bits (or characters) of storage per path is generally sufficient for the path-history memory when the maximum likelihood output rule is used. In addition, hard-decision decoding is 2 dB inferior to the unquantized case, and the eight-level quantizer utilizing three-bit metrics is 0.1 to 0.2 dB inferior to decoding on the unquantized channel. However, other shortcuts typically employed in a practical Viterbi decoder design can lead to significant performance degradations. Fortunately, though, by providing a modest increase in the path-history storage, total performance degradation can be limited to the range of 0.2 to 0.25 dB.

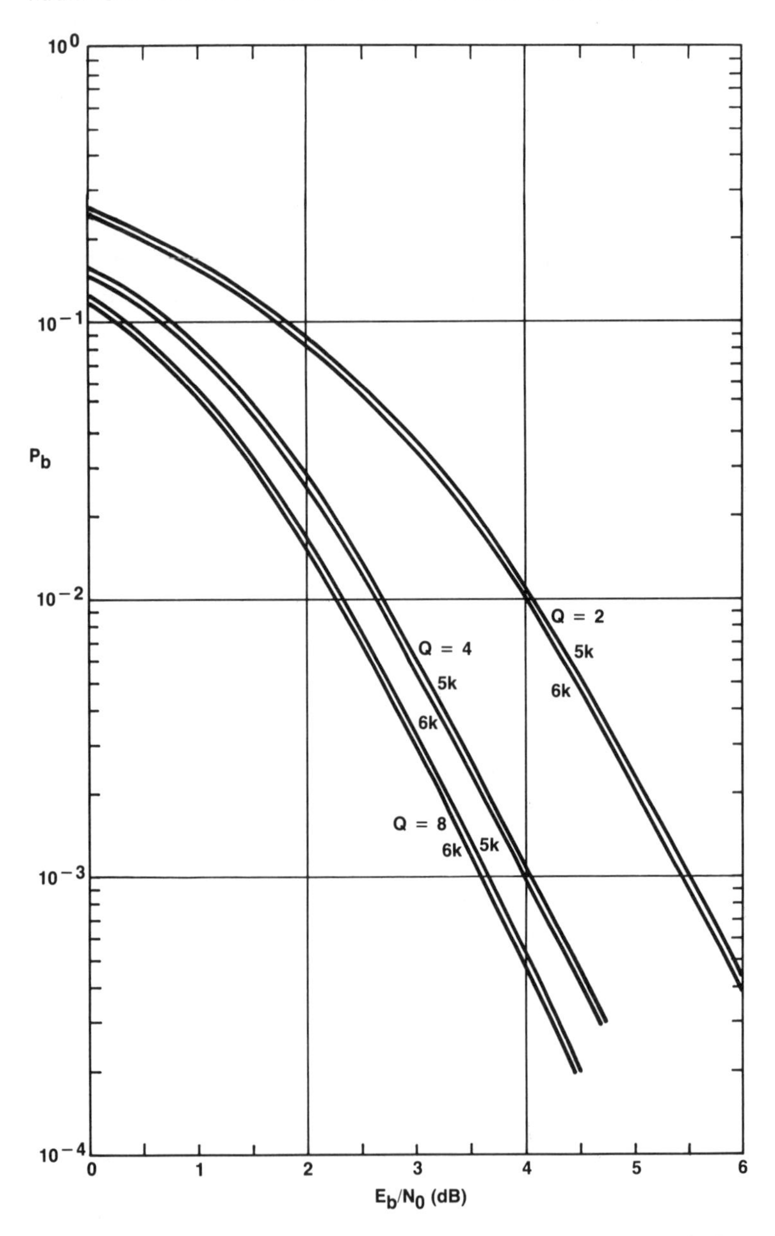

FIGURE 9.14. Post-decoding bit-error rate vs. E_b/N_0 for the $k = 4$, rate-1/2 code. Two-, four-, and eight-level quantization with $5k$ and $6k$ bits of path history per path; majority-voting output decision rule.

For example, we have assumed that the output decision rule involves first finding the path that has the smallest accumulated metric, the highest likelihood. A more easily implemented technique would be to either make a majority-voting decision on the oldest bits in the stored path memories or simply to pick a path at random to use for the output decision. Simulations have shown, however, that in order to provide comparable performance when an arbitrary path is selected, the size of the path-history storage must be approximately doubled [51].

Performance is considerably better for the majority-voting decision rule, as can be seen from Fig. 9.14. Simulation results are given for the majority-voting rule when $5k$ and $6k$ bits of history are provided for each path. Results are shown for $Q = 8$-, $Q = 4$-, and $Q = 2$-level quantization. Note that in each case the difference in performance between $5k$ and $6k$ bits of path-history storage is small, and therefore $5k$ bits will suffice for most applications. Total degradation from maximum likelihood decoding is about 0.2 dB for eight-level quantization.

9.5.4. Other Features

Some very attractive additional features can be included in a Viterbi decoder design for a modest additional cost. For example, during normal operation on a good channel, the decoder typically sees the path metrics for the correct path grow at a fairly steady rate while the other path metrics lag behind more or less as a group. The rate of increase of the correct path metric is a function of the received SNR, and after averaging the rate of increase of the metric along the path favored by the decoder, an estimate of the channel quality can be produced. In addition, should the channel SNR degrade, the Viterbi decoder can readily recognize the poorer link transmission quality.

Another attractive feature of Viterbi decoding is that it can be made self-synchronizing. That is, it is not necessary to know where a message starts or, for a $b = 1$, rate-$1/2$ code, which pair of consecutive channel bits correspond to the two bits on the same branch. One may simply make an arbitrary code-bit framing decision and attempt decoding with all path metrics set initially to the same value. Since the correct state of the encoder is included in the set of states assumed by the decoder, the correct path metric can be expected to emerge after sufficient received data has been processed if a correct sync assumption is made. If not, the assumption on code-bit framing is changed and decoding restarted, the procedure continuing until a dominant path metric finally emerges.

The use of Viterbi decoding with differential data encoding and decoding on the coherent channel also provides an important feature when used with a transparent code. A *transparent convolutional code* is a code having the property that complements of codewords are codewords. For such codes the all-ones channel sequence is a codeword, and the reader should verify that a code will be transparent if each parity generator has an odd number of taps.

The use of differential data encoding prior to the convolutional encoder and differential decoding following the Viterbi decoder then provides protection against 180-degree phase ambiguity in the channel, for should a phase reversal occur, the Viterbi decoder would still be presented with a decodable string of data. When the resulting decoded output is differentially decoded, the intended information sequence is reconstructed except for errors at the points where phase reversals occur. In the absence of phase reversals, a degradation in the output error rate can be expected. However, post-Viterbi-decoding errors tend

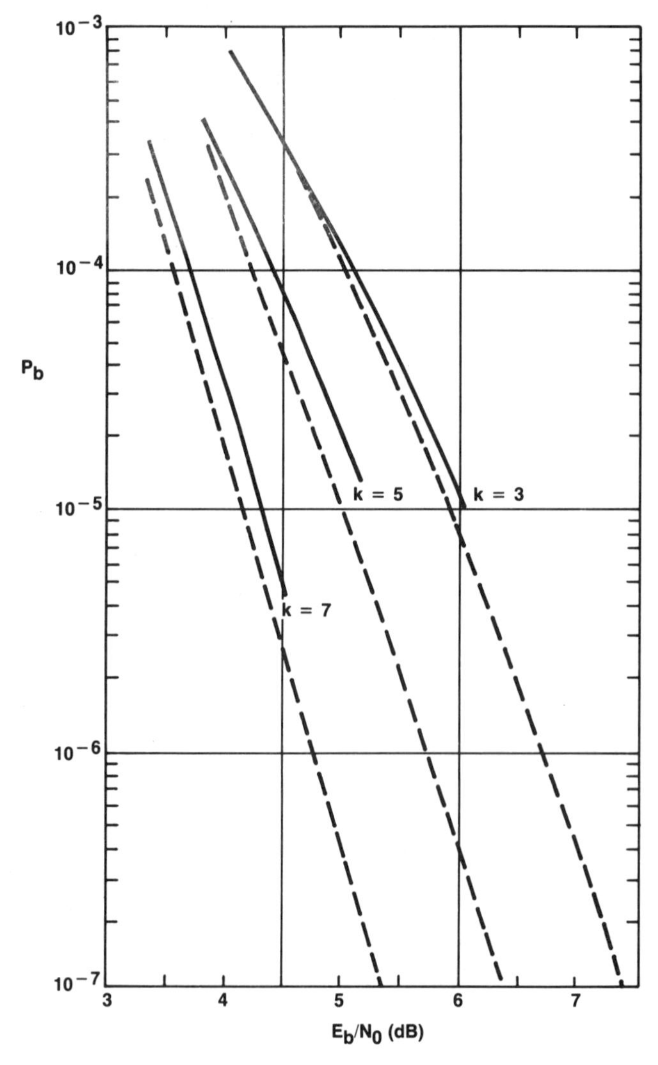

FIGURE 9.15. Post-decoding bit-error rate vs. E_b/N_0; rate-1/2, $b = 1$, $k = 3, 5$, and 7; eight-level quantization; 32 bits of path history per path. Simulation results (solid) and analytical bound (dashed).

to occur in bursts, and the increase in error rate is less than a factor of two. Typically a degradation of only 0.1 to 0.2 dB is observed.

9.6. PERFORMANCE RESULTS FOR VITERBI DECODING

Performance results for Viterbi decoding of a number of convolutional codes were given in references 51 and 64. A wide range of code parameters spanning the set of codes of interest was considered. Specifically, for the $b = 1$

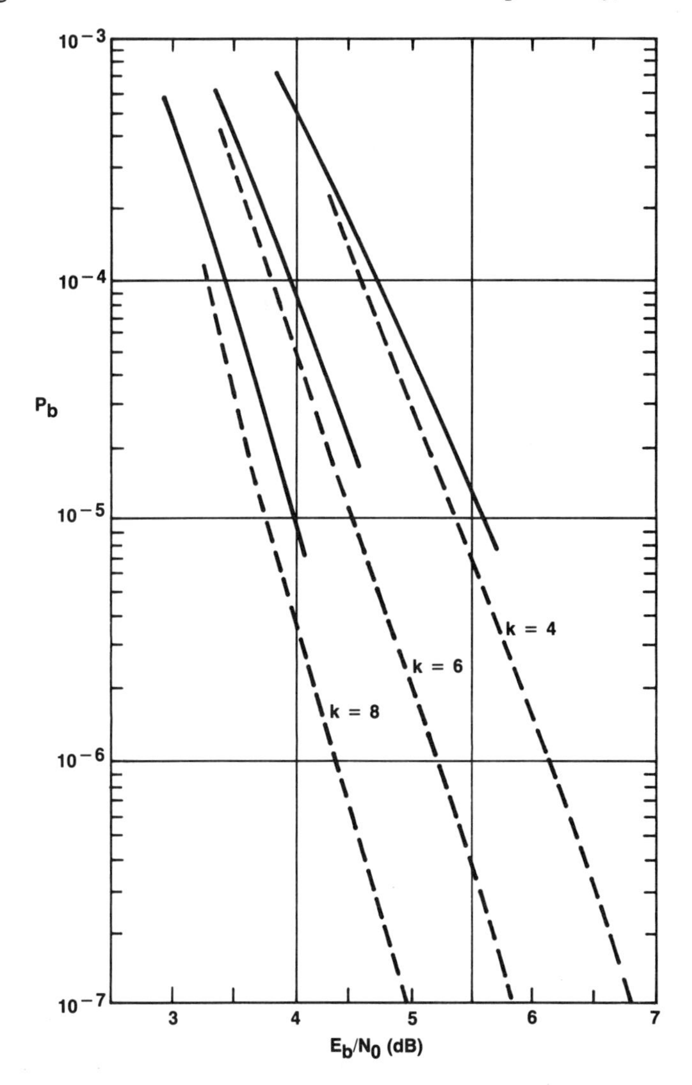

FIGURE 9.16. Post-decoding bit-error rate vs. E_b/N_0; rate-1/2, $b = 1$, $k = 4, 6$, and 8; eight-level quantization; 32 bits of path history per path. Simulation results (solid) and analytical bound (dashed).

codes, simulation results for constraint lengths up to 8 were presented. Perhaps somewhat more complex decoders may be feasible today, but the results given in that original work provide a good deal of information, and thus portions are included here.

We consider $b = 1$, rate-1/2 codes with constraint lengths between 3 and 8. Binary antipodal signaling and coherent detection are assumed, and in each case a sufficiently large path-history storage memory is provided (32 bits per path). The results of the simulations presented are shown in Figs. 9.15 and 9.16 for eight-level quantization and in Fig. 9.17 for two-level quantization. The

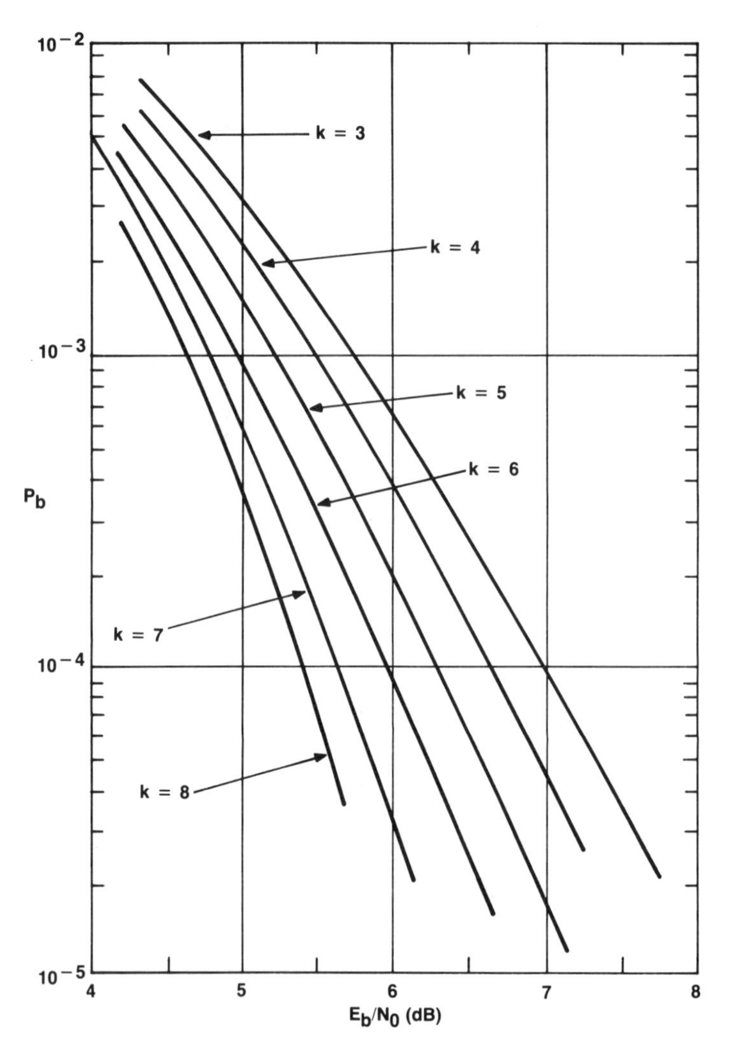

FIGURE 9.17. Post-decoding bit-error rate vs. E_b/N_0; rate-1/2, $b = 1$, $k = 3, 4, 5, 6, 7$, and 8; hard-decision decoding; 32 bits of path-history storage per path.

performance bounds are also shown in Figs. 9.15 and 9.16. The following general observations can be made:

1. In all cases, hard-decision (two-level) quantization provides performance about 1.8 dB poorer than eight-level quantization.
2. There is 0.1 to 0.2 dB degradation associated with eight-level quantization relative to unquantized operation.

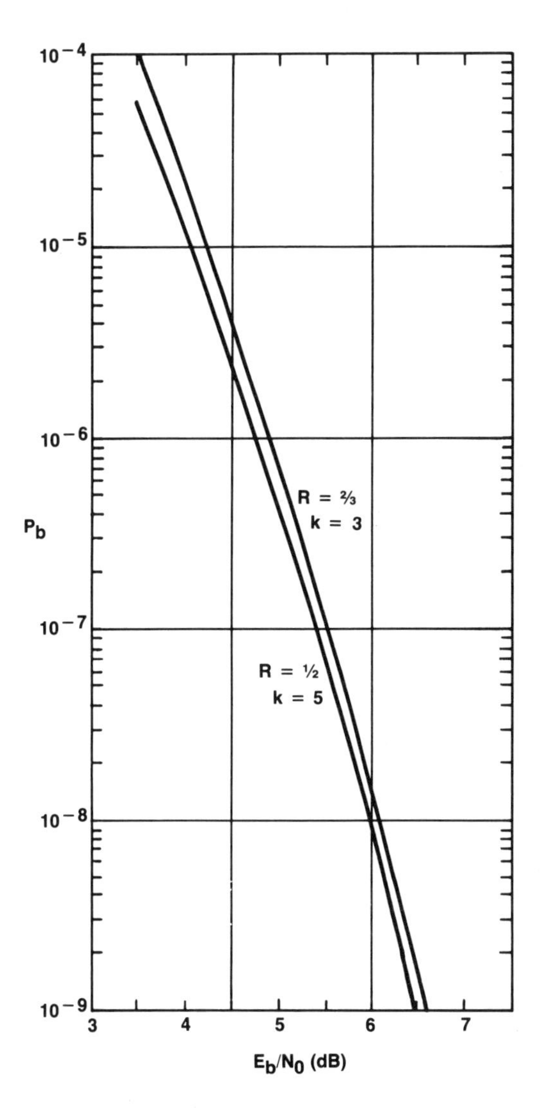

FIGURE 9.18. Bounds on post-decoding bit-error rate vs. E_b/N_0 for a $k = 5$, $b = 1$, rate-1/2 code and a $k = 3$, $b = 2$, rate-2/3 code.

3. For $b = 1$, rate-1/2 codes and low output error rates, somewhat less than 0.5 dB in performance enhancement is obtained by increasing k by 1, which doubles the complexity of the decoder.

Another interesting comparison of a rate-1/2 and a rate-2/3 code was included in reference 51. Those results are indicated in Fig. 9.18, where post-Viterbi-decoding error bounds are shown for a $k = 5$, $b = 1$, rate-1/2 code and a $k = 3$, $b = 2$, rate-2/3 code. Note that both encoders are equivalent to 16-state machines, and the respective Viterbi decoders, although logically quite different, are of comparable complexity. Thus the fact that similar performance results are observed is not surprising. This leads to a final general comment: Viterbi decoders that are of comparable complexity for well-chosen convolutional codes provide comparable performance.

9.7. GOOD CONVOLUTIONAL CODES FOR USE WITH VITERBI DECODING

A set of good convolutonal codes that can be used with Viterbi decoding has been found by Odenwalder [118]. Care must be taken in selecting a short-constraint-length convolutional code, since code performance can vary widely for a given constraint length. That is, there exist short-constraint-length codes with poor distance properties compared with others of the same constraint length. Of course, only noncatastrophic codes, codes that do not exhibit infinite error propagation, were considered, and those that provided the largest possible minimum free distance were found. Where more than one such code existed, the code that produced the fewest post-decoding errors for incorrect decodings to the minimum-distance words was chosen. If a selection could not be made on that basis, the procedure was continued for higher-weight words and error patterns. This approach led to the enumeration of a set of rate-1/2

TABLE 9.1. Good Rate-1/2, $b = 1$ Convolutional Codes for Viterbi Decoding and Binary Signaling

Constraint Length, k	Code Tap Connections (G_1, G_2)
3	111,101
4	1111,1101
5	11101,10011
6	111101,101011
7	1111001,1011011
8	11111001,10100111
9	111101011,101110001

TABLE 9.2. Good Rate-1/3, $b = 1$ Convolutional Codes for Binary Signaling and Viterbi Decoding

Constraint Length, k	Code Tap Connections (G_1, G_2, G_3)
3	111,111,101
4	1111,1101,1011
5	11111,11011,10101
6	111101,101011,100111
7	1111001,1110101,1011011
8	11110111,11011001,10010101

TABLE 9.3. Transfer Functions of the Good Rate-1/2 and -1/3 Codes for Binary Signaling

	k	T(D)
Rate-½ codes		
	3	$D^5 + 2D^6 + 4D^7 + 8D^8 + 16D^9 + 32D^{10} + 64D^{11} + 128D^{12} + 256D^{13} + 512D^{14} + 1{,}024D^{15} + 2{,}048D^{16} + 4{,}096D^{17} + \ldots$
	4	$D^6 + 3D^7 + 5D^8 + 11D^9 + 25D^{10} + 55D^{11} + 121D^{12} + 267D^{13} + 589D^{14} + 1{,}299D^{15} + 2{,}865D^{16} + 6{,}319D^{17} + 13{,}937D^{18} + \ldots$
	5	$2D^7 + 3D^8 + 4D^9 + 16D^{10} + 37D^{11} + 68D^{12} + 176D^{13} + 432D^{14} + 925D^{15} + 2{,}156D^{16} + 5{,}153D^{17} + 11{,}696D^{18} + 26{,}868D^{19} + \ldots$
	6	$D^8 + 8D^9 + 7D^{10} + 12D^{11} + 48D^{12} + 95D^{13} + 281D^{14} + 605D^{15} + 1{,}272D^{16} + 3{,}334D^{17} + 7{,}615D^{18} + 18{,}131D^{19} + 43{,}197D^{20} + \ldots$
	7	$11D^{10} + 38D^{12} + 193D^{14} + 1{,}331D^{16} + 7{,}230D^{18} + \ldots$
Rate-⅓ codes		
	3	$2D^8 + 5D^{10} + 13D^{12} + 34D^{14} + 89D^{16} + 233D^{18} + 610D^{20} + \ldots$
	4	$3D^{10} + 2D^{12} + 15D^{14} + 24D^{16} + 87D^{18} + 188D^{20} + 557D^{22} + \ldots$
	5	$5D^{12} + 3D^{14} + 13D^{16} + 62D^{18} + 108D^{20} + 328D^{22} + 1{,}051D^{24} + \ldots$
	6	$D^{13} + 3D^{14} + 6D^{15} + 4D^{16} + 5D^{17} + 12D^{18} + 14D^{19} + 33D^{20} + 66D^{21} + 106D^{22} + 179D^{23} + 317D^{24} + 513D^{25} +$
	7	$3D^{15} + 3D^{16} + 6D^{17} + 9D^{18} + 4D^{19} + 18D^{20} + 35D^{21} + 45D^{22} + 77D^{23} + 153D^{24} + 263D^{25} + 436D^{26} + 763D^{27} +$

and rate-1/3, $b = 1$ codes whose generators are given in Tables 9.1 and 9.2. The code generators are written with lowest order terms on the left. For example, the encoder specified for the $k = 4$ code in Table 9.1 is shown in Fig. 9.7. The most significant terms of the transfer functions for these codes are given in Table 9.3, and the derivatives of the transfer functions are shown in Table 9.4.

We note that in general the post-decoding bit-error rate depends on the entire weight distribution, not just the number of low-weight codewords. At high SNRs, however, the dominant error event is incorrect decoding to the low-weight words and the codes listed are the best performers in that region. At low SNRs, other codes may provide somewhat better performance.

Up to this point we have considered using convolutional codes only in conjunction with binary signaling, but convolutional codes can easily be used

TABLE 9.4. Derivatives of the Transfer Functions of the Good Rate-1 / 2 and -1 / 3 Codes for Binary Signaling

	k	$\frac{d}{dN} T(D,N) \Big\rvert_{N=1}$
Rate-½ codes	3	$D^5 + 4D^6 + 12D^7 + 32D^8 + 80D^9 + 192D^{10} + 448D^{11} + 1{,}024D^{12} + 2{,}304D^{13} + 5{,}120D^{14} + \ldots$
	4	$2D^6 + 7D^7 + 18D^8 + 49D^9 + 130D^{10} + 333D^{11} + 836D^{12} + 2{,}069D^{13} + 5{,}060D^{14} + 12{,}255D^{15} + \ldots$
	5	$4D^7 + 12D^8 + 20D^9 + 72D^{10} + 225D^{11} + 500D^{12} + 1{,}324D^{13} + 3{,}680D^{14} + 8{,}967D^{15} + 22{,}270D^{16} + \ldots$
	6	$2D^8 + 36D^9 + 32D^{10} + 62D^{11} + 332D^{12} + 701D^{13} + 2{,}342D^{14} + 5{,}503D^{15} + 12{,}506D^{16} + 36{,}234D^{17} + \ldots$
	7	$36D^{10} + 211D^{12} + 1{,}404D^{14} + 11{,}633D^{16} + 76{,}628D^{18} + 469{,}991D^{20} + \ldots$
Rate-⅓ codes		
	3	$3D^8 + 15D^{10} + 58D^{12} + 201D^{14} + 655D^{16} + 2{,}052D^{18} + \ldots$
	4	$6D^{10} + 6D^{12} + 58D^{14} + 118D^{16} + 507D^{18} + 1{,}284D^{20} + 4{,}323D^{22} + \ldots$
	5	$12D^{12} + 12D^{14} + 56D^{16} + 320D^{18} + 693D^{20} + 2{,}324D^{22} + 8{,}380D^{24} + \ldots$
	6	$D^{13} + 8D^{14} + 26D^{15} + 20D^{16} + 19D^{17} + 62D^{18} + 86D^{19} + 204D^{20} + 420D^{21} + 710D^{22} + 1{,}345D^{23} + \ldots$
	7	$7D^{15} + 8D^{16} + 22D^{17} + 44D^{18} + 22D^{19} + 94D^{20} + 219D^{21} + 282D^{22} + 531D^{23} + 1{,}104D^{24} + 1{,}939D^{25} + \ldots$

with M-ary signaling as well. For example, a $b = 1$, rate-1/2 code that produces two output parity bits for each input information bit can be used in a natural way with 4-ary signaling. Each pair of parity bits on a code branch is associated with one of the four waveforms produced by the modulator. However, in this case, a different distance measure for the design of convolutional codes is useful.

Here we are interested in the nonbinary Hamming distance between codewords in the code space, and a nonbinary transfer function must be defined. This can be done by relabeling the state diagram so that the nonbinary distances of the output symbols are indicated. That is, any state transition that produces a nonzero output symbol would include the factor D on the corresponding segment of the state diagram rather than D^i when i ones are contained in the symbol. Evaluation of the code transfer function can then proceed as in the binary case.

This approach was investigated by Trumpis [170], who found good $b = 1$ convolutional codes for use with nonbinary signaling and Viterbi decoding. For example, the generators for good rate-1/2 and rate-1/3 codes for use with 4-ary and 8-ary modulation, respectively, are indicated in Table 9.5. The most significant terms of the nonbinary transfer functions are indicated in Table 9.6, and the derivatives of the transfer functions are indicated in Table 9.7. The issue of designing good $b = 1$ convolutional codes for M-ary signaling has been addressed by other authors, for example, Modestino and Mui [113] and Lyon [99].

TABLE 9.5. Good Rate-1/2 Convolutional Codes for 4-ary Signaling and Rate-1/3 Codes for 8-ary Signaling and Viterbi Decoding

	Constraint Length, k	Code Tap Connections (G_1, G_2) and (G_1, G_2, G_3)
Rate-½ codes		
	3	111,101
	4	1101,1010
	5	11010,10101
	6	111101,101111
	7	1011011,1111110
Rate-⅓ codes		
	3	100,110,101
	4	1001,1101,1011
	5	10011,11011,10110
	6	100110,110111,101101
	7	1111110,1101101,1010111

TABLE 9.6. Transfer Functions of the Good Rate-1/2 and -1/3 Codes for 4-ary and 8-ary Signaling

k	T(D)
Rate-½ codes (4-ary signaling)	
3	$D^3 + 2D^4 + 4D^5 + 8D^6 + 16D^7 + 32D^8 + 64D^9 + 128D^{10} + 256D^{11} + 512D^{12} + 1{,}024D^{13} + 2{,}048D^{14} + 4{,}096D^{15} + \ldots$
4	$D^4 + 5D^5 + 5D^6 + 19D^7 + 45D^8 + 118D^9 + 270D^{10} + 704D^{11} + 1{,}714D^{12} + 4{,}287D^{13} + 10{,}551D^{14} + 26{,}308D^{15} + 65{,}047D^{16} + \ldots$
5	$2D^5 + 5D^6 + 5D^7 + 40D^8 + 66D^9 + 242D^{10} + 585D^{11} + 1{,}712D^{12} + 4{,}672D^{13} + 12{,}725D^{14} + 35{,}672D^{15} + 97{,}073D^{16} + 269{,}988D^{17} + \ldots$
6	$4D^6 + 5D^7 + 14D^8 + 41D^9 + 144D^{10} + 369D^{11} + 1{,}090D^{12} + 3{,}065D^{13} + 9{,}119D^{14} + 25{,}688D^{15} + 74{,}733D^{16} + 213{,}330D^{17} + 618{,}424D^{18} + \ldots$
7	$4D^7 + 10D^8 + 19D^9 + 62D^{10} + 195D^{11} + 574D^{12} + 1{,}665D^{13} + 4{,}820D^{14} + 14{,}225D^{15} + 42{,}087D^{16} + 122{,}970D^{17} + 361{,}085D^{18} + \ldots$
Rate-⅓ codes (8-ary signaling)	
3	$D^3 + D^4 + 2D^5 + 3D^6 + 5D^7 + 8D^8 + 13D^9 + 21D^{10} + 34D^{11} + 55D^{12} + 89D^{13} + 144D^{14} + 233D^{15} + \ldots$
4	$D^4 + D^5 + 3D^6 + 5D^7 + 11D^8 + 21D^9 + 43D^{10} + 85D^{11} + 171D^{12} + 341D^{13} + 683D^{14} + 1{,}365D^{15} + 2{,}731D^{16} + \ldots$
5	$D^5 + 2D^6 + 3D^7 + 7D^8 + 16D^9 + 35D^{10} + 72D^{11} + 156D^{12} + 336D^{13} + 726D^{14} + 1{,}552D^{15} + 3{,}339D^{16} + 7{,}171D^{17} +$
6	$D^6 + 2D^7 + 3D^8 + 9D^9 + 18D^{10} + 40D^{11} + 98D^{12} + 201D^{13} + 480D^{14} + 1{,}054D^{15} + 2{,}372D^{16} + 5{,}384D^{17} + 11{,}964D^{18} + \ldots$
7	$D^7 + 2D^8 + 3D^9 + 13D^{10} + 20D^{11} + 38D^{12} + 131D^{13} + 235D^{14} + 587D^{15} + 1{,}363D^{16} + 3{,}046D^{17} + 6{,}956D^{18} + 16{,}269D^{19} + \ldots$

TABLE 9.7. The Derivatives of the Transfer Functions of the Good Rate-1/2 and -1/3 Codes for 4-ary and 8-ary Signaling

	k	$\frac{d}{dN} T(D,N) \Big\|_{N=1}$
Rate-½ codes (4-ary signaling)		
	3	$D^3 + 4D^4 + 12D^5 + 32D^6 + 80D^7 + 192D^8 + 448D^9 + 1{,}024D^{10} + 2{,}304D^{11} + 5{,}120D^{12} + \ldots$
	4	$D^4 + 14D^5 + 21D^6 + 94D^7 + 261D^8 + 818D^9 + 2{,}173D^{10} + 6{,}335D^{11} + 17{,}220D^{12} + 47{,}518D^{13} + \ldots$
	5	$3D^5 + 15D^6 + 22D^7 + 196D^8 + 398D^9 + 1{,}737D^{10} + 4{,}728D^{11} + 15{,}832D^{12} + 47{,}491D^{13} + 144{,}170D^{14} + \ldots$
	6	$9D^6 + 14D^7 + 62D^8 + 212D^9 + 874D^{10} + 2{,}612D^{11} + 9{,}032D^{12} + 28{,}234D^{13} + 93{,}511D^{14} + 288{,}974D^{15} + \ldots$
	7	$7D^7 + 39D^8 + 104D^9 + 352D^{10} + 1{,}348D^{11} + 4{,}540D^{12} + 14{,}862D^{13} + 48{,}120D^{14} + 156{,}480D^{15} + 505{,}016D^{16} + \ldots$
Rate-⅓ codes (8-ary signaling)		
	3	$D^3 + 2D^4 + 5D^5 + 10D^6 + 20D^7 + 38D^8 + 71D^9 + 130D^{10} + 235D^{11} + 420D^{12} + \ldots$
	4	$D^4 + 2D^5 + 7D^6 + 16D^7 + 41D^8 + 94D^9 + 219D^{10} + 492D^{11} + 1{,}101D^{12} + 2{,}426D^{13} + \ldots$
	5	$D^5 + 5D^6 + 8D^7 + 25D^8 + 64D^9 + 170D^{10} + 392D^{11} + 958D^{12} + 2{,}270D^{13} + 5{,}406D^{14} + \ldots$
	6	$D^6 + 5D^7 + 7D^8 + 34D^9 + 76D^{10} + 200D^{11} + 557D^{12} + 1{,}280D^{13} + 3{,}399D^{14} + 8{,}202D^{15} + \ldots$
	7	$D^7 + 4D^8 + 8D^9 + 49D^{10} + 92D^{11} + 186D^{12} + 764D^{13} + 1{,}507D^{14} + 4{,}198D^{15} + 10{,}744D^{16} + \ldots$

9.8. NOTES

Currently, Viterbi decoding of short-constraint-length convolutional codes is probably the most widely used forward-error-control technique. Viterbi decoding is especially well suited for use on satellite links and other applications where performance is limited primarily by additive white Gaussian noise and a moderate improvement in communication efficiency is required.

There are a number of useful references on Viterbi decoder design and the construction of good short-constraint-length convolutional codes. A construction procedure for good rate-b/V codes and lists of code generators along with the associated transfer functions can be found in references 28 and 29. In reference 170 Trumpis gives an iterative procedure for evaluating complex code transfer functions. A good description of the original hardware design is given in reference 51, and a discussion of path-history storage and metric calculation issues can be found in reference 141. Forney [43] has shown formally that the Viterbi algorithm is maximum likelihood and considered its use on band limited channels with intersymbol interference [42].

Additional performance results are presented in Chapter 11, where the use of Viterbi decoding on other channels is considered. We note finally that H. Gibbons prepared the computer simulation programs used to obtain results for the $k = 4$ code and also obtained the transfer-function results presented.

CHAPTER TEN

Sequential Decoding

The first sequential decoding algorithm was described by Wozencraft in 1957 [184]. Unlike other decoding procedures for convolutional codes, sequential decoding can be used with complex long-constraint-length codes and can provide very dramatic performance enhancements on the AWGN channel. Since its discovery, sequential decoding has been the subject of a good deal of analytical and experimental work. The Wozencraft sequential decoder has been refined, a variety of related algorithms have been proposed, and performance analyses have been developed to estimate the coding gains that might be realized. Experimental work has verified the analytical results and has enhanced our understanding of sequential decoding.

In this chapter we consider the design and experimental evaluation of a sophisticated soft-decision sequential decoder for a long-constraint-length convolutional code. Our intention is to explain the design issues associated with sequential decoding. Experimental results are presented to illustrate how the decoder parameters can be adjusted to optimize performance and to show the performance gains that can be realized. First, we discuss sequential decoding in qualitative terms and point out an important design problem, namely, minimizing the computational burden. The issues of the choice of code rate and the design of an optimum decoder quantizer are addressed. The Fano sequential decoding algorithm, a suboptimum but computationally efficient procedure for searching any code tree, is then described in detail, and experimental results are examined for a complex soft-decision decoder design. The performance that can be obtained is presented, and the discussion of the Fano algorithm is concluded with a brief description of a hard-decision decoder. The last section of the chapter deals with another sequential decoding algorithm, the Zigangirov-Jelinek *stack decoder*.

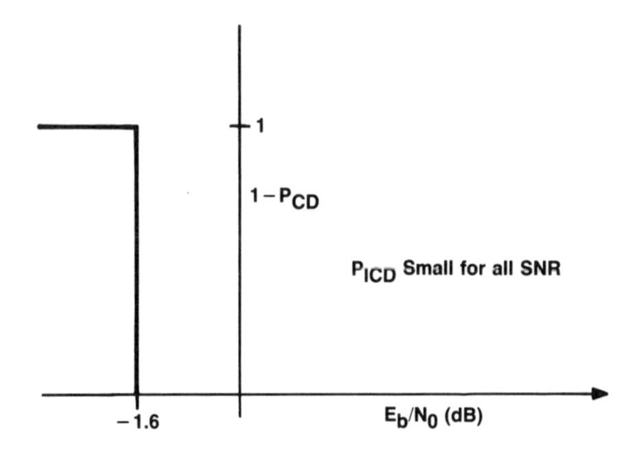

FIGURE 10.1. Ideal performance characteristic for a message-transmission system.

In the discussion of the design of a complex soft-decision Fano decoder, we shall take a different approach from that chosen by some of the early researchers. In many of the initial investigations the transmission of arbitrarily long messages was addressed. However, the messages considered through most of this chapter will be of finite length, and thus we shall consider sequential decoding of a truncated convolutional code.

It is appropriate at this point to review what might be considered an ideal operating characteristic for a message-transmission system. First, a negligibly small probability of incorrect decoding, P_{ICD}, for any SNR clearly would be desirable. In addition, we would like to have a high probability of correct decoding, P_{CD}, for SNRs above some nominal design point or equivalently a small probability of missing the message, given approximately by $1 - P_{\text{CD}}$. For the AWGN channel, the channel capacity theorem shows that the nominal design point can be at best $E_b/N_0 = -1.6$ dB, and we would like to operate as closely to this fundamental limit as possible. Thus the ideal performance characteristic takes the form shown in Fig. 10.1. It will be seen in the following sections that the use of a low-rate long-constraint-length convolutional code and soft-decision sequential decoding can provide performance within about 3 dB of this ideal operating characteristic.

10.1. A QUALITATIVE DESCRIPTION OF SEQUENTIAL DECODING

After reception and demodulation of the transmitted data, the function of any error-control decoder is to find among the set of all possible transmitted codewords the word that most closely resembles the received data. A maximum likelihood decoder, which is described in Chapter 9, effectively searches the entire code space, whereas suboptimum decoders search only a portion of the space. A sequential decoder for a convolutional code is a suboptimum decoder that continually restricts its attention to those portions of the code space that appear most likely to contain the actual message. When the noise background is low, one might expect to be able to pick out the correct message with little difficulty, but if the noise is locally severe, one can expect the choice to be quite difficult. In a sense a maximum likelihood decoder is designed for the worst-case noise events and does not take advantage of lulls in the noise background. A sequential decoder, on the other hand, does not burden itself with a large amount of work when the correct information sequence is fairly obvious but rather waits to expend the greater effort required to find the transmitted information sequence when the decision is not obvious.

We consider first how a decoder might proceed to recover the intended codeword for a nonsystematic rate-1/2, $b = 1$ convolutional code. In Fig. 10.2 the encoder and the associated code tree for a $k = 12$ nonsystematic code are

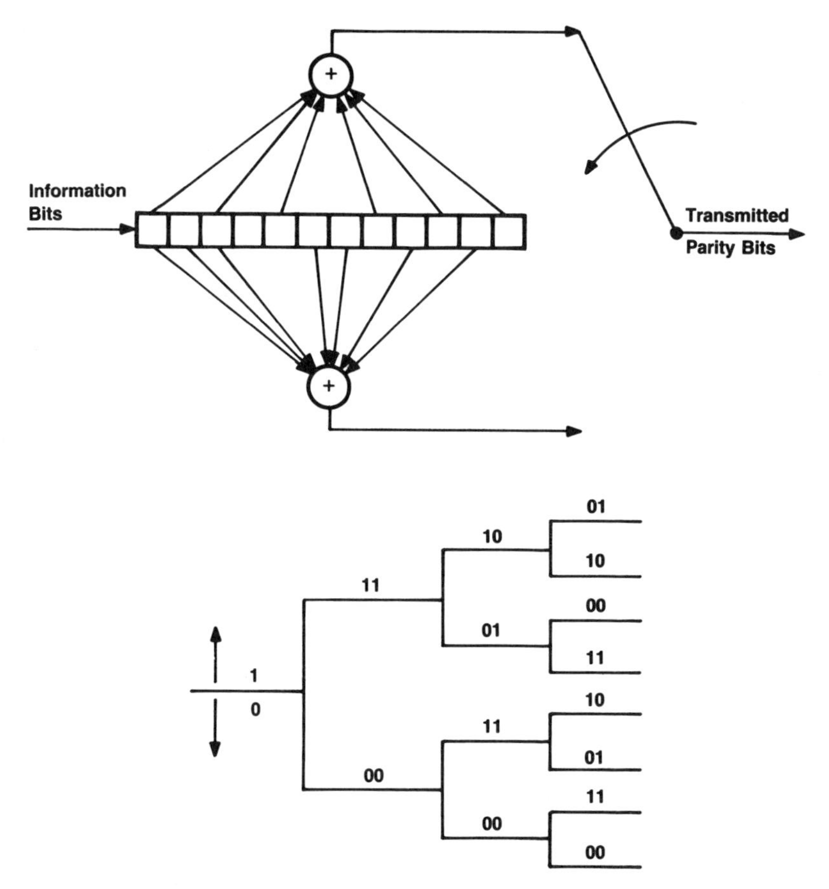

FIGURE 10.2. Encoder and the code tree for a rate-1/2 nonsystematic code with $k = 12$ and $b = 1$.

shown. We assume that the decoder contains a copy of the encoder, and we restrict the initial state of the encoder to be the all-zeros state. The channel is initially assumed to be noiseless.

On the noiseless channel it is a simple matter to determine the value of the first information bit after reception of the first two parity bits. All that is required is to compare the received parity-bit values with those on the first two branches of the code tree and choose the information bit that results in two agreements. That is, for the code indicated in Fig. 10.2, if 11 is received, surely a one was intended, and if 00 is received, we can be certain that the first information bit was a zero. Note that having made that first decision, the decoder has knowledge of the state of the encoder at the point in time when the second information bit is encoded, and the process may continue. The decoder, upon receipt of the second pair of parity bits, shifts the encoding register, computes the expected parities, and again tests for agreement with the parity bits associated with the transmission of a zero or one information bit on the second code branch. If the channel is noiseless, the second decision is also made correctly. For example, if the first information bit transmitted were a one and the parity bits 10 are received on the second code branch, the decoder concludes that the second information bit is a one.

As long as the channel is noiseless, this procedure can continue indefinitely and the decoder can reproduce the information bits exactly. There is clearly perfect correlation between what is transmitted and what is received. That is, the decoder continually steps forward one node in the code tree for each received pair of parity bits and decodes one information bit at each step. Actually, in the absence of noise, the decoder requires knowledge of only a single parity-bit sequence and the initial state of the encoder in order to reconstruct the information sequence.

But what happens if the channel is noisy? Clearly the individual decisions on information bits can no longer be made with certainty. For example, if the first pair of received parity bits for the encoder shown in Fig. 10.2 is 10 or 01, the initial decision on the value of the first information bit must be simply a guess between equally likely alternatives. It is also possible that 00 might be received when 11 was transmitted or vice versa, which would result in a wrong initial decision. Therefore, the individual bit decisions made by a sequential decoder must be considered tentative at each step in the decoding process. The decoder must have the capability to reverse any and all bit decisions at a later time if it seems necessary.

We suppose now that the decoder does make a mistake, an incorrect bit decision, and we ask what the consequences are. After an error has been made, the state of the encoder assumed by the decoder is no longer valid. There is, for example, a one hypothesized in the second shift register stage when there should be a zero. In this case, for the encoder shown in Fig. 10.2, we see that the error will affect the lower parity bit but not the upper bit. In general, long-constraint-length convolutional codes are purposely constructed so that after the first and subsequent errors are made, this characteristic continues;

half the parity bits can be expected to be correct and half incorrect. This means that the individual shift register stages are, on the average, connected to half of the parity adders. Thus we can expect relatively poor correlation between an incorrect path postulated by the decoder and the received parity bits, in contrast with relatively good correlation expected between the correct path and the received bits.

To be more specific, let us initially consider the case of the binary symmetric channel with bit-error probability p. For this channel the correlation between the path postulated by the decoder and the received data may be defined as the number of agreements minus the number of disagreements between the postulated path and the received parity-bit sequence. That is, we use this correlation as the metric of the path followed by the decoder. Then, for a rate-$1/V$ code, as the decoder moves forward along the correct path, the expected rate at which discrepancies are observed is the channel bit-error rate p, and the correct path metric can be expected to increase at the average rate $(1 - p)V - pV = (1 - 2p)V$ at each node. On the other hand, since the expected number of agreements equals the expected number of disagreements when the decoder is following an incorrect path, there will be no growth in the path metric on average when the decoder departs from the correct path. This behavior of the mean of the correct and incorrect path metrics is shown in Fig. 10.3.

The fact that the growth in the value of the path metric is expected to be different for the correct and incorrect paths suggests that by monitoring the value of the metric of the hypothesized path, the decoder will be able to determine when it has departed from the correct path. The decoder has high confidence that the hypothesized path is the correct one as long as its metric

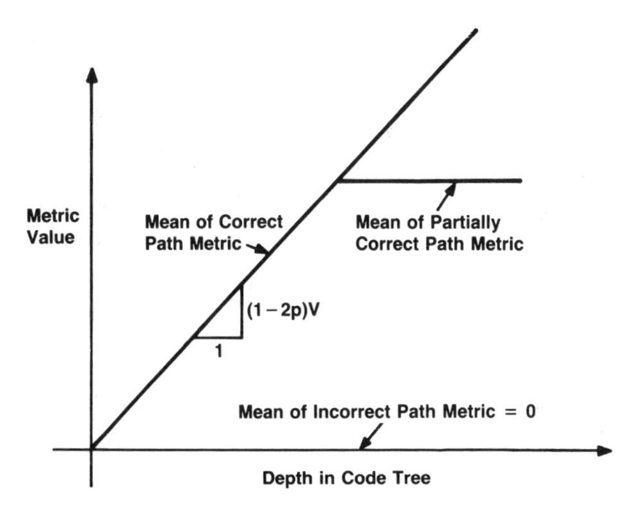

FIGURE 10.3. Behavior of the mean of the correct, incorrect, and partially correct path metrics on the binary symmetric channel with bit-error rate p (no metric bias).

continues to grow. If that growth ceases, the decoder reexamines previous decisions to see if an alternate path can be found whose metric ultimately increases. It is possible that a troublesome sequence of noise digits can cause an incorrect path metric to grow for several consecutive sets of received parity bits, but eventually the growth of metrics on all incorrect paths can be expected to cease.

From the previous discussion it is clear that a mechanism is needed to reject paths followed by the decoder when the growth in the path metric ceases. This can be implemented with a simple threshold test. For example, we could require that a next branch on the currently examined path have a metric greater than some value, say T_0. When the branch metric is too small, the current path would be discarded and a search would begin for a better one. Furthermore, we would like the decoder to recognize that it has departed from the correct path as quickly as possible, and thus we would initially choose T_0 to be relatively large. However, should some branch on the correct path have a metric smaller than T_0, the correct path would be discarded. Branch metrics on incorrect paths would also, with high probability, eventually fail the threshold test, and if the decoder is to return to the correct path and move forward again, the threshold T_0 must be lowered. Therefore, we must also provide for adjustment of the threshold value, perhaps many times.

All sequential decoding algorithms are suboptimum tree-searching procedures that share these two features. Specifically, a sequential decoder attempts to quickly recognize that it has departed from the correct path in the code tree and uses a sequence of test criteria on path metrics to determine the relative "goodness" of the path being pursued. A sequential decoder attempts to find the best path in the code tree, that is, the path with the largest metric.

The detailed set of rules that specify which path is extended, accepted, or discarded and controls the search for a better one defines a particular sequential decoding algorithm. Following the development of the original algorithm of Wozencraft, Fano [33] proposed a similar technique that has several important implementation advantages. The Fano decoder moves forward and backward in the code tree one node at a time depending on the behavior of the path metric. Backing up all the way to the beginning of the message is permitted. Later Zigangirov [188] and Jelinek [75] independently proposed a quite different approach, one that permits jumping between widely separated nodes in the tree, and more recently other algorithms have been suggested [94]. However, the most widely used sequential decoding procedure is the Fano algorithm, which we shall subsequently describe in detail.

10.2. THE COMPUTATIONAL PROBLEM

It should be apparent from the previous discussion that the time a sequential decoder takes to process a message is a random variable. This is also true for block codes where the number of computations required to decode increases with the number of errors in the received word and usually depends on the locations of the errors. However, the variability in the decoding time with sequential decoding can be very large, much larger than with typical block codes.

To begin a discussion of the variability of decoding effort, let us define a decoder computation as the process of extending the path currently being examined and deciding what to do next, that is, choosing to accept or discard the new path. It has been shown that if C is the number of computations required to decode an information bit, then the *computational distribution* for a sequential decoder is asymptotically Paretian [153, 154]. That is, the probability that C exceeds some number X is given by

$$P(C > X) = AX^{-\alpha}$$

for large X. The *Pareto exponent* α and the constant A are numbers that depend on the channel and the rate of the code used. For the purpose of this initial discussion we assume the use of a very low-rate code or equivalently a large bandwidth expansion.

The functional behavior of the computational distribution was anticipated by Wozencraft when he stated that "bad decoding situations occur with a probability that is exponentially small but when they do occur they are of exponential magnitude" [154]. What Wozencraft observed was that a string of errors that spans l received digits occurs with a probability that decreases exponentially with l but forces the decoder to examine a number of paths that increases exponentially with l. The balance between these opposing exponential tendencies results in the Pareto distribution, which is a linear function on a log-log plot. The Pareto exponent α is the slope of that function.

We now examine the behavior of the computational distribution and the Pareto exponent as a function of the quality of the channel. We assume that the channel noise is additive, white, and Gaussian. Clearly, if the SNR is high, α must be large and the slope of the computational distribution nearly vertical as shown in Fig. 10.4. This is true because the received data must closely resemble the transmitted codeword and the decoder can be expected to find the correct path with little difficulty. That is, the decoder continually moves forward in the code tree, executing one computation per information bit.

For small SNRs and low code rates there is a simple algebraic relationship between the Pareto exponent and the SNR (see for example reference 72 or 48). Specifically, for arbitrarily low-rate codes and SNRs within a few decibels of channel capacity, α is given by

$$\alpha = \eta \frac{E_b/N_0}{E_b'/N_0} - 1 \tag{10.1}$$

where E_b/N_0 is the received SNR per information bit, E_b'/N_0 is the SNR corresponding to operation at channel capacity, and η is a parameter, upper-

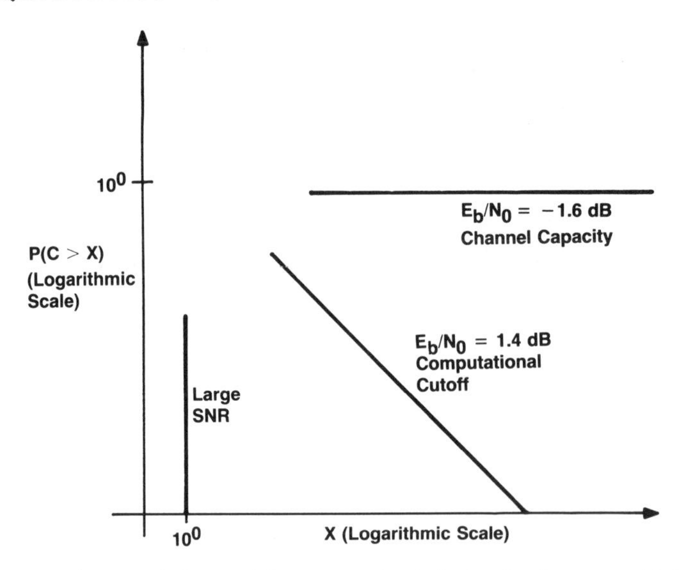

FIGURE 10.4. Asymptotic forms of the computational distribution for sequential decoding.

bounded by unity, that depends on the performance degradation due to quantization. One of the major advantages of sequential decoding is that soft-decision information can be readily accommodated and η can be made to approach unity. Then, if there is no appreciable quantization loss, we see that $\alpha = 1$ when the ratio of E_b/N_0 to E_b'/N_0 is 2 or at an SNR that is 3 dB above channel capacity, that is, at $E_b/N_0 = 1.4$ dB. At channel capacity ($E_b/N_0 = -1.6$ dB) we have $E_b'/N_0 = E_b/N_0$ and $\alpha = 0$.

The asymptotic behavior of the computational distribution for a sequential decoder operating at $E_b/N_0 = 1.4$ dB and -1.6 dB is also indicated in Fig. 10.4. It is interesting to note that as E_b/N_0 decreases from an arbitrarily large value to 1.4 dB, the slope of the computational distribution rotates through 45 degrees. As E_b/N_0 is reduced just 3 dB further, to channel capacity, the slope rotates again by 45 degrees and goes to zero. Thus we see that within a few decibels of channel capacity, small changes in SNR produce very large changes in the computational load for a sequential decoder.

In addition, the Pareto distribution has a finite mean for $\alpha > 1$, but for $0 \leq \alpha \leq 1$ the mean is unbounded [185]. Thus one might expect to be able to operate a sequential decoder at SNRs above $E_b/N_0 = 1.4$ dB but not below. In fact, the point at which $\alpha = 1$ is commonly called *computational cutoff*. However, we will consider sequential decoding of a truncated convolutional code, and an unlimited decoding time for a received code sequence will not be allowed. That is, a maximum number of computations permitted to decode a message is defined, which in turn changes the distribution and imposes a bound on the mean. When the computational limit is exceeded, the decoder simply erases the received word and quits. We shall see shortly that a sequential decoder can, in fact, operate near or perhaps slightly beyond the point of computational cutoff.

10.3. EFFECTS OF CODE RATE AND QUANTIZATION

In the discussion so far we have assumed use of a very low-rate code, large bandwidth expansion, and no quantization loss. In this section we consider the effects of a more realistic set of constraints. Key design issues will emerge and be discussed.

The communication system model that will be used is the familiar one shown in Fig. 10.5. We limit the discussion to convolutional codes with rate $R = 1/V$. The modulation and demodulation strategy assumed is binary antipodal signaling with coherent detection, for example, coherent PSK. The channel is assumed to have additive white Gaussian noise with single-sided noise-power spectral density N_0. The mean of the matched-filter output is then $\pm\sqrt{E_s}$, the sign depending on the value of the transmitted channel bit (either 0 or 1, respectively), and the variance is $N_0/2$. If y represents the matched-filter output and p_Y the associated probability density function, we have

$$p_Y(y|0) = \frac{1}{\sqrt{\pi N_0}} e^{-(y-\sqrt{E_s})^2/N_0}$$

$$p_Y(y|1) = \frac{1}{\sqrt{\pi N_0}} e^{-(y+\sqrt{E_s})^2/N_0}$$

As in the previous chapter it is convenient to normalize the matched-filter

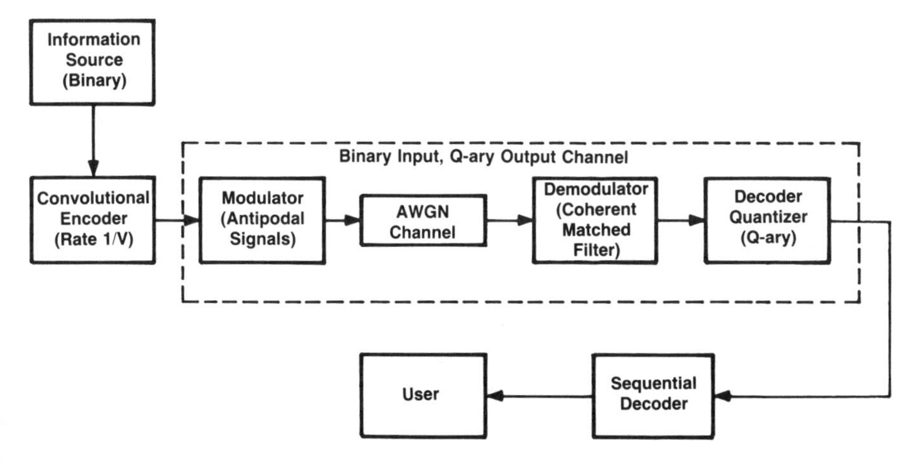

FIGURE 10.5. Communication model assumed for use with sequential decoding.

output by $\sqrt{2/N_0}$, in which case we have

$$s = \sqrt{2/N_0}\, y$$

and

$$p_S(s|0) = \frac{1}{\sqrt{2\pi}} e^{-(s-a)^2/2} \tag{10.2}$$

$$p_S(s|1) = \frac{1}{\sqrt{2\pi}} e^{-(s+a)^2/2}$$

where $a = \sqrt{2E_s/N_0}$. Thus the mean of the normalized matched-filter output s is $\pm a$ and the variance is unity.

Note that the functions enclosed in the dashed lines in Fig. 10.5 define a binary input, Q-ary output channel for the sequential decoder, and a quantizer is shown following the demodulator. The design goal for this quantizer is somewhat different from the classic quantization goal, which is to minimize the mean-squared error between input and output signals. The design goal for the decoder quantizer is, as discussed in Chapter 1, to maximize the computational cutoff rate of the discrete channel.

10.3.1. Selection of Code Rate

We have seen that for the unquantized channel and arbitrarily small code rates, computational cutoff occurs with $E_b/N_0 = 1.4$ dB. For any real system, though, arbitrarily large bandwidth expansion is not practical. Furthermore, the complexity of the arithmetic unit in the decoder can be reduced by use of a suitable decoder quantizer. Therefore we can expect that for any real system, computational cutoff will occur with $E_b/N_0 > 1.4$ dB. However, for moderately low code rates and a properly designed decoder quantizer, operation near $E_b/N_0 = 1.4$ dB is possible.

The characteristic of central interest in the design of a sequential decoder for the binary input, Q-ary output channel is the computational cutoff rate R_{COMP}, which was introduced in Chapter 1. R_{COMP} is defined as the information-transfer rate, measured in information bits per channel symbol, for which the Pareto exponent is -1. As discussed in Chapter 1, R_{COMP} for the unquantized channel is given by

$$R_{\mathrm{COMP}} = 1 - \log_2\left[1 + \int_{-\infty}^{\infty} \sqrt{p_S(s|0)\, p_S(s|1)}\, ds\right] \tag{10.3}$$

and with AWGN we have

$$R_{\mathrm{COMP}} = 1 - \log_2[1 + e^{-E_s N_0}] \tag{10.4}$$

TABLE 10.1. Signal-to-Noise Ratios Corresponding to Computational Cutoff for Several Code Rates

Code Rate, 1/V	E_b/N_0	E_b/N_0 (dB)	E_S/N_0 (dB)
1/2	1.76	2.5	−0.5
1/3	1.60	2.0	−2.7
1/4	1.53	1.9	−4.2
1/6	1.48	1.7	−6.1
1/9	1.44	1.6	−8.0
1/12	1.43	1.5	−9.2
1/96	1.39	1.4	−18.4

For a rate-$1/V$ code

$$E_b = VE_s \tag{10.5}$$

and if we assume operation of the communication system with $R = 1/V = R_{\text{COMP}}$, Eqs. (10.4) and (10.5) can be solved for the SNRs that correspond to computational cutoff. In particular, we have

$$\frac{E_b}{N_0} = -V\ln[2^{(1-1/V)} - 1] \tag{10.6}$$

and

$$\frac{E_s}{N_0} = -\ln[2^{(1-1/V)} - 1]$$

The SNRs corresponding to computational cutoff for several code rates are given in Table 10.1. Note that as V is increased (the code rate reduced), E_b/N_0 approaches 1.4 dB and the degradation due to a finite bandwidth constraint can be small. For example, with $V = 6$, a degradation of only 0.3 dB is observed, but for $V = 2$ the degradation is more than 1 dB. It is also important to note that as the code rate is reduced, E_s/N_0 decreases and modem synchronization becomes more difficult.

10.3.2. Design of the Decoder Quantizer

For the binary input, Q-ary output channel, the expression for R_{COMP}, Eq. (10.3), becomes

$$R_{\text{COMP}} = 1 - \log_2\left[1 + \sum_{j=1}^{Q} \sqrt{P(j|1)P(j|0)}\right] \tag{10.7}$$

where $P(j|i)$ is the probability of observing a matched filter output in the jth quantization interval given that symbol i was transmitted ($i = 0,1$) [106]. For coherent reception in Gaussian noise we have

$$P(j|0) = \frac{1}{\sqrt{2\pi}} \int_{I_{j-1}}^{I_j} e^{-(t-a)^2/2}\, dt$$

where I_j and I_{j-1} are the boundaries of the jth quantization interval. Note that since $P(j|i)$ depends on the particular quantization intervals chosen, R_{COMP} does also, and it is our intention to choose the quantization intervals such that R_{COMP} is maximized for a given SNR.

It should be noted that maximizing R_{COMP} for the quantized channel serves two purposes. First, the SNR at which computational cutoff occurs is minimized. Second, as we saw in Chapter 1, R_{COMP} is numerically equal to R_0, which appears in the exponent of a bound on the probability of error for coded systems. Thus, maximizing R_{COMP} also minimizes the bound on the probability of error.

Massey [106] developed an iterative procedure for finding the optimum quantization boundaries, that is, the quantization thresholds that maximize R_{COMP}. The procedure first obtains the optimum quantization intervals for the *likelihood ratio function* $l(s)$. The likelihood ratio is given by

$$l(s) \triangleq \frac{p_S(s|0)}{p_S(s|1)} \tag{10.8}$$

which is simply the ratio of probability density functions of the matched-filter outputs conditioned on the transmission of a zero and one, respectively. For the case of antipodal signaling and coherent detection, these densities are given by Eq. (10.2). Clearly we have $0 < l(s) < \infty$.

The procedure described by Massey is best illustrated with an example. We consider the design of an optimum eight-level quantizer. Our approach is to quantize $l(s)$ by specifying the eight regions indicated in Fig. 10.6. The problem is to choose the seven thresholds T_j, $1 \le j \le 7$, such that R_{COMP} is maximized. Since $l(s)$ can be any positive number, we have selected $T_0 = 0$ and $T_8 = \infty$.

Massey showed that if we let

$$\lambda(\mathscr{R}_j) = \frac{\int_{\mathscr{R}_j} p_S(s|0)\, ds}{\int_{\mathscr{R}_j} p_S(s|1)\, ds} \tag{10.9}$$

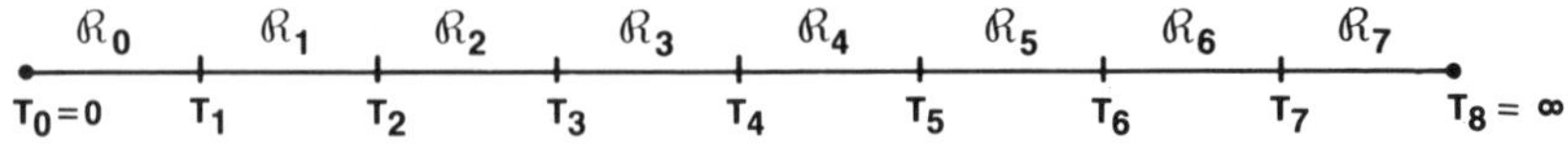

$\mathscr{R}_0$ $\mathscr{R}_1$ $\mathscr{R}_2$ $\mathscr{R}_3$ $\mathscr{R}_4$ $\mathscr{R}_5$ $\mathscr{R}_6$ $\mathscr{R}_7$

$T_0 = 0$ T_1 T_2 T_3 T_4 T_5 T_6 T_7 $T_8 = \infty$

FIGURE 10.6. Quantization regions $\mathscr{R}_j$ and thresholds T_j for the likelihood ratio function $l(s)$ to maximize R_{COMP}; eight-level quantization.

where the integrals in Eq. (10.9) are evaluated over the jth region indicated in Fig. 10.6, then the optimum values of the T_j are given by

$$
\begin{aligned}
T_1 &= \sqrt{\lambda(\mathscr{R}_0)\lambda(\mathscr{R}_1)} \\
T_2 &= \sqrt{\lambda(\mathscr{R}_1)\lambda(\mathscr{R}_2)} \\
&\vdots \\
T_7 &= \sqrt{\lambda(\mathscr{R}_6)\lambda(\mathscr{R}_7)}
\end{aligned} \tag{10.10}
$$

However, in order to calculate $\lambda(\mathscr{R}_j)$ we must know T_j and T_{j+1} or, conversely, in order to calculate T_j we need to know $\lambda(\mathscr{R}_{j-1})$ and $\lambda(\mathscr{R}_j)$. Thus the following iterative procedure was suggested to find the $\{T_j\}$ for an eight-level quantizer:

1. Choose T_1 arbitrarily.
2. Calculate

$$\lambda(\mathscr{R}_0) = \frac{\int_0^{T_1} p_S(s|0)\,ds}{\int_0^{T_1} p_S(s|1)\,ds}$$

3. Find T_2 using Eqs. (10.9) and (10.10). That is, first calculate

$$\lambda(\mathscr{R}_1) = \frac{T_1^2}{\lambda(\mathscr{R}_0)}$$

and then find T_2 from

$$\lambda(\mathscr{R}_1) = \frac{\int_{T_1}^{T_2} p_S(s|0)\,ds}{\int_{T_1}^{T_2} p_S(s|1)\,ds}$$

4. Find T_3 through T_7 in the same manner by repeating steps 2 and 3.
5. Compute

$$\lambda'(\mathscr{R}_7) = \frac{T_7^2}{\lambda(\mathscr{R}_6)}$$

and see if $\lambda'(\mathscr{R}_7) = \lambda(\mathscr{R}_7)$ that is,

$$\lambda(\mathscr{R}_7) = \frac{\int_{T_7}^{\infty} p_S(s|0)\,ds}{\int_{T_7}^{\infty} p_S(s|1)\,ds} \stackrel{?}{=} \lambda'(\mathscr{R}_7)$$

If $\lambda'(\mathscr{R}_7) > \lambda(\mathscr{R}_7)$, choose a smaller value for T_1 and return to step 2. If $\lambda'(\mathscr{R}_7) < \lambda(\mathscr{R}_7)$, choose a larger value for T_1 and return to step 2. If $\lambda'(\mathscr{R}_7) \simeq \lambda(\mathscr{R}_7)$, stop; the procedure has successfully found the set of likelihood thresholds that maximizes R_{COMP}.

Of course, the point of this exercise is to find the optimum quantization thresholds in signal space rather than in a likelihood space. That is, our intention is to compute the quantization thresholds for an eight-level quantizer. To convert the likelihood thresholds $\{T_j\}$ to quantization thresholds $\{I_j\}$ for the matched-filter outputs, note that

$$T_j = \frac{p_S(s = I_j|0)}{p_S(s = I_j|1)}$$

$$= \frac{(1/\sqrt{2\pi})e^{-(I_j-a)^2/2}}{(1/\sqrt{2\pi})e^{-(I_j+a)^2/2}}$$

$$= e^{2aI_j}, \qquad 1 \le j \le 7$$

or

$$I_j = \left(\frac{1}{2a}\right)\ln T_j, \qquad 1 \le j \le 7$$

where $I_0 = -\infty$ and $I_8 = \infty$.

TABLE 10.2. Optimum Four-Level Quantizer; $E_s/N_0 = -6.1$ dB

I_j	Threshold Value
I_0	$-\infty$
I_1	-0.994
I_2	0
I_3	0.994
I_4	∞

TABLE 10.3. Optimum Eight-Level Quantizer; $E_s/N_0 = -6.1$ dB; $I_{8-j} = -I_j$ for $0 < j < 4$

I_j	Threshold Value
I_0	$-\infty$
I_1	-1.760
I_2	-1.056
I_3	-0.503
I_4	0

This procedure generalizes easily for other values of Q. Note that with binary antipodal signals the quantization intervals must be symmetric about the origin; that is, $I_{Q-j} = -I_j$ for $0 \leq j \leq Q/2$, and one actually need compute only half the required number of thresholds. In addition, when Q is even, the boundary in the middle must be the origin, $I_{Q/2} = 0$.

It is important to note that the Q-level quantizer that maximizes R_{COMP} depends on $a = \sqrt{2E_s/N_0}$. The optimum quantizer is thus a function of SNR, although when E_s/N_0 is small the dependence is weak. Since the channel symbol SNR is generally not known a priori, one could pick a nominal system operating point and design the quantizer to maximize R_{COMP} at that SNR. For

TABLE 10.4. Optimum 16-Level Quantizer; $E_s / N_0 = -6.1$ dB; $I_{16-j} = -I_j$ for $0 < j < 8$

I_j	Threshold Value
I_0	$-\infty$
I_1	−2.411
I_2	−1.850
I_3	−1.441
I_4	−1.102
I_5	−0.802
I_6	−0.524
I_7	−0.259
I_8	0

TABLE 10.5. Optimum 32-Level Quantizer; $E_s / N_0 = -6.1$ dB; $I_{32-j} = -I_j$ for $0 < j < 16$

I_j	Threshold Value
I_0	$-\infty$
I_1	−2.983
I_2	−2.509
I_3	−2.177
I_4	−1.911
I_5	−1.684
I_6	−1.483
I_7	−1.301
I_8	−1.132
I_9	−0.973
I_{10}	−0.822
I_{11}	−0.677
I_{12}	−0.536
I_{13}	−0.399
I_{14}	−0.265
I_{15}	−0.132
I_{16}	0

SNRs above nominal, the quantizer is not optimum but the computational problem is less severe. For lower SNRs, optimization of performance is not a concern.

We now show that the nominal operating point can be chosen as the SNR corresponding to operation at R_{COMP} for the unquantized channel. As an example we consider a rate-1/6 code and assume no quantization loss. From Table 10.1 we see that R_{COMP} for the unquantized channel corresponds to $E_s/N_0 = -6.1$ dB, and the question now is whether a quantizer can be found such that R_{COMP} for the quantized channel is close to 1/6.

Use of the procedure suggested by Massey results in the quantization intervals shown in Tables 10.2 to 10.6 for $Q = 4$, 8, 16, 32, and 64 when

TABLE 10.6. Optimum 64-Level Quantizer; $E_s/N_0 = -6.1$ dB; $l_{64-j} = -l_j$ for $0 < j < 32$

l_j	Threshold Value
l_0	$-\infty$
l_1	−3.495
l_2	−3.080
l_3	−2.796
l_4	−2.573
l_5	−2.386
l_6	−2.224
l_7	−2.079
l_8	−1.948
l_9	−1.827
l_{10}	−1.714
l_{11}	−1.608
l_{12}	−1.508
l_{13}	−1.412
l_{14}	−1.321
l_{15}	−1.233
l_{16}	−1.149
l_{17}	−1.067
l_{18}	−0.987
l_{19}	−0.909
l_{20}	−0.833
l_{21}	−0.759
l_{22}	−0.686
l_{23}	−0.614
l_{24}	−0.544
l_{25}	−0.474
l_{26}	−0.405
l_{27}	−0.336
l_{28}	−0.268
l_{29}	−0.201
l_{30}	−0.134
l_{31}	−0.067
l_{32}	0

TABLE 10.7. R_{COMP} **for the Discrete Channels with Optimum Quantization**

Q	R_{COMP}
4	0.1480
8	0.1612
16	0.1652
32	0.1663
64	0.1666

$E_s/N_0 = -6.1$ dB. In Table 10.7, values of R_{COMP} for the corresponding discrete channels are listed. Note that for $Q = 32$ and 64, R_{COMP} for the quantized channel is close to $1/6$ and thus the quantization loss is small.

Clearly, the optimum two-level quantizer employs a boundary at the origin. It is easy to show that if $E_s/N_0 = -6.1$ dB, R_{COMP} for a hard-decision two-level quantizer is 0.108. Furthermore, it can be shown that the SNR must be increased 2 dB in order that R_{COMP} approach $1/6$. Thus we observe again that hard-decision quantization results in a 2-dB performance degradation.

10.4. THE FANO SEQUENTIAL DECODER

As we have mentioned previously, the most often used sequential decoding algorithm is the one developed by Fano [33]. The technique can be used for hard- or soft-decision decoding, and we shall consider soft-decision decoding first. The *Fano decoder*, in its search for the most likely transmitted message, moves forward or backward in the code tree one node at a time, with the behavior of the metric along the path followed by the decoder determining whether a forward or backward move is made. The selection of the metric to use with sequential decoding is crucial for proper operation of the decoder. Fano argued heuristically, and it was shown rigorously later by Massey [105], that the proper metric function for sequential decoding is the logarithm of the probability that a particular output is observed given that a certain symbol was transmitted, normalized by the total probability of the observed output. Specifically, for the binary input Q-ary output channel, Fano suggested that the metric associated with a particular received parity bit i, say the kth bit on the lth branch, be set equal to

$$\lambda_{k,l} = \log\left(\frac{P(j|i)}{P(j)}\right) - U$$

where $P(j)$ is the total probability of observing an output in the jth quantization level, and U is a bias term. We shall see shortly that U can be selected for computational convenience and to reduce the probability of error. For binary antipodal signaling and coherent detection, we have

$$\lambda_{k,l} = \log\left(\frac{\frac{1}{\sqrt{2\pi}}\int_{I_{j-1}}^{I_j} e^{-(x\pm a)^2/2}\,dx}{P(j)}\right) - U$$

and when zeros and ones are equally likely we have

$$P(j) = \tfrac{1}{2}[P(j|0) + P(j|1)]$$

The metric associated with the lth information bit, or with the lth branch in the code tree, is then

$$\lambda(l) = \sum_{k=1}^{V} \lambda_{k,l}$$

and the metric associated with the entire path from the origin to the lth node is

$$\Lambda(l) = \sum_{j=1}^{l} \lambda(j)$$

The metric function suggested by Fano has the property that correct path metrics tend to increase at a moderate rate but, in contrast with the example considered previously (Section 10.1), incorrect path metrics decrease relatively quickly. The Fano algorithm exploits these features of the path metrics.

The sequence of steps used by the Fano decoder to find the path through the code tree with the largest metric is straightforward, but any particular implementation may contain many subtleties. The process has been compared to driving a car through a confusing maze. The basic operations involve comparing the metric along a postulated path with a sequence of thresholds that are separated by a uniform spacing ΔT. In Fig. 10.7 a possible correct path metric is shown as the solid line, and the thresholds used by the decoder are indicated with lighter horizontal lines.

When a Fano decoder attempts to move forward from a node it computes the branch metrics corresponding to the information bits 0 and 1 for the next code branch and tentatively chooses the larger. The favored branch metric is added to the previous path metric, and the newly extended path is accepted as long as its metric is above the current threshold value. During initial forward moves the threshold is adjusted upward whenever the path metric exceeds a larger threshold value. If the extended path metric falls below the current threshold, the path extension is rejected, perhaps temporarily, and the decoder goes into a backward search for a more promising sequence of information bits.

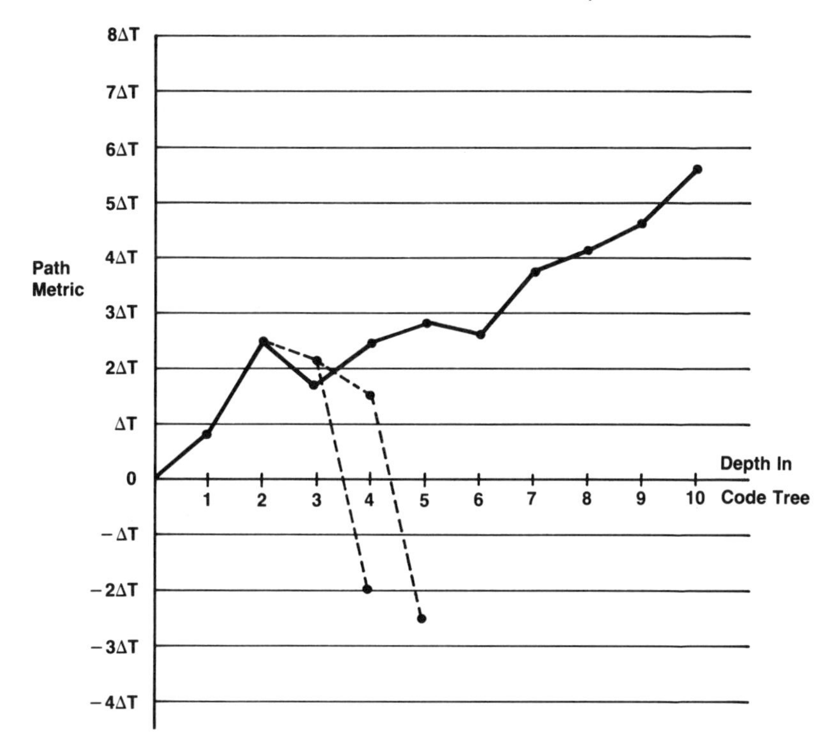

FIGURE 10.7. Behavior of a possible correct path metric (solid line) and some incorrect alternatives (dashed lines).

In the backward search the Fano decoder retreats one node at a time and examines the alternative branches along the path that was traced initially. Path metrics are compared against the current threshold T, and if an alternative is found whose metric is larger than T, forward progress is attempted again. If there are no such alternatives or if all the paths extended from the alternatives violate the current threshold eventually, the threshold value is lowered by ΔT and the initial best path is retraced. If the extensions from the initial best path do not remain above the new threshold, another backward search is started, and again the alternatives are considered. If no alternative can be found whose metric remains above $T - \Delta T$, the threshold is lowered again. This sequence of events continues until a path is found whose metric ultimately increases.

An important point to note is that the threshold cannot be permitted to increase when the decoder retraces previously examined nodes, or an infinite loop would result. Thus one implementation issue is obvious: how is the decoder to distinguish between the process of retracing an old path and reaching a new node for the first time? Since backward searches can be quite lengthy and involved, it would be costly to store information on an entire backward search. One of the principal advantages of the Fano algorithm is that this bookkeeping is accomplished in a very efficient way.

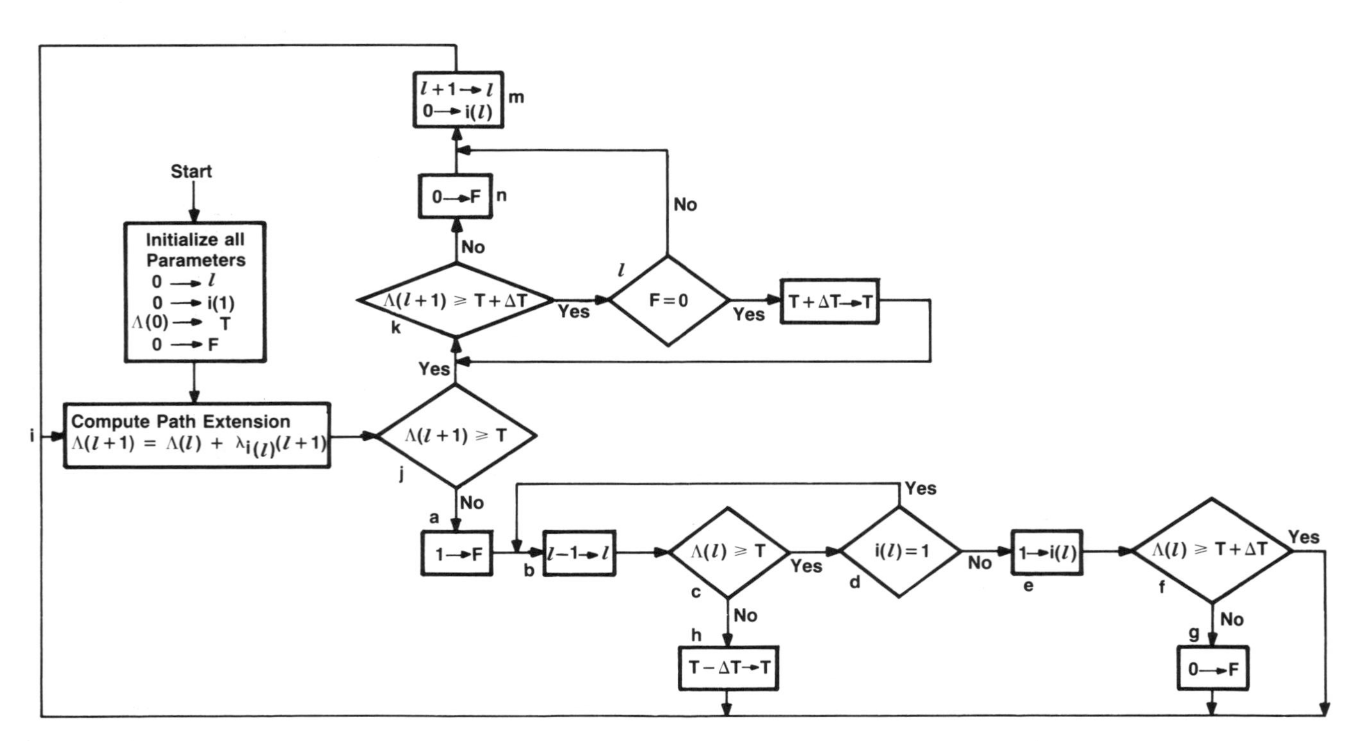

FIGURE 10.8. Flow chart for the Fano sequential decoder. From Fano [33], © 1963 IEEE, reprinted with permission.

The Fano decoder is best described in detail with a computer flow chart, and in Fig. 10.8 we show the original flow chart used in the first digital computer implementation. The algorithm makes use of a flag F and a binary vector $\boldsymbol{i}$ whose lth component is denoted $i(l)$. It is this flag and binary vector that control the adjustment of thresholds. The flag, set when a threshold violation occurs, indicates the beginning of a backward search. The vector $\boldsymbol{i}$ is used to control the choice of the most likely branch extension or the alternative. When a node is reached for the first time, the most likely branch extension is pursued, but in a backward search it is necessary to examine the alternatives. The vector $\boldsymbol{i}$ is initially set to the all-zeros vector, and when $i(l) = 0$, the path extension with the larger metric is used at the lth node; when $i(l) = 1$, the alternative branch is followed.

The other variables shown in the flow chart are defined as follows:

$$\Lambda(l) \triangleq \text{path metric to a node at depth } l \text{ in the code tree}$$

$$T \triangleq \text{current threshold value}$$

$$\Delta T \triangleq \text{spacing between thresholds (a constant)}$$

$$\lambda_{i(l)}(l+1) \triangleq \text{the metric of the } i(l)\text{th extension to the node at depth } l+1 \text{ in the code tree}$$

$$\Lambda(0) \triangleq \text{initial threshold value (usually zero)}$$

The key to understanding the dynamic behavior of the Fano algorithm is to note that once a backward search has begun, the decoder reexamines the sequence of nodes along the initially favored path whose metrics lie in the interval $(T, T + \Delta T)$. The alternative paths leaving these nodes are pursued in new attempts to move forward. If eventually all such alternatives have metrics that fall below the threshold, or if none exists, the threshold is lowered by ΔT and forward progress is attempted along the alternative paths that begin at the originally favored nodes with metrics in the new interval $(T, T + \Delta T)$ which is the old $(T - \Delta T, T)$. This process is continued, lowering the threshold by ΔT and searching for alternatives originating in the interval $(T, T + \Delta T)$, until a path is found on which threshold violations cease.

It is instructive to work through an example of the dynamics of the Fano algorithm, and so in Fig. 10.7 we have shown the values of some alternative branch metrics, connected by dashed lines, departing from the correct path. Note that the incorrect path extension from the node at depth 2 in the code tree has a larger metric than the correct extension, which forces the decoder to depart from the correct path initially. The next extension from the originally favored path to depth 4, however, falls below the current threshold ($2\Delta T$), and the decoder enters a backward search. The subsequent sequence of events is described, with reference to the flow chart in Fig. 10.8, as follows:

1. The flag is set to indicate that a threshold violation has occurred (point **a** on the flow chart).
2. The decoder backs up one node (point **b**) and determines that the previous path metric is above the threshold (**c**).
3. The decoder then tests to see if there is an alternative path available (**d**), and, since there is, $i(2)$ is set to 1 so that the alternative path will be pursued (**e**).
4. Next it is observed that the path metric to be extended lies in the interval $(T, T + \Delta T)$, and the flag remains set.

However, the alternative path metric extended from node 2 also fails the threshold and the sequence of steps listed below results:

Step in Flow Chart	Outcome
a $1 \rightarrow F$	Flag remains set.
b $1 \rightarrow l$	Back up to node 1.
c $\Lambda(1) > T$?	No.
h $T - \Delta T \rightarrow T$	Lower threshold by ΔT, now $T = \Delta T$.

Having just lowered the threshold and set the flag, the decoder is prepared to retrace the initially favored path, and the following sequence occurs:

Step in Flow Chart	Outcome
i $\Lambda(2) = \Lambda(1) + \lambda_{i(1)}(2)$	$i(1) = 0$, so the more likely extension from node 1 to 2 is used
j $\Lambda(2) \geq T$?	Yes
k $\Lambda(2) \geq T + \Delta T$?	Yes.
l $F = 0$?	No, so threshold is not increased.
m set $2 \rightarrow l$, $0 \rightarrow i(2)$	Prepare to move forward again.
i $\Lambda(3) = \Lambda(2) + \lambda_0(3)$	Initially follow the better path extension.
j $\Lambda(3) \geq T$?	Yes.
k $\Lambda(3) \geq T + \Delta T$?	Yes.
l $F = 0$?	No, so threshold is not tightened.
m $3 \rightarrow l$, $0 \rightarrow i(3)$	Prepare to move forward again.

Note that in the steps just examined, the path metrics $\Lambda(2)$ and $\Lambda(3)$ were larger than $T + \Delta T$. Therefore the flag remains set, and the decoder recognizes that a previously examined path is being retraced. In the next phase of the search a new node will be encountered, and we observe the following:

Step in Flow Chart		Outcome
i	$\Lambda(4) = \Lambda(3) + \lambda_0(4)$	Follow the better path extension.
j	$\Lambda(4) \geq T?$	Yes.
k	$\Lambda(4) \geq T + \Delta T?$	No.
n	$0 \rightarrow F$	Flag is reset.
m	$4 \rightarrow l$	Prepare to move forward.
	$0 \rightarrow i(4)$	

Since the flag is now off, new forward progress is attempted. However, on the very next move, the threshold is violated, the flag goes back on, we back up one node, and the search continues. We have $l = 3$, $F = 1$, $i(3) = i(2) = 0$, $T = \Delta T$ and are at point **c** in the flow chart. We observe:

Step in Flow Chart		Outcome
c	$\Lambda(3) \geq T?$	Yes.
d	$i(3) = 1?$	No, so there is an alternative to consider.
e	$1 \rightarrow i(3)$	
f	$\Lambda(3) \geq T + \Delta T?$	Yes, so the flag stays on.
i	$\Lambda(4) = \Lambda(3) + \lambda_1(4)$	Take the alternative path.
j	$\Lambda(4) \geq T?$	No.
a	$1 \rightarrow F$	
b	$2 \rightarrow l$	Back up again.
c	$\Lambda(2) \geq T?$	Yes.
d	$i(2) = 1?$	No, so there is an alternative.
e	$1 \rightarrow i(2)$	
f	$\Lambda(2) \geq T + \Delta T?$	Yes, flag stays on.
i	$\Lambda(3) = \Lambda(2) + \lambda_1(3)$	Take the alternative (correct) path.
j	$\Lambda(3) \geq T?$	Yes.
k	$\Lambda(3) \geq T + \Delta T?$	No, so flag is reset to zero. We are about to encounter a new node for the first time.
n	$0 \rightarrow F$	

Note that at step **n** the decoder has returned to the correct path, the flag is off, and forward progress can resume, with the threshold being tightened whenever possible.

This moderately involved search was caused by a small dip in the correct path metric between nodes 2 and 3. The search was somewhat involved not only because the correct path metric fell below an incorrect extension but also because the correct path metric violated a threshold. If the correct path metric had not violated a threshold, the decoder would have quickly returned to the correct path. However, if several consecutive thresholds are violated, the search can become very complex.

Although in this chapter we are concerned only with $b = 1$, rate-$1/V$ convolutional codes, it is a simple matter to modify the flow chart for the Fano decoder to accommodate the more general case of rate-b/V codes. With rate-b/V codes there are 2^b branches leaving each node and thus $2^b - 1$ alternatives to consider at each node in a backward search. We must then let $i(l)$ count from 0 to $2^b - 1$, and we need only modify the question at point **d** and insert a counter at point **e** in the flow chart. For the more general case, the question should be: "$i(l) = 2^b - 1$?" and the counter, "$i(l) + 1 \rightarrow i(l)$."

To minimize the computational load, the various parameters associated with the Fano decoder must be chosen carefully. For example, if the threshold spacing ΔT is too small, each dip in the correct path metric will force the decoder into a backward search, and many of these searches would be premature. On the other hand, if ΔT is too large, the decoder might not initiate a backward search until after many consecutive incorrect decisions had been made, and the resulting search would be unnecessarily long. Thus we can expect an optimum value for ΔT to exist. Its determination as well as the evaluation of the best values for other decoder parameters are easily handled experimentally.

10.5. SOME FURTHER DESIGN ISSUES AND PERFORMANCE RESULTS

In this section we consider the design of a sophisticated soft-decision Fano decoder for a powerful convolutional code. Our aim is to show the performance achievable with a complex decoder that can be realized with current technology. A processor implementation is therefore assumed.

In order to provide a large performance gain, we use a relatively low-rate code. In practice the code rate is determined by bandwidth restrictions and limitations on modem complexity, and we assume here that these constraints are not severe. A rate-$1/6$ code is considered as the example, and the nominal decoder design point will thus be $E_s/N_0 = -6.1$ dB (see Table 10.1).

It should be apparent from the earlier discussion of the Fano algorithm that the complexity of the decoder is only weakly dependent on the constraint length of the code. The constraint length affects only the encoding portion of the decoder, and as k is increased the complexity of the encoder is increased modestly. Thus long-constraint-length codes can be used with Fano decoding, and since the probability of decoding error is reduced exponentially as the constraint length is increased, very low output error rates can be provided. This is an important advantage of sequential decoding relative to the Viterbi algorithm.

For the example at hand, we pick a code with constraint length 64, which is sufficiently powerful to satisfy stringent requirements on the probability of incorrect decoding. The code tap connections are listed in octal notation in

TABLE 10.8. Generators in Octal Notation for the Rate-1 / 6, $k = 64$, $b = 1$ Convolutional Code

G_1	1746146627377550472755
G_2	1034404433262145223742
G_3	1504772236241044721165
G_4	1653226065047015140426
G_5	1123422125726014134303
G_6	1323730145241274230233

Table 10.8. In designing the code, the first five tap connections were chosen to maximize the minimum distance over the first five code branches, and the remainder of the connections were chosen randomly. Another advantage of using long-constraint-length codes is that the code design need not be optimized to provide good performance with sequential decoding.

To assure a small quantization loss, a 32-level optimum quantizer is assumed initially. As we have seen, R_{COMP} for optimum 32-level quantization is quite close to the code rate (Table 10.7). In addition, in order to avoid degradation due to limited arithmetic precision, a large metric range is assumed. Specifically, we use a nine-bit metric range.

Several authors have obtained results that provide guidance on the selection of the metric bias term U. Fano argued intuitively that the best value to use for the bias is the code rate, and Massey [105] subsequently showed that setting $U = R$ minimizes the probability of decoding error. Gallager [49] has shown that a bound on the number of computations to decode a bit is minimized when the metric bias is set equal to R_{COMP}. However, if the number of quantization levels is large and the quantizer is implemented properly, we have $R_{\text{COMP}} \approx R$. We thus use $U = 1/6$ and simultaneously minimize the probability of error and the computational load.

At this point we have, at least tentatively, specified all the parameters of a basic decoder design except the threshold spacing ΔT, which can easily be adjusted experimentally. The performance of the Fano decoder has been studied with the aid of simulation experiments that made use of finite-length messages. The experimental approach was suggested originally in a paper on the computational problem by Jacobs and Berlekamp [74], who postulated the existence of a benevolent magic genie as part of the decoder. The role of the genie was to inform the decoder when it reached the nth node in the code tree correctly for the first time and to send it back to search for the correct message when the path to the nth node was decoded incorrectly. The authors then showed that the distribution of computations for successful decoding of an n-bit message was the Pareto distribution that was described previously (Section 10.2). However, it turns out that this particular magic genie can actually be implemented! For example, if $k - 1$ tail bits, k being reasonably large, are appended to an n-bit message, and if the decoder uses the a priori knowledge of the tail-bit values while decoding the tail, there is essentially no chance that the decoder will reach the end of the tree having decoded the message incorrectly. This is a consequence of the fact that the probability of incorrect decoding can be made extremely small for sequential decoding of long-constraint-length codes. A key assumption in this discussion is that the beginning and end of the message are known.

The simulation experiments that we describe made use of a 500-bit message and 63 tail bits. The decoder was permitted to execute up to 500 computations per information bit (500^2 total computations) while decoding a message. When the computational limit was exceeded, the decoder simply quit, announced computational overflow, and went on to the next message. Computations in the tail were not counted so that the effect of the genie would not be distorted.

Performance results obtained with use of various values for the Fano threshold spacing are indicated in Table 10.9. We show the average number of computations per information bit required to decode the message, $\overline{C}$, for several values of $\Delta T/\mu_B$, where μ_B is the mean of the correct branch metric at the design point. The SNR assumed is $E_s/N_0 = -6.1$ dB. Note that for the data shown, $\overline{C}$ is minimum when the ratio of ΔT to μ_B is between 5.3 and 6.2. Generally it has been observed that the optimum value of ΔT is between $5\mu_B$ and $6\mu_B$. The probability of exceeding 500 computations per information bit is also shown in Table 10.9. Note that the asymptotic behavior of the decoder also degrades when ΔT is not optimum. However, performance is clearly not sensitive to small changes in ΔT in the vicinity of the optimum value.

The computational distribution that was measured with the soft-decision Fano decoder with $E_s/N_0 = -6.1$ dB is shown in Fig. 10.9. Note that the distribution does indeed represent a Pareto random variable for large values of X. That is, a linear relationship is evident in the plot of $\log P(C > X)$ as a function of $\log X$. In addition, the slope for large X is -1, the value of the Pareto exponent at computational cutoff. This shows again that the quantization loss is small with the 32-level optimum quantizer, since $E_s/N_0 = -6.1$ dB is the SNR corresponding to computational cutoff for the unquantized channel.

With an interest in examining the effects of simplifying the decoder quantizer, the decoder was also operated at the design point, $E_s/N_0 = -6.1$ dB, with

TABLE 10.9. Average Number of Computations to Decode a 500-Bit Message, $\overline{C}$, and the Probability of Exceeding 500 Computations per Bit as a Function of $\Delta T/\mu_B$; $E_s/N_0 = -6.1$ dB

$\Delta T/\mu_B$	$\overline{C}$	$P(C > 500)$
1.8	8.6	1.7×10^{-3}
3.5	6.0	1.0×10^{-3}
4.4	5.6	8.6×10^{-4}
5.3	5.4	8.6×10^{-4}
6.2	5.4	8.6×10^{-4}
7.1	5.5	9.2×10^{-4}
8.8	5.8	9.8×10^{-4}
10.6	6.8	1.3×10^{-3}

$Q = 4$-, $Q = 8$-, and $Q = 16$-level optimum quantizers. The computational distributions that were measured are shown in Fig. 10.10, and the associated values of the mean number of decoder computations are listed in Table 10.10. Note that the performance data achieved with the 16- and 32-level quantizers are similar, but the 16-level quantizer results in a decoder that is 7 percent slower. The degradations associated with the eight- and four-level quantizers are more severe.

As an alternative to the optimum decoder quantizers considered, for simplicity one might choose to implement a quantizer with uniform intervals. For a uniform quantizer, we have $I_0 = -\infty$, $I_Q = \infty$, and

$$I_j - I_{j-1} = \Delta, \qquad j = 2, 3, 4, \ldots, Q - 1$$

The value of Δ can be varied to maximize R_{COMP}, and the values obtained for the cutoff rate for the uniform discrete channels are, in fact, quite close to the values observed with the optimum quantizers. With $Q = 2$ and $Q = 4$, the

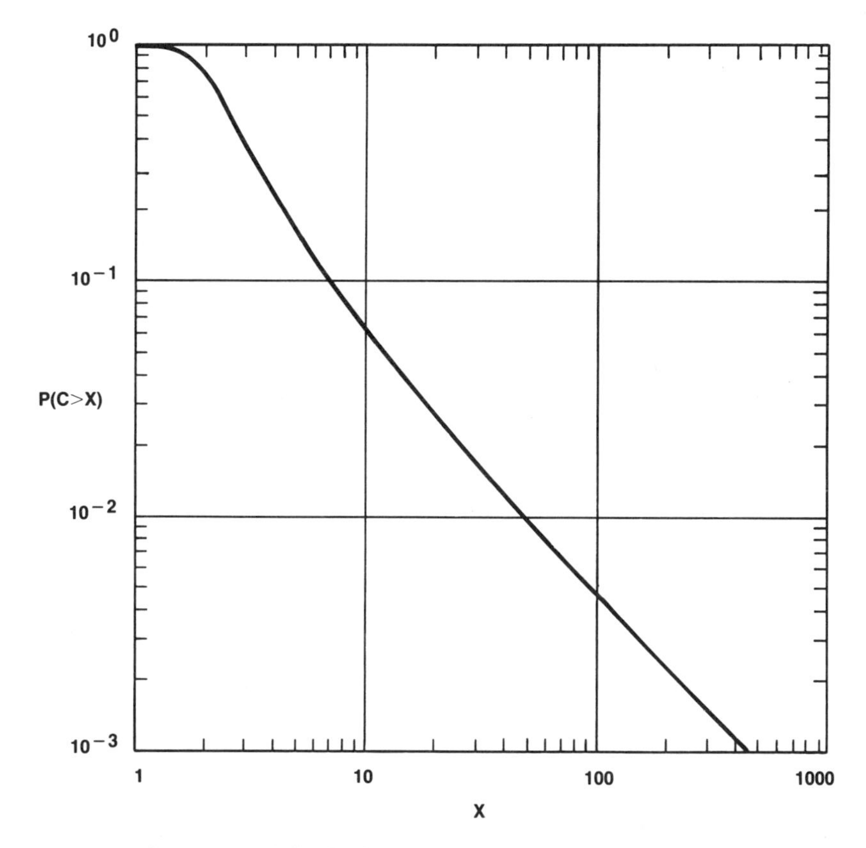

FIGURE 10.9. Computational distribution for baseline sequential decoder design; $E_s/N_0 = -6.1$ dB.

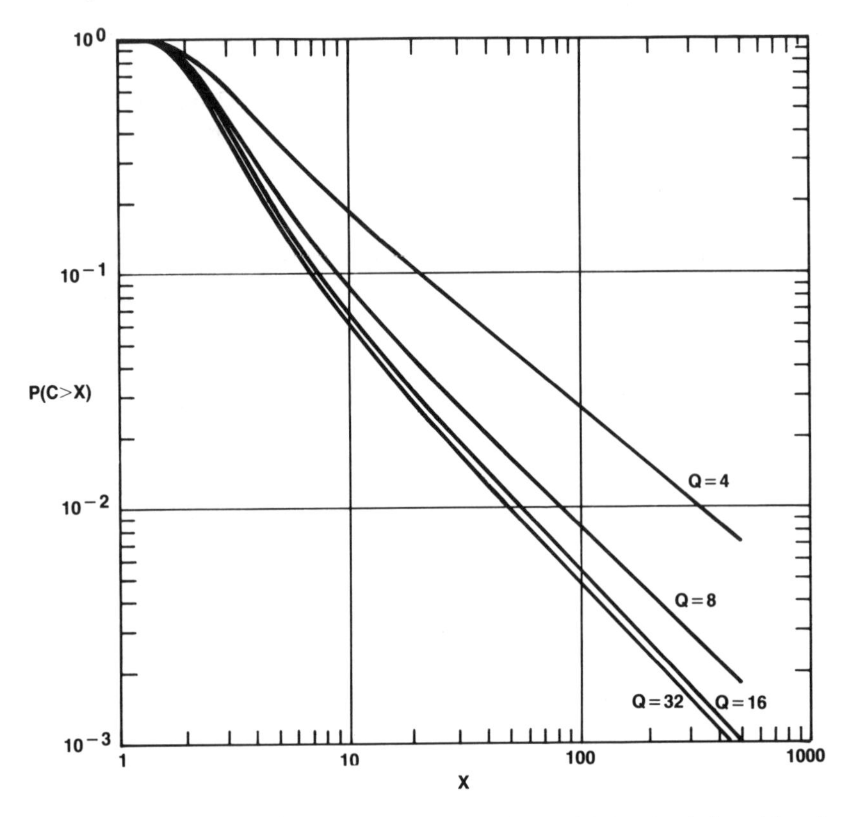

FIGURE 10.10. Computational distributions for baseline sequential decoder design with various levels of optimum quantization; $E_s/N_0 = -6.1$ dB.

uniform and optimum quantizers are identical. The performance of the Fano decoder with uniform quantization at the nominal design point was found to be essentially identical to the performance observed with the optimum intervals. Change of the decoder quantizer, of course, necessitates redesign of the decoding metrics and determination of a new value for the threshold spacing ΔT.

The effects of changing the assumed values for the metric range and bias are also of interest. Use of the 32-level optimum quantizer is assumed here. The metric range (nine bits) was intentionally chosen to be large so that performance

Table 10.10. Average Number of Decoder Computations per Information Bit for Various Q-Level Optimum Quantizers; $E_s/N_0 = -6.1$ dB

Q	$\bar{C}$
4	15.1
8	7.1
16	5.8
32	5.4

would not be inadvertently compromised. Other experiments showed that the choice of nine bits was indeed generous, since decoder performance did not degrade appreciably until the metric range was reduced to five bits. With a five-bit metric range, the average number of decoder computations increased about 10 percent. Small changes in the metric bias, though, resulted in more rapid performance degradation. For example, changing the bias by ± 10 percent increased the average computational load by 10 to 20 percent.

10.6. PERFORMANCE AS A FUNCTION OF SNR

Thus far we have considered the performance of the soft-decision Fano decoder at only one SNR, the design point $E_s/N_0 = -6.1$ dB. In this section we present performance results for somewhat larger and smaller SNRs, corresponding to operation both above and below computational cutoff. Use of a 32-level optimum quantizer, a nine-bit metric range, bias equal to $1/6$, and a 500-bit message is assumed.

For proper operation of a sequential decoder over a range of SNRs, an *automatic gain control* (AGC) must be implemented prior to the decoder quantizer. We will assume use of an AGC that scales the matched-filter outputs by g, where

$$g = \sqrt{\frac{(2E_s + N_0)_D}{2E_s + N_0}}$$

and where $\frac{1}{2}(2E_s + N_0)_D$ is the signal-plus-noise energy at the system design point and $\frac{1}{2}(2E_s + N_0)$ is the received signal-plus-noise energy.

The decoder computational distributions measured at SNRs that correspond to $R_{\text{COMP}} + 1$ dB to $R_{\text{COMP}} - 2$ dB in 0.5 dB increments are shown in Fig. 10.11. Note that each distribution is asymptotically Paretian and the slopes for SNRs between computational cutoff and channel capacity (R_{COMP} to $R_{\text{COMP}} - 3$ dB) are in good agreement with analytical predictions, found from Eq. (10.1). Note also that the slope of the computational distribution approaches zero quite rapidly as E_s/N_0 is reduced from -6.1 dB. Thus, it is feasible to operate a sequential decoder only at data rates near or slightly above R_{COMP}.

The results shown in Fig. 10.11 can be used to determine the performance of the soft-decision Fano decoder as a function of E_b/N_0. Since tail bits represent additional system overhead, E_b/N_0 depends on the specific number of tail bits used. It has been observed that for long-constraint-length codes it is not necessary to use $k - 1$ tail bits to assure a very low probability of incorrect decoding. If a shorter tail is used, information bits toward the end of the message are somewhat less reliable than those transmitted elsewhere in the message. Nonetheless, the soft-decision sequential decoder for the $k = 64$ code under consideration provides a very low incorrect message decoding probability (less than 10^{-8}) with 30 tail bits [149]. We have then

$$\frac{E_b}{N_0} \text{ dB} = \frac{E_s}{N_0} \text{ dB} + 10 \log \frac{6(500 + 30)}{500}$$

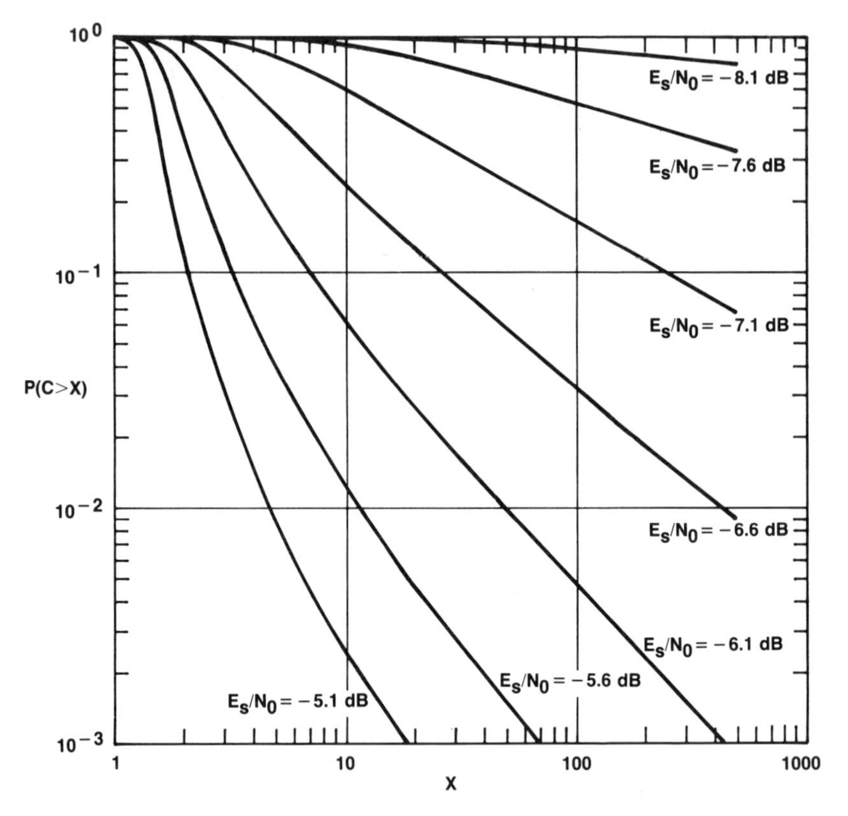

FIGURE 10.11. Computational distributions for baseline sequential decoder design and various signal-to-noise ratios near computational cutoff.

for the $k = 64$, rate-$1/6$ code utilizing 30 tail bits. The nominal decoder design point is thus $E_b/N_0 = 1.9$ dB.

If we now let C_{max} be the maximum number of decoder computations permitted in decoding an information bit, the results given in Fig. 10.11 can be used to obtain $1 - P_{\text{CD}}$ as a function of E_b/N_0, with C_{max} a parameter. For example, with $C_{\text{max}} = 10$ and $E_b/N_0 = 1.9$ dB, we have

$$P(C > 10) = 6.0 \times 10^{-2}$$

In addition,

$$\begin{aligned} 1 - P_{\text{CD}} &= P(C > C_{\text{max}}) + P_{\text{ICD}} \\ &\approx P(C > C_{\text{max}}) \end{aligned}$$

since the probability of incorrect decoding is small.

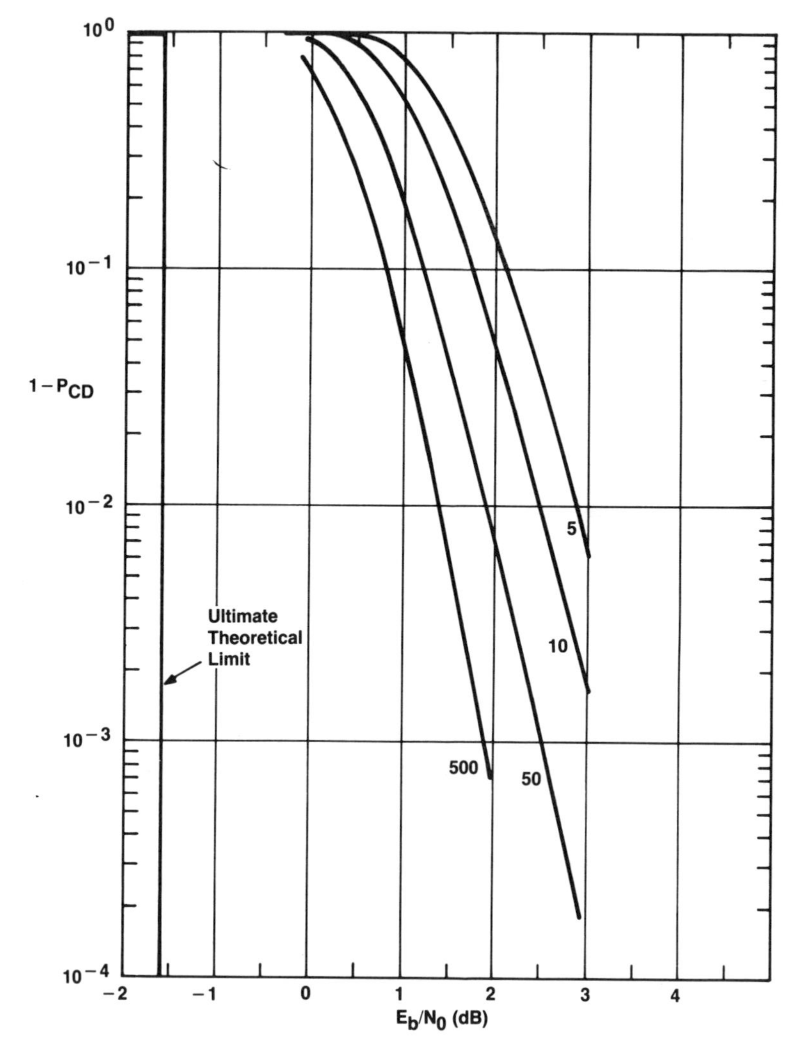

FIGURE 10.12. $1 - P_{CD}$ vs. E_b/N_0 for the rate-1/6 code and baseline sequential decoder design with 5, 10, 50, and 500 computations per bit allowed.

Figure 10.12 shows $1 - P_{CD}$ as a function of E_b/N_0 for the soft-decision Fano decoder with $C_{max} = 5$, 10, 50, and 500 computations per information bit. Note that by increasing C_{max} from 5 to 50, a 1.0-dB gain in communication efficiency results, but only 0.5 dB is obtained by increasing C_{max} from 50 to 500. A further increase in C_{max} would provide marginal performance enhancements. Finally, we point out that for large C_{max} the performance provided by the soft-decision Fano sequential decoder is within 3 to 4 dB of the ultimate design goal described in Section 10.1, that is, operation within about 3 dB of channel capacity is achieved.

10.7. A BRIEF DESCRIPTION OF A HARD-DECISION FANO DECODER DESIGN

The soft-decision Fano decoder described in the previous section represents perhaps the ultimate in performance that can currently be achieved with sequential decoding. The decoder is complex and to date has been suitable only for processor implementation, which restricts operation to data rates in the low kilobit per second region. Other implementation approaches have been proposed, and some have led to hard-wired designs capable of operating at data rates of several megabits per second while providing more modest performance gains. In this section we briefly describe one such design approach due to Forney, et al. [44, 45].

The complexity of a sequential decoder can be reduced considerably by using a systematic rate-1/2 code and hard-decision decoding. To provide comparable performance, the constraint length for a systematic code must be larger than that for a nonsystematic code, but this is a minor disadvantage with Fano decoding. Hard-decision decoding permits use of a less complex arithmetic unit.

The messages to be considered here are assumed to be arbitrarily long, and thus no consideration is given to the use of tail bits. In this case a data buffer must be provided to accommodate the variability in decoding effort associated with sequential decoding. Most of the time the decoder can be expected to keep up with the incoming data and the buffer will be nearly empty. The buffer should be large enough that when a rare difficult decoding condition exists, incoming data can be held until the decoder has worked its way through the noisy data segment. However, the buffer cannot be made sufficiently large to assure that the probability of overflow is negligibly small, since the Pareto computational distribution decreases slowly for large X. Thus, buffer overflow is analogous to exceeding the computational limit for the decoder considered in the previous section. In this case, though, provision must be made to restart the decoder without use of tail bits each time an excessive computational load results in overflow.

The decoder described by Forney uses a constant delay between input data and output bit decisions. If the decoder is stalled in a backward search when the time comes to make an output decision on a given bit, buffer overflow is declared and a resynchronization attempt begins. In the resynchronization mode the decoder simply moves forward in the code tree, away from the locally severe noise, and $k - 1$ consecutive received information bits are used to initialize the state of the decoder. This way of restarting the decoder is relatively simple, given the use of a systematic code.

If the $k - 1$ consecutive received information bits are all correct, resynchronization will be achieved and decoding can proceed. For a channel with bit error rate p, the probability of $k - 1$ correct received bits is $(1 - p)^{k-1}$, and one might expect to have to try $(1 - p)^{-(k-1)}$ times, on the average, to successfully restart the decoder. However, it is not necessary that all

TABLE 10.11. E_b/N_0 **Required for Coherent PSK to Provide** $P_b = 10^{-5}$ **with Hard-Decision Sequential Decoding for Several Computational Speed Factors** μ. **Code Rate = 1/2**

μ	E_b/N_0 (dB)
4	5.3
8	4.9
16	4.7
32	4.6
64	4.5

$k - 1$ bits be correct if the decoder is permitted to move backward or forward from the new starting point. That is, if there are just a few errors, the decoder can correct them if permitted. This was verified experimentally by Forney.

It is assumed that when buffer overflow occurs, the decoder uses the sequence of information bits as received for the output bit decisions. Thus post-decoding errors can result from either buffer overflow or undetected error events. An undetected error occurs when the decoder has accepted a string of incorrect bit decisions and yet manages to get back on the correct path before overflow occurs. For the hard-decision decoder considered, the probability of undetected error can be nonnegligible.

The probability of undetected error can be reduced by increasing the magnitude of the metric bias U. However, when the magnitude of the bias is increased, the computational load is increased, and performance can degrade due to a higher rate of buffer overflow. It was found that overall decoder performance is optimized when the bias is adjusted so that the probability of undetected error is not negligibly small.

The key parameter that limits performance of a hard-decision sequential decoder is the *computational speed factor* μ. Defined as the number of decoder computations that can be executed in the time interval required to transmit $V = 2$ channel bits, μ represents the ratio of the decoder speed to the information transfer rate. Clearly, as μ is increased, performance improves but the decoder becomes more complex. This can be seen from Table 10.11, where the SNRs that are required to provide a 10^{-5} post-decoding bit-error rate are shown for several values of μ. Coherent PSK is assumed, and we note that an uncoded transmission requires $E_b/N_0 = 9.6$ dB to provide a bit-error rate of 10^{-5}. Thus for $\mu = 64$, a performance improvement or coding gain of 5.1 dB is achieved.

10.8. THE STACK ALGORITHM FOR SEQUENTIAL DECODING

We conclude the discussion of sequential decoding with a brief description of an algorithm proposed independently by Zigangirov and Jelinek [75, 188]. Its major advantage and disadvantage are pointed out.

The *Zigangirov-Jelinek* (ZJ) decoder is conceptually simpler than the Fano decoder. The basic decoding computations for the two algorithms are similar. That is, for extension of a node in the code tree, the metrics associated with the zero and one path extension are computed. However, at this point the ZJ decoder forms the metrics for both extensions and stores the information on the new paths in a portion of memory called the *stack*. Each stack entry contains all the information necessary to resume decoding from the node in question. Entries stored in the stack are ordered by path metric value, and the best path metric observed during the decoding process is located at the top of the stack. The path chosen for extension in a decoding computation is the best path, the path at the top of the stack. The ZJ decoder is free to jump between widely separated nodes in the code tree in successive computations, while the Fano decoder is constrained to move forward or backward one node at a time. Use of the Fano path metric is assumed.

The ZJ decoding algorithm can thus be described by the following steps:

1. Initialize (clear memory, and create one stack entry corresponding to the root of the code tree).
2. Obtain the path in the stack that has the largest metric.
3. Extend the best path, and store the resulting new paths in their appropriate places in the stack.
4. Return to step 2.

There are clearly several implementation problems suggested by this brief description of the ZJ decoder. First, it has been assumed that there will always be room to insert the new path entries in the stack after each node extension. However, for any finite stack size, we can expect to encounter noisy segments of received data that will cause the decoder to examine a very large number of paths and the stack will become full. Therefore, step 3 must be modified so that when the stack is full, room is created by deleting old entries. The most appropriate entries to delete are clearly the worst, that is, those with the lowest path metric values. However, if the correct path is ever deleted from the stack, the decoder cannot recover.

In addition, while the notion of a perfectly ordered stack is helpful in describing the algorithm, its implementation is impractical. Instead, Jelinek suggested that the stack be segmented into a sequence of bins that are used to store paths having metrics within specified ranges. That is, bin j would be used to store all paths with metric values in the interval $(\Lambda_j, \Lambda_j + \Delta\Lambda)$. A constant bin width $\Delta\Lambda$ is employed. The bins are ordered by path metric value, but the entries within a bin are not ordered. Step 2 is modified so that the decoder retrieves an entry from the highest non-empty bin for extension. Any convenient path in that bin may be used, for example, the newest or oldest entry.

Numerous tricks can be employed in a ZJ decoder implementation. For example, if a path is extended and the metric associated with the preferred path has increased, there is no point in storing the new path in the stack, since it can be used immediately for the next extension. Even when the path metric decreases slightly, if the new path belongs in the highest non-empty bin, it can be used for the next extension. In addition, large savings in memory can be realized by use of high-order data structures in the stack implementation [149].

Upon its discovery, it was thought that the ZJ algorithm would be faster than the Fano but would require more memory. Since the ZJ decoder may jump between widely separated nodes in the code tree, it is better suited for decoding very noisy segments of data. This has proven to be true, but the gain in speed is relatively small considering the cost in additional memory. For example, performance comparable to that obtained with the soft-decision Fano decoder for the rate-1/6 code can be obtained with a ZJ decoder that is permitted to execute about half the number of decoder computations per message. However, the ZJ computations are more complex, requiring stores and calls from the stack, and the resulting overall speed advantage is only about 1.5:1. The ZJ decoder, on the other hand, requires 10 to 20 times more memory than the Fano.

10.9. NOTES

Much of the early work on sequential decoding was done at MIT Lincoln Laboratory. The original Wozencraft algorithm was extended by Reiffen [187], and the first Fano decoder was implemented by G. Blustein at Lincoln Laboratory [15]. It should be noted that Blustein also implemented the soft-decision Fano decoder that was used to obtain most of the performance results reported here and J. Dickson provided significant support as well. Jordan studied the performance that can be achieved with sequential decoding on channels with orthogonal signaling and noncoherent detection and gave performance results for the Fano algorithm [78]. A detailed design of a real-time processor implementation of the ZJ decoder has been reported by Richer [149], and we have followed Richer closely in portions of our presentation. Contributions have also been made by J. K. Wolf, who suggested the presentation of Massey's algorithm for finding the optimum quantizer, and S. Cheng, who computed the optimum quantization intervals. L. N. Lee [89] has generalized Massey's procedure to the case of M-ary signaling and found optimum decision regions in likelihood space. Lists of good moderate constraint length codes for use with sequential decoding can be found in reference 94.

■ ■ CHAPTER ELEVEN

Applications of Error-Control Coding

Thus far in this book, error-control coding techniques have been considered individually. The structure of particular codes as well as encoding and decoding algorithms have been described for the more important techniques. In Chapter 7, performance results for linear block codes with bounded-distance decoding were presented for both binary and nonbinary codes, and results for feedback, Viterbi, and sequential decoding of convolutional codes were given in Chapters 8 to 10. In this chapter we consider the larger problem of how one goes about selecting an error-control technique given a particular performance objective or requirement.

There is, of course, no single class of codes or error-control technique that is "best," that is, suitable for all applications, since communication channels and performance requirements vary widely. Furthermore, there is no commonly accepted single figure of merit upon which a final design decision can be made for an arbitrary application. We have observed that each error-control technique has its own set of features that may or may not be useful in a given application. The selection of an error-control scheme consists in part of matching the features of a technique with the system objectives being addressed.

We first consider modulation and coding techniques for coherent detection on the AWGN channel. We consider the use of coding with two representative types of modulation, binary PSK and M-ary FSK. The presentation is organized to address three broad ranges of performance objectives: highly efficient designs providing large coding gains (> 8 dB), moderately efficient designs providing moderate coding gains (4 to 7 dB), and designs that can offer only small improvements (1 to 3 dB) over uncoded operation. In many cases, error-control coding has been used as an applique to provide low to moderate improvements in the performance of already existing communication systems.

In other cases, where overall communication efficiency has been judged to be very important, the selection of modulation and coding techniques have been treated together in an integrated design approach. Therefore, in portions of the discussion that follows, we consider alternatives in the choice of modulation type as well as coding technique.

We also consider two typical ways that a user may choose to specify the performance of a coded communication system. The more common approach is simply to specify the post-decoding bit-error rate as a function of SNR, which is a natural extension of a modem specification. When this approach is used, one is generally not permitted to ignore post-decoding errors that result from decoding-failure events. In this case, the results presented in Chapter 7, which include only the errors associated with incorrect decoding, must be modified to include the errors that occur when the decoder detects uncorrectable error patterns.

An alternative way to specify communication system performance, which is becoming more common, imposes requirements at the message level rather than the bit level. For example, the user may specify the probability of correct message reception as a function of SNR. When communication performance is stated in this way, one generally is also required to assure that the probability of incorrect decoding is below some small specified value for all SNRs. This approach is typically used when the penalty for accepting incorrect data is high, and we assume in this case that detected but uncorrected errors can be ignored by the system. That is, messages that have been recognized as erroneous may be thrown away. We shall see in the following sections that in addressing both types of communication performance requirements, it is useful to consider combining two or more quite different error-control techniques in a single integrated design.

A variation of the message-level specification that we do not consider is often used in defining the quality of service for commercial data transmission systems. Typically, the requirement is stated in terms of the percentage of *error-free minutes* provided. However, it should be clear to the reader how this type of specification is addressed, given the results presented here.

In subsequent discussions *coding gain* is used to compare coded and uncoded system performance. Regardless of the figure of merit used, one can determine the SNR necessary to provide the performance desired. Coding gain is then simply the difference in the E_b/N_0 required for coded and uncoded operation to provide the specified level of communication performance. By comparing alternatives on the basis of E_b/N_0, the issues of message transmission time and SNR are addressed. For example, on channels where additional bandwidth can be provided a coding gain of 3 dB can be used to reduce the radiated signal power by a factor of 2, to reduce the transmission time by a factor of 2, or to simultaneously reduce the needed signal power and transmission time. However, in many applications it is difficult or impossible to increase the channel bandwidth. If this is the case, coding may still be used productively but the time required to transmit a message is necessarily increased.

For example, suppose we are using uncoded coherent PSK on a channel with $E_b/N_0 = 9$ dB and thus the delivered bit-error rate is 3.4×10^{-5}. In this case E_s/N_0 is also equal to 9 dB. Suppose further that we have found a rate-1/2 coding technique that provides a post-decoding bit-error rate of 3.4×10^{-5} with coherent PSK signaling and $E_b/N_0 = 3$ dB. We thus have a coding gain of 6 dB. Note that for the coded system, $E_s/N_0 = 0$ dB, since a rate-1/2 code is assumed, and that the required radiated signal energy per channel bit is reduced by 9 dB. Then, if it is possible to double the signaling rate in the channel, doubling the channel bandwidth, the time required to transmit a message can be the same for the coded and uncoded systems and an advantage of 6 dB in communication efficiency is achieved. If the signaling rate cannot be increased, use of the rate-1/2 code results in a doubling of the message transmission time but E_s/N_0 may still be reduced by 9 dB. In this case the message transmission time is doubled, E_b/N_0 is 3 dB, the coding gain is again 6 dB, and the required radiated signal power is reduced by 12 dB.

We consider similar design issues for the AWGN channel with noncoherent detection. The modulation types treated are DPSK and noncoherent M-ary FSK. Then coding techniques for communication channels that exhibit a mixture of burst-error and independent-error phenomena are discussed briefly. It is shown that when a feedback channel is available, *automatic repeat request* (ARQ) provides a highly efficient and robust error-control approach. Interleaving of random-error-correcting codes for application on burst-error channels is also described. Where appropriate, performance results are discussed and references cited.

11.1. COHERENT RECEPTION ON THE AWGN CHANNEL

Although seldom done in practice, the proper way to develop an error-controlled communication system is to design the modulation, demodulation, and coding functions together at the outset. However, for the AWGN channel, where coherent detection is feasible, the set of reasonable design choices need not be large. This is certainly the case for the important and representative modulations considered here: binary coherent PSK and M-ary coherent FSK.

It was shown in Chapter 1 that binary coherent PSK provides a 3-dB performance advantage over binary coherent FSK. PSK also has some clear implementation advantages, such as a smaller signaling bandwidth and the use of only one rather than two matched filters. However, for higher-order alphabets, the performance of coherent FSK improves significantly, and in the limit of very large alphabets, operation at channel capacity can theoretically be achieved. This, of course, is not the case for PSK.

Let us say, then, that we are given a coherent channel and the freedom to choose the modulation type as well as the coding technique. One reasonable set of alternatives is to use PSK or M-ary FSK modulation with a coding technique appropriate to the modulation type in each case. As we shall see shortly, from a theoretical point of view, the choice of modulation doesn't matter; that is, highly efficient communication can be provided in either case. Therefore the key issue is, which approach leads to the most economical realization within practical constraints? For M-ary coherent FSK, the modem is more complex than for binary PSK, and a larger signaling bandwidth may be required; however, a simpler error-control scheme can be employed due to the improved performance provided by M-ary FSK modulation itself. Issues of phase coherence time, codeword framing, bandwidth, and symbol synchronization, which vary with the application, must be considered. However, it has generally been true that simpler designs result when a carefully chosen code is applied with binary PSK on the coherent channel. It will be seen subsequently that the situation is somewhat different for the case of noncoherent detection.

11.1.1. High Performance Techniques, $E_b/N_0 \approx 2$ to 3 dB

In this section we consider the error-control techniques that have been developed to provide superior performance in systems using coherent detection. Although certainly not simple, these schemes are not unacceptably complex for some applications. We take as the nominal design point $E_b/N_0 \approx 2$ to 3 dB, which corresponds to operation within 3.5 to 4.5 dB of channel capacity. (Recall from Chapter 1 that the Shannon limit is -1.6 dB.) We assume that in order to achieve such highly efficient communication, large bandwidth expansion will be permitted, which will in turn allow use of low-rate codes.

Permitting use of low-rate codes is the key to obtaining highly efficient communication on the coherent channel. For instance, in Chapter 10 we saw that as one reduces the rate of a convolutional code, the SNR corresponding to operation at computational cutoff decreases monotonically to $E_b/N_0 = 1.4$ dB. This is true also for block-coded systems, although monotonicity does not necessarily apply for a particular class of block codes. The principle of using large bandwidth expansion and low-rate codes to provide efficient communication was discussed in Chapter 1 with reference to Shannon's channel coding theorem.

Soft-decision sequential decoding of long-constraint-length convolutional codes can certainly be regarded as a high-performance error-control technique. However, we first consider two variations of a scheme that use the cascade or *concatenation* of two error-control codes. The first approach to be presented was originally proposed by Forney [37].

11.1.1.1. Concatenated Block Codes. A very powerful low-rate error-control code can be constructed by concatenating two or more relatively simple codes. In Fig. 11.1, an example of the concatenation of two block codes is shown. Information to be transmitted is first encoded with an (n, k) code, which is labeled the *outer code* in the figure. The symbols in an (n, k) code block are then treated as an information stream that is encoded as a sequence of (N, K) *inner code* blocks. Clearly, if the symbol alphabets of the inner and outer codes are not the same, it is necessary to reformat the data between the encoders for the inner and outer codes. At the receive side, the demodulated data is first decoded with a decoder for the inner code, and the symbols released are then decoded with a decoder for the outer code.

Several characteristics of concatenated block codes are evident from Fig. 11.1. First, the resulting overall code block length can be large, much larger than the block length of the inner or outer code, and the resulting code rate can be quite low. Specifically, the block length of the concatenated code is the product of the block lengths of the inner and outer codes, and the rate is the product of the code rates. Thus, by use of concatenation, long low-rate codes are constructed that can be decoded in stages. In general, a concatenated code is not as powerful as the best single-stage code with the same rate and block length, but, since decoding is implemented in stages, decoder complexity is much reduced.

Several design issues are also evident from Fig. 11.1. First we note that the inner and outer codes must be carefully matched. Clearly, the job of the outer code is to correct the errors that remain after decoding of the inner code, errors that may exhibit either independent- or burst-error statistics. In addition, the rate of the outer code must be selected carefully. If the outer code rate is too high, errors leaving the inner decoder cannot be reliably corrected. However, if we fix E_b/N_0 and the inner code, then as we reduce the rate of the outer code, the SNR at the input to the inner decoder, E_s/N_0, is reduced. If the outer code rate is reduced too much, E_s/N_0 is lowered to the point where there is a coding loss rather than a gain associated with the inner code. Thus we can expect an optimum rate to exist for the outer code.

The only codes that have been used as outer codes in powerful concatenation schemes are the RS codes. This is because the RS codes, being maximum-distance-separable codes, make highly efficient use of redundancy, and block lengths and symbol sizes can be readily adjusted to accommodate a wide range of message sizes. RS codes also provide a wide range of code rates that can be chosen to optimize performance. In addition, as we saw in Chapter 6, efficient decoding techniques are available for use with RS codes.

The design issues for selection of the inner code of a concatenated code are somewhat different from those for other applications. First, what is not desired is a very steep performance characteristic, that is, a waterfall curve with a pronounced knee. If such a code were selected as the inner code and a nominal design point chosen slightly to the right of the knee, at an output error rate of,

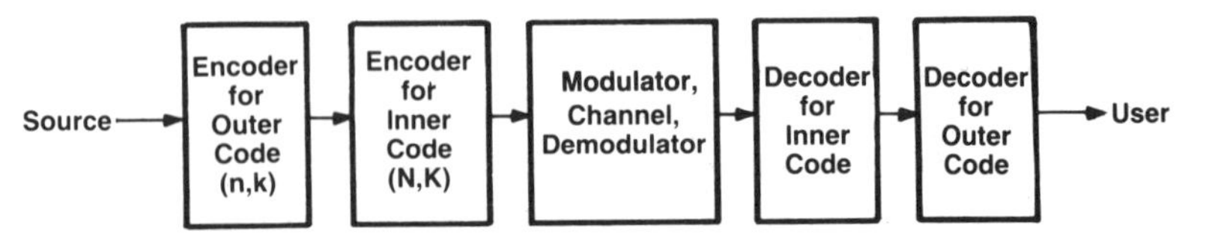

FIGURE 11.1. Concatenated block code using an (n,k) outer code and an (N,K) inner code.

say, 10^{-3}, an inefficient and unstable design would result. In this case the RS code would be designed to improve a 10^{-3} error rate to the overall level of performance required. However, should the channel quality degrade even slightly, the error rate at the input to the RS decoder would increase sharply and the RS decoder would be overloaded with errors. On the other hand, if E_s/N_0 were to increase slightly, the inner code by itself would be capable of providing the desired quality of service at the same or lower E_b/N_0, thus eliminating the need for the outer code. In this case, the outer code causes an overall loss in communication efficiency.

It can be shown that the considerations outlined above lead to the conclusion that the inner code should be designed to provide a modest gain at the system design point, that is, an improvement in error rate of one to two orders of magnitude. Powerful RS codes may then be used to reduce the moderate error rate to whatever performance level is required. The inner codes that best suit this design goal are relatively simple codes that can be decoded with maximum likelihood decoders.

In his original work on concatenated codes, Forney suggested using RS codes as outer codes and biorthogonal codes with maximum likelihood decoding as the inner coding technique. A *biorthogonal code* can be formed by adding an overall parity-check bit to a PN-sequence code and then doubling the number of codewords by including the complements. Therefore m bits of information are conveyed in a block of $N = 2^{m-1}$ bits. In general, a biorthogonal code is any code comprising a set of orthogonal sequences and their complements.

With the scheme suggested by Forney, a block of km information bits is encoded by first segmenting the data into k m-bit subblocks. The subblocks are then treated as k information symbols in a 2^m-ary alphabet and are encoded with an (n,k) RS code. The RS encoder releases n m-bit characters, each of which is encoded with a $(2^{m-1},m)$ biorthogonal code. As we shall see, concatenated codes constructed in this manner can provide excellent performance with moderate decoder complexity.

The performance of biorthogonal codes with maximum likelihood (correlation) decoding was obtained by Viterbi [171] for the case of PSK modulation and coherent detection in additive white Gaussian noise. Viterbi's results, post-decoding 2^m-ary symbol-error rate as a function of E_b/N_0, are shown in Fig. 11.2 for $m = 5$ to 10. Although not realizable, arbitrarily small error probability for all $E_b/N_0 > -1.6$ dB is theoretically provided in the limit of very large m. Similar results were presented in Chapter 1 for coherent M-ary FSK. The two sets of results are, in fact, closely related.

To see that highly efficient signaling can be achieved with the concatenated code design, we consider use of an $m = 6$ biorthogonal code with RS codes defined on a 64-ary alphabet. With $m = 6$, a biorthogonal codeword conveys six bits of information, and the inner and outer code symbol alphabets are conveniently matched. To compute performance results for this scheme, the symbol-error probabilities given in Fig. 11.2 are viewed as the input character-error rates for RS decoding, and E_b/N_0 is adjusted to include the RS code

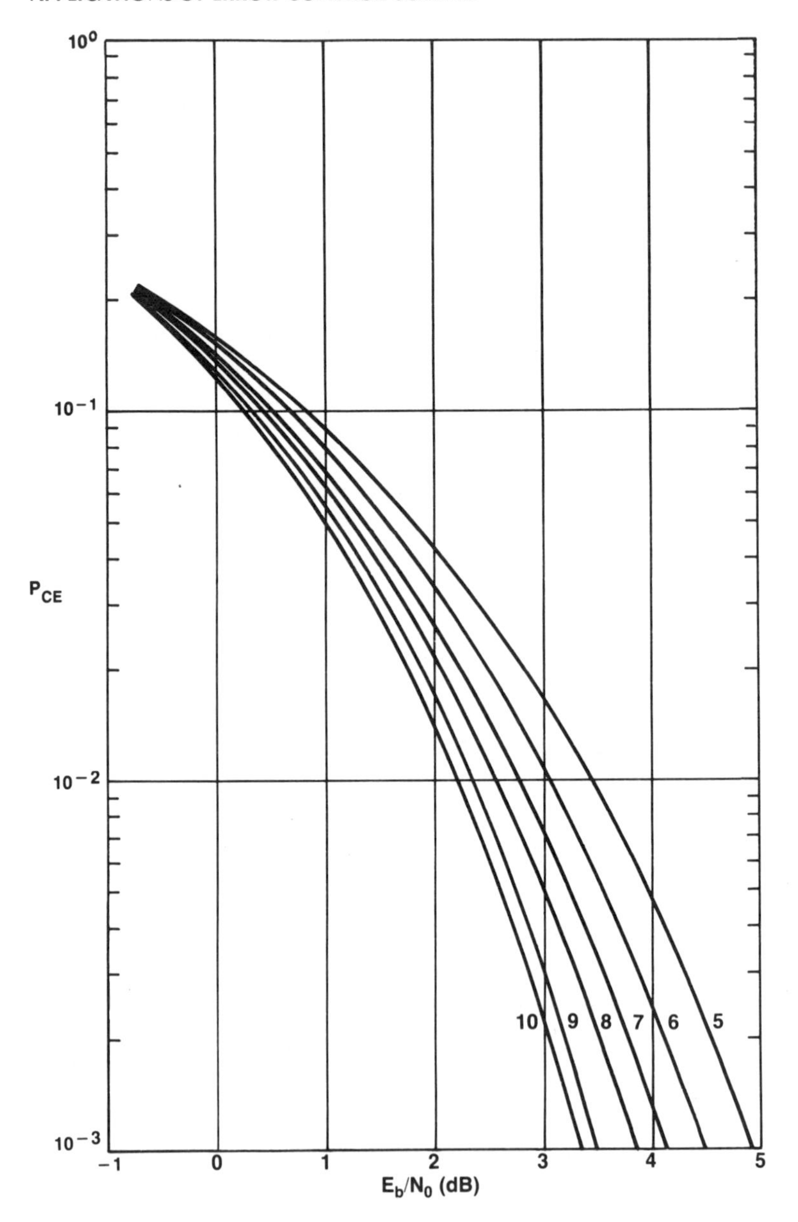

FIGURE 11.2. Performance of biorthogonal codes with correlation decoding. P_{CE} vs. E_b/N_0 for $m = 5$ to 10.

rate. Errors-and-erasures decoding of the RS code can provide enhancements of only a few tenths of a decibel relative to errors-only decoding on the AWGN channel, so we limit the discussion to bounded-distance errors-only decoding of the RS code.

We now estimate the post-RS-decoding bit-error rate including the errors associated with decoding-failure events. We make the conservative assumption that when more than t errors are present in the received word, the decoder introduces t additional errors. For the probability of post-decoding bit error we have, then,

$$P_b \le \frac{2^{m-1}}{2^m - 1} \sum_{i=t+1}^{n} \left[\frac{t+i}{n}\right] \binom{n}{i} P_{\mathrm{CE}}^i (1 - P_{\mathrm{CE}})^{n-i} \tag{11.1}$$

where n is the block length of the RS code and P_{CE} is the character-error rate observed at the output of the decoder for the biorthogonal code. The factor $2^{m-1}/(2^m - 1)$ represents the average number of bit errors per decoded RS character error.

The SNRs required to achieve a post-decoding BER of 10^{-6} obtained from Eq. (11.1) and Fig. 11.2 are given in Table 11.1 for various RS codes defined on $GF(64)$. Note that an optimum rate is observed for the RS code in the vicinity of $R = 0.7$. With $E_b/N_0 = 3.7$ dB, the (63,45) $t = 9$ RS code concatenated with an $m = 6$ biorthogonal code provides $P_b \le 10^{-6}$.

In order to achieve reliable communication with $E_b/N_0 \approx 2$ to 3 dB, it is necessary to use more complex biorthogonal and RS codes, that is, codes with longer block lengths. In Fig. 11.3, performance results are given for $m = 6$, 7, and 8 biorthogonal codes concatenated with particular RS codes defined on $GF(64)$, $GF(128)$, and $GF(256)$, respectively. Note that with $E_b/N_0 = 2.4$ dB, the $m = 8$ biorthogonal code and the (255,179) $t = 38$ RS code provide a post-decoding BER of 10^{-6}. Forney reported the same result for the $m = 9$ biorthogonal code concatenated with the (511,383) $t = 64$ RS code with $E_b/N_0 = 2.0$ dB. We note that for uncoded PSK transmission, $P_b = 10^{-6}$

TABLE 11.1. E_b/N_0 **Required to Provide** $P_b \le 10^{-6}$ **for Various RS Codes Defined on** $GF(64)$ **Concatenated with an** $m = 6$ **Biorthogonal Code**

RS Code Parameters (n,k)		E_b/N_0 (dB)
(63,57)	t=3	4.4
(63,53)	t=5	4.0
(63,49)	t=7	3.8
(63,45)	t=9	3.7
(63,41)	t=11	3.7
(63,37)	t=13	3.9
(63,33)	t=15	4.0

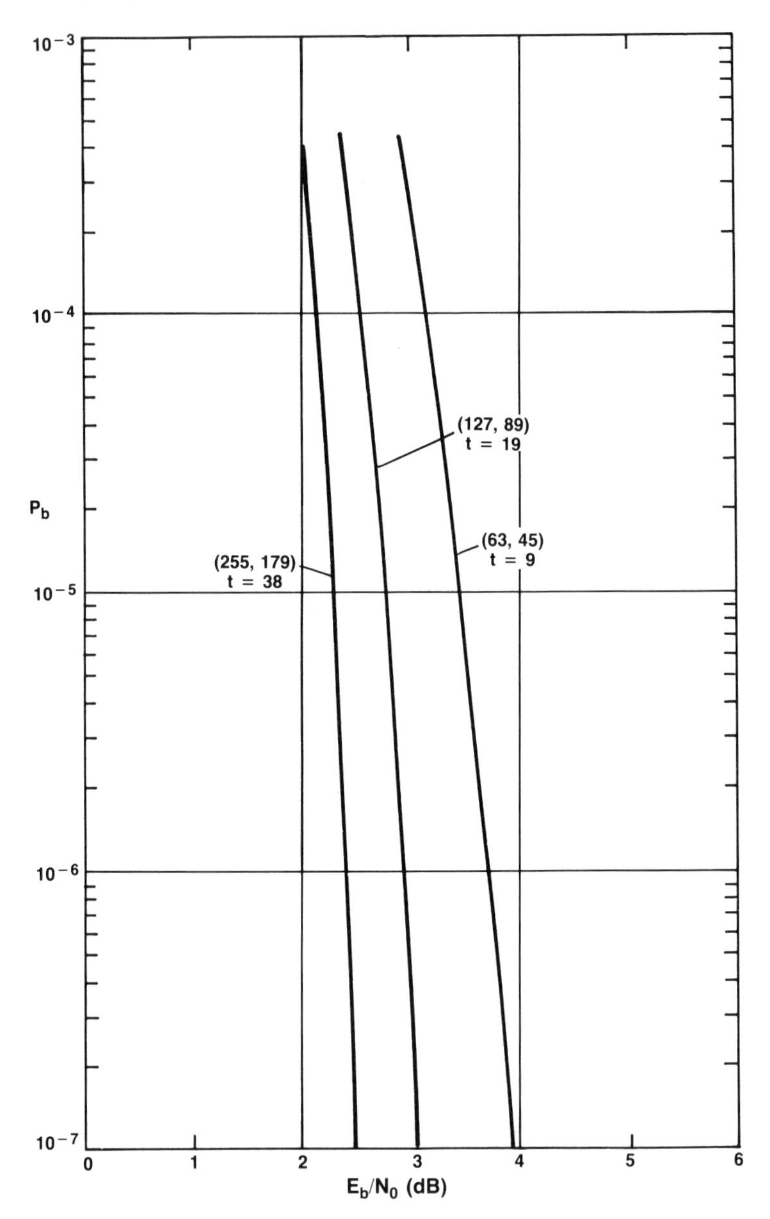

FIGURE 11.3. Upper bound on post-decoding bit-error rate as a function of E_b/N_0 for $m = 6, 7$, and 8 biorthogonal codes concatenated with RS codes defined on $GF(64)$, $GF(128)$, and $GF(256)$.

requires $E_b/N_0 = 10.5$ dB, and thus the most powerful concatenated codes considered here would provide coding gains of 8.1 and 8.5 dB, respectively, based on a comparison of post-decoding BERs.

It is also interesting to compare the performance of concatenated codes with uncoded PSK on the basis of message-level performance. We consider first the probability of not correctly decoding a message conveyed in one (255,179) RS codeword on $GF(256)$, that is, a message containing 1432 bits. Since this is a $t = 38$ code, we have

$$1 - P_{CD} = 1 - \sum_{i=0}^{38} \binom{255}{i} P_{CE}^{i}(1 - P_{CE})^{(255-i)} \tag{11.2}$$

where P_{CE} is the symbol-error probability at the output of the decoder for the $m = 8$ biorthogonal code. For the uncoded case, the message is received correctly only if each of the 1432 bits is correct. It can easily be verified that $1 - P_{CD} = 10^{-3}$ with $E_b/N_0 = 2.1$ dB for the concatenated code and with $E_b/N_0 = 10.7$ dB for uncoded PSK transmission. Thus a coding gain of 8.6 dB is provided.

For the case of message-level results, we are also interested in the maximum probability of wrong message, or, for our example, the probability of incorrect decoding of a (255,179) RS code block. Such long block codes often provide a very low probability of incorrect decoding even when used to correct the maximum number of errors per codeword. This result can be computed exactly for any value of P_{CE} using the tedious approach presented in Chapter 7. However, an important and very useful measure of performance is the probability of incorrect decoding on the *completely random channel* [the channel with no signal present and $P_{CE} = (q - 1)/q$], which can be evaluated as

$$P_{ICD}\Big|_{P_{CE}=(q-1)/q} = \sum_{i=0}^{t} \binom{n}{i} \left[q^{-r}(q-1)^i - \left(\frac{q-1}{q}\right)^i \left(1 - \frac{q-1}{q}\right)^{n-i} \right] \tag{11.3}$$

where $r = n - k$. Equation (11.3) is based on the observation that when all possible received n-tuples are equally likely, the probability of incorrect decoding is given by the ratio of the total number of points contained within the radius-t spheres centered on all the codewords to the total number of possible received n-tuples minus the probability that a received point is contained within the sphere centered on the correct word. Since Kasami and Lin [81] have shown that for RS codes P_{ICD} decreases monotonically in the interval $(q - 1)/q \geq P_{CE} \geq 0$, Eq. (11.3) provides an important bound on the probability of incorrect decoding. We have

$$P_{ICD} < q^{-r} \sum_{i=0}^{t} \binom{n}{i} (q-1)^i$$

This bound, in fact, applies for any maximum-distance-separable code. For the (255,179) code on $GF(256)$, we have then

$$P_{\mathrm{ICD}} < (256)^{-76} \sum_{i=0}^{38} \binom{255}{i} (255)^{i}, \qquad 0 \leq P_{\mathrm{CE}} \leq \frac{q-1}{q}$$

which can be shown to be quite small. For example, the largest term in the sum corresponds to $i = 38$, and its value is

$$(256)^{-76}(255)^{38}\binom{255}{38} = (9.41 \times 10^{-184})(2.81 \times 10^{91})(2.95 \times 10^{45})$$

$$= 7.8 \times 10^{-47}$$

Thus, very low probabilities of wrong message can be provided with long block codes, by virtue of the large number of syndromes that are not decodable with bounded-distance decoding.

Therefore, in principle at least, highly efficient communication can be obtained with concatenated block codes. An obvious problem with using this approach, however, is that the overall code rate is quite low and consequently E_s/N_0 is small. For example, for the $m = 8$ scheme suggested, with $E_b/N_0 =$ 2.5 dB, the channel bit SNR (E_s/N_0) at the input to the decoder for the biorthogonal code is -11.1 dB. Thus, bit synchronization, phase estimation, and codeword framing can be difficult at best. In the following section we consider a concatenated code that provides similar performance results with higher overall code rate.

We mentioned earlier in this chapter that highly efficient communication can also be provided on the coherent channel with orthogonal signaling and error-control coding. In fact, the performance of M-ary coherent FSK used with RS codes defined on an M-ary alphabet is comparable to that of the concatenated block codes. With this approach, orthogonal signaling is substituted for the biorthogonal code and PSK modulation in the block concatenation schemes.

11.1.1.2. Concatenated Block and Convolutional Codes. Since the error-control codes that are most suitable for use as inner codes are those that lend themselves to maximum likelihood decoding, short-constraint-length convolutional codes may be considered as candidates. Viterbi decoding typically provides a 10^{-2} output error rate at the SNR corresponding to R_{COMP} for the code employed, and high-rate block codes can be used to reduce the 10^{-2} error rate to whatever quality of service is required. Odenwalder [121] considered use of convolutional codes with Viterbi decoding as the inner coding technique. Again RS codes were used as the outer codes.

A block diagram of the scheme treated by Odenwalder is given in Fig. 11.4. Data to be transmitted is first encoded with an RS code and then a $b = 1$

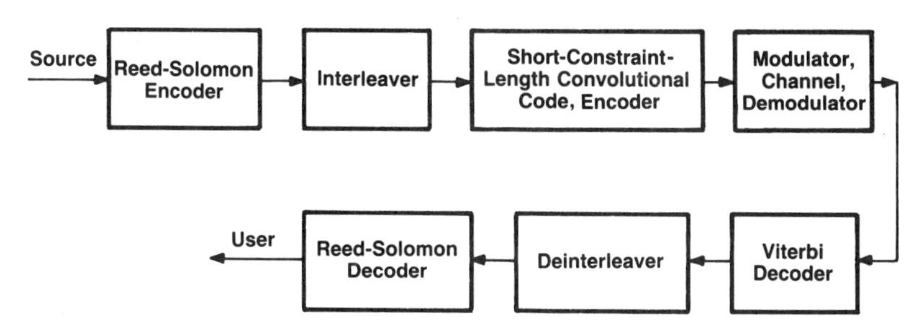

FIGURE 11.4. Concatenated code using a short-constraint-length convolutional code and an RS code.

short-constraint-length convolutional code. At the receiving end, Viterbi decoding of the convolutional code and hard-decision, bounded-distance decoding of the RS code are implemented. A major difference between the scheme proposed by Odenwalder and the approach using only block codes is that errors leaving the Viterbi decoder are not independent. That is, when errors are observed at the output of a Viterbi decoder, they tend to be clustered in bursts.

A way to make the RS code more effective in this burst-error environment is to *interleave*, or *interlace*, symbols of successive RS code blocks. With this approach, error bursts following Viterbi decoding are spread among several consecutive RS code blocks rather than being contained within a single block. Thus, an *interleaver* is shown in Fig. 11.4 following RS encoding and a *deinterleaver* is shown prior to RS decoding. More will be said shortly on the use of interleaving to combat burst disturbances, but the reader should be able to see that, in this application, the inclusion of an interleaver enables use of higher-rate RS codes than would otherwise be required. Thus, a more efficient design results with a less complex RS decoder and a less severe synchronization requirement (due to larger E_s/N_0).

In order to compute the performance afforded by a concatenated code using a convolutional inner code, Odenwalder assumed that sufficient interleaving is provided so that independent character errors are observed at the input to the RS decoder. Character-error rates following Viterbi decoding were then measured as a function of SNR, and the bound on post-decoding bit error rate, Eq. (11.1), was used to obtain performance results for a variety of code parameters. It was observed again that an optimum RS code rate exists that, in this case, is somewhat higher than for the block-coded systems.

Since the most widely used Viterbi decoding configuration employs the $b = 1$, $k = 7$, rate-1/2 code that was given in Chapter 9 (Table 9.1), we consider performance results for this code first. However, we note that this constraint-length-7 code provides the lowest output BER of all the $k = 7$ codes for high SNRs. In the concatenation scheme, of course, inner code performance at relatively low SNRs (high error rates) is of interest, and it is possible that better $k = 7$ codes exist for this region.

Performance results for the $k = 7$, rate-1/2 code concatenated with interleaved high-rate RS codes defined on six- through nine-bit alphabets are shown in Fig. 11.5. Soft-decision Viterbi decoding with three-bit quantization is assumed. Note that for RS symbol alphabets of seven or more bits, low post-decoding BERs are provided with $2.5 \leq E_b/N_0 \leq 3$ dB. Odenwalder also obtained results for somewhat more complex inner codes and Viterbi decoders. For instance, results are shown in Fig. 11.6 for some RS codes and the $k = 8$, rate-1/3 inner code (Table 9.2). These concatenated codes provide good performance with 2.0 dB $\leq E_b/N_0 \leq$ 2.5 dB. Note that the techniques using the rate-1/2 convolutional inner code require bit synchronization and phase tracking with $E_s/N_0 \approx -1.0$ dB and the rate-1/3 schemes with $E_s/N_0 \approx -3.4$ dB. These are considerably larger SNRs than are required for the concatenated block codes considered in the previous section. Thus reliable communication can be provided with the concatenation of block and convolutional codes at levels in the range 2 dB $\leq E_b/N_0 \leq$ 3 dB with a less stringent synchronization requirement.

Another difference between the concatenation scheme using a convolutional inner code and the block-coded scheme is that the symbol alphabet at the output of the inner decoder is not matched to the RS code alphabet. Thus, error bursts in the Viterbi decoder output are not necessarily lined up with the RS code symbols. Of course, one could consider use of a convolutional inner code with $b > 1$ and RS codes defined on 2^b-ary alphabets. Lee [90] proposed and analyzed the performance that can be obtained with such schemes. It was shown that, as with the technique considered here, highly reliable communication is achieved with $E_b/N_0 \approx 2$ to 3 dB.

As a final note on concatenation schemes using convolutional and RS codes, we point out results obtained by Roefs and Best [151], who chose to use an RS code with a block length of 256 for 16-error correction with the $k = 7$ rate-1/2 convolutional inner code. It was noted that 257 is a Fermat prime (a prime of the form $2^m + 1$) and that significant computational gains can be realized with fast transforms defined on fields with q elements when q is a Fermat prime [146, 147]. Roefs and Best thus used an RS code designed on $GF(257)$. Significant reductions in decoder complexity were obtained relative to a more standard code design on $GF(256)$.

11.1.1.3. Soft-Decision Sequential Decoding of Long-Constraint-Length, Low-Rate Convolutional Codes. It was shown in Chapter 10 that a long-constraint-length, low-rate convolutional code decoded with a soft-decision sequential decoder can also provide highly efficient communication. The transmission of a 500-bit message with a rate-1/6, $k = 64$, $b = 1$ nonsystematic code was considered. It was shown that a carefully designed soft-decision Fano decoder provides $1 - P_{\text{CD}} = 10^{-3}$ for the code considered, with $E_b/N_0 = 1.9$ dB when 500 computations per information bit are permitted. This represents a coding gain of 8.4 dB relative to uncoded PSK transmission. With 10 computations per bit, $E_b/N_0 = 3.1$ dB is required for the same P_{CD}.

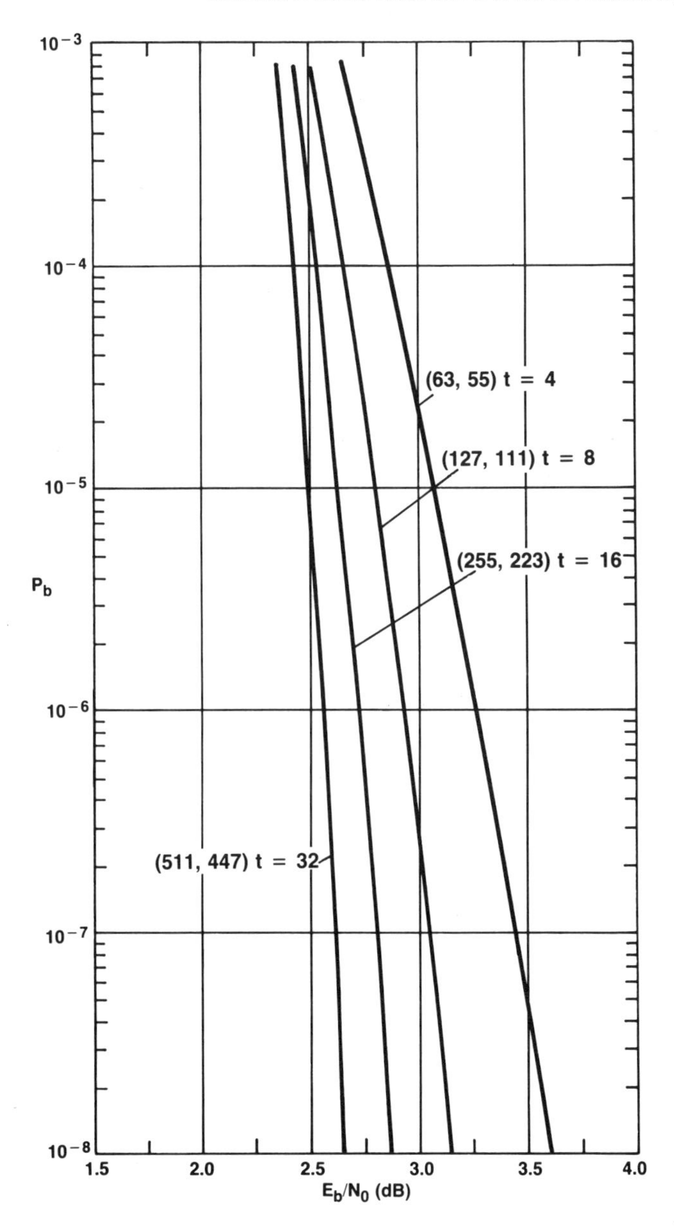

FIGURE 11.5. Post-decoding bit-error rate for several concatenated codes using the $k = 7$, rate-$1/2$ convolutional code as the inner code. Outer codes are RS codes defined on $GF(64)$, $GF(128)$, $GF(256)$, and $GF(512)$.

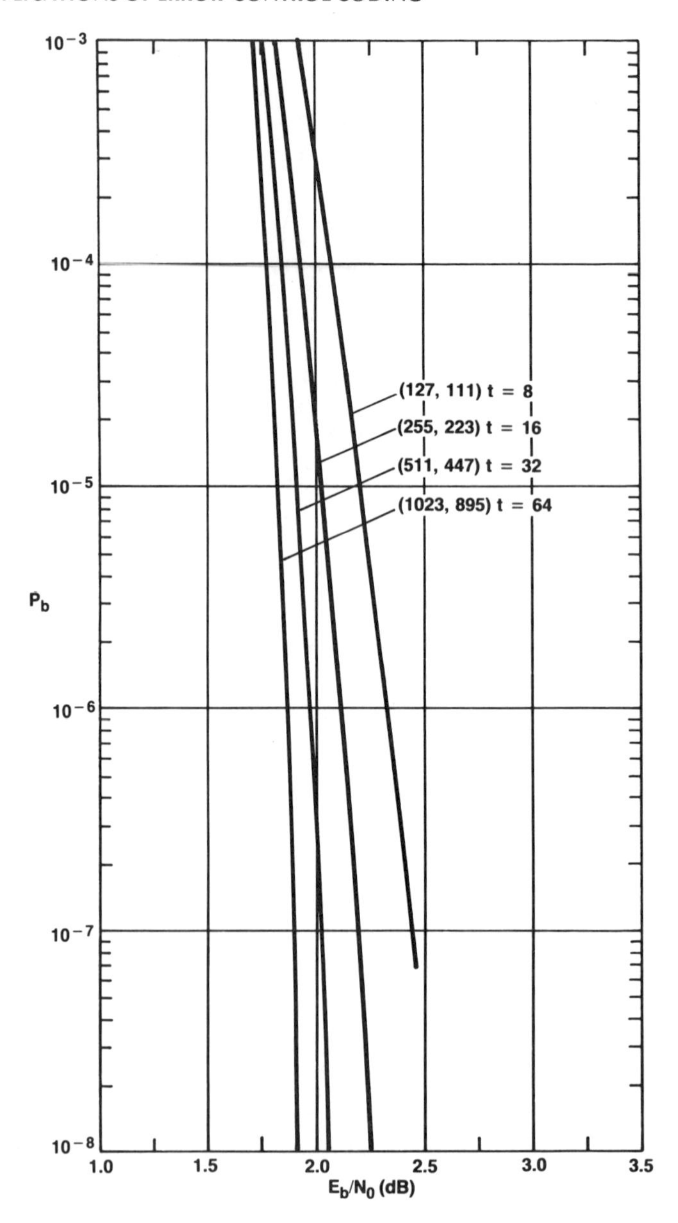

FIGURE 11.6. Post-decoding bit-error rate for several concatenated codes using the $k = 8$, rate-1/3 convolutional code as the inner code. Outer codes are RS codes defined on $GF(128)$, $GF(256)$, $GF(512)$, and $GF(1024)$.

In addition, sequential decoding can also provide a very low probability of incorrect decoding for all SNRs. It was observed that the decoder can be designed either to reach the end of the code tree and deliver the message correctly or to quit, that is, erase the received message when computational overflow occurs. One might expect then that the post-decoding bit-error rate is given by one-fourth to one-half of the probability of not decoding correctly. However, with the nonsystematic code considered, a convenient estimate of the received information bits is not available when computational overflow occurs, and post-decoding bit-error rate has no real meaning in this case.

11.1.2. Techniques That Provide Moderate Coding Gain

The high-performance schemes considered in the previous section can, of course, be designed and implemented with simpler codes and decoding strategies at the cost of reduced performance. Specifically, coding gains of 4 to 7 dB can be obtained by use of convolutional codes with hard-decision sequential decoding or by use of concatenated codes designed with simpler inner and outer codes. However, coding gains of 4 to 7 dB can also be obtained with other techniques. In the following sections we first consider binary block codes with hard-decision decoding and then convolutional codes with Viterbi decoding.

11.1.2.1. Binary BCH Codes and Hard-Decision Decoding. We consider the use of binary BCH codes on the AWGN channel where hard bit decisions are made on demodulator outputs. Bounded-distance decoding as described in Chapter 5 is assumed. In addition we assume that the performance requirement being addressed is based on a message-level specification and that delivery of correct messages with moderate communication efficiency is desired.

Coding gain in this case corresponds to the difference in the SNRs required by coded and uncoded systems to provide a particular probability of correct message reception. We assume initially that a message can be conveyed in a single code block and correct message reception occurs when we have a correct decoding. Coherent binary PSK modulation is also assumed. The probability of correct decoding with bounded-distance decoding is given by

$$P_{\mathrm{CD}} = \sum_{i=0}^{t} \binom{n}{i} p^i (1-p)^{n-i} \tag{11.4}$$

where p is the BER on the channel. For uncoded transmission of a k-bit message, the probability of correct message reception is $(1-p)^k$.

The performance provided with uncoded transmission and some binary BCH codes with block lengths 63, 127, and 255 is indicated in Figs. 11.7 to 11.11 for message lengths of 30, 92, 64, 187, and 131 bits. Note that the (63,30) code provides a gain of 4 dB when $1 - P_{\mathrm{CD}}$ is small, and the codes with block

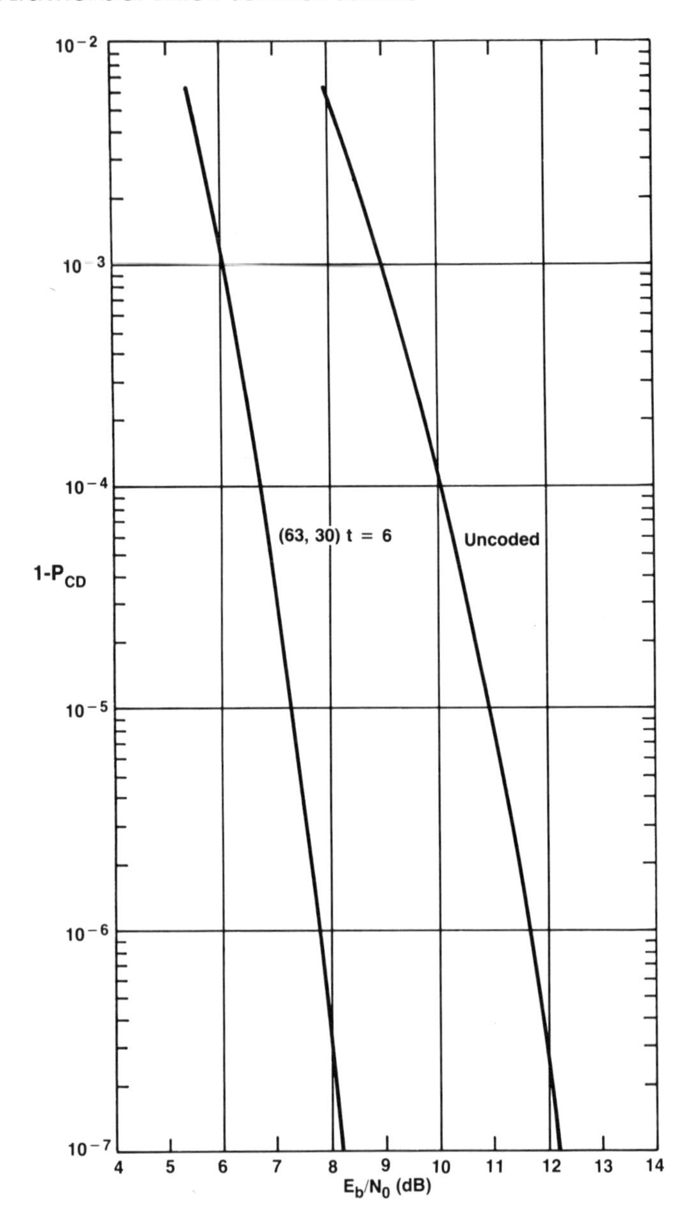

FIGURE 11.7. $1 - P_{\mathrm{CD}}$ vs. E_b/N_0 for the (63,30) $t = 6$ binary BCH code and an uncoded 30-bit transmission.

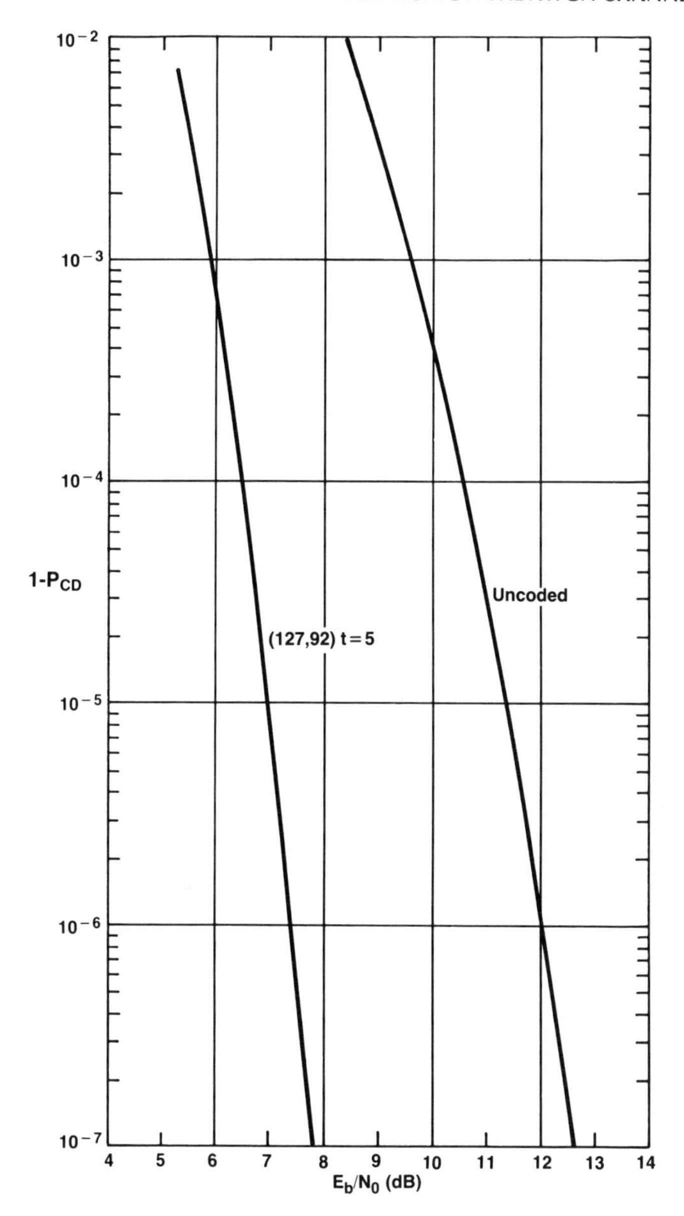

FIGURE 11.8. $1 - P_{CD}$ vs. E_b/N_0 for the (127,92) $t = 5$ binary BCH code and an uncoded 92-bit transmission.

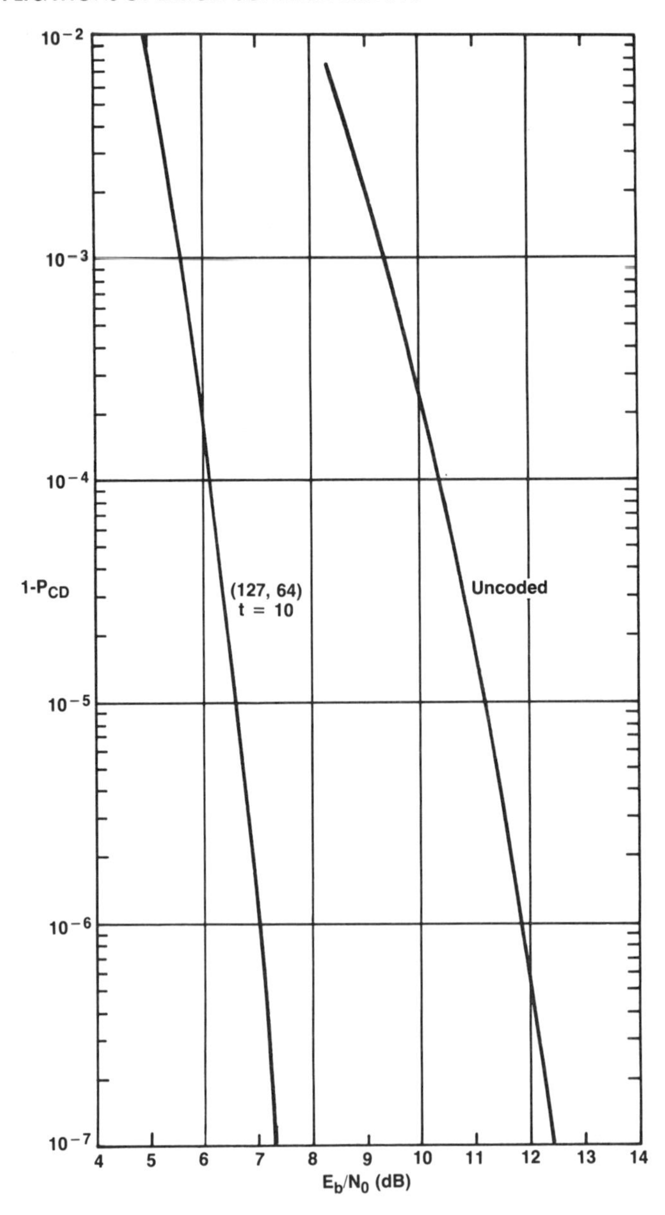

FIGURE 11.9. $1 - P_{\mathrm{CD}}$ vs. E_b/N_0 for the (127,64) $t = 10$ binary BCH code and an uncoded 64-bit transmission.

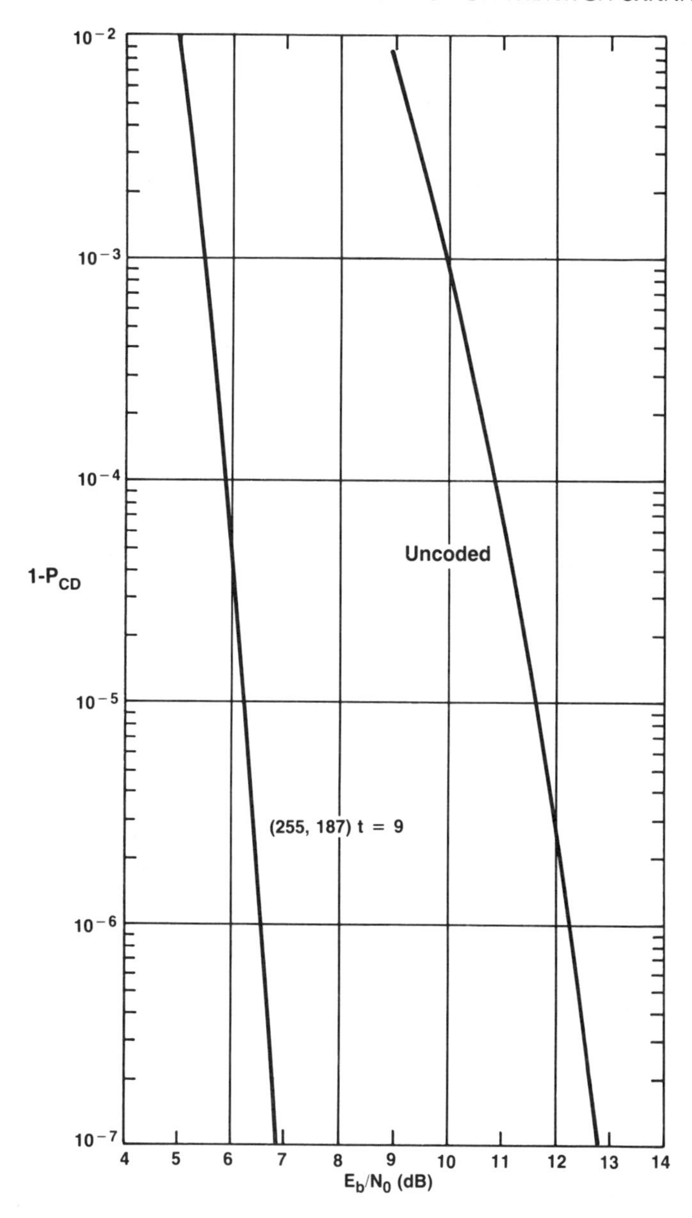

FIGURE 11.10. $1 - P_{\text{CD}}$ vs. E_b/N_0 for the (255,187) $t = 9$ binary BCH code and an uncoded 187-bit transmission.

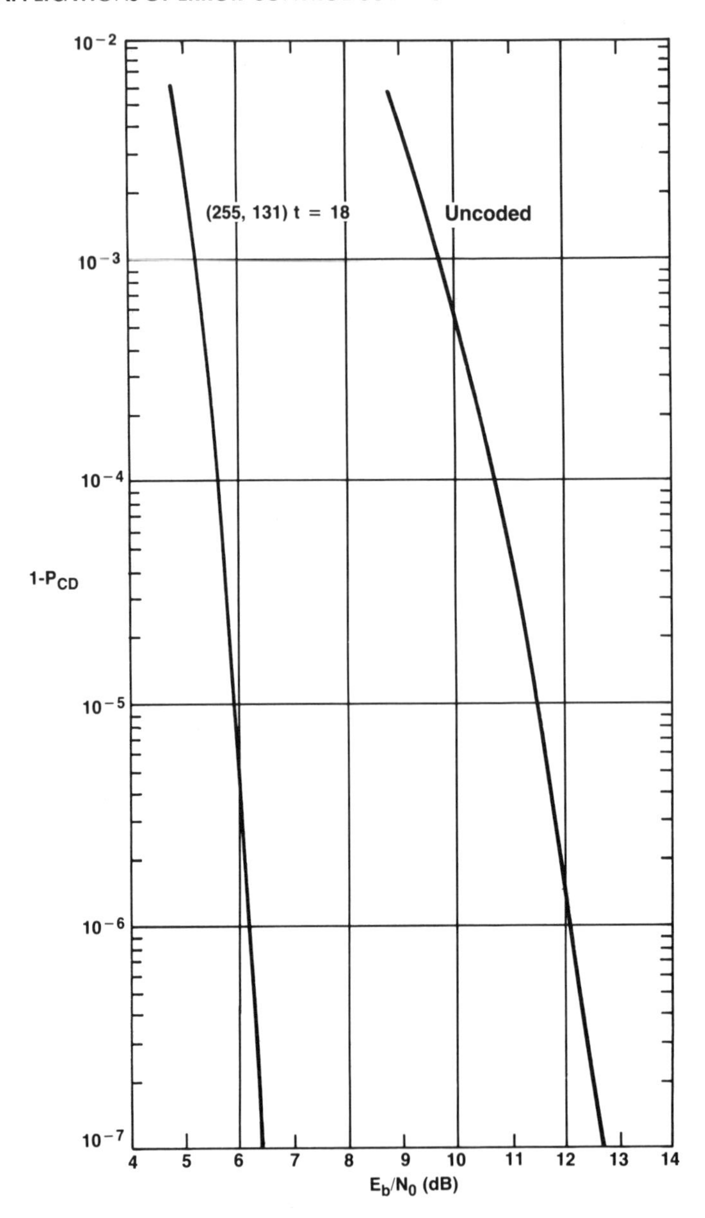

FIGURE 11.11. $1 - P_{\text{CD}}$ vs. E_b/N_0 for the (255,131) $t = 18$ binary BCH code and an uncoded 131-bit transmission.

TABLE 11.2. Probability of Incorrect Decoding for Some Binary BCH Codes on the Random Channel ($p = 1/2$)

BCH Code Parameters (n,k)		P_{ICD} for p = 1/2
(63,30)	t=6	8.8×10^{-3}
(127,92)	t=5	7.7×10^{-3}
(127,64)	t=10	2.5×10^{-5}
(255,187)	t=9	3.8×10^{-5}
(255,131)	t=18	8.9×10^{-11}

lengths 127 and 255 provide gains of approximately 5 dB and 6 dB, respectively. The probability of incorrect decoding for these codes on the random channel ($p = 1/2$) is shown in Table 11.2, and we see that for long binary BCH codes with rate approximately equal to 1/2, low incorrect-decoding probabilities can be assured. These results are obtained from Eq. (11.3) with $q = 2$.

The examples just considered are rather specialized in that the messages to be transmitted contain exactly k information bits and can each be conveyed in a single (n,k) codeword. In general, codes must be shortened, and perhaps more than one codeword must be used to transmit a single message. For transmission of a message that consists of l codewords, the probability of correct message reception is given by $P_{CM} = P_{CD}^l$ and the probability of incorrect message is given by

$$P_{ICM} = \sum_{i=1}^{l} \binom{l}{i} P_{ICD}^i P_{CD}^{l-i}$$

In general, P_{ICM} is not a monotonic function of the channel-error probability.

As an example, we consider a message consisting of 500 information bits. Clearly, the (255,131) code can be shortened to (249,125) and four codewords used to encode the message. Similarly, the (255,187) code can be shortened to (235,167) and the (127,64) code shortened to (113,50) with three and 10 code blocks, respectively, used to transmit the message. For these three schemes, $1 - P_{CM}$ is shown in Fig. 11.12 along with the performance for an uncoded 500-bit transmission. Note that for $1 - P_{CM}$ small, coding gains of about 5 to 6 dB are obtained again.

There are certainly many other configurations of moderately complex binary block codes that can be used to provide coding gains of 4 to 7 dB on the independent-error channel. Furthermore, Gore has shown that RS codes can be used effectively with binary signaling [57]. Specifically, it was demonstrated that RS codes provide better performance for low output error rates when compared with binary BCH codes with the same rate and block length. In addition, RS codes provide some protection against clustered errors or bursts as well as independent bit errors.

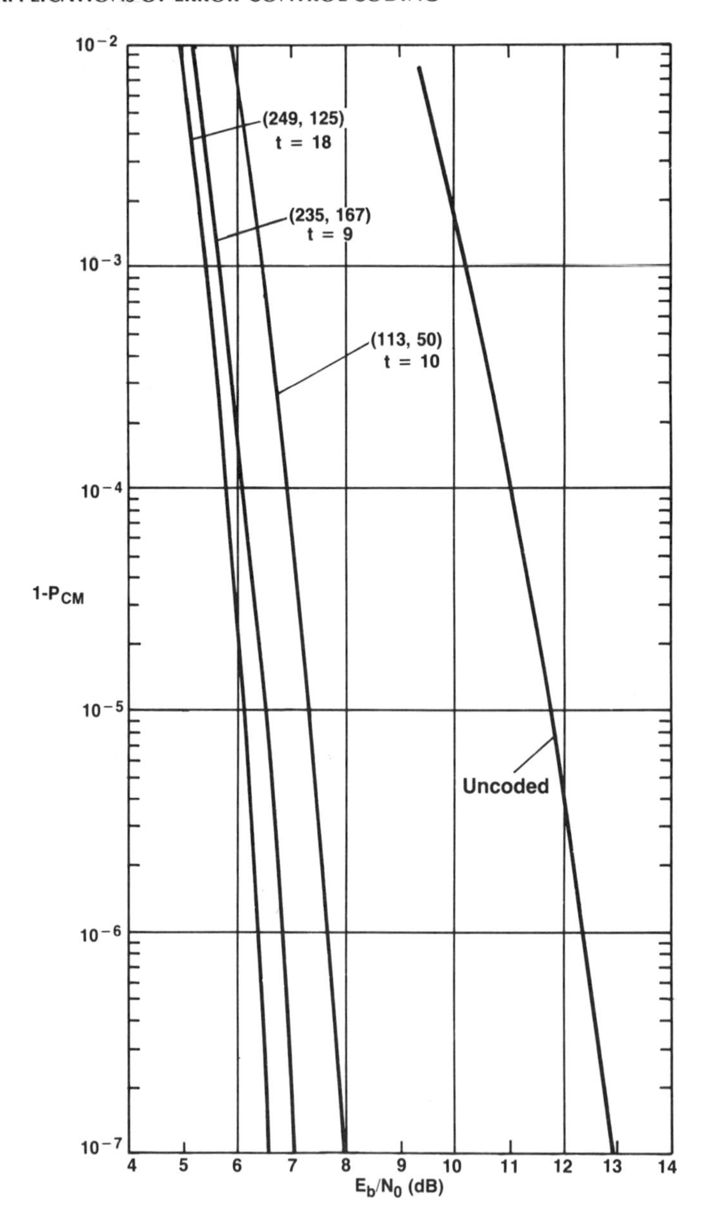

FIGURE 11.12. $1 - P_{\mathrm{CM}}$ vs. E_b/N_0 for a 500-bit message with multiple codeword transmission of some shortened binary BCH codes and an uncoded 500-bit transmission.

TABLE 11.3. Coding Gain (in dB) at a 10^{-7} Post-decoding Bit-Error Rate for a Selection of Convolutional Codes with Soft-Decision (Eight-Level) Viterbi Decoding

R	b	k	Coding Gain (dB) for $P_b = 10^{-7}$
1/3	1	7	6.2
1/2	1	7	5.8
1/2	1	5	4.9
2/3	2	4	5.2
2/3	2	3	4.7
3/4	3	3	4.8
3/4	3	2	3.9

11.1.2.2. Short-Constraint-Length Convolutional Codes and Viterbi Decoding. It is widely known that soft-decision Viterbi decoding of short-constraint-length convolutional codes can be used to provide coding gains of 4 to 7 dB with PSK signaling and coherent reception. In this case coding gain is usually stated in terms of the improvement in post-decoding BER. In Chapter 9, good codes for use with Viterbi decoding were presented and performance results described for a selection of code parameters. Table 11.3 summarizes some of the results given by Jacobs [73] for a post-decoding BER of 10^{-7}. Note that the widely used rate-1/2, $b = 1$, $k = 7$ code provides a gain of 5.8 dB, and the rate-3/4, $b = 3$, $k = 3$ code provides a gain of 4.8 dB. The rate-1/3, $k = 7$ code provides 6.2 dB of coding gain

It must be noted that with Viterbi decoding, very low error rates cannot be assured for all SNRs. Thus, short-constraint-length convolutional codes and Viterbi decoding cannot be used alone to address a system specification that imposes a stringent wrong-message-probability requirement for arbitrary SNRs. Short-constraint-length convolutional codes can, of course, be concatenated with binary BCH codes to provide low message-error probabilities. In this case the block code is used as the outer code solely for error detection. For example, a distance-5 binary BCH code with 20 parity bits can be used to protect a 1000-bit message and provides a maximum incorrect-decoding probability of 9.5×10^{-7}. When this BCH code is concatenated with a short-constraint-length convolutional code, the penalty in communication efficiency for the extra 20 parity bits is small (about 0.1 dB).

11.1.3. Techniques Providing Modest Coding Gain

Although in this book we have concentrated on the more powerful error-control schemes, some relatively simple techniques have been developed that provide coding gains in the range of 1 to 3 dB. There are both block and convolutional coding techniques that fall into this category. Examples of techniques that use convolutional codes to provide 1 to 3 dB of gain are hard-decision feedback decoding of short-constraint-length codes and majority-logic and threshold decoding. These schemes were treated briefly in Chapter 8, and more complete descriptions that include detailed performance results can be found in a number of references, such as references 27 and 94.

Many relatively simple schemes can also be used to provide coding gains of 1 to 3 dB with block codes. Perhaps the most widely known example is hard-decision Kasami decoding of the (23,12) Golay code. The reader may readily verify from the results presented in Chapter 7 that the (23,12) code provides a coding gain of 2.3 dB at a post-decoding bit-error rate of 10^{-5}.

11.2. NONCOHERENT RECEPTION ON THE AWGN CHANNEL

A good deal of the literature devoted to performance evaluations of coded communication systems deals only with coherent reception. However, there are applications for which coherent reception is not practical or is even impossible. For example, there are communication channels on which the received signal is not stable for the period of time required to estimate or track phase. Furthermore, as one uses more powerful error-control techniques, E_s/N_0 is reduced, which makes phase tracking more difficult. This is especially true for *spread-spectrum* systems, whether they use *frequency-hopping* or *direct-sequence* band spreading. Thus, the issue of applying coding to noncoherent and differentially coherent channels is of considerable interest.

We have observed for the coherent case that examination of R_{COMP}, or R_0, provides very useful guidance for the selection of a system design point. In fact, the parameter R_{COMP} is in one sense more useful than channel capacity. As we have seen, the SNR that corresponds to operation at R_{COMP} has proved to be the practical limit for operation of a coded coherent communication system. The operation of a communication system at rates above R_{COMP} results in vastly increased complexity, and attempts to realize the last 3 dB in communication efficiency have not been successful.

The same is true for noncoherent or differentially coherent reception. Jordan [78] evaluated R_{COMP} for M-ary orthogonal signaling on the AWGN channel assuming noncoherent detection. For soft-decision detection and decoding, R_{COMP} is given by

$$R_{\text{COMP}} = \log \frac{M}{1 + (M-1)\left[\int \sqrt{p_n(y)\, p_{s+n}(y)}\, dy\right]^2} \tag{11.5}$$

where $p_n(y)$ is the probability density function of the noise at the output of the matched filter with no signal present and $p_{s+n}(y)$ is the density with signal present. For the case of hard-decision detection and decoding, we have

$$R_{\text{COMP}} = -\log \frac{\left[\sqrt{q_M(M-1)} + \sqrt{(1-q_M)}\right]^2}{M} \tag{11.6}$$

where q_M is the probability that an M-ary symbol is received correctly.

Bucher [18] also evaluated the expressions (11.5) and (11.6) and has presented the results in a particularly illuminating way. The SNR that corresponds to operation at R_{COMP} for the noncoherent channel was evaluated as a function of the rate of the code employed. That is, it was assumed that the communication system was being operated with the code rate equal to R_{COMP}. The results for M-ary orthogonal signaling and soft-decision decoding are shown in Fig. 11.13 and for hard-decision decoding in Fig. 11.14. We note that for noncoherent reception, the results indicate that there is an optimum code rate near $R = 1/2$, and that for both very low and very high code rates a severe penalty in communication efficiency is paid. However, for code rates between 0.2 and 0.9, the SNR corresponding to operation at R_{COMP} changes by only 1 dB, and within these limits performance is relatively insensitive to the choice of code rate.

These results are distinctly different from those for the coherent channel, where performance theoretically improves indefinitely as the code rate is reduced and the SNR corresponding to computational cutoff approaches

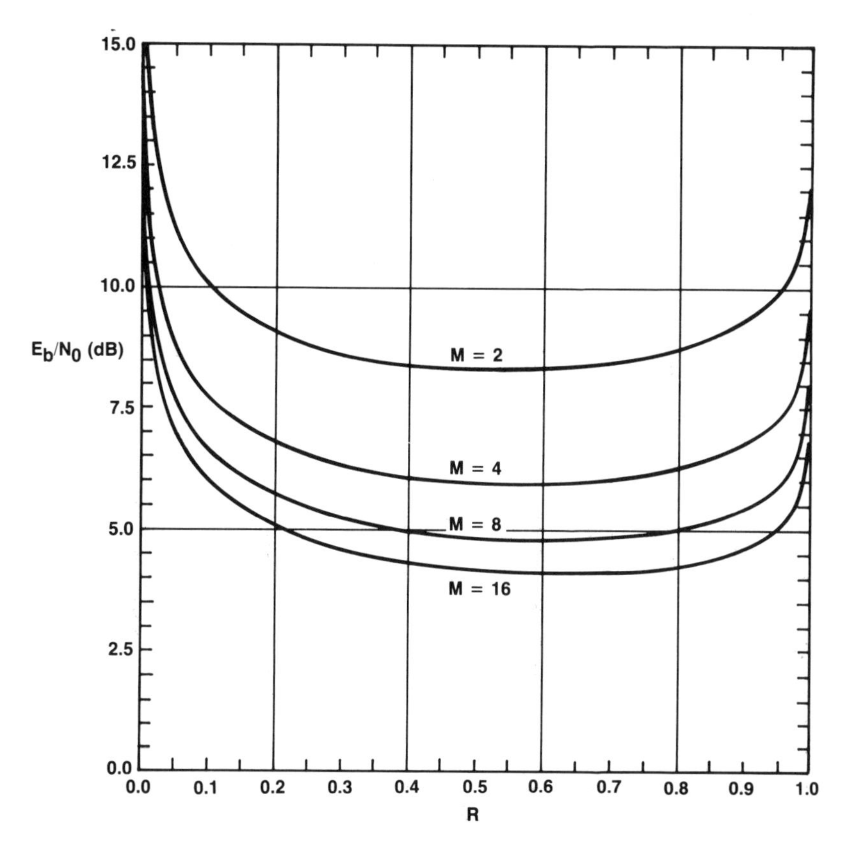

FIGURE 11.13. E_b/N_0 corresponding to operation at $R = R_{\text{COMP}}$ for binary, 4-ary, 8-ary, and 16-ary orthogonal signaling with noncoherent detection and soft-decision decoding. From Bucher [18], © 1980 IEEE, reprinted with permission.

$E_b/N_0 = 1.4$ dB monotonically. One is tempted to say, then, that for noncoherent reception moderate to high code rates should be favored, that is, $0.5 \leq R \leq 0.8$. For these rates, optimum or near-optimum performance can be realized with less complex decoders and higher E_s/N_0, which results in less severe synchronization and codeword framing requirements. A note of caution is in order, however, since Bucher also evaluated R_{COMP} for some nonstationary channels and found that the performance of high-rate schemes degraded appreciably. Examples of the nonstationary channels considered by Bucher and others [165] include both Rayleigh and Rician fading and pulsed interference. In each case, the low-rate error-control schemes showed promise of more robust performance.

Several other points are also apparent from Figs. 11.13 and 11.14. First, as in the coherent case, hard-decision decoding is 2 dB inferior to soft-decision decoding. However, for nonstationary channels the difference can be much larger. In addition, for soft-decision decoding, a nominal design point can be

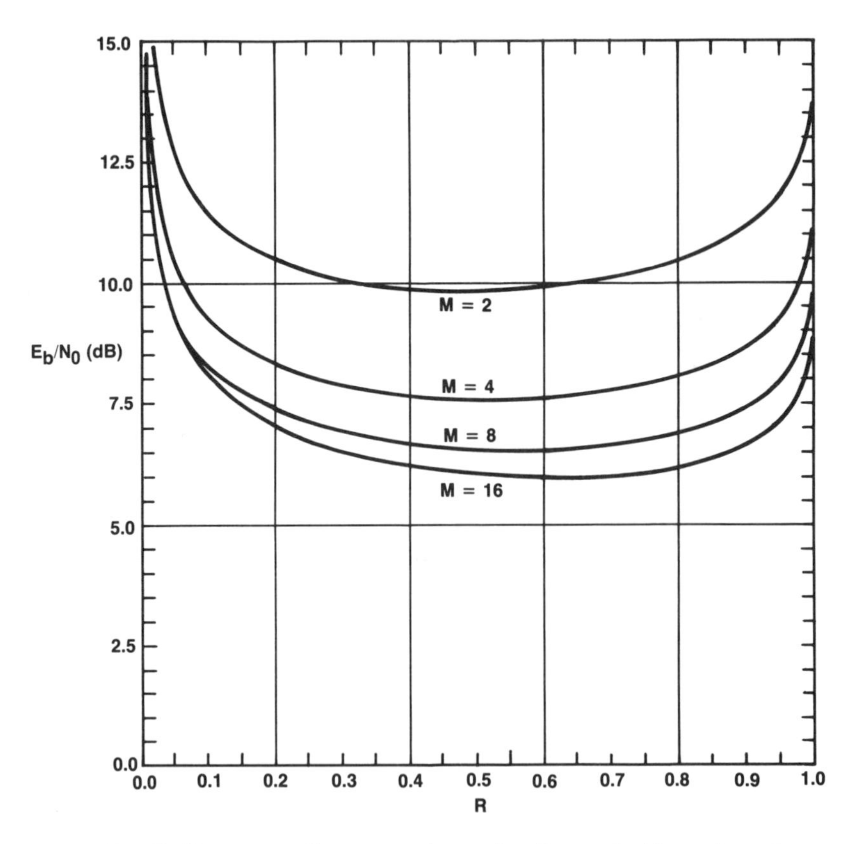

FIGURE 11.14. E_b/N_0 corresponding to operation at $R = R_{\text{COMP}}$ for binary, 4-ary, 8-ary, and 16-ary orthogonal signaling with noncoherent detection and hard-decision decoding. From Bucher [18], © 1980 IEEE, reprinted with permission.

selected from about $E_b/N_0 = 4.0$ dB to 8.0 dB depending on the complexity of the modem and error-control scheme utilized. Thus, even for large M, a penalty in communication efficiency of at least 2 dB is expected for noncoherent reception compared with the coherent case. Finally, since differentially coherent PSK is 3 dB superior to binary noncoherent FSK, Fig. 11.13 provides a design goal for DPSK as well, namely $E_b/N_0 \approx 5.3$ dB for $R = 1/2$.

The design choices available in developing a coded noncoherent communication system are thus somewhat different than those for the coherent case. In the following sections we consider some examples of the application of error-control coding to noncoherent and differentially coherent channels.

11.2.1. M-ary Orthogonal Signaling and Reed-Solomon Coding

Remarkably good performance can be achieved with relatively straightforward error-control schemes and orthogonal signaling on the noncoherent channel. We first consider hard-decision (errors-only) decoding of RS codes used with M-ary orthogonal signaling. In Figs. 11.15 and 11.16, $1 - P_{CD}$ is shown as a function of E_b/N_0 for rate $\approx 1/2$ RS codes defined on 32-ary and 64-ary alphabets with 32-ary and 64-ary orthogonal signaling, respectively. $1 - P_{CD}$ is very small with $E_b/N_0 \geq 6.5$ dB for the 32-ary scheme, and with $E_b/N_0 \geq 5.4$ dB for the 64-ary case.

The message-level performance provided by uncoded 32-ary and 64-ary transmission is also indicated in Figs. 11.15 and 11.16. Note that although these error-control schemes are operating near R_{COMP} for the hard-decision channel (only small improvements over 16-ary signaling are obtained with larger signaling alphabets), relatively modest coding gains are provided with respect to uncoded transmission. Specifically, for the 32-ary scheme, a coding gain of 2.9 dB is observed, and for the 64-ary case a 3.4-dB gain is achieved. This is a result of the fact that by simply using M-ary rather than binary orthogonal signaling on the noncoherent channel, significant enhancements in communication efficiency are realized. These schemes provide large coding gains compared with uncoded binary FSK.

11.2.2. Convolutional Codes on Noncoherent Channels

The convolutional coding schemes considered previously for the coherent channel can also be used on the noncoherent or differentially coherent channels. For example, Jordan [78] showed that soft-decision sequential decoding can be used with 16-ary signaling on the noncoherent channel to provide reliable communication with $E_b/N_0 \approx 4.5$ dB. The concatenation schemes utilizing block and convolutional codes can also be used on the differentially coherent channel with binary antipodal signaling. However, in this section we consider somewhat simpler techniques.

Here we consider the performance obtained with Viterbi decoding of some short-constraint-length convolutional codes used with binary, 8-ary, and 32-ary

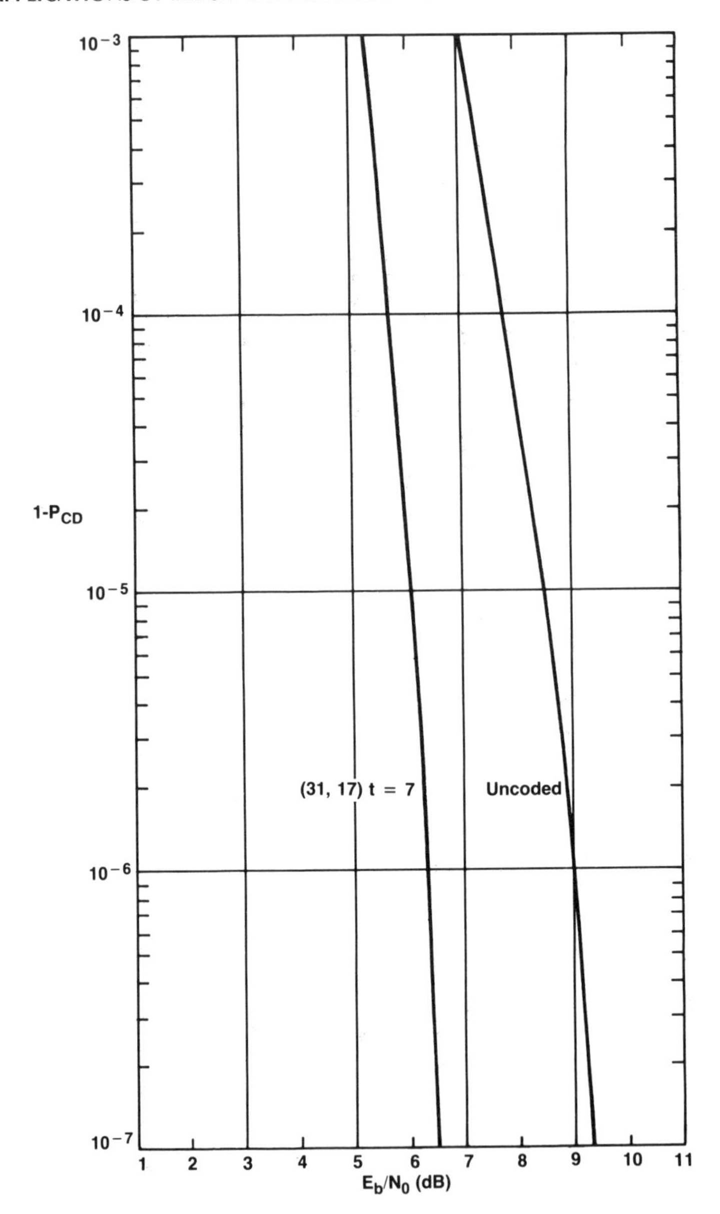

FIGURE 11.15. $1 - P_{CD}$ vs. E_b/N_0 for the (31,17) $t = 7$ RS code defined on $GF(32)$ with 32-ary orthogonal signaling and noncoherent detection; uncoded 32-ary transmission of a 17-character message also shown.

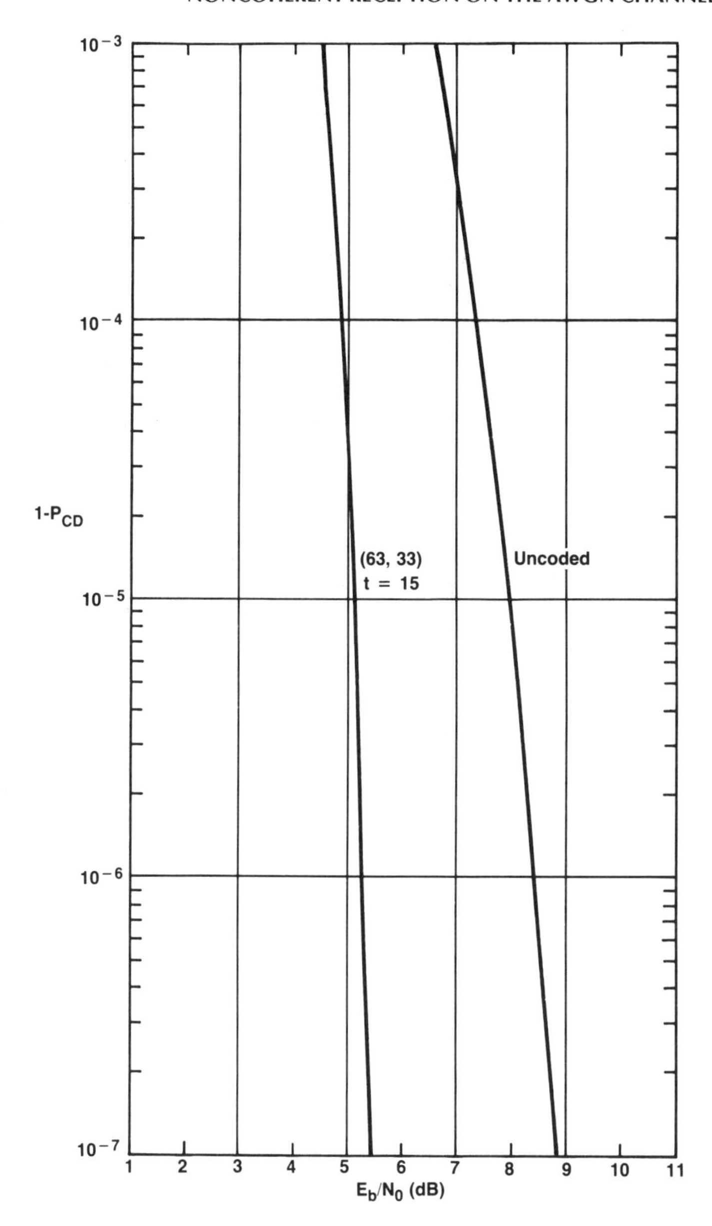

FIGURE 11.16. $1 - P_{\text{CD}}$ vs. E_b/N_0 for the (63,33) $t = 15$ RS code defined on $GF(64)$ with 64-ary orthogonal signaling and noncoherent detection; uncoded 64-ary transmission of a 33-character message also shown.

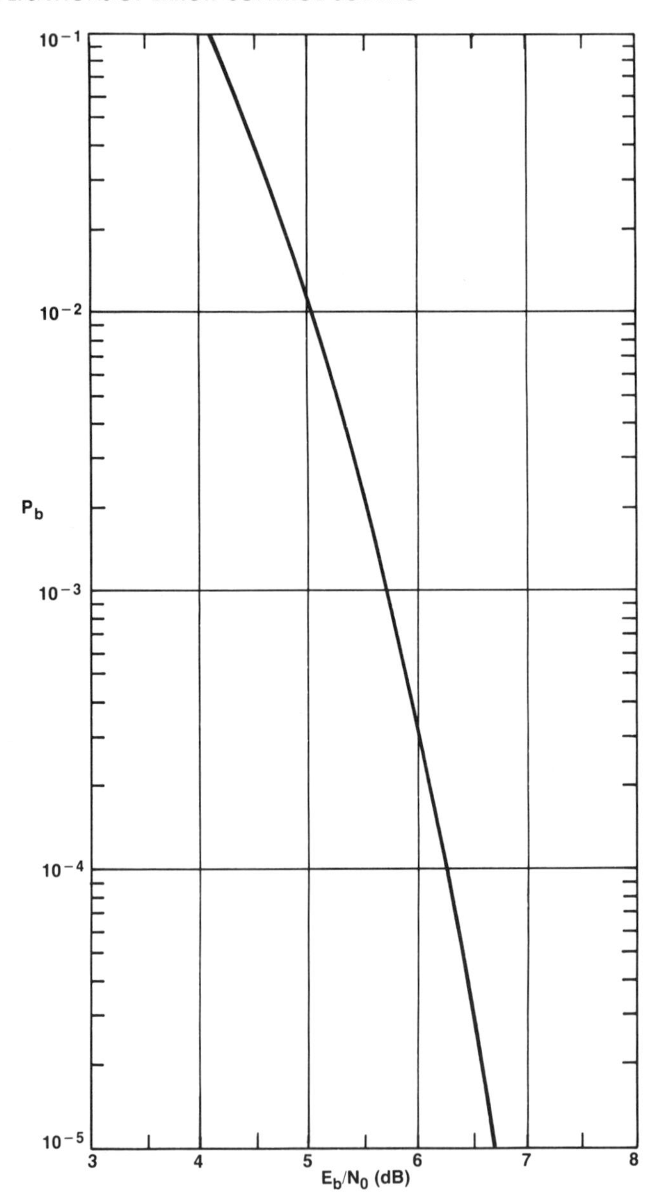

FIGURE 11.17. Post-decoding bit-error rate vs. E_b/N_0 for the $k = 7$, rate-$1/2$ convolutional code; three-bit soft-decision Viterbi decoding on the DPSK channel with interleaving.

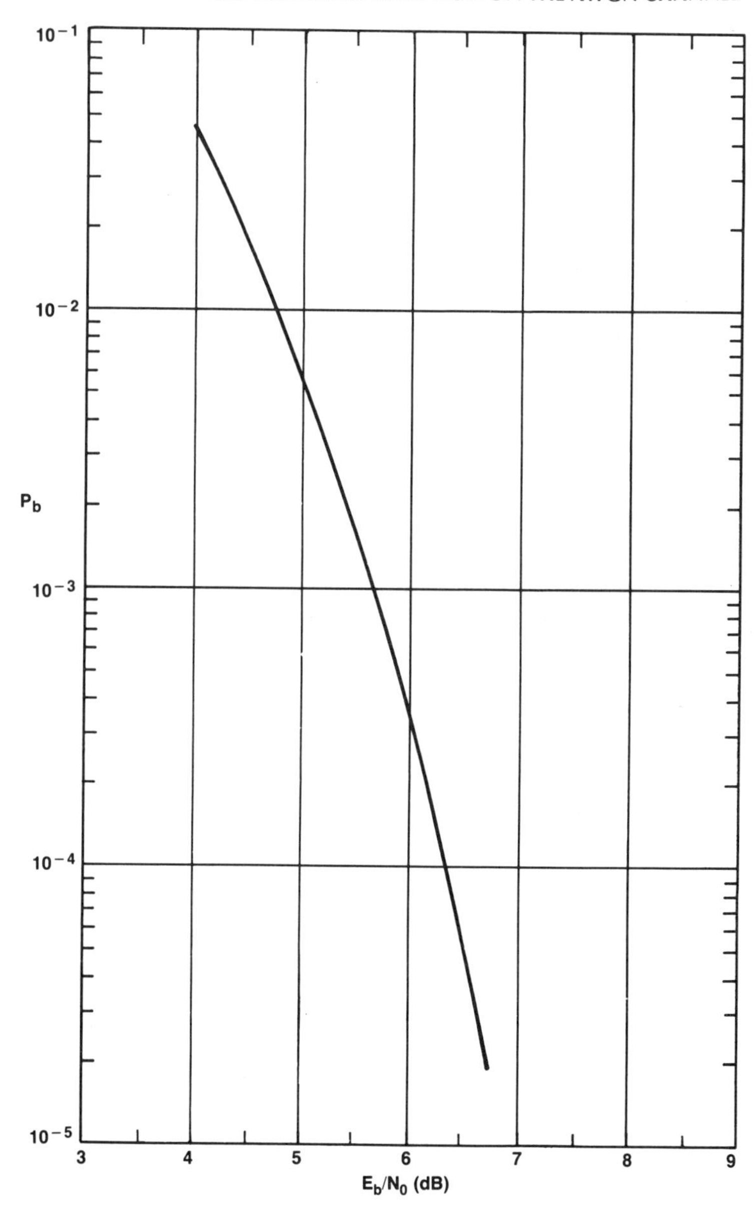

FIGURE 11.18. Post-decoding bit-error rate vs. E_b/N_0 for the $k = 7$ rate-1/3 convolutional code given by Trumpis with 8-ary orthogonal signaling and noncoherent detection.

signaling. Odenwalder [120] has obtained performance results for the $b = 1$, $k = 7$ convolutional code on the DPSK channel. The decoding metric used was the differential detector output quantized to three bits. Since Viterbi decoders are sensitive to input error bursts, and paired errors are more likely on the DPSK channel than on the independent-error channel, use of an interleaver was assumed. Figure 11.17 presents simulation results for the $b = 1$, $k = 7$ code decoded with a Viterbi decoder, and we note that for $E_b/N_0 \geq 7$ dB, low post-decoding BERs are observed. With $P_b = 10^{-5}$, a coding gain of 3.5 dB over uncoded DPSK is realized.

In terms of communication efficiency, Trumpis [170] showed that comparable performance can be obtained with 8-ary orthogonal signaling and the $k = 7$, rate-1/3 code given in Table 9.5. Simulation results were presented for soft-decision Viterbi decoding of the $k = 7$ code, and these results are shown in Fig. 11.18. The output of a *square-law detector* is assumed for the decoding metric. With $P_b = 10^{-5}$, a coding gain of about 2.2 dB relative to uncoded 8-ary signaling is obtained.

Odenwalder [119] has also described a class of nonbinary convolutional codes with a constraint length of 2 that can be conveniently used with orthogonal signaling and Viterbi decoding. These codes are called *dual-k* codes. The information and parity symbols are taken as elements of a finite field $GF(q)$, where $q = 2^k$, and $GF(q)$ arithmetic is used for encoding. As an example we consider the rate-1/2 dual-5 code, which is encoded as follows.

$$p_{1j} = i_j + i_{j-1}$$

$$p_{2j} = i_j + \alpha i_{j-1}$$

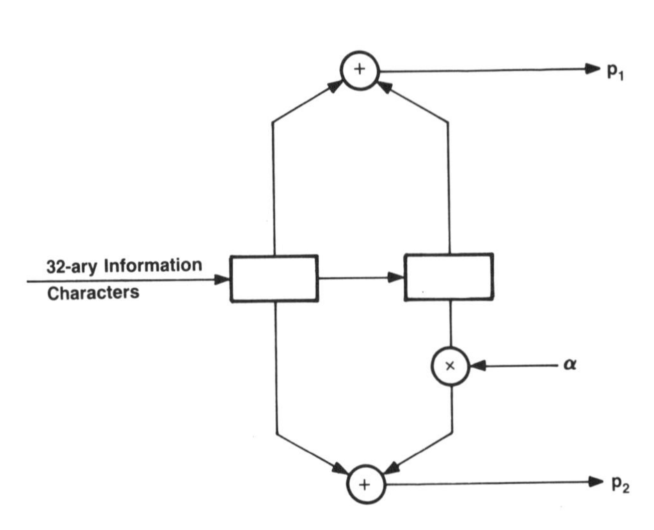

FIGURE 11.19. Encoder for the rate-1/2 dual-5 code.

where the i's and p's are the 32-ary information and parity symbols, respectively, and α is a primitive element of $GF(32)$.

An encoder for the dual-5 code is shown in Fig. 11.19. Each of the two delay elements indicated holds a 32-ary character, and one 32-ary information character is inserted in the register for each cycle of the encoder. The performance of the dual-5 code with Viterbi decoding and 32-ary orthogonal signaling is conveniently obtained with computer simulations for high output

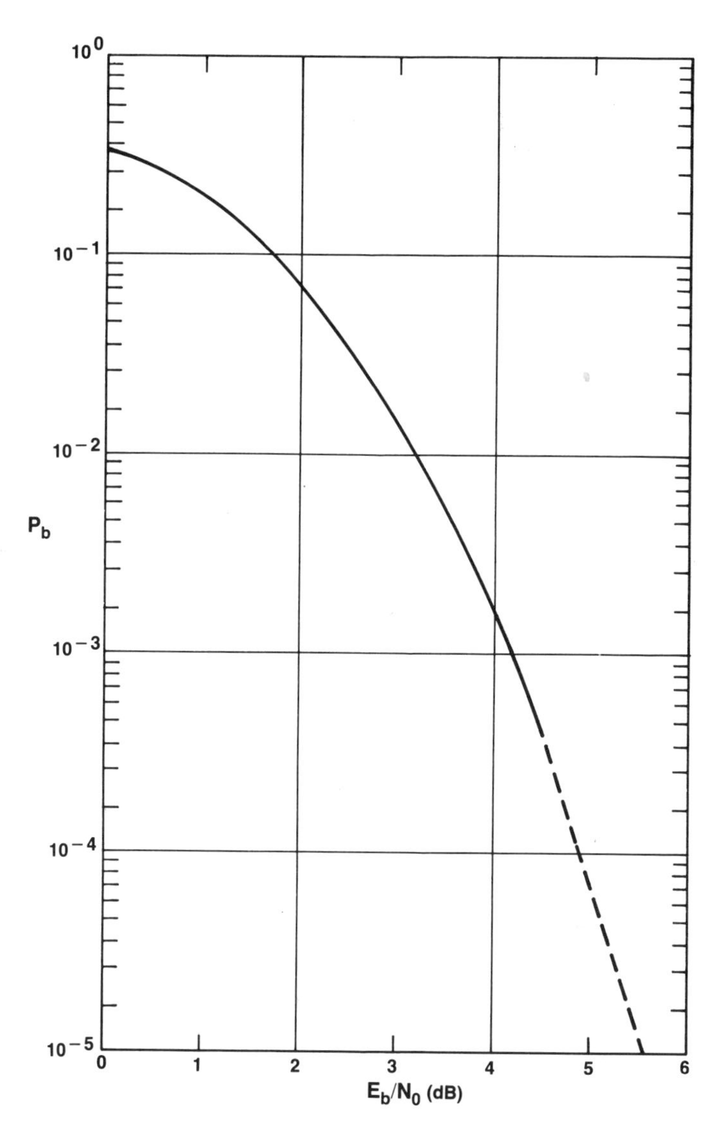

FIGURE 11.20. Post-decoding bit-error rate vs. E_b/N_0 for 32-ary orthogonal signaling, noncoherent detection, dual-5 convolutional code, and Viterbi decoding.

error rates and with bounding techniques for low error rates. An approach for obtaining the error-rate bound for Viterbi decoding was described in Chapter 9. It is assumed that the square-law detector output (unquantized) is used as the decoding metric.

Performance results for Viterbi decoding of the dual-5 code with 32-ary signaling and noncoherent detection are shown in Fig. 11.20. Post-decoding bit-error probability is shown as a function of E_b/N_0. The solid portion of the curve represents simulation results, and the dashed portion the bounding result. At a post-decoding BER of 10^{-5}, a coding gain of 1.5 dB is provided over uncoded 32-ary signaling.

11.3. CODING FOR COMPOUND-ERROR CHANNELS

Throughout this book we have been concerned principally with coding for independent-error channels, that is, channels for which the probability of error is constant and errors occur independently from one signaling interval to the next. However, as was pointed out in Chapter 1, the error events observed on many real communication channels exhibit a mixture of independent- and burst-error statistics. Such channels are called *compound-error channels*.

Examples of phenomena that produce bursts of errors are varied. On telephone channels, lightning, transients in central office switching equipment, and even maintenance activity can result in impulsive noise on circuits. On radio channels, atmospherics, multipath, fading, and interference from other users of the band are well-known phenomena that result in nonindependent error events. Error bursts can result from other sources in special situations, for example, a noisy engine ignition system on vehicle-mounted equipment, collocation of receiving antennas with high-power pulsed-radar systems, and even arcing across faulty welds on ships that have been at sea for a prolonged time. Many other examples can be cited.

We have seen that highly efficient error-controlled communication can be provided for the strictly independent-error channel. On channels that exhibit predominantly burst-error statistics, effective error-control schemes can also be devised. For example, consider a channel for which bursts of errors span at most B bits and each error burst is followed by a *guard space* of G error-free bits. When G is sufficiently large relative to B, several error-control schemes can be used quite effectively.

An example of such a scheme, called *burst trapping*, was proposed by Tong [169]. With this technique, which uses block codes, information bits are transmitted twice, once in the information bit positions of a codeword, and later added (mod 2) to the check set of a subsequent codeword. Should the initial codeword be decoded correctly on the first transmission, the information bits are recovered and are added to the check set of the appropriate word to be received later, and normal decoding of that subsequent word can proceed.

However, if the initial word is not successfully decoded due to an error burst, the assumption is made that the second transmission is received error-free, hence the need for an error-free guard space. In this case, the information bits of the subsequent word are encoded and the calculated check bits are added to the received check bits. This provides an estimate of the initial information bits that were lost due to the error burst.

The basic scheme proposed by Tong permits random-error correction during the normal decoding attempts. A generalization of burst trapping has been described [20, 127] that also provides for some limited random-error correction during the guard time. However, these burst-trapping schemes are effective only when the channel can be accurately characterized as having two well-defined states, a relatively good state and a bad state. Furthermore, the channel is required to remain in the good state for a specified minimum period of time and in the bad state for a specified maximum period of time.

Other schemes have been proposed for use on two-state channels. Examples are given by Gallager [85] and Sullivan [167]. Strict-sense burst-error-correcting codes have also been developed [34, 71]. RS codes, of course, provide some burst-error protection when used with binary signaling, since the effect of B consecutive bit errors is spread over a smaller number of RS code characters, and when error bursts can be identified, errors-and-erasures decoding can be used effectively. However, none of these schemes has been widely applied on real compound-error channels. Each approach is based on certain assumptions about the statistics of the communication channel, assumptions that are not likely to always be true. In the application of error-control coding on compound-error channels, accurate modeling of the rare events is important because it is these events that generally limit communication performance. If the assumed channel model does not accurately account for the rare occurrences, error-control schemes based on the model are not likely to be effective. Forney [41] has discussed these issues in detail.

There are two quite effective approaches to providing error protection on real channels exhibiting mixed random- and burst-error statistics. Neither approach depends on a detailed statistical model of the channel; therefore, should channel conditions change somewhat, performance levels do not change drastically. That is, robust performance is achieved. We briefly consider these two techniques in the following sections.

11.3.1. Automatic Repeat Request (ARQ)

Up to this point we have considered only *forward-error-control* techniques. That is, we assume that a message is encoded, transmitted over the communication channel and received; then an attempt is made to decode and deliver whatever data is available. That attempt may be successful or unsuccessful, but in either case no further processing is done. If, however, a feedback channel exists between the receiver and transmitter, requests can be

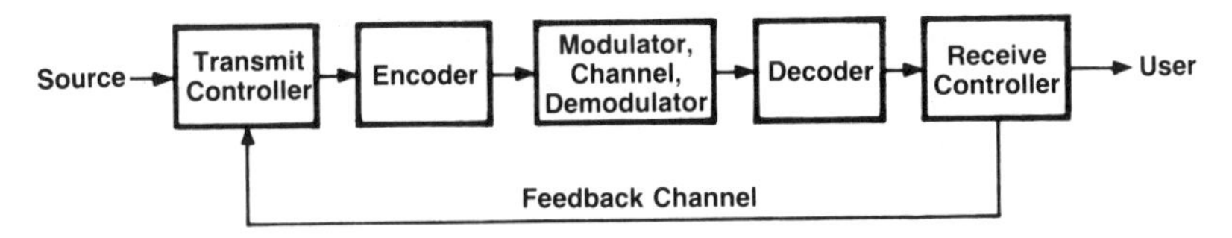

FIGURE 11.21. Basic block diagram of an ARQ system.

made for retransmission of data not successfully decoded. When feasible, such schemes can be used on compound-error channels to provide highly reliable and efficient communication.

The technique that utilizes a feedback channel to request message retransmission is called *automatic repeat request* (ARQ). It is by far the oldest and most widely applied error-control scheme in use today. A block diagram of a basic ARQ system is shown in Fig. 11.21. As usual, encoded digits are passed along to a modulator for transmission over the communication channel, and received digits are demodulated and decoded. The equipment shown in the figure includes transmit and receive controllers that exchange information by way of the feedback channel. The ARQ strategy is implemented within the controllers.

We shall consider ARQ systems that utilize block codes, although convolutional codes can be used as well. The simplest ARQ scheme is called the *stop-and-wait* strategy. With this approach, a block of data is encoded into a codeword and is transmitted over the channel. The transmitter then stops and waits until correct receipt of the codeword is acknowledged or a request for retransmission is received on the feedback channel. Generally, only error detection rather than detection and correction is implemented at the receiver. A noiseless feedback channel is usually assumed, but this is not a severe restriction, since the rate of information transfer in the feedback channel is low and a substantial amount of redundancy can be used to ensure reliable feedback transmission.

Several important aspects of ARQ can now be pointed out. First, since only error-detection decoding is to be employed, efficient high-rate block codes can be used, and the associated penalty in E_b/N_0 is small. As we have seen, there are many good high-rate block codes that can be used effectively for error detection and require simple encoders and decoders. A decoder can be implemented with an encoder and a circuit to compare received parity bits with those computed. Clearly, transmission can be made highly reliable by using a sufficient number of parity bits.

More important, though, ARQ is well suited for operation on many real compound-error channels. Specifically, should a channel have a set of relatively good states and a set of bad states, ARQ can be quite effective. When the channel is in a good state, most of the transmissions are decoded correctly on the first try, and when the channel switches to a bad state the link is, in effect, shut down. That is, attempts to transmit data are almost always unsuccessful and the communication system waits until the channel reverts to a good state to continue the communication process. For channels that are truly either good or bad, very efficient communication can be provided, and there are no restrictions on how long the channel must remain in either a good or bad state.

The application of ARQ is not limited to compound-error channels. ARQ can be considered for any communication channel provided that a feedback link exists and system requirements permit a variable information-transfer rate. A key figure of merit for an ARQ system is its *throughput efficiency*, which is defined as the ratio of the average number of information bits per second accepted at the receiver to the maximum data transmission rate on the channel. An obvious problem with stop-and-wait ARQ is that while the transmitter is idling, waiting for acknowledgments, transmission time is wasted and throughput suffers. When round-trip delays are long, throughput suffers appreciably. Thus, use of long code blocks is favored, but as the block length of the code increases, the probability of detected errors in a received word, and hence the probability of request for retransmission, also increases.

The problem of idling with stop-and-wait ARQ can be alleviated with the use of a slightly more complex strategy called *continuous ARQ* or *go-back-N*. With this approach, the transmitter does not wait for acknowledgments but rather continually transmits successive code blocks until a request for a retransmission is received. Then, the transmitter stops, backs up to the code block that was not successfully decoded, and restarts the transmission with that code block. All N blocks that were transmitted in the time interval between the original transmission and the receipt of the request for the retransmission are sent again in sequence.

The enhancement in throughput associated with continuous ARQ can be pronounced. However, many of the blocks that are retransmitted may have already been successfully received, as many as all $N - 1$ blocks following the one that was received with detected errors. Thus, additional gains in throughput can be realized if only those blocks that contain detected errors are retransmitted. This scheme is appropriately called *continuous ARQ with selected repeats*.

Continuous ARQ with selected repeats is the most complex of the three basic ARQ strategies, requires the most involved controllers and data buffers, and provides the best performance. Continuing interest in ARQ has produced numerous variations of the three basic ARQ schemes described thus far. For example, it has been proposed that forward error correction be implemented at the receiver to reduce the number of retransmissions that are required. Such schemes are called *hybrid ARQ*. A price is paid in code rate and complexity, but if the number of retransmissions required is reduced significantly, throughput is increased. In addition, other schemes have been proposed that use different modes of retransmission.

However, all ARQ systems have the same general performance characteristic. For good channel conditions, throughput is high, and for very poor channels, throughput is low. The use of forward error correction and other modifications increases the range of channel BER for which throughput remains high. These schemes are useful when the channel BER is expected to remain in this intermediate region for long periods of time. The detailed calculation of throughput as a function of BER can be complex, since it depends on many factors, including the channel, the size of the data buffers, round-trip time delay, and the detailed rules of the particular ARQ strategy.

This discussion is intended simply to introduce the basic concept of ARQ and identify the design issues. More detailed treatments can be found in several references, for example, references 19 and 94.

It is interesting to note that sequential decoding of convolutional codes provides features that are well suited for use in a hybrid ARQ system. Namely, we have seen that a sequential decoder can be designed to either decode a message correctly or quit and erase the received message. Very low incorrect decoding probability can be provided. Thus sequential decoding can be considered for use as the forward-error-control technique in an ARQ system. In reference 155, Shacham considers a hybrid ARQ technique that uses sequential decoding for a *packet radio* channel.

11.3.2. Interleaving

A straightforward and effective way to apply coding on a compound-error channel is to use *interleaving*. With this approach, a bursty channel is transformed into an independent-error channel, for which, as we have seen, many effective coding techniques are applicable. It is not necessary to know the detailed burst-error statistics of the channel to use interleaving effectively, but care must be taken when disturbances that produce periodic errors may exist.

A block diagram of an error-control system that uses interleaving is shown in Fig. 11.22. At the transmit end of the link, following error-control encoding, an interleaver, which scrambles the encoded data stream, is indicated. On the receive side the transmitted data is deinterleaved or descrambled prior to decoding. The intention is that successive bits (or symbols) transmitted over the communication channel be widely separated in the data sequence to be decoded. Thus, after deinterleaving, error bursts that occur on the communication channel are spread out in the data sequence to be decoded and may span many received codewords.

A simple interleaver structure for use with block codes can be visualized as a rectangular array having I rows and n columns. Encoded data are read into the array by rows and out by columns. Each element of the array is used to

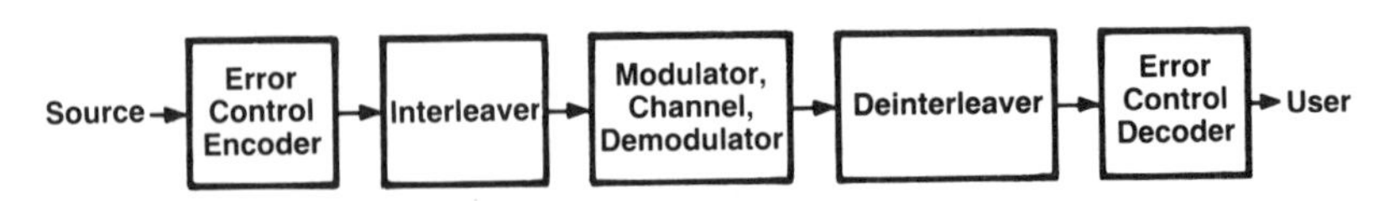

FIGURE 11.22. Typical structure for an error-control system using interleaving.

store one code symbol. Ordinarily the size of a row, n, is the block length of the code used, so that each row contains one codeword. The vertical dimension of the array, I, is called the *interleaving factor* or *interleaving degree*. It is common to refer to the entire array as an *interleaving frame*. It can be seen that an error burst spanning $l \leq I$ code symbols produces at most a single error in l consecutive codewords. Selection of I, therefore, depends on the expected error-burst lengths.

At the receiving end, data is read into an $I \times n$ array by columns and out by rows. Although bursts of channel errors are spread among successive codewords, it is important to note that periodic disturbances can produce many errors in a single word. For example, if the channel produces periodic transmission errors at a rate that is a multiple of I, many errors can be inserted in a single row or codeword.

The use of interleaved block codes has been studied for application on several real compound-error channels. For example, Pierce et al. [133] investigated the suitability of interleaved binary BCH codes on the high-frequency (HF) radio channel. HF suffers from a number of phenonema that produce bursts of errors, for example, frequency-selective and nonselective fading, atmospherics, and interference from other users of the band. In the study by Pierce, a sizable amount of recorded bit-error data collected with multitone HF modems was analyzed. It was shown that the (127,64) $t = 10$ binary BCH code with a 511×127 rectangular interleaver can provide up to 5 orders of magnitude improvement in post-decoding bit-error rate on poor-quality HF circuits. A key conclusion of Pierce's work was that in order for interleaving to be effective on fading HF radio circuits, the total size of the interleaving frame (In) should correspond to the duration of a typical fade period. Brayer [17] has also given results for interleaved code designs for HF and other real compound-error channels.

Although the performance improvements achievable with interleaved block codes can be dramatic for some real channels, two important system design issues must be considered. First, an additional level of framing or synchronization is required, namely, the synchronization of the interleaver and deinterleaver. In addition, for most real communication channels, a large interleaver is required to provide significant enhancements, and there can be a sizable associated time delay. In some applications, such as two-way digitized voice communication, a long time delay is unacceptable. It is only when the required time delay can be tolerated that interleaving is feasible.

Interleaving can also be used effectively with convolutional codes on compound-error channels. With convolutional codes a different interleaver structure is convenient. Ramsey [142] described several interleavers for convolutional codes, but we will consider one discussed by Forney [41]. Again encoded data to be transmitted are read into an array, but in this case the shape of the array is triangular. This type of interleaver is usually called a *convolutional interleaver*. Convolutional interleavers can be used with either convolutional or block codes.

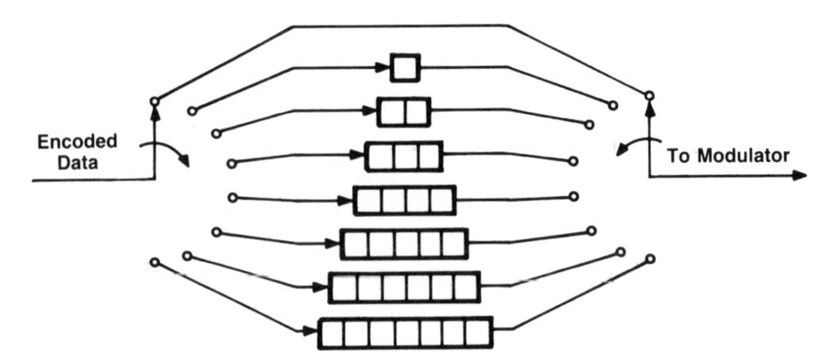

FIGURE 11.23. Convolutional interleaver structure.

A particular convolutional interleaver is indicated in Fig. 11.23. Each storage register holds a single bit, and encoded data are fed in one bit at a time. Inspection of the figure will show that adjacent bits in the encoded data sequence are separated by at least eight bits in the transmitted data. This is a consequence of the structure of the interleaver and the fact that there are eight tap connections on the commutators.

The circuit shown in Fig. 11.23 can also serve as a deinterleaver if the commutators are rotated in the reverse direction. It is important to note that a deinterleaver for a soft-decision decoder requires storage and shifting of nonbinary data. Other interleaver structures for convolutional codes have been described by Forney [41].

11.4. CONCLUDING REMARKS

We have only touched upon the issue of coding for compound-error channels here. In general, the basic techniques that have proved most useful are the same as those that provide good performance on independent-error channels, although much tailoring is required. At this point we would like to add that the coding gains realizable on compound-error channels can be much larger than those for independent-error channels. For example, as was mentioned in Chapter 1, some communication links have the property that error-rate curves "flatten out." That is, an irreducible error rate can exist for high SNRs, and no matter how much the transmitted signal power is increased the probability of error cannot be reduced further. In some such cases, a carefully chosen error-control technique can be used to reduce that error rate to whatever level of communication performance is required. One can argue that for these applications an infinite coding gain is provided.

It should also be pointed out that we have not thoroughly discussed some of the important issues that are closely related to making a good coding design work, for example, the registration of received codewords or codeword framing. In our discussion of decoding throughout the book, we have usually assumed that the beginning and end of a codeword are known; however, this cannot always be assumed. In some applications, multiple decoding attempts are used to accomplish framing and decoding simultaneously, but a price is paid in decoder complexity, since faster decoders are required. In addition, multiple decode attempts increase the probability of incorrect decoding.

Codeword framing, modem synchronization, AGC, and phase tracking are communication system design issues that must be addressed on a case-by-case basis. As a result, it is not possible to say unequivocally that one error-control scheme is always better than another. Consider, for example, concatenated codes and sequential decoding which can provide comparable performance. The features of these techniques are quite different and may or may not be useful in a particular application. Sequential decoding requires unambiguous codeword framing, while the framing requirements for the concatenation scheme that uses a convolutional inner code and Viterbi decoding are less severe. RS codewords must be framed, but this takes place after Viterbi decoding and consequently at an improved error rate. However, the concatenation scheme requires an interleaver, which must be synchronized, and there is an associated time delay. It is even difficult at times to directly compare the complexity of various coding alternatives because the computations each performs can be quite different. In addition, the dependence on the particular characteristics of the rapidly changing hardware technology is obvious.

In Chapter 1, the Shannon view of digital communication was introduced. It was argued that by using error-control coding, arbitrarily reliable communication can be provided for suitably chosen rates of transmission. It was pointed out that the noise on a communication channel does not limit the reliability that can be provided, but rather imposes a limit on the rate at which information can be reliably transmitted. Since Shannon's pioneering work, a good deal of effort has been devoted to finding techniques that achieve the efficient reliable communication performance promised by the channel coding theorem.

In this book, we have described many of the error-control techniques that have been developed to provide reliable digital communication. We have focused on the schemes that in our experience have proved to be the most useful. In this chapter we have seen that many error-control techniques can be used to provide a wide range of performance levels on the AWGN channel, improvements of a few to over 8 dB in communication efficiency. We have also seen that very low incorrect-decoding probabilities can be provided for all SNRs and that communication techniques exist today that deliver messages reliably at SNRs within 3 to 4 dB of channel capacity.

In hindsight it is easy to see that error-control coding was initially viewed improperly. Coding was thought by some to be a panacea for any and all communication system problems. As the first results were obtained, there was understandable disappointment, since the techniques were complex and did not achieve the promised levels of performance. However, the state of the art in hardware technology has advanced rapidly, and with the advent of the Viterbi algorithm and the subsequent decoder designs, real decoders were developed that actually justified their cost. Coding then began to be properly viewed as an engineering discipline that can be applied productively in the design and development of efficient digital communication systems.

What is important to recognize today is that a great variety of techniques exist that provide a wide range of coding gains. Designers can choose from a set of proven techniques with varying levels of complexity in addressing a particular performance requirement. The focus now is on applying the excellent techniques that have already been developed. Error-control coding has become a mature field and has taken its proper place as an engineering discipline.

■ ■ APPENDIX A

Matrix Notation and Terminology

Matrix notation provides a means for writing sets of equations in compact forms. This appendix provides a brief summary of commonly used matrix notation and terminology. Detailed presentations of the theory of matrices can be found in a number of textbooks, including Hildebrand [66] and Shields [160]. This appendix is adapted from a similar appendix in Wozencraft and Jacobs [185] (© 1965 Wiley) by permission of John Wiley & Sons, Inc.

BASIC DEFINITIONS

1. An $m \times n$ (read "m by n") *matrix* is an m-row, n-column rectangular array of elements such as

$$\mathbf{A} = \begin{bmatrix} a_{11} & a_{12} & \cdots & a_{1n} \\ a_{21} & a_{22} & \cdots & a_{2n} \\ \vdots & \vdots & & \vdots \\ a_{m1} & a_{m2} & \cdots & a_{mn} \end{bmatrix}$$

If the elements in $\mathbf{A}$ are in a finite field, say $GF(q)$, then $\mathbf{A}$ is called an $m \times n$ matrix over $GF(q)$.

2. The (i, j)th *element* a_{ij} of a matrix $\mathbf{A}$ is the element located in ith row and jth column of $\mathbf{A}$.

3. A matrix having every element equal to zero is called a *zero matrix* and is usually denoted by $\mathbf{0}$. A matrix having an equal number of rows and

columns is called a *square matrix*. The *main diagonal* of a square matrix is the set of elements a_{ii} extending from the top left corner to the bottom right corner of the matrix. A square matrix is called a *diagonal matrix* if all elements off the main diagonal are equal to zero, that is, $a_{ij} = 0$ for $i \neq j$. A diagonal matrix having every element along the main diagonal equal to unity is called an *identity matrix* and is usually denoted by **I**. When it is necessary to indicate the number k of rows or columns in **I**, the symbol $\mathbf{I}_k$ is used in place of **I**.

4. The *transpose* of an $m \times n$ matrix **A** is the $n \times m$ matrix, denoted by $\mathbf{A}^T$, obtained by interchanging the rows and columns of **A**. An example is as follows:

$$\mathbf{A} = \begin{bmatrix} 1 & 2 & 3 \\ 4 & 5 & 6 \end{bmatrix} \qquad \mathbf{A}^T = \begin{bmatrix} 1 & 4 \\ 2 & 5 \\ 3 & 6 \end{bmatrix}$$

5. A $1 \times n$ (single-row) matrix is called a *row vector*, while an $m \times 1$ (single-column) matrix is called a *column vector*. It is conventional to use a lowercase boldface letter such as **a** to denote a row or column vector; the orientation of the vector is usually obvious from the context in which it is used. It is sometimes convenient to denote a matrix in terms of vectors corresponding to its rows or columns. For example, we can express an $m \times n$ matrix **A** as $\mathbf{A} = (\mathbf{a}_1, \mathbf{a}_2, \ldots, \mathbf{a}_n)$, where $\mathbf{a}_1, \mathbf{a}_2, \ldots, \mathbf{a}_n$ are the $m \times 1$ column vectors constituting the respective columns of **A**.

Another notational convenience is provided by partitioning a matrix into submatrices. If any rows or columns, or both, of a matrix **A** are deleted, the rectangular array that remains is called a *submatrix* of **A**. A trivial example is the description of an element a_{ij} as a submatrix of **A** formed by deleting all rows except row i and all columns except column j. A *partitioned matrix* is a matrix expressed as a rectangular array of submatrices, where each submatrix is itself a matrix of elements in a contiguous set of rows and a contiguous set of columns. For example, the matrix

$$\mathbf{A} = \begin{bmatrix} 1 & 2 & 3 & 4 \\ 5 & 6 & 7 & 8 \\ 9 & 0 & 1 & 2 \end{bmatrix}$$

may be written in partitioned forms such as

$$\mathbf{A} = [\mathbf{A}_1; \mathbf{A}_2] = \begin{bmatrix} \mathbf{A}_{11} & \mathbf{A}_{12} \\ \mathbf{A}_{21} & \mathbf{A}_{22} \end{bmatrix}$$

where

$$\mathbf{A}_1 = \begin{bmatrix} 1 & 2 \\ 5 & 6 \\ 9 & 0 \end{bmatrix}, \qquad \mathbf{A}_2 = \begin{bmatrix} 3 & 4 \\ 7 & 8 \\ 1 & 2 \end{bmatrix}$$

and

$$\mathbf{A}_{11} = [1], \qquad \mathbf{A}_{12} = [2 \quad 3 \quad 4]$$

$$\mathbf{A}_{21} = \begin{bmatrix} 5 \\ 9 \end{bmatrix}, \qquad \mathbf{A}_{22} = \begin{bmatrix} 6 & 7 & 8 \\ 0 & 1 & 2 \end{bmatrix}$$

Submatrices appearing in partitioned forms such as the examples given above are also called *blocks*.

6. Two matrices with the same dimensions are said to be *equal* if and only if all corresponding elements are equal. That is, $\mathbf{A} = \mathbf{B}$ if and only if $a_{ij} = b_{ij}$ for all i, j.

OPERATIONS

1. The *sum* of two $m \times n$ matrixes $\mathbf{A}$ and $\mathbf{B}$ is an $m \times n$ matrix $\mathbf{C}$ obtained by adding corresponding elements in $\mathbf{A}$ and $\mathbf{B}$, that is, $c_{ij} = a_{ij} + b_{ij}$ for all i, j. This is written in matrix notation as $\mathbf{C} = \mathbf{A} + \mathbf{B}$. The addition of matrices is associative and commutative, as in the addition of numbers in ordinary arithmetic. Thus matrices of a given shape and size form a commutative group under addition (see Section 3.1). The sum of two matrices having different dimensions is not defined.

2. The *scalar product* or *scalar multiple* of an $m \times n$ matrix $\mathbf{A}$ by a constant c is a new $m \times n$ matrix $\mathbf{B}$ formed by multiplying each element of $\mathbf{A}$ by c, that is, $b_{ij} = ca_{ij}$ for all i, j. We write this as $\mathbf{B} = c\mathbf{A}$. The laws of addition and multiplication for ordinary numbers also hold for scalar multiples of matrices. That is, for any two constants c and c' and any two $m \times n$ matrices $\mathbf{A}$ and $\mathbf{A}'$, we have

$$(cc')\mathbf{A} = c(c'\mathbf{A})$$

$$c(\mathbf{A} + \mathbf{A}') = c\mathbf{A} + c\mathbf{A}'$$

$$(c + c')\mathbf{A} = c\mathbf{A} + c'\mathbf{A}$$

3. If $\mathbf{a}$ is a row vector and $\mathbf{b}$ is a column vector having the same number of elements (say n), then the *product* of $\mathbf{a}$ times $\mathbf{b}$, denoted by $c = \mathbf{ab}$, is a single number equal to the sum of the products of corresponding elements in $\mathbf{a}$ and $\mathbf{b}$, that is,

$$c = a_1 b_1 + a_2 b_2 + \cdots + a_n b_n$$

In conventional matrix notation, the product of two row vectors (or *row matrices*) or two column vectors (*column matrices*) is not defined. However, if we had let $\mathbf{a}$ and $\mathbf{b}$ both be row vectors just above, we could have written

$c = \mathbf{a}\mathbf{b}^T$. The reader familiar with vector analysis will see that c is algebraically identical to the *dot product* of vectors **a** and **b**, where the association of the vectors with rows or columns of matrices is superfluous.

The product of a $k \times m$ matrix **A** and an $m \times n$ matrix **B** is a $k \times n$ matrix **C** in which each element c_{ij} is the product of the ith row in **A** times the jth column in **B**. That is, denoting the *matrix product* as

$$\mathbf{C} = \mathbf{AB}$$

we have

$$c_{ij} = \sum_{l=1}^{m} a_{il} b_{lj}$$

where l is a dummy index that ranges over the column dimension of **A** and the (identical) row dimension of **B**. In computing a matrix product with pencil and paper, it is convenient to position the matrices in a way that makes it easy to visualize the row and column to be multiplied together in forming each element in the product. This is done in the following example, where the 2×3 matrix **A** is multiplied by the 3×4 matrix **B** to form the 2×4 matrix **C**, multiplications and additions being done in ordinary arithmetic.

$$\mathbf{B} \rightarrow \begin{bmatrix} 1 & 2 & 3 & 4 \\ 5 & 6 & 7 & 8 \\ 9 & 0 & 1 & 2 \end{bmatrix}$$

$$\mathbf{A} \rightarrow \begin{bmatrix} 1 & 2 & 3 \\ 4 & 5 & 6 \end{bmatrix} \qquad \begin{bmatrix} 38 & 14 & 20 & 26 \\ 83 & 38 & 53 & 68 \end{bmatrix} \leftarrow \mathbf{C}$$

In order for the product of two matrices (say **A** times **B**) to be defined, the matrices must be *conformable*, that is, the number of columns in the first matrix (**A**) must equal the number of rows in the second matrix (**B**). If this relationship does not hold, the two matrices are said to be *nonconformable* and the product **AB** is not defined. The number of rows in **A** and the number of columns in **B** are arbitrary; in the special case where each is just one, the product **AB** is simply the product of a single row vector times a single column vector, and **C** contains just one element. It is straightforward to form a triple product, say $\mathbf{D} = (\mathbf{AB})\mathbf{C}$, as long as **A** and **B** are conformable and **B** and **C** are conformable. For example, we might multiply a 3×7 matrix **A** by a 7×5 matrix **B** and then multiply the resulting 3×5 matrix product by a 5×4 matrix **C**, yielding a 3×4 matrix **D**. It is readily seen that we can therefore multiply an arbitrarily long string of matrices, so long as each adjacent pair is conformable, and the dimensions of the final product are the row dimension of the first matrix and the column dimension of the last matrix in the string.

It is sometimes advantageous in multiplying matrices to express the matrices in partitioned form. Let **A** and **B** be written in partitioned form as

$$\mathbf{A} = \begin{bmatrix} \mathbf{A}_{11} & \cdots & \mathbf{A}_{1r} \\ \vdots & & \vdots \\ \mathbf{A}_{q1} & \cdots & \mathbf{A}_{qr} \end{bmatrix}, \qquad \mathbf{B} = \begin{bmatrix} \mathbf{B}_{11} & \cdots & \mathbf{B}_{1t} \\ \vdots & & \vdots \\ \mathbf{B}_{r1} & \cdots & \mathbf{B}_{rt} \end{bmatrix}$$

If the matrix products $\mathbf{A}_{ij}\mathbf{B}_{jk}$ are defined for all values of i, j, and k, the product **AB** may be computed in terms of submatrices as follows:

$$\mathbf{AB} = \mathbf{C} = \begin{bmatrix} \mathbf{C}_{11} & \cdots & \mathbf{C}_{1t} \\ \vdots & & \vdots \\ \mathbf{C}_{q1} & \cdots & \mathbf{C}_{qt} \end{bmatrix}$$

where

$$\mathbf{C}_{ij} = \mathbf{A}_{i1}\mathbf{B}_{1j} + \mathbf{A}_{i2}\mathbf{B}_{2j} + \cdots + \mathbf{A}_{ir}\mathbf{B}_{rj}$$

Each product $\mathbf{A}_{ij}\mathbf{B}_{jk}$ is, of course, defined as long as the number of columns in $\mathbf{A}_{ij}$ equals the number of rows in $\mathbf{B}_{jk}$, which is to say that $\mathbf{A}_{ij}$ is conformable to $\mathbf{B}_{jk}$. Thus we see that if **A** and **B** are expressed with a conformable set of partitionings, the product **AB** may be computed using the "row-by-column" rule of multiplication as if the submatrices were ordinary matrix elements using the definitions of matrix addition and multiplication. Matrix multiplication performed in this manner is called *block multiplication*.

PROPERTIES OF MATRIX MULTIPLICATION

The definitions given above imply certain key properties of matrix multiplication. All these properties can be verified directly from the definitions of matrix multiplication and addition.

1. Matrix multiplication is *distributive*, that is,

$$\mathbf{A}(\mathbf{B} + \mathbf{C}) = \mathbf{AB} + \mathbf{AC}, \quad \text{and} \quad (\mathbf{B} + \mathbf{C})\mathbf{D} = \mathbf{BD} + \mathbf{CD}$$

2. Matrix multiplication is *associative*, that is, $(\mathbf{AB})\mathbf{C} = \mathbf{A}(\mathbf{BC})$. This can be verified by showing that an element of either triple product, say d_{ij}, is given by

$$d_{ij} = \sum_m \sum_n a_{im} b_{mn} c_{nj}$$

regardless of which multiplication is carried out first.

3. Matrix multiplication is not generally *commutative*. That is, $\mathbf{AB} \neq \mathbf{BA}$ in general, although equality may hold in special instances. It is clear that two matrices conformable in one order need not be conformable in the other order. However, even in the case of two $m \times m$ matrices, multiplication is not necessarily commutative, and in fact is generally not. For example, if

$$\mathbf{A} = \begin{bmatrix} 1 & 2 \\ 3 & x \end{bmatrix} \quad \text{and} \quad \mathbf{B} = \begin{bmatrix} 3 & 1 \\ 2 & y \end{bmatrix}$$

we have

$$\mathbf{AB} = \begin{bmatrix} 7 & 1 + 2y \\ 9 + 2x & 3 + xy \end{bmatrix}, \qquad \mathbf{BA} = \begin{bmatrix} 6 & 6 + x \\ 2 + 3y & 4 + xy \end{bmatrix}$$

and by looking at the (1, 1) element in each matrix product, we see that $\mathbf{AB} \neq \mathbf{BA}$ regardless of the values of x and y.

INVERSE MATRICES

Let $\mathbf{I}$ denote the $n \times n$ identity matrix, defined earlier. The *inverse* of an $n \times n$ matrix $\mathbf{A}$ is another $n \times n$ matrix, denoted by $\mathbf{A}^{-1}$, such that

$$\mathbf{AA}^{-1} = \mathbf{A}^{-1}\mathbf{A} = \mathbf{I}$$

Not every square matrix has an inverse; if $\mathbf{A}$ has an inverse, then $\mathbf{A}$ is said to be *nonsingular*. If $\mathbf{A}$ has no inverse, it is said to be *singular*. Let the inverse of $\mathbf{A}$ be denoted also by $\mathbf{B}$, that is, $\mathbf{B} = \mathbf{A}^{-1}$. It can be readily shown that the (i, j)th element of $\mathbf{B}$ is computed as

$$b_{ij} = (-1)^{i+j} \frac{|\mathbf{A}_{ji}|}{|\mathbf{A}|} \tag{A.1}$$

where $\mathbf{A}_{ji}$ is the submatrix formed by deleting the jth row and ith column from $\mathbf{A}$, and $|\mathbf{A}_{ji}|$ is its *determinant*, to be defined presently. [Note that the subscripts i and j are in reversed order on the two sides of Eq. (A.1).] The denominator $|\mathbf{A}|$ is the determinant of $\mathbf{A}$. The determinant of an $n \times n$ matrix is a sum of signed terms, each an n-fold product of the form

$$(-1)^t a_{1i_1} a_{2i_2} \cdots a_{ni_n}$$

where the set $(i_1, \ldots, i_n)$ is a permutation of $(1, \ldots, n)$, so that there is one element from each row and column of $\mathbf{A}$ in each product. There are $n! = n(n-1) \cdots 1$ such permutations of elements from $\mathbf{A}$, and the determinant of $\mathbf{A}$ is defined as the sum of all $n!$ products, with signs $(-1)^t$ appropriately chosen. The value of t, and hence the sign, for each product is determined by the particular permutation of elements in the product. A simple statement of the rule for setting t is as follows. Let the elements in a given product be joined in pairs by line segments. The total number of such segments sloping upward to the right is the value to be assigned to t. It is readily seen that odd values of t produce negative products, while even values (including zero) produce positive products. As a simple example, we compute the determinant of a 2×2 matrix as follows.

$$|\mathbf{A}| = \begin{vmatrix} a_{11} & a_{12} \\ a_{21} & a_{22} \end{vmatrix} = a_{11}a_{22} - a_{12}a_{21}$$

A convenient method for calculation of a determinant is provided by the *Laplace expansion* formula,

$$|\mathbf{A}| = \sum_{j=1}^{n} a_{ij}(-1)^{i+j}|\mathbf{M}_{ij}| = \sum_{i=1}^{n} a_{ij}(-1)^{i+j}|\mathbf{M}_{ij}|$$

for any i or j, where $\mathbf{M}_{ij}$ is the submatrix of **A** formed by deleting the ith row and jth column. The quantity

$$(-1)^{i+j}|\mathbf{M}_{ij}|$$

is called the *cofactor* of the matrix element a_{ij}. The alternate forms of the Laplace expansion formula denote the fact that the expansion may be performed with any row or column of **A**. If the matrix **A** contains zeros, advantage can be taken by expanding about a row or column containing as many zeros as possible, so as to minimize the number of products to be computed.

Returning now to the definition of the inverse matrix, as given by Eq. (A.1), we see that if the determinant $|\mathbf{A}|$ is zero, the inverse of **A** is undefined, since its calculation would require division by zero. Thus a square matrix having its determinant equal to zero is a singular matrix. It is possible to show that an $n \times n$ matrix has a nonzero determinant if, and only if, its n rows (or n columns) are *linearly independent*, that is, if no sum of scalar products of the rows (or columns), except with the set of all-zero scalars, sums to the all-zeros row (or column) vector. In the context of using matrix methods to solve sets of simultaneous linear equations, this is completely equivalent to saying that we use n equations to find a unique solution for n unknowns if, and only if, the n equations are linearly independent. The maximum number of linearly independent rows of a matrix **A** is called the *row rank* of **A**. It can be shown that this value is equal to the maximum number of linearly independent columns of **A**, called the *column rank* of **A**. The value is usually referred to simply as the *rank* of **A**. The rank of **A** can be defined equivalently as the size of the largest square submatrix of **A** whose determinant is nonzero.

Tables of Irreducible Polynomials over *GF*(2)

This appendix is reprinted, with minor editorial changes, from Peterson and Weldon, *Error-Correcting Codes*, 2nd ed., MIT Press (© 1972 The Massachusetts Institute of Technology) by permission of the MIT Press.

From Table B.2 all irreducible polynomials of degree 16 or less over $GF(2)$ can be found, and certain of their properties and relations among them are given. A primitive polynomial with a minimum number of nonzero coefficients and polynomials belonging to all possible exponents are given for each degree 17 through 34.

Polynomials are given in an octal representation. Each digit in the table represents three binary digits according to the following code:

0	000	2	010	4	100	6	110
1	001	3	011	5	101	7	111

The binary digits then are the coefficients of the polynomial, with the high-order coefficients at the left. For example, 3525 is listed as a tenth-degree polynomial. The binary equivalent of 3525 is 011101010101, and the corresponding polynomial is $X^{10} + X^9 + X^8 + X^6 + X^4 + X^2 + 1$.

The reciprocal polynomial of an irreducible polynomial is also irreducible, and the reciprocal polynomial of a primitive polynomial is primitive. Of any pair consisting of a polynomial and its reciprocal polynomial, only one is listed in the table. Each entry that is followed by a letter in the table is an irreducible polynomial of the indicated degree. For degree 2 through 16, these polynomials along with their reciprocal polynomials comprise all irreducible polynomials of that degree.

The letters following the octal representation give the following information:

A, B, C, D	Not primitive.
E, F, G, H	Primitive.
A, B, E, F	The roots are linearly dependent.
C, D, G, H	The roots are linearly independent.
A, C, E, G	The roots of the reciprocal polynomial are linearly dependent.
B, D, F, H	The roots of the reciprocal polynomial are linearly independent.

The other numbers in the table tell the relation between the polynomials. For each degree, a primitive polynomial with a minimum number of nonzero coefficients was chosen, and this polynomial is the first in the table of polynomials of this degree. Let α denote one of its roots. Then the entry following j in the table is the minimum polynomial of α^j. The polynomials are included for each j unless for some $i < j$ either α^i and α^j are roots of the same irreducible polynomial or α^i and α^{-j} are roots of the same polynomial. The minimum polynomial of α^j is included even if it has smaller degree than is indicated for that section of the table; such polynomials are not followed by a letter in the table.

EXAMPLES. The primitive polynomial (103), or $X^6 + X + 1 = p(X)$ is the first entry in the table of sixth-degree irreducible polynomials. If α designates a root of $p(X)$, then α^3 is a root of (127) and α^5 is a root of (147). The minimum polynomial of α^9 is (015) $= X^3 + X^2 + 1$ and is of degree 3 rather than 6.

There is no entry corresponding to α^{17}. The other roots of the minimum polynomial of α^{17} are α^{34}, $\alpha^{68} = \alpha^5, \alpha^{10}, \alpha^{20}$, and α^{40}. Thus the minimum polynomial of α^{17} is the same as the minimum polynomial of α^5, or (147). There is no entry corresponding to α^{13}. The other roots of the minimum polynomial $p_{13}(X)$ of α^{13} are α^{26}, α^{52}, $\alpha^{104} = \alpha^{41}$, $\alpha^{82} = \alpha^{19}$, and α^{38}. None of these is listed. The roots of the reciprocal polynomial $p_{13}^*(X)$ of $p_{13}(X)$ are $\alpha^{-13} = \alpha^{50}$, $\alpha^{-26} = \alpha^{37}$, $\alpha^{-52} = \alpha^{11}$, $\alpha^{-41} = \alpha^{22}$, $\alpha^{-19} = \alpha^{44}$ and $\alpha^{-38} = \alpha^{25}$. The minimum polynomial of α^{11} is listed as (155) or $X^6 + X^5 + X^3 + X^2 + 1$. The minimum polynomial of α^{13} is the reciprocal polynomial of this, or $p_{13}(X) = X^6 + X^4 + X^3 + X + 1$.

The exponent to which a polynomial belongs can be found as follows: If α is a primitive element of $GF(2^m)$, then the order e of α^j is

$$e = \frac{(2^m - 1)}{\text{GCD}(2^m - 1, j)}$$

and e is also the exponent to which the minimum function of a^j belongs.

Thus, for example, in $GF(2^{10})$, α^{55} has order 93, since

$$93 = \frac{1023}{\text{GCD}(1023,55)} = \frac{1023}{11}$$

Thus the polynomial (3453) belongs to 93. In this regard Table B.1 is useful.

Marsh [100] has published a table of all irreducible polynomials of degree 19 or less over $GF(2)$. In Table B.2 the polynomials are arranged in lexicographical order; this is the most convenient form for determining whether or not a given polynomial is irreducible.

For degree 19 or less, the minimum-weight polynomials given in this table were found in Marsh's tables. For degree 19 through 34, the minimum-weight polynomial was found by a trial-and-error process in which each polynomial of weight 3, then 5, was tested. The following procedure was used to test whether a polynomial $f(X)$ of degree m is primitive:

1. The residues of 1, X, X^2, $X^4, \ldots, X^{2m-1}$ are formed modulo $f(X)$.
2. These are multiplied and reduced modulo $f(X)$ to form the residue of $X^{2^m} - 1$. If the result is not 1, the polynomial is rejected. If the result is 1, the test is continued.
3. For each factor r of $2^m - 1$, the residue of X^r is formed by multiplying together an appropriate combination of the residues formed in step 1. If none of these is 1, the polynomial is primitive.

Each other polynomial in the table was found by solving for the dependence relations among its roots by the method illustrated at the end of Section 8.1 in Peterson and Weldon [131].

TABLE B.1. Factorization of $2^m - 1$ into Primes

$2^3 - 1 = 7$	$2^{19} - 1 = 524287$
$2^4 - 1 = 3 \times 5$	$2^{20} - 1 = 3 \times 5 \times 5 \times 11 \times 31 \times 41$
$2^5 - 1 = 31$	$2^{21} - 1 = 7 \times 7 \times 127 \times 337$
$2^6 - 1 = 3 \times 3 \times 7$	$2^{22} - 1 = 3 \times 23 \times 89 \times 683$
$2^7 - 1 = 127$	$2^{23} - 1 = 47 \times 178481$
$2^8 - 1 = 3 \times 5 \times 17$	$2^{24} - 1 = 3 \times 3 \times 5 \times 7 \times 13 \times 17 \times 241$
$2^9 - 1 = 7 \times 73$	$2^{25} - 1 = 31 \times 601 \times 1801$
$2^{10} - 1 = 3 \times 11 \times 31$	$2^{26} - 1 = 3 \times 2731 \times 8191$
$2^{11} - 1 = 23 \times 89$	$2^{27} - 1 = 7 \times 73 \times 262657$
$2^{12} - 1 = 3 \times 3 \times 5 \times 7 \times 13$	$2^{28} - 1 = 3 \times 5 \times 29 \times 43 \times 113 \times 127$
$2^{13} - 1 = 8191$	$2^{29} - 1 = 233 \times 1103 \times 2089$
$2^{14} - 1 = 3 \times 43 \times 127$	$2^{30} - 1 = 3 \times 3 \times 7 \times 11 \times 31 \times 151 \times 331$
$2^{15} - 1 = 7 \times 31 \times 151$	$2^{31} - 1 = 2147483647$
$2^{16} - 1 = 3 \times 5 \times 17 \times 257$	$2^{32} - 1 = 3 \times 5 \times 17 \times 257 \times 65537$
$2^{17} - 1 = 131071$	$2^{33} - 1 = 7 \times 23 \times 89 \times 599479$
$2^{18} - 1 = 3 \times 3 \times 3 \times 7 \times 19 \times 73$	$2^{34} - 1 = 3 \times 43691 \times 131071$

TABLE B.2. Irreducible Polynomials of Degree ≤ 34 over $GF(2)$

DEGREE 2	1 7H				
DEGREE 3	1 13F				
DEGREE 4	1 23F	3 37D	5 07		
DEGREE 5	1 45E	3 75G	5 67H		
DEGREE 6	1 103F	3 127B	5 147H	7 111A	9 015
11 155E	21 007				
DEGREE 7	1 211E	3 217E	5 235E	7 367H	9 277E
11 325G	13 203F	19 313H	21 345G		
DEGREE 8	1 435E	3 567B	5 763D	7 551E	9 675C
11 747H	13 453F	15 727D	17 023	19 545E	21 613D
23 543F	25 433B	27 477B	37 537F	43 703H	45 471A
51 037	85 007				
DEGREE 9	1 1021E	3 1131E	5 1461G	7 1231A	9 1423G
11 1055E	13 1167F	15 1541E	17 1333F	19 1605G	21 1027A
23 1751E	25 1743H	27 1617H	29 1553H	35 1401C	37 1157F
39 1715E	41 1563H	43 1713H	45 1175E	51 1725G	53 1225E
55 1275E	73 0013	75 1773G	77 1511C	83 1425G	85 1267E
DEGREE 10	1 2011E	3 2017B	5 2415E	7 3771G	9 2257B
11 2065A	13 2157F	15 2653B	17 3515G	19 2773F	21 3753D
23 2033F	25 2443F	27 3573D	29 2461E	31 3043D	33 0075C
35 3023H	37 3543F	39 2107B	41 2745E	43 2431E	45 3061C
47 3177H	49 3525G	51 2547B	53 2617F	55 3453D	57 3121C
59 3471G	69 2701A	71 3323H	73 3507H	75 2437B	77 2413B
83 3623H	85 2707E	87 2311A	89 2327F	91 3265G	93 3777D
99 0067	101 2055E	103 3575G	105 3607C	107 3171G	109 2047F
147 2355A	149 3025G	155 2251A	165 0051	171 3315C	173 3337H
179 3211G	341 0007				
DEGREE 11	1 4005E	3 4445E	5 4215E	7 4055E	9 6015G
11 7413H	13 4143F	15 4563F	17 4053F	19 5023F	21 5623F
23 4757B	25 4577F	27 6233H	29 6673H	31 7237H	33 7335G
35 4505E	37 5337F	39 5263F	41 5361E	43 5171E	45 6637H
47 7173H	49 5711E	51 5221E	53 6307H	55 6211G	57 5747F
59 4533F	61 4341E	67 6711G	69 6777D	71 7715G	73 6343H
75 6227H	77 6263H	79 5235E	81 7431G	83 6455G	85 5247F
87 5265E	89 5343B	91 4767F	93 5607F	99 4603F	101 6561G
103 7107H	105 7041G	107 4251E	109 5675E	111 4173F	113 4707F
115 7311C	117 5463F	119 5755E	137 6675G	139 7655G	141 5531E
147 7243H	149 7621G	151 7161G	153 4731E	155 4451E	157 6557H
163 7745G	165 7317H	167 5205E	169 4565E	171 6765G	173 7535G
179 4653F	181 5411E	183 5545E	185 7565G	199 6543H	201 5613F
203 6013H	205 7647H	211 6507H	213 6037H	215 7363H	217 7201G
219 7273H	293 7723H	299 4303B	301 5007F	307 7555G	309 4261E
331 6447H	333 5141E	339 7461G	341 5253F		
DEGREE 12	1 10123F	3 12133B	5 10115A	7 12153B	9 11765A
11 15647E	13 12513B	15 13077B	17 16533H	19 16047H	21 10065A
23 11015E	25 13377B	27 14405A	29 14127H	31 17673H	33 13311A
35 10377B	37 13565E	39 13321A	41 15341G	43 15053H	45 15173C
47 15621E	49 17703C	51 10355A	53 15321G	55 10201A	57 12331A
59 11417E	61 13505E	63 10761A	65 00141	67 13275E	69 16663C
71 11471E	73 16237E	75 16267D	77 15115C	79 12515E	81 17545C
83 12255E	85 11673B	87 17361A	89 11271E	91 10011A	93 14755C
95 17705A	97 17121G	99 17323D	101 14227H	103 12117E	105 13617A
107 14135G	109 14711G	111 15415C	113 13131E	115 13223A	117 16475C
119 14315C	121 16521E	123 13475A	133 11433B	135 10571A	137 15437G
139 12067F	141 13571A	143 12111A	145 16535C	147 17657D	149 12147F
151 14717F	153 13517B	155 14241C	157 14675G	163 10663F	165 10621A

TABLE B.2. (*Continued*)

```
DEGREE 12--CONTINUED
  167 16115G   169 16547C   171 10213B   173 12247E   175 16757D   177 16017C
  179 17675E   181 10151E   183 14111A   185 14037A   187 14613H   189 13535A
  195 00165    197 11441E   199 10321E   201 14067D   203 13157B   205 14513D
  207 10603A   209 11067F   211 14433F   213 16457D   215 10653B   217 13563B
  219 11657B   221 17513C   227 12753F   229 13431E   231 10167B   233 11313F
  235 11411A   237 13737B   239 13425E   273 00023    275 14601C   277 16021G
  279 16137D   281 17025G   283 15723F   285 17141A   291 15775A   293 11477F
  295 11463B   297 17073C   299 16401C   301 12315A   307 14221E   309 11763B
  311 12705E   313 14357F   315 17777D   325 00163    327 17233D   329 11637B
  331 16407F   333 11703A   339 16003C   341 11561E   343 12673B   345 14537D
  347 17711G   349 13701E   355 10467B   357 15347C   359 11075E   361 16363F
  363 11045A   365 11265A   371 14043D   397 12727F   403 14373D   405 13003B
  407 17057G   409 10437F   411 10077B   421 14271G   423 14313D   425 14155C
  427 10245A   429 11073B   435 10743B   437 12623F   439 12007F   441 15353D
  455 00111    585 00013    587 14545G   589 16311G   595 13413A   597 12265A
  603 14411C   613 15413H   619 17147F   661 10605E   683 10737F   685 16355C
  691 15701G   693 12345A   715 00133    717 16571C   819 00037   1365 00007

DEGREE 13        1 20033F     3 23261E     5 24623F     7 23517F     9 30741G
   11 21643F    13 30171G    15 21277F    17 27777F    19 35051G    21 34723H
   23 34047H    25 32535G    27 31425G    29 37505G    31 36515G    33 26077F
   35 35673H    37 20635E    39 33763H    41 25745E    43 36575G    45 26653F
   47 21133F    49 22441E    51 30417H    53 32517H    55 37335G    57 25327F
   59 23231E    61 25511E    63 26533F    65 33343H    67 33727H    69 27271E
   71 25017F    73 26041E    75 21103F    77 27263F    79 24513F    81 32311G
   83 31743H    85 24037F    87 30711G    89 32641G    91 24657F    93 32437H
   95 20213F    97 25633F    99 31303H   101 22525E   103 34627H   105 25775E
  107 21607F   109 25363F   111 27217F   113 33741G   115 37611G   117 23077F
  119 21263F   121 31011G   123 27051E   125 35477H   131 34151G   133 27405E
  135 34641G   137 32445G   139 36375G   141 22675E   143 36073H   145 35121G
  147 36501G   149 33057H   151 36403H   153 35567H   155 23167F   157 36217H
  159 22233F   161 32333H   163 24703F   165 33163H   167 32757H   169 23761E
  171 24031E   173 30025G   175 37145G   177 31327H   179 27221E   181 25577F
  183 22203F   185 37437H   187 27537F   189 31035G   195 24763F   197 20245E
  199 20503F   201 20761E   203 25555E   205 30357H   207 33037H   209 34401G
  211 32715G   213 21447F   215 27421E   217 20363F   219 33501G   221 20425E
  223 32347H   225 20677F   227 22307F   229 33441G   231 33643H   233 24165E
  235 27427F   237 24601E   239 36721G   241 34363H   243 21673F   245 32167H
  247 21661E   265 33357H   267 26341E   269 31653H   271 37511G   273 23003F
  275 22657F   277 25035E   279 23267F   281 34005G   283 34555G   285 24205E
  291 26611E   293 32671G   295 25245E   297 31407H   299 33471G   301 22613F
  303 35645G   305 32371G   307 34517H   309 26225E   311 35561G   313 25663F
  315 24043F   317 30643H   323 20157F   325 37151G   327 24667F   329 33325G
  331 32467H   333 30667H   335 22631E   337 26617F   339 20275E   341 36625G
  343 20341E   345 37527H   347 31333H   349 31071G   355 23353F   357 26243F
  359 21453F   361 36015G   363 36667H   365 34767H   367 34341G   369 34547H
  371 35465G   373 24421E   375 23563F   377 36037H   391 31267H   393 27133F
  395 30705G   397 30465G   399 35315G   401 32231G   403 32207H   405 26101E
  407 22567F   409 21755E   411 22455E   413 33705G   419 37621G   421 21405E
  423 30117H   425 23021E   427 21525E   429 36465G   431 33013H   433 27531E
  435 24675E   437 33133H   439 34261G   441 33405G   443 34655G   453 32173H
  455 33455G   457 35165G   459 22705E   461 37123H   463 27111E   465 35455G
  467 31457H   469 23055E   471 30777H   473 37653H   475 24325E   477 31251G
  547 35163H   549 33433H   551 37243H   553 27515E   555 32137H   557 26743F
  563 30277H   565 20627F   567 35057H   569 24315E   571 24727F   581 30331G
  583 34273H   585 23207F   587 31113H   589 36023H   595 27373F   597 20737F
  599 36235G   601 21575E   603 26215E   605 21211E   611 20311E   613 34003H
  615 34027H   617 20065E   619 22051E   621 22127F   627 23621E   629 24465E
  651 26457F   653 31201G   659 34035G   661 27227F   663 22561E   665 21615E
  667 22013F   669 23365E   675 26213F   677 26775E   679 32635G   681 33631G
  683 32743H   685 31767H   691 34413H   693 22037F   695 30651G   697 26565E
  711 22141E   713 22471E   715 35271G   717 37445G   723 22717F   725 26505E
  727 24411E   729 24575E   731 23707F   733 25173F   739 21367F   741 25161E
  743 24147F   793 36307H   795 24417F   805 20237F   807 36771G   809 37327H
  811 27735E   813 31223H   819 36373H   821 33121G   823 32751G   825 33523H
```

DEGREE 13--CONTINUED

839 26415E 841 23737F 843 25425E 845 34603H 851 31047H 853 37305G
855 21315E 857 35777H 859 32725G 869 20571E 871 30301G 873 34757H
875 21067F 877 25151E 1171 27513F 1173 33721G 1179 34775G 1189 23571E
1195 27411E 1197 20457F 1203 21557F 1205 30177H 1227 26347F 1229 27477F
1235 34243H 1237 27235E 1323 25175E 1325 31231G 1331 31131G 1333 25503F
1355 33045G 1357 24253F 1363 35351G 1365 26053F

DEGREE 14 1 42103F 3 40547B 5 43333E 7 51761E 9 54055A
11 40503F 13 77141G 15 47645A 17 62677G 19 44103F 21 46425A
23 45145E 25 76303G 27 62603D 29 64457G 31 57231E 33 52737B
35 64167F 37 60153F 39 62115C 41 55753F 43 72427D 45 64715A
47 70423H 49 47153F 51 67653D 53 53255E 55 41753F 57 74247D
59 40725E 61 42667F 63 65301A 65 67517H 67 45653F 69 72501C
71 67425G 73 42163F 75 73757D 77 45555E 79 74561G 81 60523B
83 53705E 85 40123E 87 41403B 89 56625E 91 70311E 93 75547C
95 45627F 97 67335G 99 56733A 101 53253F 103 66411E 105 57745A
107 65551G 109 43017F 111 62125A 113 71073E 115 67333H 117 70677C
119 52215E 121 44177F 123 70535C 125 46327F 127 71747D 129 00203
131 61335G 133 43161E 135 46047B 137 60645G 139 40317F 141 47727A
143 65001G 145 54335E 147 76175C 149 65153H 151 50351E 153 42711A
155 41625E 157 44435E 159 41163A 161 47667F 163 41441E 165 54175A
167 45713F 169 75267H 171 72051C 173 64223H 175 42337F 177 51275A
179 65155E 181 63015E 183 57521A 185 67173H 187 50661E 189 41735A
191 50645E 193 72433F 195 47043B 197 65133H 199 53543F 201 62431A
203 42777F 205 47203F 207 46605A 209 64377H 211 73725G 213 43611A
215 42301A 217 51145E 219 44307B 221 73647H 223 74427H 225 53747A
227 45511E 229 42637F 231 63117D 233 40363E 235 75201G 237 63155C
239 72717G 241 56557F 243 75363D 245 70553F 247 66675G 249 55501A
251 60263H 261 53043B 263 75303F 265 74315E 267 66031A 269 62505G
271 60057H 273 54473A 275 60253F 277 45671E 279 71525C 281 61443E
283 64635G 285 64475C 287 67401G 289 44203F 291 50343A 293 77747H
295 54101E 297 65645A 299 41177F 301 65661A 303 42361A 305 43047F
307 45563F 309 50717A 311 53233E 313 67101G 315 62251C 317 64251E
323 40635E 325 46113E 327 44367B 329 40665E 331 63331G 333 71545C
335 73107H 337 42727F 339 43775A 341 65667E 343 61677H 345 53525A
347 52723F 349 42323F 351 41433B 353 43173E 355 46305E 357 45663B
359 71315E 361 44031E 363 73457B 365 52577E 367 52621E 369 40063B
371 52027F 373 45201E 375 77001C 377 45737E 379 64035G 381 52225A
387 00253 389 60765G 391 66545G 393 71323A 395 62767G 397 73137H
399 40145A 401 63265G 403 47551E 405 71711C 407 40353F 409 76055G
411 70065C 413 73527F 415 67201G 417 43723B 419 61251E 421 47357F
423 62261C 425 50575E 427 61267H 429 40511A 431 71721G 433 65121G
435 61053D 437 45371E 439 54627E 441 77703A 443 65057H 445 76225E
451 73071G 453 52553B 455 60025E 457 60471G 459 53513B 461 67303H
463 42763F 465 52261A 467 53657F 469 75443F 471 67267D 473 53373B
475 65165E 477 44037B 479 54737F 481 61175E 483 65031A 485 51707E
487 57627F 489 57251A 491 44073F 493 45761E 495 63463C 529 65277F
531 55247B 533 56171E 535 63513H 537 43377B 539 45641E 541 63227H
547 54243F 549 62055C 551 53061E 553 46321E 555 51431A 557 71147H
559 64053D 561 41551A 563 75521E 565 46701E 567 53763B 569 56463F
571 77057G 573 41105A 579 41171A 581 41307F 583 70425E 585 74117D
587 50135E 589 67737H 591 47615A 593 53057F 595 55103F 597 54443B
599 53051E 601 61555G 603 64157D 605 57407F 611 64653F 613 65513H
615 73603D 617 47525E 619 55165E 621 64215C 623 76377H 625 57365E
627 50557B 629 45725E 631 71301G 633 56465A 635 51745A 645 00217
647 47233F 649 53015E 651 53361A 653 46215E 655 50613E 657 77211C
659 46565E 661 44141E 663 55771A 665 71263G 667 41315E 669 62225C
675 51565A 677 76267H 679 62467H 681 64003C 683 71645G 685 76223G
687 52627A 689 70665G 691 45773F 693 64033D 695 45533E 697 50007F
699 45257B 701 45311E 707 44023F 709 72153G 711 60117D 713 46617E
715 70461G 717 47513B 719 65575E 721 56435E 723 67157C 725 71403G
727 46107F 729 65007A 731 50667B 733 55331E 739 52017F 741 51317B
743 66163F 745 70767G 747 70215C 749 76401G 751 63043H 753 63753D
755 43317F 781 77031G 783 45617B 785 52603F 787 57503F 789 63667D
791 75761G 793 60075G 795 72307B 797 51633F 803 57475E 805 61533G

TABLE B.2. (*Continued*)

DEGREE 14--CONTINUED

807	60561C	809	53575E	811	62027H	813	64633C	815	67123F	817	43445A
819	73655C	821	54003F	823	62347F	825	63271C	827	71337F	837	57715A
839	54635E	841	46505E	843	64407C	845	57017E	847	54751E	849	42417A
851	57033F	853	54077F	855	42567B	857	50455E	859	62533H	861	42411A
867	74133D	869	72441G	871	43577F	873	52353B	875	55325E	877	67527G
879	75605C	881	52467F	883	61757F	885	66105C	887	51261E	889	62723D
903	00375	905	63537H	907	52457E	909	44735A	911	62413H	913	51671E
915	41001A	917	70773H	919	56031E	921	60227D	923	71345G	925	46125E
931	40655E	933	44221A	935	55323F	937	76005E	939	55435A	941	42531E
943	62671E	945	74277D	947	64617G	949	52137F	951	56637B	953	47753F
955	46773F	1093	72155G	1095	56067A	1097	63007E	1099	47111E	1101	54021A
1107	44523B	1109	54257F	1111	63567H	1113	43215A	1115	73665G	1117	45335E
1123	44147E	1125	62731C	1127	41657F	1129	77235G	1131	65643B	1133	51055E
1139	47637F	1141	40071E	1143	47771A	1161	00271	1163	57541E	1165	57107F
1171	61621G	1173	51511A	1175	57201E	1177	70251G	1179	43633B	1181	53315E
1187	44343F	1189	55705E	1191	40413B	1193	64641E	1195	44567E	1197	46451A
1203	60241C	1205	65705E	1207	71117H	1209	66703D	1211	53477F	1221	45355A
1223	74531G	1225	74607H	1227	71763C	1229	76707H	1235	60235G	1237	47673F
1239	54321A	1241	75571G	1243	77515G	1245	57611A	1251	55643B	1253	46175E
1255	74357H	1257	70267D	1259	46461E	1301	77345G	1303	51243F	1305	76151C
1307	56061E	1309	66427G	1315	54517F	1317	72465C	1319	50733F	1321	74045G
1323	71057D	1325	73143F	1331	51231E	1333	70201C	1335	77631C	1337	64021G
1351	72643H	1353	41777B	1355	71675G	1357	63073H	1363	47537E	1365	61261A
1367	65227H	1369	55073F	1371	77727B	1373	61363H	1379	43701E	1381	65147H
1383	52267B	1385	63153F	1387	72337G	1389	56607A	1395	40371A	1397	42721A
1419	00211	1421	75273F	1427	73555G	1429	67225G	1431	76617C	1433	74711E
1435	50325E	1437	70713C	1443	72513D	1445	57737F	1447	61333G	1449	40327A
1451	55111E	1453	40633F	1459	61641G	1461	65315C	1463	43647F	1465	67621G
1479	62745C	1481	41755E	1483	65727F	1485	74263D	1587	41573B	1589	55631E
1591	66405A	1593	60121C	1607	71615E	1609	77615G	1611	41447B	1613	46437F
1619	70633H	1621	65615G	1623	64605C	1625	55075E	1627	73151G	1637	75033H
1639	57327F	1641	66277D	1643	56007F	1645	55703F	1651	77277D	1677	00345
1683	57743A	1685	42645E	1687	50045E	1689	74255C	1691	53623E	1701	50477B
1703	52071E	1705	61237H	1707	67533B	1709	55417F	1715	45173E	1717	61461G
1719	43731A	1721	56717E	1735	54041E	1737	44613A	1739	70341G	1741	52065E
1747	56345E	1749	44441A	1751	76663H	1753	50777F	1755	70443D	2341	55471E
2347	53727F	2349	65637C	2355	57143B	2357	44741E	2379	67627D	2381	77177G
2387	51213E	2389	70273H	2395	62101G	2405	50241E	2411	65263H	2413	41241A
2451	00357	2453	76047H	2459	75723F	2469	73145C	2475	61377D	2477	41357F
2643	56421A	2645	76213H	2667	64213D	2709	00313	2731	41235E	2733	67605C
2739	44537B	2741	76505G	2763	65375C	2765	50721E	2771	75517H	2861	65357G
2867	47121E	5461	00007								

DEGREE	15	1	100003F	3	102043F	5	110013F	7	125253B
9	102067F	11	104307F	13	100317F	15	177775E	17	103451E
19	110075E	21	127701A	23	102061E	25	114725E	27	103251E
29	163005G	31	103437A	33	112611E	35	137733B	37	120265E
39	117423F	41	106341E	43	161007H	45	174003E	47	113337E
49	125263B	51	126007E	53	105257E	55	114467E	57	177207G
59	147047F	61	111511E	63	127635A	65	114633E	67	133663F
69	102171E	71	170465G	73	131427E	75	161615E	77	136143A
79	115155E	81	123067F	83	102561E	85	170057H	87	125235E
89	173117E	91	125747B	93	124677B	95	134531E	97	125507F
99	171737G	101	152417F	103	142305G	105	146255C	107	120043F
109	136173F	111	122231E	113	164705G	115	177757F	117	146637E
119	177535C	121	102643F	123	103145E	125	112751E	127	151537G
129	115135E	131	137067E	133	122707A	135	174443E	137	100541E
139	112273F	141	145573F	143	114273F	145	124511E	147	122563B
149	140703F	151	101361A	153	103125E	155	150451C	157	147303G
159	123023F	161	103751A	163	154463H	165	177541G	167	101561E
169	144473G	171	162375G	173	131013F	175	117767A	177	160521G
179	164727G	181	102367E	183	147363F	185	132367E	187	172431E
189	133627B	191	156333E	193	114505E	195	176561G	197	152235G
199	127143F	201	176133E	203	123075A	205	173357G	207	117143E
209	144461E	211	151447G	213	173661E	215	151043F	217	142327B

DEGREE 15--CONTINUED

219	166775E	221	153143G	223	172213F	225	105213E	227	156053H
229	156745G	231	170623B	233	140373G	235	152361G	237	142157H
239	117633F	241	103605E	243	116361E	245	137523A	247	101705E
249	116135E	251	102337E	253	173515G	259	136321A	261	120447F
263	117511E	265	115141E	267	173613F	269	131735E	271	114225E
273	121125A	275	136577F	277	113227E	279	114533B	281	166151E
283	112231E	285	165033E	287	120177B	289	117547F	291	126051E
293	111335E	295	177101G	297	143703G	299	106047E	301	137427B
303	110427F	305	131211E	307	110037F	309	160511G	311	153731G
313	144275G	315	151513C	317	133775E	319	134447E	321	127347E
323	163767H	325	110717E	327	175001E	329	100377A	331	125121E
333	136237F	335	132103F	337	171035G	339	132651E	341	134105A
343	100261A	345	170227H	347	101233F	349	100445E	351	144707G
353	165355E	355	150243H	357	163353C	359	114041E	361	113025E
363	104447F	365	143301G	367	165011G	369	137361E	371	117201A
373	141655G	375	160113G	377	106715E	379	140575E	381	112123E
387	140733F	389	124243E	391	116073E	393	147321E	395	123721E
397	150225G	399	134741A	401	157111G	403	134411A	405	172317G
407	153327E	409	140573H	411	113625E	413	101673B	415	170543F
417	176735E	419	115307F	421	141635E	423	157241G	425	153005E
427	167051A	429	177175G	431	146331G	433	166541G	435	102513F
437	123121E	439	162463G	441	134037B	443	174571E	445	123433F
447	150167H	449	175465E	451	113255E	453	137325A	455	123045A
457	133571E	459	135215E	461	110221E	463	157435E	465	121437A
467	177707G	469	143501C	471	161667F	473	157427G	475	150671G
477	112407F	479	165563E	481	112053E	483	135363B	485	130617F
487	125613F	489	114713F	491	165113G	493	143733G	495	162155E
497	135017B	499	126753F	501	137765E	503	106577E	521	112113F
523	105555E	525	153425C	527	115313A	529	105761E	531	132165E
533	176147H	535	114621E	537	135751E	539	152763C	541	124757F
543	112245E	545	123221E	547	141757G	549	160547F	551	101331E
553	156065C	555	156725G	557	113373E	559	137643F	561	156237G
563	141151G	565	126015E	567	171335C	569	146717H	571	130305E
573	121355E	579	166021G	581	145361C	583	134325E	585	157155E
587	124647E	589	163761C	591	114457E	593	155243G	595	153137D
597	137253F	599	151551G	601	113645E	603	150305G	605	163745G
607	165473F	609	113057B	611	160173H	613	177663F	615	161117H
617	144115E	619	156635G	621	150633H	623	115061A	625	143253H
627	165451G	629	160305E	631	146025E	633	106751E	635	132625E
637	160553D	643	123561E	645	116637F	647	111423E	649	117107E
651	166761C	653	153555G	655	132127F	657	112333E	659	135267F
661	146727H	663	132753F	665	143343A	667	131705E	669	141005E
671	113147F	673	125323F	675	123235E	677	103653F	679	173025C
681	120661E	683	154545G	685	133553F	687	132001E	689	153773G
691	175241G	693	160237B	695	171131E	697	172415E	699	145111G
701	122603F	707	170507C	709	160757G	711	171207G	713	147553B
715	112365E	717	146111E	719	122003F	721	121273B	723	122005E
725	135401E	727	102441E	729	175515G	731	132507E	733	130223F
735	142713C	737	102615E	739	105713F	741	134241E	743	173643F
745	163617G	747	175043E	749	132051A	751	104217F	753	115523F
755	120247B	757	164447H	759	173667F	761	137051E	775	104073B
777	177065C	779	117071E	781	115537E	783	135201E	785	146643F
787	113465E	789	152263G	791	177617D	793	104755E	795	147415G
797	126001E	799	170307F	801	174425E	803	112475E	805	173263C
807	176643H	809	130303F	811	125471E	813	173711G	815	165547E
817	163723G	819	116075A	821	150677G	823	175227G	825	166407H
827	152447H	829	126205E	835	120557E	837	160335A	839	125543E
841	144377H	843	100713E	845	121251E	847	141123D	849	174517F
851	106251E	853	116277F	855	106611E	857	174563H	859	140023H
861	132037A	863	147767G	865	164531G	867	155065E	869	146263F
871	160401G	873	102057F	875	146133C	877	117021E	879	147003F
881	127723F	883	120471E	885	162455G	887	130627F	889	152135C
891	157057H	901	162153F	903	151755C	905	170277H	907	165633H
909	173105E	911	102507F	913	176037H	915	171627G	917	162171C
919	130745E	921	177517H	923	114327F	925	127167F	927	133113E

TABLE B.2. (*Continued*)

DEGREE 15--CONTINUED									
929	160461E	931	117137B	933	134323F	935	123361E	937	105237F
939	166737F	941	147571G	943	127743F	945	116351A	947	157315E
949	162645G	951	162403G	953	105335E	955	124767E	957	175301E
963	134755E	965	116645E	967	143307G	969	124125E	971	155261G
973	104163A	975	167753F	977	127423F	979	115667F	981	140171E
983	133041E	985	156767H	987	116037A	989	142267G	991	130635E
1057	000057	1059	104427F	1061	113075E	1063	162133H	1065	120717F
1067	144713F	1069	121605E	1071	122225A	1073	134657E	1075	130125E
1077	177621G	1079	110741E	1081	136745E	1083	152531G	1085	115455A
1091	161235G	1093	144137G	1095	140675E	1097	145277G	1099	114303B
1101	101507E	1103	115271E	1105	151735E	1107	157205G	1109	114011E
1111	171125E	1113	147071A	1115	134721E	1117	122123F	1123	104735E
1125	133011E	1127	162337A	1129	105261E	1131	101427E	1133	156563F
1135	103663E	1137	146043H	1139	151403H	1141	100157A	1143	163653E
1145	105413F	1147	143651C	1157	156157E	1159	102463F	1161	151025G
1163	176657H	1165	166425G	1167	103617E	1169	160021A	1171	161277H
1173	165565G	1175	152153F	1177	111243E	1179	165655G	1181	134165E
1187	171467H	1189	150161E	1191	122011E	1193	125403F	1195	170007H
1197	167765C	1199	103415E	1201	137703E	1203	111563F	1205	147305G
1207	156257F	1209	175177B	1211	141317B	1213	177467H	1219	140421G
1221	127071E	1223	142457F	1225	122021A	1227	146771E	1229	110211E
1231	134567F	1233	156321G	1235	114335E	1237	111603E	1239	121275A
1241	110103E	1243	127161E	1245	163273H	1251	144533F	1253	173135C
1255	155445E	1257	140441E	1259	103761E	1261	173523F	1263	167307F
1265	127457F	1267	102205A	1269	112251E	1291	106311E	1293	141633F
1295	135151A	1297	106641E	1299	102265E	1301	164453G	1303	163071G
1305	111641E	1307	134403E	1309	102667A	1315	177055E	1317	115373F
1319	150231G	1321	175651G	1323	160377B	1325	136063E	1327	101073F
1329	165303G	1331	116675E	1333	140221A	1335	100201E	1337	103223B
1339	105415E	1341	122445E	1347	143631E	1349	137441E	1351	104421A
1353	154023H	1355	127225E	1357	176427H	1359	151265C	1361	150215E
1363	144225G	1365	115205A	1367	123307E	1369	133437E	1371	166653E
1373	101515E	1379	126023B	1381	166553H	1383	172701E	1385	140271G
1387	121143E	1389	111577E	1391	132747E	1393	143057C	1395	111137B
1397	127401E	1399	150317E	1401	177731G	1415	155335G	1417	123057F
1419	117715E	1421	162657B	1423	171745G	1425	130527F	1427	144467G
1429	115045E	1431	177115G	1433	155751G	1435	103767A	1437	115127E
1443	176741E	1445	141475G	1447	112553E	1449	154307D	1451	105621E
1453	170051G	1455	147707F	1457	160445A	1459	161031E	1461	131405E
1463	164121A	1465	111003F	1467	167331E	1469	165311G	1475	157405G
1477	140557A	1479	156655G	1481	164561G	1483	114231E	1485	106407F
1487	111033F	1489	172123G	1491	146667D	1493	143523G	1495	170765G
1497	105725E	1499	132155E	1501	150261G	1507	122517E	1509	107567E
1511	166267E	1561	153461C	1563	166011G	1565	133445E	1571	156365G
1573	176111G	1575	137331A	1577	165407G	1579	106445E	1581	145551C
1583	124341E	1585	127215E	1587	135005E	1589	117731A	1591	110141E
1593	152345G	1595	164441G	1605	172621G	1607	143567G	1609	153443H
1611	146203E	1613	120417F	1615	103553F	1617	110567A	1619	126067F
1621	140747F	1623	107037F	1625	135503E	1627	126735E	1629	172445G
1635	117131E	1637	105173F	1639	105071E	1641	174167G	1643	114745A
1645	133407A	1647	136215E	1649	153113H	1651	141321E	1653	132523F
1655	136335E	1657	167255E	1671	146301G	1673	131265A	1675	120133F
1677	157557E	1679	107711E	1681	174751E	1683	133257F	1685	151217G
1687	144653C	1689	176203H	1691	155213H	1693	135207F	1699	131367F
1701	146543C	1703	130033F	1705	166311A	1707	150213G	1709	143227F
1711	176013G	1713	147751G	1715	131543B	1717	131111E	1719	111267F
1721	144151G	1723	110433F	1733	171173F	1735	116367F	1737	115421E
1739	112223F	1741	111635E	1743	157165C	1745	135223F	1747	106143F
1749	176015G	1751	142461G	1753	154233E	1755	114677F	1757	103363A
1763	150327F	1765	126325E	1767	126105A	1769	111713F	1771	172303B
1773	170763G	1775	124175E	1777	176357F	1807	164667E	1809	136611E
1811	163123E	1813	151037D	1815	121431E	1817	110165E	1819	172005G
1821	104265E	1827	154763A	1829	152703D	1831	163555G	1833	135021E
1835	124071E	1837	164247H	1839	166113H	1841	101625A	1843	145427H
1845	106633F	1847	155437E	1849	174633H	1851	161657H	1861	174605G

DEGREE 15--CONTINUED

1863 136701E 1865 144425E 1867 126747F 1869 157441C 1871 167015E
1873 142737H 1875 152301E 1877 131727E 1879 120221E 1881 102147E
1883 106457B 1885 152253H 1891 157645A 1893 141541G 1895 170325E
1897 141677C 1899 102733E 1901 135443F 1903 124251E 1905 150731G
1907 127137F 1909 100347F 1911 130415A 2185 147161G 2187 154247F
2189 161205G 2195 101313E 2197 175203F 2199 154507G 2201 121055A
2203 113061E 2205 170211C 2211 102763E 2213 167367H 2215 106503F
2217 133641E 2219 160175C 2221 161061E 2227 103035E 2229 173037F
2231 130737F 2233 166137C 2235 130017F 2245 122213F 2247 144577D
2249 117027F 2251 106273F 2253 107217F 2259 146373F 2261 153445C
2263 145727D 2265 121451A 2267 146607F 2269 113543F 2275 161013A
2277 177131G 2279 112633E 2281 137545E 2283 140227F 2285 112377F
2323 123163F 2325 100725A 2327 162315G 2329 155027G 2331 173551C
2333 132357F 2339 141231E 2341 117457F 2343 143403H 2345 124005A
2347 137601E 2349 143271G 2355 143727F 2357 107447F 2359 136401A
2361 157711G 2363 170337E 2373 166257D 2375 131733E 2377 176453H
2379 116057F 2381 156773H 2387 114371A 2389 155505G 2391 100641E
2393 151573E 2395 106713F 2397 177751G 2403 175601G 2405 177563G
2407 155175G 2409 170367G 2411 132015E 2413 126375E 2419 170433F
2421 151747G 2443 173153B 2445 111505E 2451 127243F 2453 107323F
2455 106745E 2457 165327B 2459 153577H 2461 150341G 2467 155737H
2469 150005G 2471 146007A 2473 146155E 2475 117655E 2477 101023E
2483 126227F 2485 173163B 2487 103175E 2489 105143F 2491 174743G
2501 101433F 2503 155757H 2505 121017F 2507 100425E 2509 126657E
2515 172363H 2517 120463E 2519 154561G 2601 126771E 2603 156161E
2605 147725G 2611 177527D 2613 121641E 2615 111365E 2617 125057E
2631 142611G 2633 110435E 2635 104575A 2637 164313G 2643 126163E
2645 112347F 2647 126155E 2649 131667F 2651 141365G 2653 116307B
2659 143531E 2661 141445E 2663 104141E 2665 167001G 2667 110343A
2669 111047F 2675 107121E 2677 106125E 2699 167203G 2701 175337F
2707 165201G 2709 106767B 2711 152351G 2713 144731G 2715 161043G
2717 113171E 2723 133533A 2725 175405G 2727 177231G 2729 127653E
2731 165535G 2733 114701E 2739 146177H 2741 121327E 2743 132277F
2745 153175G 2759 155407A 2761 145433H 2763 167463H 2765 104263A
2771 127437F 2773 176255E 2775 134435E 2777 124335E 2779 143373D
2781 170501G 2787 126711E 2789 103257E 2791 120601E 2793 155773B
2839 134255E 2841 103737F 2843 164001G 2845 161147F 2851 135565E
2853 110573E 2855 175711E 2857 116631E 2859 131623E 2861 155725G
2867 154537F 2869 114347B 2871 140755G 2873 113515E 2887 120155E
2889 160137E 2891 163647B 2893 121725E 2899 157255G 2901 141401G
2903 141125G 2905 107337A 2907 117125E 2909 144603H 2915 147635E
2917 154331G 2919 115607A 2921 154411E 2923 154155E 2925 122275E
2931 136457F 2957 126433F 2963 154515E 2965 150371G 2967 173331E
2969 146753E 2971 132741E 2973 145477H 3171 000073 3173 124115E
3175 127365E 3177 107645E 3179 117443F 3181 163335E 3187 115675E
3213 131651A 3219 170523H 3221 167313H 3223 137127F 3225 140205E
3227 102357B 3237 163365G 3239 172027H 3241 131165A 3243 162241E
3245 142223G 3251 164155G 3253 176753H 3255 152433B 3257 125271E
3271 177377G 3273 100647E 3275 121101E 3277 142751E 3283 115721A
3285 144437G 3287 177443H 3289 101613F 3291 142633H 3301 156527H
3355 165725E 3365 110405E 3367 107675A 3369 115133E 3371 101551E
3373 133213E 3379 155621C 3381 114363A 3383 161253F 3385 160413F
3399 127077E 3401 136213E 3403 171115E 3405 121553E 3411 140007G
3413 116601E 3415 147437H 3417 100223E 3419 126643E 3429 133231E
3431 162037H 3433 141027E 3435 125255E 3437 166275A 3475 171621G
3477 107373E 3479 125337A 3481 110255E 3483 114611E 3493 114055A
3495 110501E 3497 104111E 3499 146375G 3501 126557F 3507 125361A
3509 121617F 3511 103333F 3513 103053E 3527 171371E 4681 000013
4683 133261A 4685 123735E 4691 142175G 4693 131645E 4699 167637G
4709 155303H 4715 160215G 4717 163275G 4755 124053F 4757 133201E
4763 141115G 4773 161105E 4779 100021E 4781 116567B 4787 145675G
4789 123471E 4811 137613F 4813 105701E 4819 121305E 4821 146705E
4907 124621A 4909 122443E 4915 123537E 4917 124317F 4939 106677E
4941 160723H 4947 131601E 4949 113405A 4955 155517G 5285 000045
5291 155707H 5293 134277F 5299 140513C 5301 111041A 5323 127273F

TABLE B.2. (*Continued*)

DEGREE 15--CONTINUED									
5325	117243F	5331	141707H	5333	134205E	5419	107417F	5421	122401E
5427	170037E	5429	107127E	5451	161465E	5453	171027C	5459	174707H
5461	145453E								
DEGREE	16	1	210013F	3	215435A	5	227215A	7	234313F
9	225657B	11	233303F	13	307107H	15	311513D	17	336523D
19	307527H	21	363501C	23	306357H	25	353573D	27	357333D
29	201735E	31	272201E	33	310327D	35	304341C	37	242413F
39	327721C	41	270155E	43	302157H	45	374111C	47	210205E
49	305667H	51	237403B	53	236107F	55	212113B	57	314061C
59	271055E	61	313371G	63	333575C	65	267313B	67	311405G
69	323527D	71	346355G	73	350513H	75	237421A	77	203213F
79	233503F	81	261105A	83	306221G	85	267075A	87	235063B
89	244461E	91	204015E	93	327421C	95	226455A	97	202301E
99	351641C	101	376311G	103	201637F	105	365705C	107	352125G
109	273435E	111	202545A	113	243575E	115	251645A	117	277535A
119	327277D	121	250723F	123	340047D	125	274761A	127	226135E
129	357047D	131	214443F	133	277213F	135	315633D	137	300205G
139	367737H	141	230535A	143	342567H	145	265157B	147	371771C
149	217137F	151	262367F	153	301663D	155	370565C	157	201045E
159	304731C	161	303657H	163	212653F	165	245351A	167	347433H
169	260237F	171	311651C	173	256005E	175	206353B	177	362053D
179	352603H	181	310017H	183	333013D	185	256415A	187	376175C
189	243513B	191	312301G	193	260475E	195	347211C	197	215345E
199	201551E	201	362555C	203	333643H	205	304261C	207	230541A
209	250311E	211	333117H	213	274317B	215	301425C	217	247353F
219	254601A	221	212063B	223	207661E	225	317171C	227	214215E
229	322661G	231	274635A	233	326035G	235	200215A	237	324127D
239	230653F	241	342105G	243	305471C	245	242437B	247	363637H
249	330561C	251	211473F	253	266663F	255	361617D	257	000717
259	255517F	261	344733D	263	311155G	265	340207D	267	273211A
269	366421G	271	221257F	273	207753B	275	226315A	277	250017F
279	243111A	281	242225E	283	204703F	285	323563D	287	230451E
289	323341C	291	271725A	293	353263H	295	306575C	297	271251A
299	335227H	301	213375E	303	340333D	305	232013B	307	312405G
309	233017B	311	266701E	313	262351E	315	324141C	317	365221G
319	213651E	321	200365A	323	215613B	325	207221A	327	323077D
329	274627F	331	302335G	333	251211A	335	262421A	337	360667H
339	223133B	341	356255G	343	337553H	345	215015A	347	221213F
349	276531E	351	325413D	353	362737H	355	240171A	357	241173B
359	274353F	361	222563F	363	231753B	365	227065A	367	217451E
369	254471A	371	356221G	373	235275E	375	372075C	377	357527H
379	241341E	381	335263D	383	311515G	385	202155A	387	254241A
389	370137H	391	300405C	393	227157B	395	237733B	397	207717F
399	303375C	401	257051E	403	245367F	405	324631C	407	274621E
409	211101E	411	324755C	413	326261G	415	236555A	417	341343D
419	220625E	421	332745G	423	374163D	425	264255A	427	234015E
429	206635A	431	320731G	433	243631E	435	325757D	437	241677F
439	217473F	441	366373D	443	230355E	445	301653D	447	264433B
449	302321G	451	333323H	453	344045C	455	317163D	457	265401E
459	325033D	461	341667H	463	276645E	465	346725C	467	301535G
469	342325G	471	202265A	473	247617F	475	325475C	477	343213D
479	237351E	481	341741G	483	361353D	485	260665A	487	276727F
489	273141A	491	233743F	493	252023B	495	272423B	497	265617F
499	273015E	501	267421A	503	351353H	505	377171C	507	317357D
517	202703F	519	241245A	521	356057H	523	217633F	525	277215A
527	257643B	529	267507F	531	311661C	533	235145E	535	202411A
537	205003B	539	366155G	541	212115E	543	375437D	545	354377D
547	236511E	549	277745A	551	241251E	553	211571E	555	245733B
557	362633H	559	201031E	561	371643D	563	340311G	565	200751A
567	232211A	569	341345G	571	374721G	573	310745C	575	227063B
577	271161E	579	322367D	581	375213H	583	330073H	585	273007B
587	341147H	589	371427H	591	200451A	593	251741E	595	345267D
597	205143B	599	212355E	601	252623F	603	331627D	605	241175A
607	355507H	609	261177B	611	317203H	613	361541G	615	363211C

DEGREE 16--CONTINUED

617	366345G	619	337521G	621	362745C	623	366171G	625	204227B
627	222473B	629	233725A	631	346101G	633	261253B	635	354723D
637	262073F	643	206603F	645	317531C	647	215343F	649	311203H
651	221245A	653	324747H	655	301065C	657	223561A	659	232643F
661	363271G	663	253723B	665	260145A	667	337071G	669	273361A
671	224611E	673	267615E	675	377373D	677	316431G	679	237337F
681	214143B	683	272071E	685	364225C	687	230371A	689	240675E
691	306643H	693	366537D	695	233521A	697	325173D	699	241647B
701	244333F	703	311733H	705	222123B	707	234777F	709	304535G
711	202141A	713	256461E	715	374343D	717	341061C	719	375761G
721	323175G	723	236041A	725	260725A	727	222017F	729	352077D
731	231253B	733	330233H	735	210447B	737	324073H	739	306015G
741	316757D	743	302115G	745	322031C	747	226255A	749	351225G
751	344623H	753	211125A	755	337017D	757	302063H	759	375223D
761	303361G	763	313751G	765	366557D	771	000573	773	326137H
775	256553B	777	223463B	779	302577H	781	234667F	783	225405A
785	201717B	787	230257F	789	357617D	791	367333H	793	346243H
795	272445A	797	325723H	799	311103D	801	310267D	803	330177H
805	302635C	807	372301C	809	246613F	811	264507F	813	341043D
815	275357B	817	301101G	819	262135A	821	350403H	823	367033H
825	301347D	827	201607F	829	202607F	831	212737B	833	232315A
835	201367B	837	222003B	839	223121E	841	200475E	843	221151A
845	316261C	847	245265E	849	226447B	851	234155E	853	305235G
855	222267B	857	335105G	859	227475E	861	362577D	863	224671E
865	356471C	867	223255A	869	301213H	871	321453H	873	341645C
875	350277D	877	240315E	879	220343B	881	343503H	883	366673H
885	337063D	887	225733F	889	221101E	891	343547D	893	231265E
899	315737H	901	300733D	903	270403B	905	271347B	907	356741G
909	260775A	911	204343F	913	225051E	915	332655C	917	276241E
919	244251E	921	311165C	923	201771E	925	305263D	927	337547D
929	234545E	931	261141E	933	374765C	935	335205C	937	303463H
939	356233D	941	256243F	943	373053H	945	204025A	947	346467H
949	256653F	951	310671C	953	274757F	955	247275A	957	277047B
959	332663H	961	367231G	963	233035A	965	355155C	967	352653H
969	213625A	971	320225G	973	323547H	975	276031A	977	213253F
979	226073F	981	201153B	983	373363H	985	352123D	987	367065C
989	301451G	991	262233F	993	373553D	995	270253B	997	263737F
999	214267B	1001	217237F	1003	257507B	1005	365501C	1007	205535E
1041	343055C	1043	344651G	1045	211245A	1047	306573D	1049	264001E
1051	343655G	1053	201515A	1055	370743D	1057	313415G	1059	307713D
1061	320445G	1063	222425E	1065	243043B	1067	214371E	1069	370321G
1071	265231A	1073	365405G	1075	305301C	1077	364355C	1079	312615G
1081	300155G	1083	333177D	1085	341703D	1091	370275G	1093	267205E
1095	325731C	1097	376443H	1099	332033H	1101	266167B	1103	326461G
1105	244547B	1107	212647B	1109	322171G	1111	206257F	1113	277641A
1115	310517D	1117	312247H	1119	365307D	1121	310437H	1123	344513H
1125	302167D	1127	337245G	1129	247743F	1131	275141A	1133	216607F
1135	317567D	1137	255355A	1139	353153D	1141	222633F	1143	254543B
1145	211377B	1147	243135E	1149	377147D	1155	253207B	1157	337311G
1159	272175E	1161	222541A	1163	226367F	1165	324433D	1167	360623D
1169	315713H	1171	337503H	1173	326065C	1175	207307B	1177	232045E
1179	337517D	1181	353733H	1183	372435G	1185	333515C	1187	213523F
1189	200535E	1191	261263B	1193	273073F	1195	264463B	1197	347463D
1199	364201G	1201	240411E	1203	274167B	1205	362715C	1207	253603B
1209	262615A	1211	360141G	1213	315571G	1219	303045G	1221	362161C
1223	301407H	1225	251705A	1227	215615A	1229	316505G	1231	373237H
1233	214317B	1235	370541C	1237	313437H	1239	275651A	1241	361701C
1243	214663F	1245	313407D	1247	216313F	1249	271655E	1251	265663B
1253	376415G	1255	213325A	1257	355771C	1259	306235G	1261	214157F
1263	256401A	1265	272627B	1267	216777F	1269	313627D	1271	206173F
1273	361521G	1275	333733D	1285	000433	1287	264637B	1289	326317H
1291	276441E	1293	273253B	1295	341037D	1297	326715G	1299	216007B
1301	217041E	1303	222237F	1305	224107B	1307	202277F	1309	256063B
1311	240323B	1313	260655E	1315	266671A	1317	273765A	1319	377755G
1321	264037F	1323	370611C	1325	300643D	1327	335675G	1329	350057D

TABLE B.2. (*Continued*)

DEGREE 16--CONTINUED

1331 353531G	1333 303367H	1335 331751C	1337 335127H	1339 354413H
1341 314651C	1347 372705C	1349 346047H	1351 325647H	1353 255113B
1355 277341A	1357 252657F	1359 226075A	1361 353025G	1363 340065G
1365 375517D	1367 334347H	1369 225575E	1371 324711C	1373 245025E
1375 371227D	1377 225551A	1379 343145G	1381 242167F	1383 243411A
1385 247027B	1387 230365E	1389 321165C	1391 254515E	1393 367671G
1395 242115A	1397 220217F	1399 361563H	1401 301553D	1403 352365G
1405 204351A	1411 263047B	1413 261551A	1415 375627D	1417 205273F
1419 227577B	1421 331353H	1423 200677F	1425 205237B	1427 317441G
1429 377405G	1431 205335A	1433 213631E	1435 263003B	1437 330543D
1439 243375E	1441 303013H	1443 255655A	1445 203207B	1447 245255E
1449 353045C	1451 353135G	1453 215007F	1455 302131C	1457 205423F
1459 331577H	1461 327471C	1463 245313F	1465 245247B	1467 323715C
1469 255527F	1475 256311A	1477 337755G	1479 247377B	1481 224725E
1483 355403H	1485 315407D	1487 262471E	1489 324523H	1491 302737D
1493 374471G	1495 207675A	1497 330343D	1499 330023H	1501 261765E
1503 376451C	1505 237271A	1507 265553F	1509 251325A	1511 203303F
1513 374277D	1515 374573D	1517 250737F	1519 266745E	1521 342177D
1523 255505E	1549 241317F	1551 277347B	1553 255653F	1555 224655A
1557 246455A	1559 277565E	1561 367743H	1563 240233B	1565 307563D
1567 277053F	1569 254651A	1571 213075E	1573 263407F	1575 307533D
1577 275247F	1579 202753F	1581 256517B	1583 323113H	1585 216551A
1587 371051C	1589 254551E	1591 302257H	1593 342001C	1595 252313B
1597 316145G	1603 272207F	1605 253317B	1607 223043F	1609 355513H
1611 210233B	1613 277663F	1615 335645C	1617 355147D	1619 225573F
1621 366057H	1623 243217B	1625 337225C	1627 354503H	1629 337577D
1631 364445G	1633 370145G	1635 216177B	1637 330673H	1639 271341E
1641 317373D	1643 334555G	1645 255751A	1647 312411C	1649 212423B
1651 354047H	1653 241767B	1655 351157D	1657 342721G	1659 214053B
1669 267221E	1671 307541C	1673 334533H	1675 323145C	1677 211767B
1679 236423F	1681 325275G	1683 372057D	1685 327373D	1687 277505E
1689 272713B	1691 317361G	1693 347361G	1695 220433B	1697 354231G
1699 216477F	1701 311337D	1703 220741E	1705 324015C	1707 264271A
1709 231361E	1711 254323F	1713 302505C	1715 245417B	1717 212137B
1719 360721C	1721 217671E	1723 301021G	1725 321573D	1731 363525C
1733 272267F	1735 263401A	1737 215545A	1739 204141E	1741 276117F
1743 322515C	1745 213165A	1747 244425E	1749 212065A	1751 300135C
1753 277145E	1755 332415C	1757 360545G	1759 376475G	1761 252015A
1763 261455E	1765 333525C	1767 346335C	1769 310215G	1771 210435E
1773 201557B	1775 201373B	1777 210733F	1779 204235A	1781 203603F
1783 201155E	1785 214461A	1799 000703	1801 244035E	1803 261117B
1805 271563B	1807 242305E	1809 225735A	1811 240741E	1813 232435E
1815 262253B	1817 227647F	1819 352577D	1821 215171A	1823 343011G
1825 375333D	1827 313647D	1829 226117F	1831 326571G	1833 341523D
1835 341117D	1837 235151E	1839 344435C	1841 351021G	1843 330523H
1845 321507D	1847 347111G	1849 355323H	1851 367701C	1853 322643D
1859 341337H	1861 312471G	1863 307521C	1865 321735C	1867 303435G
1869 344203D	1871 202463F	1873 262645E	1875 233215A	1877 256353F
1879 344153H	1881 330235C	1883 247511E	1885 354415C	1887 316365C
1889 322111G	1891 314507H	1893 251765A	1895 226613B	1897 346145G
1899 333061C	1901 327337H	1903 225523F	1905 221173B	1907 205745E
1909 273235E	1911 371625C	1913 274571E	1915 227353B	1925 253215A
1927 331333H	1929 273703B	1931 270557F	1933 370467H	1935 350763D
1937 335717H	1939 211123F	1941 207163B	1943 322601G	1945 265101A
1947 377623D	1949 234675E	1951 200627F	1953 344615C	1955 372351C
1957 300025G	1959 216455A	1961 215411E	1963 333221G	1965 320151C
1967 277707F	1969 201435E	1971 333265C	1973 343335G	1975 300557D
1977 305567D	1979 355735G	1981 240477F	2115 331173D	2117 272425E
2119 234711E	2121 361707D	2123 213067F	2125 262271A	2127 363543D
2129 241757F	2131 326423H	2133 253125A	2135 342643D	2137 321433H
2139 251271A	2141 346173H	2147 211305E	2149 361371G	2151 246721A
2153 356561G	2155 222715A	2157 315225C	2159 361055C	2161 323055G
2163 214773B	2165 271215A	2167 266347F	2169 301321C	2171 232405E
2181 323157D	2183 324661G	2185 215233B	2187 215103B	2189 310003H
2191 311427H	2193 351701C	2195 365051C	2197 335477H	2199 227311A

DEGREE 16--CONTINUED

2201	337027H	2203	303417H	2205	202721A	2211	201165A	2213	227107F
2215	203365A	2217	306373D	2219	211213F	2221	326533H	2223	344357D
2225	361437D	2227	312357D	2229	340577D	2231	254513F	2233	345363H
2235	232177B	2237	303067H	2243	310107H	2245	342421C	2247	255541A
2249	311763H	2251	323627H	2253	220037B	2255	300741C	2257	335513H
2259	206655A	2261	374255C	2263	341233H	2265	227627B	2267	260413F
2269	337647H	2275	312763D	2277	334137D	2279	270127F	2281	320055G
2283	347265C	2285	272161A	2287	336013H	2289	310443D	2291	233715E
2293	233251E	2295	374477D	2313	000771	2315	251037B	2317	271137F
2319	253123B	2321	235533F	2323	217161E	2325	256335A	2327	210177F
2329	256731A	2331	334423D	2333	271621E	2339	234037F	2341	356031G
2343	364761C	2345	332461C	2347	215127F	2349	374735C	2351	303661G
2353	341121G	2355	221603B	2357	274077F	2359	365573H	2361	356613D
2363	272215A	2365	364617D	2371	267233F	2373	366265C	2375	306405C
2377	315017H	2379	271677B	2381	345431G	2383	346365G	2385	303255C
2387	372217H	2389	200077F	2391	222443B	2393	255267F	2395	307633D
2397	346757D	2403	357451C	2405	307413D	2407	320365G	2409	205757B
2411	335405G	2413	254717F	2415	344277D	2417	215211E	2419	244737F
2421	331145C	2423	227305E	2425	364553D	2427	215135A	2437	277131E
2439	255017B	2441	374515G	2443	362363H	2445	347477D	2447	201241E
2449	370115G	2451	336675C	2453	253521E	2455	202045A	2457	364563D
2459	367527H	2461	200457F	2467	236461E	2469	356631C	2471	254643F
2473	236547F	2475	307775C	2477	353001G	2479	312507H	2481	222605A
2483	312073H	2485	260473B	2487	275435A	2489	200071E	2491	200275E
2493	264673B	2499	310527D	2501	277355E	2503	276265E	2505	235475A
2507	366733H	2509	311061G	2511	323363D	2513	314615G	2515	204567B
2517	335515C	2519	320407H	2521	257345E	2523	211327B	2525	306667D
2531	354071G	2533	233161A	2535	370371C	2537	301605G	2539	252455E
2581	314613H	2583	321247D	2585	254373B	2587	332231G	2589	350007D
2595	271547B	2597	332157H	2599	317307H	2601	303203D	2603	306645G
2605	351731C	2607	305255C	2609	356625G	2611	243771E	2613	266761A
2615	364577D	2617	301737H	2619	251675A	2621	275367F	2627	207733F
2629	371607H	2631	277327B	2633	325437H	2635	316451C	2637	343473D
2639	250267F	2641	352101G	2643	207747B	2645	251403B	2647	245057F
2649	357225C	2651	245155E	2653	344337H	2659	242017F	2661	343415C
2663	224413F	2665	363343D	2667	361635C	2669	221027B	2671	335661G
2673	214125A	2675	215575A	2677	224161E	2679	345101C	2681	314013H
2695	263425A	2697	336623D	2699	375715G	2701	247113F	2703	301365C
2705	211223B	2707	201345E	2709	351533D	2711	244745E	2713	261163F
2715	356515C	2717	302415G	2723	363455G	2725	363477D	2727	312253D
2729	337457H	2731	246515E	2733	227713B	2735	370125C	2737	313151C
2739	371007D	2741	277461E	2743	253053F	2745	243753B	2747	352363H
2749	270263F	2755	253017B	2757	256737B	2759	336615G	2761	225315E
2763	342135C	2765	273623B	2767	367161G	2769	237013B	2771	250565A
2773	306227H	2775	350523D	2777	212641E	2779	324513H	2781	202765A
2787	350351C	2789	264111E	2791	366147H	2793	300575C	2795	217635A
2797	371253H	2799	312255C	2801	252203F	2803	336007H	2805	254465A
2827	000453	2829	302151C	2831	345223H	2833	311375G	2835	215121A
2837	231651E	2839	375751C	2841	262721A	2843	222535E	2845	327323D
2851	315157H	2853	363227D	2855	371025C	2857	232561E	2859	363557D
2861	355763H	2863	312001G	2865	220747B	2867	317271G	2869	217527F
2871	324001C	2873	204013B	2875	351413D	2877	255363B	2883	361027D
2885	376347D	2887	363657H	2889	320437D	2891	320317H	2893	241161E
2895	225557B	2897	354175G	2899	237127F	2901	216241A	2903	353265G
2905	330051C	2907	263461A	2909	224305E	2915	335135C	2917	304655G
2919	237765A	2921	254727F	2923	323267H	2925	310257D	2927	332017H
2929	224035E	2931	264721A	2933	221047F	2935	346231C	2937	336545C
2951	365455G	2953	243163F	2955	202013B	2957	334213H	2959	232233F
2961	344171C	2963	320775G	2965	276463B	2967	314447D	2969	251447F
2971	236407F	2973	353711C	2979	333345C	2981	375713H	2983	257253F
2985	371551C	2987	334565G	2989	373017H	2991	366763D	2993	263641E
2995	205621A	2997	207631A	2999	345145G	3001	247461E	3003	243667B
3005	237651A	3011	371105G	3013	253055E	3015	312075C	3017	306675G
3019	373335G	3021	207613B	3023	271671E	3121	235505E	3123	220253B
3125	333171C	3127	240631E	3129	227737B	3131	311171G	3141	337571C

TABLE B.2. (*Continued*)

DEGREE 16--CONTINUED

3143	253237F	3145	222215A	3147	316401C	3149	343071G	3151	205305E
3153	220163B	3155	202375A	3157	354565G	3159	207227B	3161	317631G
3163	216545E	3165	362243D	3171	332363D	3173	305163H	3175	344025C
3177	320461C	3179	373407D	3181	254637F	3183	240255A	3185	357641C
3187	363057H	3189	334041C	3191	204373F	3193	240235E	3207	350325C
3209	264427F	3211	305721G	3213	301167D	3215	313577D	3217	207247F
3219	317675C	3221	256577F	3223	220233F	3225	340175C	3227	273513F
3229	370647H	3235	376055C	3237	335343D	3239	332751G	3241	340115G
3243	302267D	3245	272353B	3247	270217B	3249	260027B	3251	364473H
3253	202735E	3255	245277B	3257	206225E	3259	344023H	3269	202551E
3271	306313H	3273	345257D	3275	345677D	3277	347023H	3279	355525C
3281	241005A	3283	243337F	3285	220543B	3287	234537F	3289	230415E
3291	356177D	3293	233321E	3299	206035E	3301	277257F	3303	350705C
3305	367347D	3307	223075E	3309	260011A	3311	272021E	3313	335375G
3315	272535A	3341	000551	3343	225177F	3345	355233D	3347	242243F
3349	252661A	3351	316671C	3353	221123F	3355	300645C	3357	304161C
3363	355067D	3365	217547B	3367	240763F	3369	207441A	3371	343401G
3373	265347F	3375	343225C	3377	206151E	3379	371631G	3381	216755A
3383	344227D	3385	351611C	3387	230773B	3397	300367H	3399	254735A
3401	370151G	3403	244377F	3405	340605C	3407	316123H	3409	346415G
3411	204007B	3413	226163F	3415	350453D	3417	203175A	3419	340565G
3421	363103H	3427	212321E	3429	217353B	3431	313671G	3433	257325E
3435	226567B	3437	277035E	3439	305345G	3441	230333B	3443	271451E
3445	233571A	3447	363255C	3449	250131E	3463	366331G	3465	252755A
3467	263767F	3469	214377F	3471	213463B	3473	310633H	3475	232031A
3477	304503D	3479	340731G	3481	376125G	3483	324025C	3485	306551C
3491	231075E	3493	300453H	3495	227251A	3497	325167H	3499	346527H
3501	366117D	3503	215727F	3505	327163D	3507	323461C	3509	214251E
3511	327777H	3513	215447B	3515	240343B	3525	266461A	3527	225073F
3529	354767H	3531	332275C	3533	220075E	3535	302517D	3537	217423B
3539	266611E	3541	335061G	3543	210075A	3545	220573B	3547	302555G
3549	327727D	3555	243067B	3613	345251G	3619	245337F	3621	257453B
3623	237373F	3625	334371C	3627	355263D	3629	226741E	3631	204613F
3633	375225C	3635	203173B	3637	255125E	3639	304353D	3641	257675E
3643	354773H	3653	302033H	3655	204253B	3657	361445C	3659	205723F
3661	375555G	3663	305615C	3665	375653D	3667	252447F	3669	302445C
3671	230047F	3673	302451G	3675	206621A	3677	267227F	3683	211201E
3685	350435C	3687	377301C	3689	275747B	3691	271257F	3693	336307D
3695	241267B	3697	327623H	3699	216023B	3701	234337F	3703	355645G
3721	212033F	3723	360235C	3725	203311A	3727	237645E	3729	217563B
3731	262747F	3733	377465G	3735	332725C	3737	213727F	3739	355727H
3741	242161A	3747	343063D	3749	350031G	3751	231471E	3753	310113D
3755	242745A	3757	314531C	3759	343633D	3761	250407F	3763	344543H
3765	277065A	3767	206243F	3769	220607F	3771	223233B	3781	301033H
3783	232247B	3785	363367D	3787	271633F	3789	242321A	3791	346517D
3793	237435E	3795	337745C	3797	222377F	3799	360427H	3801	247707B
3803	226153F	3805	334603D	3811	334767H	3813	244021A	3815	254133B
3817	231057F	3819	340475C	3821	266635E	3823	324463H	3825	275675A
3855	000471	4369	000023	4371	207177B	4373	217565E	4375	214467B
4377	205317B	4379	311121G	4381	250201E	4387	254045E	4389	270271A
4391	375041G	4393	237103F	4395	267441A	4397	224173F	4403	252227B
4405	310653D	4407	237501A	4409	306043H	4411	317545G	4421	264727F
4423	304757H	4425	276573B	4427	332677H	4429	254211E	4435	310723D
4437	233327B	4439	241773F	4441	224707F	4443	356435C	4445	227611A
4451	234111E	4453	200517F	4455	311045C	4457	332707H	4459	323671G
4461	367267D	4467	235603B	4469	267135E	4471	367571C	4489	375113H
4491	231313B	4493	244077F	4499	373125G	4501	233231E	4503	323365C
4505	371557D	4507	222231E	4509	351577D	4515	327117D	4517	316767H
4519	351757H	4521	333625C	4523	352231G	4525	364715C	4531	322301G
4533	274347B	4535	345107D	4537	235167F	4539	320423D	4549	377351G
4551	235473B	4553	322351G	4555	365301C	4557	372421C	4563	232277B
4565	354123D	4567	231165E	4569	220771A	4571	215507F	4645	334437D
4647	251153B	4649	321523H	4651	263273F	4653	314435C	4659	213111A
4661	272633F	4663	317645G	4665	342153D	4667	322147H	4677	321255C
4679	277723F	4681	330271G	4683	340273D	4685	353447D	4691	253305E

DEGREE 16--CONTINUED

4693 340431G 4695 300247D 4697 272647F 4699 250641E 4701 321771C
4707 224457B 4709 305343D 4711 354505G 4713 250377B 4715 232637B
4717 330741G 4723 311465G 4725 317765C 4747 206723F 4749 211521A
4755 270573B 4757 371203H 4759 347277H 4761 374567D 4763 226343F
4765 231221A 4771 267177F 4773 365433D 4775 334401C 4777 373677D
4779 232441A 4781 241413F 4787 221527F 4789 204471E 4791 357511C
4793 357173H 4795 307457D 4805 263745A 4807 375603H 4809 322165C
4811 363623D 4813 367767H 4819 261545E 4821 321537D 4823 201561E
4825 364321C 4827 212131A 4829 306711G 4835 235237B 4837 352677H
4839 316053D 4841 216523F 4843 245717F 4845 267755A 4883 000543
4885 241207B 4887 215777B 4889 371623H 4891 313167H 4893 300515C
4899 366471C 4901 352655G 4903 311705G 4905 275177B 4907 337137H
4909 255737F 4915 211527B 4917 352275C 4919 245711E 4921 372447H
4923 357321C 4933 301571G 4935 202543B 4937 360067H 4939 276323F
4941 312477D 4947 244107B 4949 345443H 4951 213233F 4953 236371A
4955 344507D 4957 361505G 4963 231235E 4965 260757B 4967 245023F
4969 366447H 4971 220265A 4973 265653F 4979 376663H 4981 235273B
5003 365753H 5005 340751C 5011 302605G 5013 316475C 5015 345053D
5017 335447H 5019 265275A 5021 260167F 5027 255707F 5029 226373F
5031 230005A 5033 243703F 5035 363433D 5037 304327D 5203 202051E
5205 345611C 5207 373065G 5209 333551G 5211 234501A 5213 346751G
5219 353073D 5221 230303F 5223 317703D 5225 211107B 5227 233355E
5229 200617B 5235 252315A 5237 203761E 5259 332201C 5261 211363F
5267 313743H 5269 372727H 5271 350471C 5273 347247H 5275 230145A
5277 341727D 5283 254753B 5285 265003B 5287 255425A 5289 305153D
5291 362113H 5293 261217F 5299 236443F 5301 275753B 5303 322375G
5305 203773B 5319 200767B 5321 234217B 5323 232513F 5325 234601A
5331 237205A 5333 246753F 5335 364757D 5337 235257B 5339 212257F
5341 216427F 5347 246551E 5349 203577B 5351 201643F 5353 302021G
5355 232531A 5397 000567 5399 356433H 5401 254453F 5403 263027B
5405 372211C 5411 302751G 5413 370257H 5415 364627D 5417 210117F
5419 367257H 5421 222451A 5427 347323D 5429 342501G 5431 233033F
5433 264705A 5447 254615E 5449 242635E 5451 311031C 5453 250555E
5459 315757H 5461 206745E 5463 230173B 5465 263201A 5467 312703H
5469 232507B 5475 334665C 5477 216351E 5479 340363H 5481 306271C
5483 221743F 5485 375511C 5491 356717D 5493 272663B 5515 375707D
5517 216127B 5523 264051A 5525 360755C 5527 310503H 5529 230743B
5531 204657F 5533 360177H 5539 302541G 5541 241443B 5543 354517H
5545 262507B 5547 211553B 5549 246747F 5555 307251C 5557 225411E
5559 226007B 5561 304137H 5575 266363B 5577 261613B 5579 347441G
5581 223531E 5587 332125G 5677 263431E 5683 235125E 5685 301035C
5687 271317F 5689 201615E 5703 242173B 5705 223763B 5707 273031E
5709 367053D 5715 374331C 5717 261227F 5719 220121E 5721 235611A
5723 273127F 5725 275463B 5731 226171E 5733 372241C 5735 250663B
5737 217173F 5739 356343D 5741 223705E 5747 215657F 5773 252513F
5779 244773F 5781 217305A 5783 236715E 5785 253251A 5787 306177D
5789 334215G 5795 337775C 5797 205341A 5799 345021C 5801 227017F
5803 216141E 5805 267607B 5811 263503B 5813 353315G 5815 365367D
5817 201323B 5831 355755C 5833 371177H 5835 321235C 5837 221223F
5843 206363F 5845 334451C 5847 362317D 5849 273645E 5851 374361G
5853 236665A 5859 235743B 5861 242633F 5863 273075E 5865 326553D
5911 000435 5913 255021A 5915 230267B 5917 210763F 5923 321717H
5925 325617D 5927 363007H 5929 203571E 5931 343725C 5933 220045A
5939 231433F 5941 304457H 5943 247553B 5945 257023B 6343 377645G
6345 220551A 6347 325303H 6349 306747H 6355 377241C 6357 372433D
6359 230731E 6361 322405G 6363 362731C 6373 247527F 6375 352527D
6425 000637 6427 341575G 6437 376677H 6439 204337F 6441 234241A
6443 257255A 6445 301767D 6451 333075G 6453 213765A 6455 271207B
6457 243441E 6471 315641C 6473 241577F 6475 330255C 6477 360733D
6483 216513B 6485 352303D 6487 210711E 6489 210241A 6491 254075E
6501 365325C 6503 336763H 6505 373547D 6507 304115C 6509 331617H
6515 221425A 6541 223755E 6547 256621E 6549 210471A 6551 377203H
6553 222527F 6555 215061A 6565 366625C 6567 376563D 6569 262227F
6571 253653F 6573 377375C 6579 262015A 6581 241113F 6583 216537F
6585 326647D 6599 270133F 6601 340521G 6603 230445A 6709 370713H

TABLE B.2. (*Continued*)

DEGREE 16--CONTINUED

6711	375261C	6713	242371E	6727	345425G	6729	363411C	6731	275661E
6733	372545G	6739	341711G	6741	223155A	6743	250353F	6745	251435A
6747	356417D	6757	352457H	6759	377537D	6761	342575G	6763	373773H
6765	301145C	6803	343565G	6805	232661A	6807	237211A	6809	330307H
6811	335025G	6821	353427H	6823	251543F	6825	325327D	6827	224505E
6829	252765E	6835	273421A	6837	350601C	6839	324657H	6841	270235E
6855	303075C	6857	315065G	6859	245165E	6861	217275A	6867	352137D
6869	277457F	6871	326701G	6873	304017D	6875	244365A	6885	321613D
6939	000643	6949	212371E	6951	360253D	6953	305051C	6955	361063D
6957	275343B	6963	355465C	6965	240367B	6967	242767F	6969	267625A
6983	272225E	6985	234117B	6987	264073B	6989	304633H	6995	304713D
6997	315467H	6999	355331C	7001	263161E	7003	314057H	7013	357145G
7015	277563B	7017	352437D	7019	217317F	7021	267601A	9363	342631C
9365	316027D	9371	255161E	9381	267357B	9387	307143D	9389	365345G
9395	245043B	9397	236503F	9419	257067F	9421	200641E	9427	235011E
9429	245147B	9435	300403D	9509	000515	9515	322315C	9517	261153F
9523	276277F	9525	225361A	9547	354407H	9549	262433B	9555	244611A
9557	221505E	9563	241201E	9573	235731A	9579	276057B	9581	335203H
9619	314777H	9621	256055A	9627	312545C	9637	223203F	9643	347127H
9645	322717D	9651	260541A	9805	310751C	9811	304773H	9813	335601C
9819	201263B	9829	262453F	9835	265757B	9877	260243B	9883	217603F
9893	372323H	9899	273433F	9901	360463H	9907	365637H	9909	263071A
10571	320157H	10573	204645E	10579	314043H	10581	366135C	10603	360515G
10645	241125A	10667	264323F	10837	317515G	10923	247167B	10925	312537D
10931	211737B	10933	362553H	10955	275221A	10957	323305G	10963	265773F
10965	220411A	11051	000537	11053	242341E	11059	240747F	11061	343555C
11083	204325E	11085	253401A	11443	254331E	11469	234133B	11475	340407D
11565	000727	13107	000037	21845	000007				

DEGREE 17									
		1	400011E	3	400017F	5	400431E	7	525251E
9	410117F	11	400731E	13	411335E	15	444257F	17	600013H
19	403555E	21	525327F	23	411077F	25	404525E	27	401523F
29	466273F	31	642015G	33	446613F	35	527427F	37	414347F
39	414443F	41	501353F	43	445141E	45	663013H	47	414663F
49	535013F	51	610215G	53	403063F	55	530765E	57	460377F
59	626653H	61	405473F	63	504671E	65	771353H	67	444611E
69	422273F	71	442571E	73	612537H	75	572325E	77	564225E
79	561175E	81	447773F	83	470337F	85	640635G	87	646775G
89	532617F	91	537773F	93	510473F	95	413651E	97	514573F
99	601335G	101	606155G	103	540041E	105	525535E	107	714303H
109	506741E	111	514045E	113	750413H	115	642433H	117	551757F
119	735207H	121	662527H	123	530645E	125	431601E	127	775325G
129	671075G	131	443043F	133	607115G	135	432265E	137	454067F
139	402545E	141	547163F	143	742377H	145	430161E	147	504505E
149	461331E	151	454433F	153	724003H	155	562467F	157	774563H
159	430053F	161	557517F	163	652243H	165	455655E	167	727113H
169	521367F	171	743537H	173	424443F	175	421215E	177	655241G
179	424107F	181	675215G	183	541625E	185	771231G	187	734703H
189	470147F	191	413557F	193	424761E	195	656765G	197	756341G
199	722571G	201	600657H	203	554347F	205	543457F	207	414115E
209	674557H	211	630043H	213	734763H	215	402253F	217	542447F
219	706353H	221	751471G	223	636031G	225	755535G	227	771451G
229	435251E	231	620071G	233	752045G	235	742003H	237	717553H
239	734727H	241	740227H	243	633051G	245	676521G	247	567321E
249	705503H	251	763237H	253	544613E	255	524057F	257	774503H
259	477743F	261	477633F	263	755723H	265	447503F	267	411347F
269	417611E	271	404121E	273	521531E	275	611073H	277	432625E
279	501275E	281	446417F	283	671763H	285	455647F	287	575245E
289	476601E	291	437631E	293	404607F	295	466457F	297	652563H
299	461471E	301	531071E	303	772047H	305	477225E	307	457601E
309	667401G	311	625321G	313	510071E	315	573213F	317	755547H
319	472617F	321	511337F	323	511373F	325	603777H	327	424505E
329	504641E	331	425601E	333	466055E	335	565765E	337	545603F
339	603205G	341	512211E	343	715147H	345	541211E	347	565171E
349	465521E	351	640407H	353	655007H	355	555477F	357	575535E

DEGREE 18	1 1000201E	3 1010301A	5 1002241E
7 1000377B	9 1200205A	11 1703601G	13 1025711E
15 1015721A	17 1115701E	19 1023141A	21 1070477B
27 1223215A	63 1313133B	63 1313133B	63 1313133B
189 1623075C	57 1160435A	57 1160435A	171 1626367D
57 1160435A	171 1626367D	171 1626367D	133 1300565A
399 1514245C	399 1514245C	1197 1052465A	399 1514245C
1197 1052465A	1197 1052465A	73 1642365C	219 1252555A
219 1252555A	657 1253607A	219 1252555A	657 1253607A
657 1253607A	1971 1334325A	511 1231145A	1533 1055321A
1533 1055321A	4599 1341035A	1533 1055321A	4599 1341035A
4599 1341035A	13797 1777777D	1387 1011011A	9709 1001001A

DEGREE 19	1 2000047F	3 2020471E	5 2013211E
7 2570103F	9 2561427F	11 2227023F	13 2001711E
15 2331067F	17 3146455G	19 3610353H	21 2766447F

DEGREE 20	1 4000011E	3 4000017B	5 4200031A
7 4001051E	9 4040217B	11 4030071A	13 4004515E
15 4221037B	17 6000031G	19 4442235E	21 4103307B
25 4307165A	75 4266075A	33 4036267B	55 4346037B
165 5145217B	55 4346037B	165 5145217B	275 6027135C
825 6044073D	31 4034755A	93 4627377B	155 4367471A
465 5057137B	155 4367471A	465 5057137B	775 6505453D
2325 4504241A	341 4510031A	1023 7552557D	1705 6406005C
5115 5327265A	1705 6406005C	5115 5327265A	8525 5746331A
25575 6647133D	41 4027577B	123 4761757B	205 5541427A
615 7113055C	205 5541427A	615 7113055C	451 7544237D
1353 7602777D	2255 5017111A	6765 5521623B	2255 5017111A
6765 5521623B	1271 7050457D	3813 4345543B	6355 6130725C
19065 7164555C	6355 6130725C	19065 7164555C	13981 4100001A
41943 4102041A			

DEGREE 21	1 10000005E	3 10040205E	5 10020045E
7 11111115A	9 10040315E	11 10000635E	13 10103075E
15 10050335E	17 10002135E	19 17000075G	21 14600067D
49 11105347A	127 10225077A	889 11166743A	889 11166743A
6223 17155161C	337 10264425A	2359 16260075C	2359 16260075C
42799 10040001A			

DEGREE 22	1 20000003F	3 20100403B	5 20001043F
7 22222223F	9 20100453B	11 25200127F	13 20401207F
15 20110517B	17 20430607F	19 20070217F	21 31400147D
23 20005611A	69 20465307B	89 20603715A	267 24146477B
2047 22404051A	6141 36544657D	683 34230073D	15709 21774413B
60787 34603145C			

DEGREE 23	1 40000041E	3 40404041E	5 40000063F
7 40010061E	9 50000241E	11 40220151E	13 40006341E
15 40405463F	17 40103271E	19 41224445E	21 40435651E
47 44636045A	178481 43073357B		

DEGREE 24	1 100000207F	3 100205645A
5 100305143B	7 100315361A	9 102746675A
11 125245661E	13 113646571A	15 112432273B
17 140775753D	19 113763063E	21 116636645A
45 170736335C	63 164260065C	35 113206017A
105 151255377B	105 151255377B	315 105404647B
39 156267123C	39 156267123C	117 131307443B
65 150051747D	195 132365525A	195 132365525A
585 157653375C	91 160503563D	273 137240727A
273 137240727A	819 144534331C	455 165330327D
1365 103446341A	1365 103446341A	4095 120652605A
51 155212435A	51 155212435A	153 130633327B
85 141720423C	255 172634307C	255 172634307C

TABLE B.2. (*Continued*)

DEGREE 24--CONTINUED			
	765 146537231C	119 123426525A	357 105732145A
	357 105732145A	1071 133125511A	595 155513755A
	1785 121720647B	1785 121720647B	5355 102474621A
	221 100466513A	663 100006161A	663 100006161A
	1989 101312015A	1105 126751351A	3315 104313243A
	3315 104313243A	9945 116055567A	1547 156652045C
	4641 124430435A	4641 124430435A	13923 112630407A
	7735 127617123A	23205 133033563B	23205 133033563B
	69615 161676707A	241 174317125C	723 171224435C
	723 171224435C	2169 154423127D	1205 132001371A
	3615 145363733B	3615 145363733B	10845 155353415A
	1687 165365701C	5061 106342635A	5061 106342635A
	15183 100605077B	8435 133567111A	25305 161276343B
	25305 161276343B	75915 100140053B	3133 101332157B
	9399 131342727B	9399 131342727B	28197 162047171C
	15665 112155405A	46995 164117115C	46995 164117115C
	140985 124055647B	21931 110001101A	109655 100011011A
DEGREE 25		1 200000011E	3 200000017F
	5 204000051E	7 200010031E	9 200402017F
	11 252001251E	13 201014171E	15 204204057F
	17 200005535E	19 200014731E	21 201015517F
	31 200523477B	601 353551603D	18631 277267355A
	1801 341573647D	55831 253566335A	
DEGREE 26		1 400000107F	3 401007131A
	5 430216473F	7 402365755E	9 410004563B
	11 426225667F	13 510664323F	15 475477275A
	17 473167545E	19 411335571E	21 433315447B
	2731 656536753D	8191 614326143D	24573 600777003D
DEGREE 27		1 1000000047E	3 1001007071E
	5 1020024171E	7 1004462703B	9 1102210617E
	11 1250025757F	13 1257242631E	15 1020560103F
	17 1112225171E	19 1037530241E	21 1006524347B
	73 1215076703A	511 1745602367D	
DEGREE 28		1 2000000011E	3 2000000017B
	5 2040000411A	7 2104210431E	9 2002004017B
	11 2000025051E	13 2020006031E	15 2040410417B
	17 2002502115E	19 2001601071E	21 2104213577A
	29 2010141305A	87 2010073021A	145 2112310701A
	435 2256267705A	43 2043450123B	129 2232610673B
	215 3417321145C	645 2507013341A	1247 2036150345A
	3741 2742450341A	6235 2052124143B	18705 2307251163B
	113 2065561561A	339 2550100465A	565 3662526717D
	1695 3655737253B	3277 2752435573B	9831 3521653421A
	4859 2313475717B	14577 3513705403A	24295 2517460277B
	72885 2037216263B	140911 2336561121A	422733 3043320155A
	127 2243345037B	381 3540233367A	635 3664406015C
	1905 3757051033D	3683 2322031441A	11049 3157336171C
	18415 2226143443B	55245 3033453267C	5461 3305002225C
	16383 2273447351A	27305 3061505731C	81915 3052445243D
	158369 3533324373D	475107 2337253731A	791845 2145723745A
	2375535 2330160331A	14351 3316136351C	43053 2374475053A
	71755 3175417143A	215265 3514237073C	416179 2755450655A
	1248537 3657555473D	617093 3614772157D	1851279 3572445367D
	3085465 3706175715C	9256395 3777777777D	
DEGREE 29		1 4000000005E	3 4004004005E
	5 4000010205E	7 4010000045E	9 4400000045E
	11 4002200115E	13 4001040115E	15 4004204435E
	17 4100060435E	19 4040003075E	21 4004064275E
	233 4125377665A	1103 4663771561A	256999 7260572607D
	2089 6202672631C	486737 6276417701C	2304167 4334123375A

DEGREE 30

		1	10040000007F	3	10045207405A
5	10104264207F	7	17254401747D	9	10466404155A
11	10421106467B	13	10115131333F	15	12531150265A
17	11326212703F	19	10343244533E	21	14340746005C
63	15671207425A	33	10617013661A	33	10617013661A
99	10231077101A	77	10347066511A	231	12551521353B
231	12551521353B	693	12363365205A	31	10537567431A
93	13104273407B	93	13104273407B	279	17565561725C
217	13063776443B	651	14475010377C	651	14475010377C
1953	16217747517D	341	15312176137D	1023	13005472403B
1023	13005472403B	3069	15027200513D	2387	17327131755A
7161	17273014127A	7161	17273014127A	21483	15222475661C
151	11732145645A	453	15642307235C	453	15642307235C
1359	13137001367A	1057	17576155211A	3171	14046056527C
3171	14046056527C	9513	15362114071A	1661	16275156545A
4983	11747625331A	4983	11747625331A	14949	14262504223C
11627	12305126253B	34881	11274077671A	34881	11274077671A
104643	16671210137D	4681	11346765601A	14043	15727555211C
14043	15727555211C	42129	11154174627A	32767	14271111643D
98301	17313775157D	98301	17313775157D	294903	17667776677D
51491	15116464137C	154473	10170400463B	154473	10170400463B
463419	13637044253B	360437	13726766575A	1081311	14437537423D
1081311	14437537423D	3243933	17657537277D	331	13214207735A
993	15100727503B	993	15100727503B	2979	11115104367B
2317	10737311047B	6951	12374572221A	6951	12374572221A
20853	11567732701A	3641	14707036127B	10923	16076273661C
10923	16076273661C	25487	10403615303A	76461	10221305567A
76461	10221305567A	10261	16150525151C	30783	10363607103A
30783	10363607103A	92349	12553152637A	71827	14221266525C
215481	17473760245C	215481	17473760245C	646443	17070134445A
112871	12527647623A	338613	12670030647A	338613	12670030647A
790097	12105065527A	2370291	10545323161A	2370291	10545323161A
49981	10400014607B	149943	10502035235A	149943	10502035235A
449829	12240170427B	349867	10101010111A	549791	11303560025A
1649373	15735076321C	1649373	15735076321C	3848537	11010100111A
1549411	12135356633B	4648233	11274767701A	4648233	11274767701A
13944699	16471647235C	10845877	11000100011A		

DEGREE 31

		1	20000000011E	3	20000000017E
5	20000020411E	7	21042104211E	9	20010010017E
11	20005000251E	13	20004100071E	15	20202040217E
17	20000200435E	19	20060140231E	21	21042107357E

DEGREE 32

		1	40020000007F	3	40001114005A
5	50521021747B	7	40460216667F	9	40220536125A
11	40035532523F	13	42003247143F	15	42644424505A
17	44165166133B	19	41760427607F	21	56032357221A
51	73274317525C	85	55255004227B	255	60537314115C
257	52213142567B	771	46633742135A	1285	53046115123B
3855	47254550703B	4369	45052437233B	13107	71265756301C
21845	65636126613D	65535	57410204175A		

DEGREE 33

		1	100000020001E	3	100020024001E
5	104000420001E	7	100000260001A	9	100020224401E
11	111100021111E	13	100000031463F	15	104020466001E
17	100502430041E	19	100601431001E	21	100034327001A
23	100021260105A	161	107167672771A	89	100123140475A
623	124155341567B	2047	142560223461C	14329	150052442055C
599479	125725100311A	13788017	101534661265A	53353631	107753475213B

DEGREE 34

		1	201000000007F	3	201051003005A
5	201472024107F	7	377000007527H	9	203123311035A
11	225213433257F	13	227712240037F	15	213753015051A
17	251132516577F	19	211636220473F	21	377235535321C
43691	327304565547D	131071	331706543633D	393213	226405640551A

References

1. Abramson, N. M., "A Class of Systematic Codes for Non-Independent Errors," *IRE Trans. Inf. Theory*, **IT-5**, 150–157 (1959).
2. Abramson, N. M., and B. Elspas, "Double-Error-Correcting Codes and Decoders for Non-Independent Binary Errors," presented at the UNESCO Inf. Proc. Conf., Paris, 1959.
3. Anderson, P. H., F. M. Hsu, and M. N. Sandler, "A New Adaptive Modem for Long Haul HF Digital Communications at Data Rates Greater Than 1 BPS/Hz," *IEEE Military Comm. Conf. Rec.*, 29.2-1 – 29.2-7, 1982.
4. Arthurs, E., and H. Dym, "On the Optimum Detection of Digital Signals in the Presence of White Gaussian Noise," *IRE Trans. Comm. Sys.*, **CS-10**, 336–372 (1962).
5. Artin, E., *Galois Theory*, 2nd ed., University of Notre Dame Press, Notre Dame, IN, 1966.
6. Berger, T., *Rate Distortion Theory: A Mathematical Basis for Data Compression*, Prentice-Hall, Englewood Cliffs, NJ, 1971.
7. Berlekamp, E. R., *Algebraic Coding Theory*, McGraw-Hill, New York, 1968.
8. Berlekamp, E. R., "Long Primitive Binary BCH Codes Have Distance $d \sim 2n \ln R^{-1} / \log n \ldots$," *IEEE Trans. Inf. Theory*, **IT-18**, 415–426 (1972).
9. Berlekamp, E. R., "Goppa Codes," *IEEE Trans. Inf. Theory*, **IT-19**, 590–592 (1973).
10. Berlekamp, E. R. (Ed.), *Key Papers in the Development of Coding Theory*, IEEE Press, New York, 1974.
11. Berlekamp, E. R., "Algebraic Codes for Improving the Reliability of Tape Storage," *Proc. Nat. Comp. Conf.*, 1975.
12. Birkhoff, G., and S. MacLane, *A Survey of Modern Algebra*, 3rd ed., Macmillan, New York, 1965.
13. Blahut, R. E., "Transform Techniques for Error-Control Codes," *IBM J. Res. Dev.*, **23**, 299–315 (1979).
14. Blahut, R. E., *Theory and Practice of Error Control Codes*, Addison-Wesley, Reading, MA, 1983.
15. Blustein, G., and K. L. Jordan, "An Investigation of the Fano Sequential Decoding Algorithm by Computer Simulation," Group Report 62G-3, MIT Lincoln Laboratory, 1963.

16. Bose, R. C., and D. K. Ray-Chaudhuri, "On a Class of Error Correcting Binary Group Codes," *Inf. Control*, **3**, 68–79 (1960).

17. Brayer, K., "Error Correction Code Performance on HF, Troposcatter, and Satellite Channels," *IEEE Trans. Comm. Tech.*, **COM-19**, 781–789 (1971).

18. Bucher, E. A., "Coding Options for Efficient Communications on Non-Stationary Channels," *Rec. IEEE Int. Conf. Comm.*, 4.1.1–4.1.7, 1980.

19. Burton, H. O., and D. D. Sullivan, "Errors and Error Control," *Proc. IEEE*, **60**, 1293–1301 (1972).

20. Burton, H. O., D. D. Sullivan, and S. Y. Tong, "Generalized Burst Trapping Codes," *IEEE Trans. Inf. Theory*, **IT-17**, 736–742 (1971).

21. Bussgang, J. J., "Some Properties of Binary Convolutional Code Generators," *IEEE Trans. Inf. Theory*, **IT-11**, 90–100 (1965).

22. Chase, D., "A Class of Algorithms for Decoding Block Codes with Channel Measurement Information," *IEEE Trans. Inf. Theory*, **IT-18**, 170–182 (1972).

23. Chase, D., "A Combined Coding and Modulation Approach for Communication Over Dispersive Channels," *IEEE Trans. Comm.*, **COM-21**, 159–174 (1973).

24. Chien, R. T., "Cyclic Decoding Procedures for Bose-Chaudhuri-Hocquenghem Codes," *IEEE Trans. Inf. Theory*, **IT-10**, 357–363 (1964).

25. Chien, R. T., "Memory Error Control: Beyond Parity," *IEEE Spectrum*, **10**, No. 7, 18–23 (July 1973).

26. Church, R., "Tables of Irreducible Polynomials for the First Four Prime Moduli," *Ann. Math.*, **36**, 198–209 (1935).

27. Clark, G. C., Jr., and J. B. Cain, *Error-Correction Coding for Digital Communications*, Plenum, New York, 1981.

28. Daut, D. G., "Code Construction for Selected Rational Rate Convolutional Codes," ECSE Report TR81-4, RPI, Troy, NY, June 1981.

29. Daut, D. G., J. W. Modestino, and L. D. Wismer, "New Short Constraint Length Convolutional Code Constructions for Selected Rational Rates," Internal ECSE Department Report, RPI, Troy, NY, March 1981.

30. Dickson, L. E., *Linear Groups with an Exposition of the Galois Field Theory*, Dover, New York, 1958.

31. Dixon, R. C. (Ed.), *Spread Spectrum Techniques*, IEEE Press, New York, 1976.

32. Elias, P., "Coding for Noisy Channels," *IRE Conv. Rec.*, Part 4, 37–46, 1955.

33. Fano, R. M., "A Heuristic Discussion of Probabilistic Decoding," *IEEE Trans. Inf. Theory*, **IT-9**, 64–74 (1963).

34. Fire, P., "A Class of Multiple-Error-Correcting Binary Codes for Non-Independent Errors," Report RSL-E-2, Sylvania Reconnaissance Systems Laboratory, Mountain View, CA, 1959.

35. Folts, H. C. (Ed.), *Data Communications Standards*, 2nd ed., McGraw-Hill, New York, 1982.

36. Forney, G. D., Jr., "On Decoding BCH Codes," *IEEE Trans. Inf. Theory*, **IT-11**, 549–557 (1965).

37. Forney, G. D., Jr., *Concatenated Codes*, MIT Press, Cambridge, MA, 1966.

38. Forney, G. D., Jr., "Generalized Minimum Distance Decoding," *IEEE Trans. Inf. Theory*, **IT-12**, 125–131 (1966).

39. Forney, G. D., Jr., "Study of Correlation Coding," Technical Report RADC-TR-67-410, Rome Air Development Center, Griffis Air Force Base, NY, September 1967.

40. Forney, G. D., Jr., "Convolutional Codes I: Algebraic Structure," *IEEE Trans. Inf. Theory*, **IT-16**, 720–738 (1970).

41. Forney, G. D., Jr., "Burst Correcting Codes for the Classic Bursty Channel," *IEEE Trans. Comm. Tech.*, **COM-19**, 772–781 (1971).
42. Forney, G. D., Jr., "Maximum-Likelihood Sequence Estimation of Digital Sequences in the Presence of Intersymbol Interference," *IEEE Trans. Inf. Theory*, **IT-18**, 363–378 (1972).
43. Forney, G. D., Jr., "Convolutional Codes II: Maximum-Likelihood Decoding," *Inf. Control*, **25**, 222–226 (1974).
44. Forney, G. D., Jr., and E. K. Bower, "A High Speed Sequential Decoder: Prototype Design and Test," *IEEE Trans. Comm. Tech.*, **COM-19**, 821–835 (1971).
45. Forney, G. D., Jr., and R. M. Langelier, "A High Speed Sequential Decoder for Satellite Communications," *Rec. IEEE Int. Conf. Comm.*, 39-9 – 39-17, 1969.
46. Freeman, R. L., *Telecommunication System Engineering*, Wiley-Interscience, New York, 1980.
47. Freeman, R. L., *Telecommunication Transmission Handbook*, 2nd ed., Wiley-Interscience, New York, 1981.
48. Gallager, R. G., "A Simple Derivation of the Coding Theorem and Some Applications," *IEEE Trans. Inf. Theory.*, **IT-11**, 3–18 (1965).
49. Gallager, R. G., *Information Theory and Reliable Communication*, Wiley, New York, 1968.
50. Gentleman, W. M., "Matrix Multiplication and Fast Fourier Transforms," *Bell Sys. Tech. J.*, **47**, 1099–1103 (1968).
51. Gilhousen, K. S., et al., "Coding Systems Study for High Data Rate Telemetry Links," Final Contract Report, N71-27786, Contract No. NAS2-6024, Linkabit Corporation, La Jolla, CA, 1971.
52. Golay, M. J. E., "Notes on Digital Coding," *Proc. IRE*, **37**, 657 (1949).
53. Golomb, S. W. (Ed.), *Digital Communications with Space Applications*, Prentice-Hall, Englewood Cliffs, NJ, 1964.
54. Golomb, S. W., *Shift Register Sequences*, Holden-Day, San Francisco, 1967. (Revised edition published by Aegean Park Press, Laguna Hills, CA, 1982.)
55. Goppa, V. D., "A New Class of Linear Error-Correcting Codes," *Probl. Peredachi Inf.*, **6**, 24–30 (1970). (Reprinted in ref. 10.)
56. Goppa, V. D., "Rational Representation of Codes and (L,g)-Codes," *Probl. Peredachi Inf.*, **7**, 41–49 (1971). (Reprinted in ref. 10.)
57. Gore, W. C., "Transmitting Binary Symbols with Reed-Solomon Codes," *Proc. Princeton Conf. Inf. Sci. Sys.*, 495–497, 1973. (Also published as Report 73-5, The Johns Hopkins University, Department of Electrical Engineering, April 1973.)
58. Green, D. H., and I. S. Taylor, "Irreducible Polynomials over Composite Galois Fields and Their Applications in Coding Techniques," *Proc. IEE*, **121**, 935–939 (1974).
59. Hagelbarger, D. W., "Recurrent Codes: Easily Mechanized, Burst-Correcting, Binary Codes," *Bell Sys. Tech. J.*, **38**, 969–984 (1959).
60. Hamming, R. W., "Error Detecting and Error Correcting Codes," *Bell Sys. Tech. J.*, **29**, 147–160 (1950).
61. Hartley, R. V. L., "Transmission of Information," *Bell Sys. Tech. J.*, **7**, 535–563 (1928).
62. Hartmann, C. R. P., and L. D. Rudolph, "An Optimum Symbol-by-Symbol Decoding Rule for Linear Codes," *IEEE Trans. Inf. Theory*, **IT-22**, 514–517 (1976).
63. Heller, J. A., "Feedback Decoding of Convolutional Codes," in A. J. Viterbi (Ed.), *Advances in Communication Systems*, Vol. 4, Academic Press, New York, 1975.
64. Heller, J. A., and I. M. Jacobs, "Viterbi Decoding for Satellite and Space Communication," *IEEE Trans. Comm. Tech.*, **COM-19**, 835–848 (1971).
65. Heller, R. M., "Forced-Erasure Decoding and the Erasure Reconstruction Spectra of Group Codes," *IEEE Trans. Comm. Tech.*, **COM-15**, 390–397 (1967).

66. Hildebrand, F. B., *Methods of Applied Mathematics*, Prentice-Hall, Englewood Cliffs, NJ, 1952.

67. Hocquenghem, A., "Codes correcteurs d'erreurs," *Chiffres*, **2**, 147–156 (1959).

68. Holmes, J. K., *Coherent Spread Spectrum Systems*, Wiley-Interscience, New York, 1981.

69. Huntoon, Z. McC., and A. M. Michelson, "On the Computation of the Probability of Post-Decoding Error Events for Block Codes," *IEEE Trans. Inf. Theory.*, **IT-23**, 399–403 (1977).

70. IEEE Special Issue on Spread Spectrum Communications, *IEEE Trans. Comm.*, **COM-30**, 817–1072 (1982).

71. Iwadare, Y., "On Type-B1 Burst-Error-Correcting Convolutional Codes," *IEEE Trans. Inf. Theory*, **IT-14**, 577–583 (1968).

72. Jacobs, I. M., "Sequential Decoding for Efficient Communications from Deep Space," *IEEE Trans. Comm. Tech.*, **COM-15**, 492–501 (1967).

73. Jacobs, I. M., "Practical Applications of Coding," *IEEE Trans. Inf. Theory*, **IT-20**, 305–310 (1974).

74. Jacobs, I. M., and E. R. Berlekamp, "A Lower Bound to the Distribution of Computation for Sequential Decoding," *IEEE Trans. Inf. Theory*, **IT-13**, 167–174 (1967).

75. Jelinek, F., "A Fast Sequential Decoding Algorithm Using a Stack," *IBM J. Res. Dev.*, **13**, 675–685 (1969).

76. Johannesson, R., "Robustly Optimum Rate One-Half Binary Convolutional Codes," *IEEE Trans. Inf. Theory*, **IT-21**, 464–468 (1975).

77. Johannesson, R., "Some Rate 1/3 and 1/4 Binary Convolutional Codes with an Optimum Distance Profile," *IEEE Trans. Inf. Theory*, **IT-23**, 281–283 (1977).

78. Jordan, K. L., Jr., "The Performance of Sequential Decoding in Conjunction with Efficient Modulation," *IEEE Trans. Comm. Tech.*, **COM-14**, 283–297 (1966).

79. Justesen, J., "A Class of Constructive Asymptotically Good Algebraic Codes," *IEEE Trans. Inf. Theory*, **IT-18**, 652–656 (1972).

80. Kasami, T., "A Decoding Procedure for Multiple Error-Correcting Cyclic Codes," *IEEE Trans. Inf. Theory*, **IT-10**, 134–138 (1964).

81. Kasami, T., and S. Lin, "On the Probability of Undetected Error for Maximum Distance Separable Codes," *IEEE Trans. Comm.*, **COM-32**, 998–1006 (1984).

82. Kasami, T., S. Lin, and T. Klove, "On the Probability of Undetected Error of Linear Block Codes," *GLOBECOM Conf. Rec.*, 1111–1113, 1982.

83. Klieber, E. J., "Some Difference Triangles for Constructing Self-Orthogonal Codes," *IEEE Trans. Inf. Theory*, **IT-16**, 237–238 (1970).

84. Kohavi, Z., *Switching and Finite Automata Theory*, McGraw-Hill, New York, 1970.

85. Kohlenberg, A., and G. D. Forney, Jr., "Convolutional Coding for Channels with Memory," *IEEE Trans. Inf. Theory*, **IT-14**, 618–626 (1968).

86. Kotel'nikov, V. A., *The Theory of Optimum Noise Immunity* (translated by R. A. Silverman), McGraw-Hill, New York, 1959.

87. Lawton, J. G., "Theoretical Error Rates of 'Differentially Coherent' Binary and 'Kineplex' Data Transmission Systems," *Proc. IRE*, **47**, 333–334 (1959).

88. Lawton, J. G., A. M. Michelson, and Z. Huntoon, "A Performance Analysis of a Family of Codes Constructed from Simple Binary Codes," *Rec. Can. Conf. Comm. Power*, 1978.

89. Lee, L. N., "On Optimum Soft-Decision Demodulation," *IEEE Trans. Inf. Theory*, **IT-22**, 437–444 (1976).

90. Lee, L. N., "Concatenated Coding Systems Employing a Unit-Memory Convolutional Code and a Byte-Oriented Decoding Algorithm," *IEEE Trans. Comm.*, **COM-25**, 1064–1074 (1977).

91. Leung, C., "Evaluation of the Undetected Error Probability of Single Parity-Check Product Codes," *IEEE Trans. Comm.*, **COM-31**, 250–253 (1983).

92. Leung-Yan-Cheong, S. K., and M. Hellman, "Concerning a Bound on Undetected Error Probability," *IEEE Trans. Inf. Theory*, **IT-22**, 235–237 (1976).

93. Leung-Yan-Cheong, S. K., E. R. Barnes, and D. U. Friedman, "On Some Properties of the Undetected Error Probability of Linear Codes," *IEEE Trans. Inf. Theory*, **IT-25**, 110–112 (1979).

94. Lin, S., and D. J. Costello, Jr., *Error Control Coding: Fundamentals and Applications*, Prentice-Hall, Englewood Cliffs, NJ, 1983.

95. Lin, S., and H. Lyne, "Some Results on Binary Convolutional Code Generators," *IEEE Trans. Inf. Theory*, **IT-13**, 134–139 (1967).

96. Lin, S., and E. J. Weldon, Jr., "Long BCH Codes are Bad," *Inf. Control*, **11**, 445–451 (1967).

97. Lindsey, W. C., and M. K. Simon, *Telecommunication Systems Engineering*, Prentice-Hall, Englewood Cliffs, NJ, 1973.

98. Lucky, R. W., J. Salz, and E. J. Weldon, Jr., *Principles of Data Communication*, McGraw-Hill, New York, 1968.

99. Lyon, R. F., "Convolutional Codes for *M*-ary Orthogonal and Simplex Channels," The Deep Space Network, Progress Report No. 42-24—Sept. and Oct., JPL, Pasadena, CA, December 15, 1974, pp.60–65.

100. Marsh, R. W., "Table of Irreducible Polynomials Over $GF(2)$ Through Degree 19," Office of Technical Services, U. S. Department of Commerce, Washington, DC, October 24, 1957.

101. Martin, J., *Teleprocessing Network Organization*, Prentice-Hall, Englewood Cliffs, NJ, 1970.

102. Mason, S. J., and H. J. Zimmermann, *Electric Circuits, Signals and Systems*, Wiley, New York, 1960.

103. Massey, J. L., *Threshold Decoding*, MIT Press, Cambridge, MA, 1963.

104. Massey, J. L., "Shift-Register Synthesis and BCH Decoding," *IEEE Trans. Inf. Theory*, **IT-15**, 122–127 (1969).

105. Massey, J. L., "Variable-Length Codes and the Fano Metric," *IEEE Trans. Inf. Theory*, **IT-18**, 196–198 (1972).

106. Massey, J. L., "Coding and Modulation in Digital Communications," *Proc. Int. Zurich Seminar Digital Comm.*, Zurich, Switzerland, 1974.

107. Massey, J. L., and M. K. Sain, "Inverses of Linear Sequential Circuits," *IEEE Trans. Comp.*, **C-17**, 330–337 (1968).

108. MacWilliams, F. J., "A Theorem on the Distribution of Weights in a Systematic Code," *Bell Sys. Tech. J.*, **42**, 79–94 (1963).

109. MacWilliams, F. J., and N. J. A. Sloane, *The Theory of Error-Correcting Codes*, North-Holland, Amsterdam, 1977.

110. McNamara, J. E., *Technical Aspects of Data Communication*, 2nd ed., Digital Equipment Corporation Press, Bedford, MA, 1982.

111. Michelson, A. M., "A Fast Transform in Some Galois Fields and an Application to Decoding Reed-Solomon Codes," presented at IEEE Int. Symp. Inf. Theory, Ronneby, Sweden, 1976.

112. Michelson, A. M., "The Calculation of Post-Decoding Bit-Error Probabilities for Binary Block Codes," *Nat. Telecomm. Conf. Rec.*, 24.3-1–24.3-4, 1976.

113. Modestino, J. W., and S. Y. Mui, "Performance of Convolutional Codes on Fading Channels Typical of Planetary Entry Missions," Second Annual Report, Contract No. NGR 33-018-188, Electrical and Systems Engineering Department, RPI, Troy, NY, June 1975.

114. Monsen, P., "Feedback Equalization for Fading Dispersive Channels," *IEEE Trans. Inf. Theory*, **IT-17**, 56–64 (1971).

115. Moore, J. B., "Constant-Ratio Code and Automatic-RQ on Transoceanic HF Radio Services," *IRE Trans. Comm. Sys.*, **CS-8**, 72–75 (1960).

116. Muller, D. E., "Application of Boolean Algebra to Switching Circuit Design and to Error Detection," *IRE Trans. Electronic Comp.*, **EC-3**, No. 3, 6–12 (1954).

117. Nyquist, H., "Certain Factors Affecting Telegraph Speed," *Bell Sys. Tech. J.*, **3**, 324–346 (1924).

118. Odenwalder, J. P., "Optimum Decoding of Convolutional Codes," Ph.D. dissertation, Eng. and Appl. Science, University of California at Los Angeles, 1970.

119. Odenwalder, J. P., "Dual-*k* Convolutional Codes for Noncoherently Demodulated Channels," *Int. Telemetering Conf. Rec.*, 165–174, 1976.

120. Odenwalder, J. P., "Error Control Coding Handbook," Final Report, Contract No. F44620-76-0056, Linkabit Corporation, La Jolla, CA, July 1976.

121. Odenwalder, J. P., et al., "Hybrid Coding Systems Study," Final Report, Contract NAS2-6722, Linkabit Corporation, La Jolla, CA, September 1972.

122. Oldham, I. B., R. T. Chien, and D. T. Tang, "Error Detection and Correction in a Photo-Digital Storage System," *IBM J. Res. Dev.*, **12**, 422–430 (1968).

123. Oppenheim, A. V., and R. W. Schafer, *Digital Signal Processing*, Prentice-Hall, Englewood Cliffs, NJ, 1975.

124. Oppenheim, A. V., et al. (Eds.), *Selected Papers in Digital Signal Processing*, Vol. II, IEEE Press, New York, 1976.

125. Padovani, R., and J. K. Wolf, "Data Transmission Using Error Detection Codes," *GLOBECOM Conf. Rec.*, 626–631, 1982.

126. Pasupathy, S., "Minimum Shift Keying: A Spectrally Efficient Modulation," *IEEE Comm. Mag.*, **17**, No. 4, 14–22 (July 1979).

127. Pehlert, W. K., Jr., "Design and Evaluation of a Generalized Burst-Trapping Error Control System," *IEEE Trans. Comm. Tech.*, **COM-19**, 863–868 (1971).

128. Peterson, W. W., "Encoding and Error-Correction Procedures for the Bose-Chaudhuri Codes," *IRE Trans. Inf. Theory*, **IT-6**, 459–470 (1960).

129. Peterson, W. W., *Error-Correcting Codes*, MIT Press, Cambridge, MA, 1961.

130. Peterson, W. W., "On the Weight Structure and Symmetry of BCH Codes," Sci. Rep. AFCRL-65-515, Air Force Cambridge Research Labs., Bedford, MA, July 1965.

131. Peterson, W. W., and E. J. Weldon, Jr., *Error-Correcting Codes*, 2nd ed., MIT Press, Cambridge, MA, 1972.

132. Pieper, J., et al., "Design of Efficient Coding and Modulation for a Rayleigh Fading Channel," *IEEE Trans. Inf. Theory*, **IT-24**, 457–468 (1978).

133. Pierce, A., et al., "Effective Application of Forward-Acting Error-Control to Multichannel HF Data Modems," *IEEE Trans. Comm. Tech.*, **COM-18**, 281–294 (1970).

134. Pless, V., *Introduction to the Theory of Error-Correcting Codes*, Wiley-Interscience, New York, 1982.

135. Pollard, J. M., "The Fast Fourier Transform in a Finite Field," *Math. Computation*, **25**, 365–374 (1971).

136. Posner, E., "Combinatorial Structures in Planetary Reconnaissance," in H. B. Mann (Ed.), *Error Correcting Codes*, Wiley, New York, 1969, 15–46.

137. Prange, E., "Cyclic Error-Correcting Codes in Two Symbols," Tech. Note AFCRC-TN-57-103, Air Force Cambridge Research Center, Cambridge, MA, September 1957.

138. Proakis, J. G., "Bounds on the Performance of Linear Block Codes and Convolutional Codes on the Additive White Gaussian Noise Channel and on the Rayleigh Fading Channel," Systems Engr. Tech. Memo. No. 92, GTE Sylvania, Eastern Division, Needham, MA, June 24, 1977.

139. Proakis, J. G., *Digital Communications*, McGraw-Hill, New York, 1983.

140. Rabiner, L. R., and C. M. Rader (Eds.), *Digital Signal Processing*, IEEE Press, New York, 1972.

141. Rader, C. M., "Memory Management in a Viterbi Decoder," *IEEE Trans. Comm.*, **COM-29**, 1399–1401 (1981).

142. Ramsey, J. L., "Realization of Optimum Interleavers," *IEEE Trans. Inf. Theory*, **IT-16**, 338–345 (1970).

143. Rao, T. R. N., *Error Coding for Arithmetic Processors*, Academic Press, New York, 1974.

144. Reed, I. S., "A Class of Multiple-Error-Correcting Codes and the Decoding Scheme," *IRE Trans. Inf. Theory*, **PGIT-4**, 38–49 (1954).

145. Reed, I. S., and G. Solomon, "Polynomial Codes over Certain Finite Fields," *J. SIAM*, **8**, 300–304 (1960).

146. Reed, I. S., T. K. Truong, and L. R. Welch, "The Fast Decoding of Reed-Solomon Codes Using Fermat Transforms," *IEEE Trans. Inf. Theory*, **IT-24**, 497–499 (1978).

147. Reed, I. S., et al., "The Fast Decoding of Reed-Solomon Codes Using Fermat Theoretic Transforms and Continued Fractions," *IEEE Trans. Inf. Theory*, **IT-24**, 100–106 (1978).

148. Rice, S. O., "Mathematical Analysis of Random Noise," *Bell Sys. Tech. J.*, **23**, 283–332 (1944); **24**, 46–156 (1945).

149. Richer, I., "Sequential Decoding with a Small Digital Computer," Tech. Rep. 491, MIT Lincoln Laboratory, January 24, 1972.

150. Robinson, J. P., and A. J. Bernstein, "A Class of Binary Recurrent Codes with Limited Error Propagation," *IEEE Trans. Inf. Theory*, **IT-13**, 106–113 (1967).

151. Roefs, H. F. A., and M. R. Best, "Concatenated Coding on a Spacecraft-to-Ground Telemetry Channel: Performance," *Rec. IEEE Int. Conf. Comm.*, 2.6.1–2.6.5, 1981.

152. Salz, J., and B. R. Saltzberg, "Double Error Rates in Differentially Coherent Phase Systems," *IEEE Trans. Comm. Sys.*, **CS-12**, 202–205 (1964).

153. Savage, J. E., "The Distribution of Sequential Decoding Time," *IEEE Trans. Inf. Theory*, **IT-12**, 143–147 (1966).

154. Savage, J. E., "Sequential Decoding—The Computational Problem," *Bell Sys. Tech. J.*, **45**, 149–175 (1966).

155. Shacham, N., "Performance of ARQ with Sequential Decoding Over One-Hop and Two-Hop Radio Links," *IEEE Trans. Comm.*, **COM-31**, 1172–1180 (1983).

156. Schonhoff, T. A., "Symbol Error Probabilities for M-ary CPFSK: Coherent and Noncoherent Detection," *IEEE Trans. Comm.*, **COM-24**, 644–652 (1976).

157. Shannon, C. E., "A Mathematical Theory of Communication," *Bell Sys. Tech. J.*, **27**, 379–423, 623–656 (1948). (Reprinted in book form with postscript by W. Weaver, University of Illinois Press, Urbana, IL, 1949.)

158. Shannon, C. E., "Communication in the Presence of Noise," *Proc. IRE*, **37**, 10–21 (1949).

159. Shannon, C. E., "Probability of Error for Optimal Codes in a Gaussian Channel," *Bell Sys. Tech. J.*, **38**, 611–656 (1959).

160. Shields, P. C., *Elementary Linear Algebra*, 3rd ed., Worth, New York, 1980.

161. Silverman, R. A., and M. Balser, "Coding for Constant-Data-Rate Systems," *IRE Trans. Inf. Theory*, **PGIT-4**, 50–63 (1954).

162. Singleton, R. C., "Maximum Distance q-nary Codes," *IEEE Trans. Inf. Theory*, **IT-10**, 116–118 (1964).

163. Slepian, D. (Ed.), *Key Papers in the Development of Information Theory*, IEEE Press, New York, 1974.

164. Solomon, G., "A Note on Alphabet Codes and Fields of Computation," *Inf. Control*, **25**, 395–398 (1974).

165. Stark, W. E., and R. J. McEliece, "Capacity and Coding in the Presence of Fading and Jamming," *Nat. Telecomm. Conf. Rec.*, B7.4.1–B7.4.5, 1981.

166. Stenbit, J. P., "Table of Generators for Bose-Chaudhuri Codes," *IEEE Trans. Inf. Theory*, **IT-10**, 390–391 (1964).

167. Sullivan, D. D., "A Generalization of Gallager's Adaptive Error Control Scheme," *IEEE Trans. Inf. Theory*, **IT-17**, 727–735 (1971).

168. Tietavainen, A., "On the Nonexistence of Perfect Codes over Finite Fields," *SIAM J. Appl. Math.*, **24**, 88–96 (1973).

169. Tong, S. Y., "Burst-Trapping Techniques for a Compound Channel," *IEEE Trans. Inf. Theory*, **IT-15**, 710–715 (1969).

170. Trumpis, B. D., "Convolutional Coding for *M*-ary Channels," Ph.D. dissertation, University of California at Los Angeles, 1975.

171. Viterbi, A. J., "Phase-Coherent Communication Over the Continuous Gaussian Channel," in S. W. Golomb (Ed.), *Digital Communication With Space Applications*, Prentice-Hall, Englewood Cliffs, NJ, 1964, Chapter 7.

172. Viterbi, A. J., *Principles of Coherent Communication*, McGraw-Hill, New York, 1966.

173. Viterbi, A. J., "Error Bounds for Convolutional Codes and an Asymptotically Optimum Decoding Algorithm," *IEEE Trans. Inf. Theory*, **IT-13**, 260–269 (1967).

174. Viterbi, A. J., "Convolutional Codes and Their Performance in Communication Systems," *IEEE Trans. Comm. Tech.*, **COM-19**, 751–772 (1971).

175. Viterbi, A. J., and J. K. Omura, *Principles of Digital Communication and Coding*, McGraw-Hill, New York, 1979.

176. Watson, E. J., "Primitive Polynomials (Mod 2)," *Math. Comp.*, **16**, 368–369 (1962).

177. Weldon, E. J., Jr., "Decoding Binary Block Codes on *Q*-ary Output Channels," *IEEE Trans. Inf. Theory*, **IT-17**, 713–718 (1971).

178. Wiener, N., *The Extrapolation, Interpolation, and Smoothing of Stationary Time Series, with Engineering Applications*, Wiley, New York, 1949. (Original work appeared as an MIT Radiation Laboratory Report in 1942.)

179. Wolf, J. K., "Decoding of Bose-Chaudhuri-Hocquenghem Codes and Prony's Method for Curve Fitting," *IEEE Trans. Inf. Theory*, **IT-13**, 608 (1967).

180. Wolf, J. K., "Adding Two Information Symbols to Certain Nonbinary BCH Codes and Some Applications," *Bell Sys. Tech. J.*, **48**, 2405–2424 (1969).

181. Wolf, J. K., A. M. Michelson, and A. H. Levesque, "The Determination of the Probability of Undetected Error for Linear Binary Block Codes," *Rec. IEEE Int. Conf. Comm.*, 65.1.1–65.1.5, 1981.

182. Wolf, J. K., A. M. Michelson, and A. H. Levesque, "On the Probability of Undetected Error for Linear Block Codes," *IEEE Trans. Comm.*, **COM-30**, 317–324 (1982).

183. Woodward, P. M., *Probability and Information Theory, with Applications to Radar*, McGraw-Hill, New York, 1955.

184. Wozencraft, J. M., "Sequential Decoding for Reliable Communications," Tech. Rep. 325, Research Laboratories of Electronics, MIT, Cambridge, MA, 1957.

185. Wozencraft, J. M., and I. M. Jacobs, *Principles of Communication Engineering*, Wiley, New York, 1965.

186. Wozencraft, J. M., and R. S. Kennedy, "Modulation and Demodulation for Probabilistic Coding," *IEEE Trans. Inf. Theory*, **IT-12**, 291–297 (1966).

187. Wozencraft, J. M., and B. Reiffen, *Sequential Decoding*, MIT Press, Cambridge, MA, 1961.

188. Zigangirov, K. Sh., "Some Sequential Decoding Procedures," *Probl. Peredachi Inf.*, **2**, 13–25 (1966). (Reprinted in ref. 10.)

Index

ABOUT THE AUTHORS

Arnold M. Michelson is a Senior Member of the Technical Staff at GTE Government Systems Corporation, Communication Systems Division, Needham Heights, Massachusetts. He has more than 15 years of experience with GTE, working in the areas of modulation, error-control coding, and digital signal processing. He has been involved in the design and development of several advanced communication systems for radio and satellite channels. His current technical interests include the application of error-control coding techniques in processor-based communication systems. He has authored or coauthored a number of journal and conference papers on the theory and applications of error-control coding.

Mr. Michelson received the BES degree from the Johns Hopkins University and the MSEE degree from the University of Rochester, and did further graduate work at the Polytechnic Institute of Brooklyn. He is a Senior Member of the IEEE, has served as a Chapter Officer, and reviews papers for the IEEE Transactions on Communications and Information Theory.

Allen H. Levesque is a Senior Scientist at GTE Government Systems. He has been with GTE for more than 20 years, with responsibility for many projects in advanced communication systems development. Currently, he leads a group of senior specialists in communication theory and techniques. Much of his recent work has been concerned with the development of new spread-spectrum communication systems. He has authored or coauthored a number of journal and conference papers in the area of error-control coding.

Dr. Levesque earned his BSEE degree at Worcester Polytechnic Institute and received the M.Eng. and D.Eng. degrees from Yale University. He is a Senior Member of the IEEE and a former Chairman of the Boston Chapter, IEEE Group on Information Theory. He is also a registered Professional Engineer in the Commonwealth of Massachusetts.